# Computability in Context

## Computation and Logic in the Real World

# Computability in Context

## Computation and Logic in the Real World

*editors*

**S Barry Cooper**
*University of Leeds, UK*

**Andrea Sorbi**
*Università degli Studi di Siena, Italy*

Imperial College Press

ICP

*Published by*

Imperial College Press
57 Shelton Street
Covent Garden
London WC2H 9HE

*Distributed by*

World Scientific Publishing Co. Pte. Ltd.
5 Toh Tuck Link, Singapore 596224
*USA office:* 27 Warren Street, Suite 401-402, Hackensack, NJ 07601
*UK office:* 57 Shelton Street, Covent Garden, London WC2H 9HE

**Library of Congress Cataloging-in-Publication Data**
Computability in context : computation and logic in the real world / edited by S. Barry Cooper & Andrea Sorbi.
p. cm.
Includes bibliographical references.
ISBN-13: 978-1-84816-245-7 (hardcover : alk. paper)
ISBN-10: 1-84816-245-6 (hardcover : alk. paper)
1. Computable functions. 2. Computational intelligence. 3. Set theory.
4. Mathematics--Philosophy. I. Cooper, S. B. (S. Barry) II. Sorbi, Andrea, 1956–
QA9.59.C655 2011
511.3'52--dc22

2010039227

**British Library Cataloguing-in-Publication Data**
A catalogue record for this book is available from the British Library.



Printed in Singapore by World Scientific Printers.

## Preface

Computability has played a crucial role in mathematics and computer science – leading to the discovery, understanding and classification of decidable/undecidable problems, paving the way to the modern computer era and affecting deeply our view of the world. Recent new paradigms of computation, based on biological and physical models, address in a radically new way questions of efficiency and even challenge assumptions about the so-called Turing barrier.

This book addresses various aspects of the ways computability and theoretical computer science enable scientists and philosophers to deal with mathematical and real world issues, ranging through problems related to logic, mathematics, physical processes, real computation and learning theory. At the same time it focuses on different ways in which computability emerges from the real world, and how this affects our way of thinking about everyday computational issues.

But the title *Computability in Context* has been carefully chosen. The contributions to be found here are not strictly speaking 'applied computability'. The literature directly addressing everyday computational questions has grown hugely since the days of Turing and the computer pioneers. The *Computability in Europe* conference series and association is built on the recognition of the complementary role that mathematics and fundamental science plays in progressing practical work; and, at the same time, of the vital importance of a sense of context of basic research. This book positions itself at the interface between applied and fundamental research, prioritising mathematical approaches to computational barriers.

For us, the conference *Computability in Europe 2007: Computation and Logic in the Real World* was a hugely exciting – and taxing – experience. It brought together a remarkable assembly of speakers, and a level of participation around issues of computability that would surely have astounded Turing and those other early pioneers of 'computing with understanding'. All of the contributions here come from invited plenary speakers or Programme Committee members of CiE 2007. Many of these articles are likely to become key contributions to the literature of computability and its real-world significance. The authors are all world leaders in their fields, all much in demand as speakers and writers. As editors, we very much appreciate their work.

*Barry Cooper and Andrea Sorbi*

## Contents

# Chapter 1

## Computation, Information, and the Arrow of Time

Pieter Adriaans & Peter van Emde Boas
*Adriaans ADZA Beheer B.V., and*
*FNWI, University of Amsterdam,*
*1098 XG Amsterdam, The Netherlands*
*E-mail: pieter@pieter-adriaans.com*

*Bronstee.com B.V., Heemstede, and*
*ILLC, FNWI, University of Amsterdam*
*1090 GE Amsterdam, The Netherlands*
*E-mail: peter@bronstee.com*

In this chapter we investigate the relation between information and computation under time symmetry. We show that there is a class of non-deterministic automata, the *quasi-reversible automata* (QRTM), that is the class of classical deterministic Turing machines operating in negative time, and that computes all the languages in NP. The class QRTM is isomorphic to the class of standard deterministic Turing machines TM, in the sense that for every $M \in TM$ there is a $M^{-1}$ in QRTM such that each computation on $M$ is mirrored by a computation on $M^{-1}$ with the arrow of time reversed. This suggests that non-deterministic computing might be more aptly described as deterministic computing in negative time. If $M_i$ is deterministic then $M_i^{-1}$ is non deterministic. If $M$ is information discarding then $M^{-1}$ "creates" information. The two fundamental complexities involved in a deterministic computation are Program Complexity and Program Counter Complexity. Programs can be classified in terms of their "information signature" with pure counting programs and pure information discarding programs as two ends of the spectrum. The chapter provides a formal basis for a further analysis of such diverse domains as learning, creative processes, growth, and the study of the interaction between computational processes and thermodynamics.

## Contents



## 1.1. Introduction

The motivation behind this research is expressed in a childhood memory of one of the authors: "When I was a toddler my father was an enthusiastic 8-mm movie amateur. The events captured in these movies belong to my most vivid memories. One of the things that fascinated me utterly was the fact that you could reverse the time. In my favorite movie I was eating a plate of French fries. When played forward one saw the fries vanish in my mouth one by one, but when played backward a miracle happened. Like a magician pulling a rabbit out of a hat I was pulling undamaged fries out of my mouth. The destruction of fries in positive time was associated with the creation of fries in negative time."

This is a nice example of the kind of models we have been discussing when we were working on the research for this paper. It deals with computation and the growth and destruction of information. Deterministic computation seems to be incapable of creating new information. In fact most recursive functions are non-reversible. They discard information. If one makes a calculation like $a + b = c$ then the input contains roughly $(\log a + \log b)$ bits of information whereas the answer contains $\log(a + b)$ bits which is in general much less. Somewhere in the process of transforming the input to the output we have lost bits. The amount of information we have lost is exactly the information needed to separate $c$ in to $a$ and $b$. There are many ways to select two numbers $a$ and $b$ that add up to $c$. So there are many inputs that could create the output. The information about the exact history of the computation is discarded by the algorithm. This leaves us with an interesting question: *If there is so much information in the world and computation does not generate information, then where does the information come from?*

Things get more fascinating if we consider the Turing machine version of the French fries example above. Suppose we make a Turing machine that only erases its input and we make a movie of its execution and play it backward. What would we see? We see a machine creating information out of nothing, just the same way the toddler in the reversed movie was pulling neat French fries out of his mouth. So also in this case, if we reverse the arrow of time, destruction of information becomes creation and vice versa. In previous papers the first author has investigated the relation between learning and data compression ([2, 4]). Here we are interested in the converse problem: How do data-sets from which we can learn something emerge in the world? What processes grow information?

There is a class of deterministic processes that discard or destroy information. Examples are: simple erasure of bits, (lossy) data compression, and learning. There is another class of processes that seems to create information: coin flipping, growth, evolution. In general, stochastic processes create information, exactly because we are uncertain of their future, and deterministic processes discard information, precisely because the future of the process is known. The basic paradigm of a stochastic information generating process is coin flipping. If we flip a coin in such a way that the probability of heads is equal to the probability of tails, and we note the results as a binary string, then with high probability this string is random and incompressible. The string will then have maximal Kolmogorov complexity, i.e. a program that generates the string on a computer will be at least as long as the string itself ([8]). On the other hand if we generate a string by means of a simple deterministic program (say "`For x = 1 to k print("1")`") then the string is highly compressible and by definition has a low Kolmogorov complexity which approximates $\log k$ for large enough $k$. In the light of these observations one could formulate the following research question: *Given the fact that creation and destruction of information seem to be symmetrical over the time axis, could one develop a time-invariant description of computational processes for which creation of information is the same process as destruction of information with the time arrow reversed?* A more concise version of the same question is: *Are destruction and creation of information computationally symmetrical in time?* The main part of this paper is dedicated to a positive answer to this question.

*Prima facie* it seems that we compute to get new information. So if we want to know what the exact value of 10! is, then the answer 3628800 really contains information for us. It tells us something we did not know. We also have the intuition, that the harder it is to compute a function, the more value (i.e. information) the answer contains. So 10! in a way contains more information than $10^2$. Yet from a mathematical point of view 10! and 3628800 are just different descriptions of the same number. The situation becomes even more intriguing if we turn our intention to the simulation of processes on a computer that really seem to create new information like the growth of a tree, game playing, or the execution of a genetic algorithm. What is happening here if computation cannot generate information? What is the exact relation between information generating processes that we find in our world and our abstract models of computation?

In most curricula, theories about information and computation are treated in isolation. That is probably the reason why the rather fundamental question studied in this paper up till now has received little attention in computer science: *What is the interaction between information and computation?* Samson Abramsky has posed this question in a recent publication with some urgency (without offering a definitive answer): *We compute in order to gain information, but how is this possible logically or thermodynamically? How can it be reconciled with the point of view of Information Theory? How does information increase appear in the various extant theories?* ([1], p. 487). Below we will formulate a partial answer to this question by means of an analysis of time invariant descriptions of computational processes.

## 1.2. A Formal Framework: Meta-computational Space

In order to study the interplay between entropy, information, and computation we need to develop a formal framework. For this purpose we develop the notion of meta-computational space in this section: formally, the space of the graphs of all possible computations of all possible Turing machines. The physical equivalent would be the space of all possible histories of all possible universes.

$C(x)$ will be the classical Kolmogorov complexity of a binary string $x$, i.e. the length of the shortest program $p$ that computes $x$ on a reference universal Turing machine $U$. Given the correspondence between natural numbers and binary strings, $\mathcal{M}$ consists of an enumeration of all possible self-delimiting programs for a preselected arbitrary universal Turing machine $U$. Let $x$ be an arbitrary bit string. The shortest program that produces $x$ on $U$ is $x^* = argmin_{M \in \mathcal{M}}(U(M) = x)$ and the Kolmogorov complexity of $x$ is $C(x) = |x^*|$. The conditional Kolmogorov complexity of a string $x$ given a string $y$ is $C(x|y)$, this can be interpreted as the length of a program for $x$ given input $y$. A string is defined to be *random* if

$C(x) \geq |x|$. $I(x)$ is the classical integer complexity function that assigns to each integer $x$ another integer $C(x)$ [8].

We will follow the standard textbook of Hopcroft, Motwani, and Ullman for the basic definitions ([7]). A *Turing machine* (TM) is described by a 7-tuple

$$M = (Q, \Sigma, \Gamma, \delta, q_0, B, F).$$

Here, as usual, $Q$ is the finite set of states, $\Sigma$ is the finite set of input symbols with $\Sigma \subset \Gamma$, where $\Gamma$ is the complete set of tape symbols, $\delta$ is a transition function such that $\delta(q, X) = (p, Y, D)$, if it is defined, where $p \in Q$ is the next state, $X \in \Gamma$ is the symbol read in the cell being scanned, $Y \in \Gamma$ is the symbol written in the cell being scanned, $D \in \{L, R\}$ is the direction of the move, either left or right, $q_0 \in Q$ is the start state, $B \in \Gamma - \Sigma$ is the blank default symbol on the tape, and $F \subset Q$ is the set of accepting states. A *move* of a TM is determined by the current content of the cell that is scanned and the state the machine is in. It consists of three parts:

(1) Change state;
(2) Write a tape symbol in the current cell;
(3) Move the read-write head to the tape cell on the left or right.

A *non-deterministic Turing machine* (NTM) is equal to a deterministic TM with the exception that the range of the transition function consists of sets of triples:

$$\delta(q, X) = \{(p_1, Y_1, D_1), (p_2, Y_2, D_2), ..., (p_k, Y_k, D_k)\}.$$

A TM is a *reversible Turing machine* (RTM) if the transition function $\delta(q, X) = (p, Y, D)$ is one-to-one, with the additional constraint that the movement $D$ of the read-write head is uniquely determined by the target state $p$.

**Definition 1.1.** An *Instantaneous Description* (ID) of a TM during its execution is a string $X_1X_2...X_{i-1}qX_iX_{i+1}...X_n$ in which $q$ is the state of the TM, the tape head is scanning the $i$-th head from the left, $X_1X_2...X_n$ is the portion of the tape between the leftmost and the rightmost blank. Given an Instantaneous Description $X_1X_2...X_{i-1}qX_iX_{i+1}...X_n$ it will be useful to define an *Extensional Instantaneous Description* (EID) $X_1X_2...X_{i-1}X_iX_{i+1}...X_n$, that only looks at the content of the tape and ignores the internal state of the machine and an *Intensional Instantaneous Description* (IID) $qX_iD$, that only looks at the content of the current cell of the tape, the internal state of the machine, and the direction $D$ in which the head will move.

We make the jump from an object- to a meta-level of descriptions of computations by means of considering the set of all possible transitions between instantaneous descriptions.

**Definition 1.2.** Let $<\mathcal{ID}_M, \vdash_M>$ be *the configuration graph* of all possible transformations of a machine $M$, i.e. $\mathcal{ID}_M$ is the countable set of all possible instantaneous descriptions and for $ID_{i,j} \in \mathcal{ID}_M$:

$$ID_i \vdash_M ID_j$$

if and only if TM can reach $ID_j$ in one move from $ID_i$. $ID_m$ is *reachable* from $ID_i$ iff there exists a sequence of transformations from one to the other:

$$(ID_i \vdash_M^* ID_m) \Leftrightarrow ID_i \vdash_M ID_j \vdash_M ID_k ... ID_l \vdash_M ID_m.$$

The intensional description of such a transformation will be: $(IID_i \vdash_M^* IID_m)$. The extensional description will be: $(EID_i \vdash_M^* EID_m)$.

Note that two machines can perform computations that are extensionally isomorphic without intensional isomorphism and vice versa. We refer here to transformations rather than computations since, in most cases, only a subpart of the configuration graph represents valid computations that begin with a start state and end in an accepting state. Note that the class of all possible instantaneous descriptions for a certain machine contains for each possible tape configuration, at each possible position of the head on the tape, an instance for each possible internal state. Most of these configurations will only be the result, or lead to, fragments of computations. On the other hand, all valid computations that begin with a start state and either continue forever or end in an accepting state, will be represented in the configuration graph.

Note that there is a strict relation between the structure of the transition function $\delta$ and the configuration graph: For a deterministic machine the configuration graph has only one outgoing edge for each configuration, for a non-deterministic machine the configuration graph can have multiple outgoing edges per $ID$, for a reversible machine the graph consists only of a number of linear paths without bifurcations either way.

**Lemma 1.1.** *Let $M$ be a Turing machine. We have $C(<\mathcal{ID}_M, \vdash_M>) < C(M) + O(1)$.*

***Proof.*** Given $M$ the graph $< \mathcal{ID}_M, \vdash_M >$ can be constructed by the following algorithm: Create $\mathcal{ID}_M$ by enumerating the language of all possible IDs, at each step of this process run $M$ for one step on all IDs created so far and add appropriate edges to $\vdash_M$ when $M$ transforms $ID_i$ in $ID_j$. □

The finite object $M$ and the infinite object $< \mathcal{ID}_M, \vdash_M >$ identify the same structure. We use here two variants of the Kolmogorov complexity: The complexity of the finite object $M$ is defined by the smallest program that computes the object on a universal Turing machine and then halts; the complexity of $< \mathcal{ID}_M, \vdash_M >$ is given by the shortest program that creates the object in an infinite run.

**Definition 1.3.** Given an enumeration of Turing machines the *meta-computational space* is defined as the disjunct sum of all configuration graphs $< \mathcal{ID}_{M_i}, \vdash_{M_i} >$ for $i \in \mathbb{N}$.

The meta-computational space is a very rich object in which we can study a number of fundamental questions concerning the interaction between information and computation. We can also restrict ourselves to the study of either extensional or intensional descriptions of computations and this will prove useful, e.g. when we want to study the class of all computational histories that have descriptions with isomorphic pre- or suffixes. For the moment we want to concentrate on time symmetries in meta-computational space.

## 1.3. Time Symmetries in Meta-computational Space

In this paragraph we study the fact that some well-known classes of computational processes can be interpreted as each others' symmetrical images in time, i.e. processes in one class can be described as processes in the other class with the time arrow reversed, or, to say it differently, as processes taking place in negative time. We can reverse the time arrow for all possible computations of a certain machine by means of reversing all the edges in the computational graph. This motivates the following notation:

**Definition 1.4.**

$$(ID_i \vdash ID_j) \Leftrightarrow (ID_j \dashv ID_i)$$

$$(ID_i \vdash^* ID_k) \Leftrightarrow (ID_k \dashv^* ID_i).$$

The analysis of valid computations of $TM$ can now be lifted to the study of reachability in the configuration graph. The introduction of such a meta-computational model allows us to study a much more general class of computations in which the arrow of time can be reversed. We will introduce the following shorthand notation that allows us to say that $M^{-1}$ is the same machine as $M$ with the arrow of time reversed:

$$M = <\mathcal{ID}_M, \vdash_M> \Leftrightarrow M^{-1} = <\mathcal{ID}_M, \dashv_M> .$$

Intuitively the class of languages that is "computed" in negative time by a certain Turing machine is the class of accepting tape configurations that can be reached from a start state. We have to stress however, that moving back in time in the configuration graph describes a process that is fundamentally different from the standard notion of "computation" as we know it. We give some differences:

- The standard definition of a Turing machine knows only one starting state and possibly several accepting states. Computing in negative time will trace back from several accepting states to one start state.
- The interpretation of the $\delta$-function or relation is different. In positive time we use the $\delta$-function to decide which action to take given a certain state-symbol combination. In negative time this situation is reversed: We use the $\delta$-function to decide which state-symbol-move combination could have led to a certain action.
- At the start of a computation there could be a lot of rubbish on the tape that is simply not used during the computation. All computations starting with arbitrary rubbish are in the configuration graph. We want to exclude these from our definitions and stick to some minimal definition of the input of a computation in negative time.

In order to overcome these difficulties the following lemma will be useful:

**Lemma 1.2. (Minimal Input-Output Reconstruction)** *If an intensional description of a fragment of a (deterministic or non-deterministic) computation of a machine $M$: ($IID_i \vdash_M^* IID_m$) can be interpreted as the trace of a valid computation then there exist a* minimal input configuration $ID_i$ *and a* minimal output configuration $ID_m$ *for which $M$ will reach $ID_m$ starting at $ID_i$. Otherwise the minimal input and output configuration are undefined.*

***Proof.*** The proof first gives a construction for the minimal output in a positive sweep and then the minimal input in a negative sweep.

Positive sweep: Note that $(IID_i \vdash_M^* IID_m)$ consists of a sequence of descriptions: $q_iX_iD_i \vdash q_{i+1}X_{i+1}D_{i+1} \vdash ... \vdash q_mX_mD_m$. Reconstruct a computation in the following way: Start with an infinite tape for which all of the symbols are unknown. Position the read-write head at an arbitrary cell and perform the following interpretation operation: Interpret this as the state-symbol-move configuration $q_iX_iD_i$. Now we know the contents of the cell $X_i$, the state $q_i$, and the direction $D$ of the move of the read-write head. The action will consist of writing a symbol in the current cell and moving the read-write head left or right. Perform this action. The content of one cell is now fixed. Now there are two possibilities:

(1) We have the read-write head in a new cell with unknown content. From the intensional description we know that the state-symbol combination is $q_{i+1}X_{i+1}D_{i+1}$, so we can repeat the interpretation operation for the new cell.
(2) We have visited this cell before in our reconstruction and it already contains a symbol. From the intensional description we know that the state-symbol combination should be $q_{i+1}X_{i+1}D_{i+1}$. If this is inconsistent with the content of the current cell, the reconstruction stops and the minimal output is undefined. If not, we can repeat the interpretation operation for the new cell.

Repeat this operation till the intensional description is exhausted. Cells on the tape that still have unknown content have not been visited by the computational process: We may consider them to contain blanks. We now have the minimal output configuration on the tape $ID_m$.

Negative sweep: start with the minimal output configuration $ID_m$. We know the current location of the read-write head and the content of the cell. From the intensional description $(IID_i \vdash_M^* IID_m)$ we know which state-symbol combination $q_mX_mD_m$ has led to $ID_m$: from this we can construct $ID_{m-1}$. Repeat this process till the intensional description is exhausted and we read $ID_i$, which is the minimal input configuration. □

Lemma 1.2 gives us a way to tame the richness of the configuration graphs: We can restrict ourselves to the study of computational processes that are intensionally equivalent, specifically intensionally equivalent processes that start with a starting state and end in an accepting state. This facilitates the following definition:

**Definition 1.5.** If $(IID_i \vdash_M^* IID_m)$ is an intensional description of a computation then

$$\texttt{INPUT}(IID_i \vdash_M^* IID_m) = x$$

gives the minimal input $x$ and

$$\texttt{OUTPUT}(IID_i \vdash_M^* IID_m) = y$$

gives the minimal output $y$. With some abuse of notation we will also apply these functions to histories of full IDs.

**Definition 1.6.** Given a Turing machine $M$ the language recognized by its counterpart $M^{-1}$ in negative time is the set of minimal output configurations associated with intensional descriptions of computations on $M$ that begin in a start state and end in an accepting state.

**Definition 1.7.** The class $P^{-1}$ is the class of languages that are recognized by an $M_i^{-1}$ with $i \in \mathbb{N}$ in time polynomial to the length of minimal input configuration.

Note that, after a time reversal operation, the graph of a deterministic machine is transformed into a specific non-deterministic graph with the characteristic that each ID has only one incoming edge. We will refer to such a model of computation as *quasi-reversible*. The essence of this analysis is that, given a specific machine $M$, we can study its behavior under reversal of the arrow of time.

We can use the symmetry between deterministic and quasi-reversible computing in proofs. Whatever we prove about the execution of a program on $M$ also holds for $M^{-1}$ with the time reversed and vice versa.

Let $QRTM$ be the class of quasi-reversible non-deterministic machines that are the mirror image in time of the class of deterministic machines $TM$, and $QRP$ be the class of languages that can be recognized by $QRTM$ in polynomial time. The lemma below is at first sight quite surprising. The class of languages that we can recognize non-deterministically in polynomial time is the same class as the class of polynomial quasi-reversible languages:

**Lemma 1.3.** *The class $L_{QRP}$ of languages recognized by a $QRTM$ in polynomial time is NP.*

***Proof.*** 1) $L_{QRP} \subseteq NP$: The class of languages recognized by quasi-reversible machines is a subclass of the class of languages recognized by non-deterministic machines. This is trivial since there is a non-deterministic machine that produces any $\{0,1\}^{\leq k}$ in time $k$.

2) $NP \subseteq L_{QRP}$: The class NP is defined in a standard way in terms of a checking relation $R \subseteq \Sigma^* \times \Sigma_1^*$ for some finite alphabets $\Sigma^*$ and $\Sigma_1^*$. We associate with each such relation $R$ a language $L_R$ over $\Sigma^* \cup \Sigma_1^* \cup \#$ defined by

$$L_R = \{w\#y|R(w,y)\}$$

where the symbol $\#$ is not in $\Sigma$. We say that $R$ is polynomial-time iff $L_R \in P$. Now we define the class NP of languages by the condition that a language $L$ over $\Sigma$ is in NP iff there is $k \in \mathbb{N}$ and a polynomial-time checking relation $R$ such that for all $w \in \Sigma^*$,

$$w \in L \Leftrightarrow \exists y(|y| < |w|^k \ \& \ R(w,y))$$

where $|w|$ and $|y|$ denote the lengths of $w$ and $y$, respectively. Suppose that $M$ implements a polynomial-time checking relation for $R$. Adapt $M$ to form $M'$ that takes $R(w,y)$ as input and erases $y$ from the tape after checking the relation, the transformation of $M$ to $M^{-1}$ is polynomial. The corresponding QRTM $M'^{-1}$ will start with guessing a value for $y$ non-deterministically and will finish in a configuration for which the relation $R(w,y)$ holds in polynomial time since $|y| < |w|^k$ and the checking relation $R$ is polynomial. □

We can formulate the following result:

**Theorem 1.1.** $NP = P^{-1}$

***Proof.*** Immediate consequence of Lemma 1.3 and Definition 1.7. □

NP is the class of languages that can be recognized by deterministic Turing machines in negative time. This shows that quasi-reversible computing is in a way a more natural model of non-deterministic computing than the classical full-blown non-deterministic model. The additional power is unnecessary.

## 1.4. The Interplay of Computation and Information

We now look at the interplay between information and computation. The tool we use will be the study of the changes in $C(ID_t)$, i.e. changes in the Kolmogorov complexity of instantaneous descriptions over time. We make some observations:

- If $ID_i \vdash_M ID_j$ then the information distance between the instantaneous descriptions $ID_i$ and $ID_j$ is $\log k + 1$ at most where $k$ is the number of internal states of $M$.
- If $EID_i \vdash_M EID_j$ then the information distance between the extensional descriptions $EID_i$ and $EID_j$ is 1 bit at most.
- If $IID_i \vdash_M IID_j$ then the information distance between the intensional descriptions $IID_i$ and $IID_j$ is $\log k + 2$ at most where $k$ is the number of internal states of $M$.
- Let $x$ be the minimal input of a computational fragment $(IID_i \vdash_M^* IID_m)$ and let $y$ be the minimal output. We have

$$C(x|IID_i \vdash_M^* IID_m) = C(y|IID_i \vdash_M^* IID_m) = O(1).$$

  This is an immediate consequence of Lemma 1.2.

We can now identify some interesting typical machines:

- No machine can produce information faster than 1 bit per computational step. There is indeed a non-deterministic machine that reaches this "speed": the non-deterministic "coin-flip" automaton that writes random bits. For such an automaton we have with high probability $C(ID_t) \approx t$. In negative time this machine is the maximal eraser. It erases information with the maximum "speed" of 1 bit per computational step.
- A unary counting machine produces information with a maximum speed of $\log t$. Note that $C(t) = I(t)$, i.e. the complexity at time $t$ is equal to the value of the integer complexity function. The function $I(x)$ has indefinite "dips" in complexity, i.e. at those places where it approaches a highly compressible number. When $t$ approaches such a dip the information produced by a unary counting machine will drop as the machine continues to write bits. The counter part of the unary counter in negative time is the unary eraser. It erases information with the maximal speed of $\log t$, although at times it will create information by erasing bits.
- The slowest information producer for its size is the busy-beaver function. When it is finished it will have written an enormous number of bits with a conditional complexity of $O(1)$. Its counterpart in negative time will be a busy-glutton automaton that "eats" an enormous number of bits of an exact size.

These insights allow us to draw a picture that tells us how information and computation are intertwined in a deterministic process.

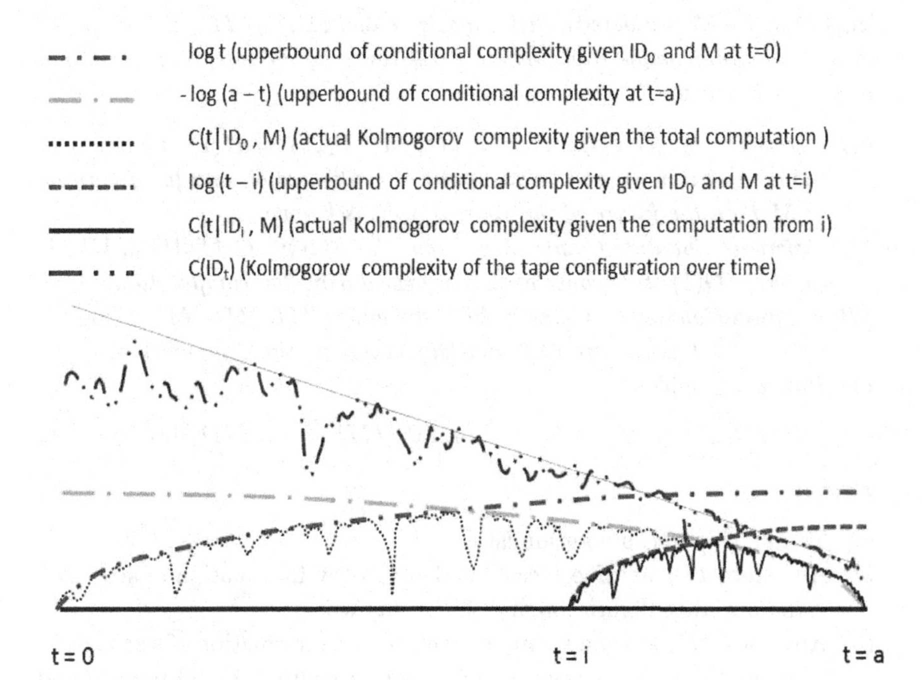


Figure 1.1. Schematic representation of the various types of complexity estimates involved in a deterministic computation.

The complexity of the history of a computation is related to the complexity of the input given the output. There are two forms of complexity involved in a deterministic computation:

- Program Complexity: This is the complexity of the input and its subsequent configurations during the process. It cannot grow during the computation. Most computations reduce program complexity.
- Program Counter Complexity: This is the descriptive complexity of the program counter during the execution of the process. It is 0 at the beginning, grows to $\log a$ in the middle, and reduces to 0 again at the end of the computation.

The relationship between these forms of complexity is given by the following theorem:

**Theorem 1.2. (Information exchange in Deterministic Computing)** *Suppose $M$ is a deterministic machine and $ID_i \vdash_M ID_a$ is a fragment of an accepting computation, where $ID_m$ contains an accepting state. For every $i \leq k \leq a$ we have:*

*(1) Determinism: $C(ID_{i+k+1} \vdash_M ID_a|M, ID_{i+k}) = O(1)$, i.e. at any moment of time if we have the present configuration and the definition of $M$ then the future of the computation is known.*
*(2) Program Counter Complexity from the start: $C(ID_t|ID_0, M) < (\log k) + O(1)$, this constraint is known during the computation.*
*(3) Program Counter Complexity from the end: $C(ID_t|ID_0, M) < (\log a - k) + O(1)$, this constraint is not known during the computation.*
*(4) Program complexity:*

$$C((IID_{i+k} \vdash_M^* IID_a)|M) = C(\mathit{INPUT}(IID_{i+k} \vdash_M^* IID_a)|M) + O(1).$$

***Proof.***

(1) Trivial, since $M$ is deterministic.
(2) Any state $ID_k$ at time $k$ can be identified by information of size $\log k$ if the initial configuration and $M$ are known.
(3) Any state $ID_k$ at time $k$ can be identified by information of size $\log(a - k)$ if the total description of the accepting computational process and $M$ are known.
(4) By the fact that the computation is deterministic it can be reconstructed from the minimal input, given $M$. By Lemma 1.2, given $M$, the minimal input can be reconstructed from $(IID_i \vdash_M^* IID_a)$. This gives the equality modulo $O(1)$. □

We cannot prove such a nice equality for the minimal output. Note that even if the following inequality holds:

$$C((IID_i \vdash_M^* IID_a)|M) \geq C((IID_{i+k} \vdash_M^* IID_a)|M) + O(1)$$

this does not imply that:

$$C(\mathtt{OUTPUT}(IID_i \vdash_M^* IID_a)|M) \geq C(\mathtt{OUTPUT}(IID_{i+k} \vdash_M^* IID_a)|M) + O(1).$$

As a counterexample, observe that a program that erases a random string has a string of blanks as minimal output. A longer string still can have a lower Kolmogorov complexity.

In computations that use counters, Program Complexity and Program Counter Complexity are mixed up during the execution. In fact one can characterize various types of computations by means of their "information signature". Informally, at extremes of the spectrum, one could distinguish:

- *Pure Information Discarding Processes*: in such processes the program counter does not play any role. They reach an accepting state by means of systematically reducing the input. Summation of a set of numbers, or erasing of a string are examples.
- *Pure Counting Processes*: `For x=1 to i write("1")`: The conditional complexity of the tape configuration grows from 0 to $\log i$ and then diminishes to 0 again.
- *Pure Search Processes*: In such processes the input is not reduced but is kept available during the whole process. The information in the program counter is used to explore the search space. Standard decision procedures for NP-hard programs, where the checking function is tested on an enumeration of all possible solutions, are an example.

A deeper analysis of various information signatures of computational processes and their consequences for complexity theory is a subject of future work.

## 1.5. Discussion

We can draw some conclusions and formulate some observations on the basis of the analysis given above.

1) Erasing and creating information are indeed, as suggested in the introduction, from a time invariant computational point of view the same processes: The quasi-reversible machine that is associated with a simple deterministic machine that erases information is a non-deterministic machine writing arbitrary bit-strings on the tape. This symmetry also implies that creation of information in positive time involves destruction of information in negative time.

2) The class of quasi-reversible machines indeed describes the class of data-sets from which we can learn something in the following way: If $L$ is the language accepted by $M$ then $M^{-1}$ generates $L$. $M^{-1}$ is an *informer* for $L$ in the sense of [6], every sentence in $L$ will be non-deterministically produced by $M^{-1}$ in the limit. $QRTM$ is the class of all informers for type-0 languages.

3) These insights suggests that we can describe stochastic processes in the real world as deterministic processes in negative time: e.g. throwing a dice in positive time is erasing information about its "future" in negative time, the evolution of species in positive time could be described as the "deterministic" computation of their ancestor in negative time. A necessary condition for the description of such growth processes as computational processes is that the number of bits that can be produced per time unit is restricted. A stochastic interpretation of a QRTM can easily be developed by assigning a set of probabilities to each split in the $\delta$ relation. The resulting stochastic-QRTM is a *sufficient statistic* for the data-sets that are generated.

4) The characterization of the class NP in terms of quasi-reversible computing seems to be more moderate than the classical description in terms of full non-deterministic computing. The full power of non-deterministic computing is never realized in a system with only one time direction.

5) Processes like game playing and genetic algorithms seem to be meta-computational processes in which non-deterministic processes (throwing a dice, adding mutations) seem to be intertwined with deterministic phases (making moves, checking the fitness function).

6) Time-symmetry has consequences for some philosophical positions. The idea that the evolution of our universe can be described as a deterministic computational process has been proposed by several authors (Zuse, Bostrom, Schmidthuber, Wolfram [10], Lloyd [9], etc.). Nowadays it is referred to as pancomputationalism [5]. If deterministic computation is an information discarding process then it implies that the amount of information in the universe rapidly decreases. This contradicts the second law of thermodynamics. On the other hand, if the universe evolves in a quasi-reversible way, selecting possible configurations according to some quasi-reversible computational model, it computes the Big Bang in negative time. The exact implications of these observations can only be explained by means of the notion of facticity [3], but that is another discussion. The concept of quasi-reversible computing seems to be relevant for these discussions [2].

## 1.6. Conclusion

Computing is moving through meta-computational space. For a fixed Turing machine $M_i$ such movement is confined to one local infinite graph $< \mathcal{ID}_{M_i}, \vdash_{M_i} >$. If $M_i$ is deterministic then $M_i^{-1}$ is non-deterministic. If $M$ is information discarding then $M^{-1}$ "creates" information. The two fundamental complexities involved in a deterministic computation are Program Complexity and Program Counter Complexity. Programs can be classified in terms of their "information signature" with pure counting programs and pure information discarding programs as two ends of the spectrum. The class $NP$ is simply the class of polynomial deterministic time calculations in negative time. Thinking in terms of meta-computational space allows us to conceptualize computation as movement in a certain space and is thus a source of new intuitions to study computation. Specifically a deeper analysis of various information signatures of computational (and other) processes is a promising subject for further study.

## References

[1] S. Abramsky. Information, Processes and Games. In eds. P. W. Adriaans and J. F. A. K. van Benthem, *Handbook of the Philosophy of Information*, In Handbooks of the Philosophy of Science, series edited by D. M. Gabbay, P. Thagard and J. Woods, pp. 483–550. Elsevier, (2008).

[2] P. W. Adriaans and J. F. A. K. van Benthem, eds., *Handbook of the Philosophy of Information.* In Handbooks of the Philosophy of Science, series edited by D. M. Gabbay, P. Thagard and J. Woods. Elsevier, (2008).

[3] P. W. Adriaans, Between order and chaos: The quest for meaningful information, *Theor. Comp. Sys.* **45**(4), (2009).

[4] P. W. Adriaans and P. Vitányi, Approximation of the two-part MDL code, *IEEE Transactions on Information Theory.* **55**(1), 444–457, (2009).

[5] L. Floridi. Trends in the philosophy of information. In eds. P. W. Adriaans and J. F. A. K. van Benthem, *Handbook of the Philosophy of Information*, In Handbooks of the Philosophy of Science, series edited by D. M. Gabbay, P. Thagard and J. Woods, pp. 113–132. Elsevier, (2008).

[6] E. M. Gold, Language identification in the limit, *Information and Control.* **10**(5), 447–474, (1967).

[7] J. E. Hopcroft, R. Motwani, and J. D. Ullman, *Introduction to Automata Theory, Languages, and Computation.* Addison-Wesley, (2001), second edition.

[8] M. Li and P. Vitányi, *An Introduction to Kolmogorov Complexity and its Applications.* Springer-Verlag, (2008), third edition.

[9] S. Lloyd, Ultimate physical limits to computation, *Nature.* **406**, 1047–1054, (2000).

[10] S. Wolfram, *A New Kind of Science.* Wolfram Media Inc., (2002).

# Chapter 2

# The Isomorphism Conjecture for NP

Manindra Agrawal *

*Indian Institute of Technology*
*Kanpur, India*
*E-mail: manindra@iitk.ac.in*

In this chapter, we survey the arguments and known results for and against the Isomorphism Conjecture.

## Contents



## 2.1. Introduction

The Isomorphism Conjecture for the class NP states that all polynomial-time many-one complete sets for NP are polynomial-time isomorphic to each other. It was made by Berman and Hartmanis [21][a], inspired in part by a corresponding result in computability theory for computably enumerable sets [50], and in part by the observation that all the existing NP-complete

*N Rama Rao Professor, Indian Institute of Technology, Kanpur. Research supported by J C Bose Fellowship FLW/DST/CS/20060225.

[a]The conjecture is also referred as *Berman–Hartmanis Conjecture* after the proposers.

sets known at the time were indeed polynomial-time isomorphic to each other. This conjecture has attracted a lot of attention because it predicts a very strong structure of the class of NP-complete sets, one of the fundamental classes in complexity theory.

After an initial period in which it was believed to be true, Joseph and Young [40] raised serious doubts against the conjecture based on the notion of *one-way* functions. This was followed by investigation of the conjecture in relativized worlds [27, 33, 46] which, on the whole, also suggested that the conjecture may be false. However, disproving the conjecture using one-way functions, or proving it, remained very hard (either implies $\mathsf{DP} \neq \mathsf{NP}$). Hence research progressed in three distinct directions from here.

The first direction was to investigate the conjecture for complete degrees of classes bigger than NP. Partial results were obtained for classes EXP and NEXP [20, 29].

The second direction was to investigate the conjecture for degrees other than complete degrees. For degrees within the 2-*truth-table-complete degree* of EXP, both possible answers to the conjecture were found [41, 43, 44].

The third direction was to investigate the conjecture for reducibilities weaker than polynomial-time. For several such reducibilities it was found that the isomorphism conjecture, or something close to it, is true [1, 2, 8, 16].

These results, especially from the third direction, suggest that the Isomorphism Conjecture for the class NP may be true contrary to the evidence from the relativized world. A recent work [13] shows that if all one-way functions satisfy a certain property then a non-uniform version of the conjecture is true.

An excellent survey of the conjecture and results related to the first two directions is in [45].

## 2.2. Definitions

In this section, we define most of the notions that we will need.

We fix the alphabet to $\Sigma = \{0, 1\}$. $\Sigma^*$ denotes the set of all finite strings over $\Sigma$ and $\Sigma^n$ denotes the set of strings of size $n$. We start with defining the types of functions we use.

**Definition 2.1.** Let $r$ be a resource bound on Turing machines. Function $f$, $f : \Sigma^* \mapsto \Sigma^*$, is *r-computable* if there exists a Turing machine (TM, in short) $M$ working within resource bound of $r$ that computes $f$. We also refer to $f$ as an *r-function.*

Function $f$ is *size-increasing* if for every $x$, $|f(x)| > |x|$. $f$ is *honest* if there exists a polynomial $p(\cdot)$ such that for every $x$, $p(|f(x)|) > |x|$.

For function $f$, $f^{-1}$ denotes a function satisfying the property that for all $x$, $f(f^{-1}(f(x))) = f(x)$. We say $f$ is *$r$-invertible* if some $f^{-1}$ is $r$-computable.

For function $f$, its range is denoted as: range$(f) = \{y \mid (\exists x)\ f(x) = y\}$.

We will be primarily interested in the resource bound of *polynomial-time*, and abbreviate it as $p$. We now define several notions of reducibilities.

**Definition 2.2.** Let $r$ be a resource bound. Set $A$ *$r$-reduces* to set $B$ if there exists an $r$-function $f$ such that for every $x$, $x \in A$ iff $f(x) \in B$. We also write this as $A \leq_m^r B$ via $f$. Function $f$ is called an *$r$-reduction* of $A$ to $B$.

Similarly, $A \leq_1^r B$ ($A \leq_{1,si}^r B$; $A \leq_{1,si,i}^r B$) if there exists a 1-1 (1-1 and size-increasing; 1-1, size-increasing and $r$-invertible) $r$-function $f$ such that $A \leq_m^r B$ via $f$.

$A \equiv_m^r B$ if $A \leq_m^r B$ and $B \leq_m^r A$. An *$r$-degree* is an equivalence class induced by the relation $\equiv_m^r$.

**Definition 2.3.** $A$ is *$r$-isomorphic* to $B$ if $A \leq_m^r B$ via $f$ where $f$ is a 1-1, onto, $r$-invertible $r$-function.

The definitions of complexity classes DP, NP, PH, EXP, NEXP etc. can be found in [52]. We define the notion of completeness we are primarily interested in.

**Definition 2.4.** Set $A$ is *$r$-complete for* NP if $A \in$ NP and for every $B \in$ NP, $B \leq_m^r A$. For $r = p$, set $A$ is called NP*-complete* in short. The class of $r$-complete sets is also called the *complete $r$-degree* of NP.

Similarly one defines complete sets for other classes.

The *Satisfiability problem* (SAT) is one of the earliest known NP-complete problems [25]. SAT is the set of all satisfiable propositional Boolean formulas.

We now define one-way functions. These are p-functions that are not p-invertible on most of the strings. One-way functions are one of the fundamental objects in cryptography.

Without loss of generality (see [30]), we can assume that one-way functions are honest functions $f$ for which the input length determines the output length, i.e., there is a *length function* $\ell$ such that $|f(x)| = \ell(|x|)$ for all $x \in \Sigma^*$.

**Definition 2.5.** Function $f$ is a *$s(n)$-secure one-way function* if (1) $f$ is a p-computable, honest function and (2) the following holds for every polynomial-time randomized Turing machine $M$ and for all sufficiently large $n$:

$$\Pr_{x \in_U \Sigma^n} [f(M(f(x))) = f(x)] < \frac{1}{s(n)}.$$

In the above, the probability is also over random choices of $M$, and $x \in_U \Sigma^n$ mean that $x$ is uniformly and randomly chosen from strings of size $n$.

We impose the property of honesty in the above definition since a function that shrinks length by more than a polynomial is trivially one-way.

It is widely believed that $2^{n^\epsilon}$-secure one-way functions exist for some $\epsilon > 0$. We give one example. Start by defining a modification of the multiplication function:

$$\text{Mult}(x, y) = \begin{cases} 1z & \text{if } x \text{ and } y \text{ are both prime numbers} \\ & \text{and } z \text{ is the product } x * y \\ 0xy & \text{otherwise.} \end{cases}$$

In the above definition, $(\cdot, \cdot)$ is a *pairing function*. In this paper, we assume the following definition of $(\cdot, \cdot)$: $(x, y) = xy\ell$ where $|\ell| = \lceil \log |xy| \rceil$ and $\ell$ equals $|x|$ written in binary. With this definition, $|(x, y)| = |x| + |y| + \lceil \log |xy| \rceil$. This definition is easily extended for $m$-tuples for any $m$.

Mult is a p-function since testing primality of numbers is in DP [11]. Computing the inverse of Mult is equivalent to factorization, for which no efficient algorithm is known. However, Mult is easily invertible on most of the inputs, e.g., when any of $x$ and $y$ is not prime. The density estimate for prime numbers implies that Mult is p-invertible on at least $1 - \frac{1}{n^{O(1)}}$ fraction of inputs. It is believed that Mult is $(1 - \frac{1}{n^{O(1)}})$-secure, and it remains so even if one lets the TM $M$ work for time $2^{n^\delta}$ for some small $\delta > 0$. From this assumption, one can show that arbitrary concatenation of Mult:

$$\text{MMult}(x_1, y_1, x_2, y_2, \ldots, x_m, y_m) = \text{Mult}(x_1, y_1) \cdot \text{Mult}(x_2, y_2) \cdots \text{Mult}(x_m, y_m)$$

is a $2^{n^\epsilon}$-secure one-way function [30](p. 43).

One-way functions that are $2^{n^\epsilon}$-secure are not p-invertible almost anywhere. The weakest form of one-way functions are *worst-case* one-way functions:

**Definition 2.6.** Function $f$ is a *worst-case one-way function* if (1) $f$ is a p-computable, honest function, and (2) $f$ is not p-invertible.

## 2.3. Formulation and Early Investigations

The conjecture was formulated by Berman and Hartmanis [21] in 1977. Part of their motivation for the conjecture was a corresponding result in computability theory for computably enumerable sets [50]:

**Theorem 2.1. (Myhill)** *All complete sets for the class of computably enumerable sets are isomorphic to each other under computable isomorphisms.*

The non-trivial part in the proof of this theorem is to show that complete sets for the class of computably enumerable sets reduce to each other via 1-1 reductions. It is then easy to construct isomorphisms between the complete sets. In many ways, the class NP is the resource bounded analog of the computably enumerable class, and polynomial-time functions the analog of computable functions. Hence it is natural to ask if the resource bounded analog of the above theorem holds.

Berman and Hartmanis noted that the requirement for p-isomorphisms is stronger. Sets reducing to each other via 1-1 p-reductions does not guarantee p-isomorphisms as p-functions do not have sufficient time to perform exponential searches. Instead, one needs p-reductions that are 1-1, size-increasing, and p-invertible:

**Theorem 2.2. (Berman–Hartmanis)** *If $A \leq^p_{1,si,i} B$ and $B \leq^p_{1,si,i} A$ then $A$ is p-isomorphic to $B$.*

They defined the *paddability* property which ensures the required kind of reductions.

**Definition 2.7.** Set $A$ is *paddable* if there exists a p-computable *padding function* $p$, $p : \Sigma^* \times \Sigma^* \mapsto \Sigma^*$, such that:

- Function $p$ is 1-1, size-increasing, and p-invertible,
- For every $x$, $y \in \Sigma^*$, $p(x, y) \in A$ iff $x \in A$.

**Theorem 2.3. (Berman–Hartmanis)** *If $B \leq^p_m A$ and $A$ is paddable, then $B \leq^p_{1,si,i} A$.*

***Proof.*** Suppose $B \leq^p_m A$ via $f$. Define function $g$ as: $g(x) = p(f(x), x)$. Then, $x \in B$ iff $f(x) \in A$ iff $g(x) = p(f(x), x) \in A$. By its definition and the fact that $p$ is 1-1, size-increasing, and p-invertible, it follows that $g$ is also 1-1, size-increasing, and p-invertible. □

Berman and Hartmanis next showed that the known complete sets for NP at the time were all paddable and hence p-isomorphic to each other. For example, the following is a padding function for SAT:

$$p_{SAT}(x, y_1 y_2 \cdots y_m) = x \wedge \bigwedge_{i=1}^{m} z_i \bigwedge_{i=1}^{m} c_i$$

where $c_i = z_{m+i}$ if bit $y_i = 1$ and $c_i = \bar{z}_i$ if $y_i = 0$ and the Boolean variables $z_1, z_2, \ldots, z_{2m}$ do not occur in the formula $x$.

This observation led them to the following conjecture:

**Isomorphism Conjecture.** *All* NP*-complete sets are p-isomorphic to each other.*

The conjecture immediately implies DP $\neq$ NP:

**Proposition 2.1.** *If the Isomorphism Conjecture is true then* DP $\neq$ NP.

***Proof.*** If DP = NP then all sets in DP are NP-complete. However, DP has both finite and infinite sets and there cannot exist an isomorphism between a finite and an infinite set. Hence the Isomorphism Conjecture is false. □

This suggests that proving the conjecture is hard because the problem of separating DP from NP has resisted all efforts so far. A natural question, therefore, is: Can one prove the conjecture assuming a reasonable hypothesis such as DP $\neq$ NP? We address this question later in the paper. In their paper, Berman and Hartmanis also asked a weaker question: Does DP $\neq$ NP imply that no *sparse* set can be NP-complete?

**Definition 2.8.** Set $A$ is *sparse* if there exist constants $k, n_0 > 0$ such that for every $n > n_0$, the number of strings in $A$ of length $\leq n$ is at most $n^k$.

This was answered in the affirmative by Mahaney [49]:

**Theorem 2.4. (Mahaney)** *If* DP $\neq$ NP *then no sparse set is* NP*-complete.*

***Proof Sketch.*** We give a proof based on an idea of [9, 19, 51]. Suppose there is a sparse set $S$ such that SAT $\leq_m^p S$ via $f$. Let $F$ be a Boolean formula on $n$ variables. Start with the set $T = \{F\}$ and do the following:

Replace each formula $\hat{F} \in T$ by $\hat{F}_0$ and $\hat{F}_1$ where $\hat{F}_0$ and $\hat{F}_1$ are obtained by setting the first variable of $\hat{F}$ to 0 and 1 respectively. Let $T = \{F_1, F_2, \ldots, F_t\}$. If $t$ exceeds a certain threshold $t_0$, then let $G_j = F_1 \vee F_j$ and $z_j = f(G_j)$ for $1 \leq j \leq t$. If all $z_j$'s are distinct then drop $F_1$ from $T$. Otherwise, $z_i = z_j$ for some $i \neq j$. Drop $F_i$ from $T$ and repeat until $|T| \leq t_0$. If $T$ has only formulas with no variables, then output Satisfiable if $T$ contains a True formula else output Unsatisfiable. Otherwise, go to the beginning of the algorithm and repeat.

The invariant maintained during the entire algorithm is that $F$ is satisfiable iff $T$ contains a satisfiable formula. It is true in the beginning, and remains true in each iteration after replacing every formula $\hat{F} \in T$ with $\hat{F}_0$ and $\hat{F}_1$. The threshold $t_0$ must be such that $t_0$ is a upper bound on the number of strings in the set $S$ of size $\max_j |f(G_j)|$. This is a polynomial in $|F|$ since $|G_j| \leq 2|F|$, $f$ is a p-function, and $S$ is sparse. If $T$ has more than $t_0$ formulas at any stage then the algorithm drops a formula from $T$. This formula is $F_1$ when all $z_j$'s are distinct. This means there are more than $t_0$ $z_j$'s all of size bounded by $\max_j |f(G_j)|$. Not all of these can be in $S$ due to the choice of $t_0$ and hence $F_1 \notin \mathsf{SAT}$. If $z_i = z_j$ then $F_i$ is dropped. If $F_i$ is satisfiable then so is $G_i$. And since $z_i = z_j$ and $f$ is a reduction of SAT to $S$, $G_j$ is also satisfiable; hence either $F_1$ or $F_j$ is satisfiable. Therefore dropping $F_i$ from $T$ maintains the invariant.

The above argument shows that the size of $T$ does not exceed a polynomial in $|F|$ at any stage. Since the number of iterations of the algorithm is bounded by $n \leq |F|$, the overall time complexity of the algorithm is polynomial. Hence $\mathsf{SAT} \in \mathsf{DP}$ and therefore, $\mathsf{DP} = \mathsf{NP}$. □

The "searching-with-pruning" technique used in the above proof has been used profitably in many results subsequently. The Isomorphism Conjecture, in fact, implies a much stronger density result: All NP-complete sets are *dense.*

**Definition 2.9.** Set $A$ is *dense* if there exist constants $\epsilon, n_0 > 0$ such that for every $n > n_0$, the number of strings in $A$ of length $\leq n$ is at least $2^{n^\epsilon}$.

Buhrman and Hitchcock [22] proved that, under a plausible hypothesis, every NP-complete set is dense infinitely often:

**Theorem 2.5. (Buhrman–Hitchcock)** *If* PH *is infinite then for any* NP*-complete set $A$, there exists $\epsilon > 0$ such that for infinitely many $n$, the number of strings in $A$ of length $\leq n$ is at least $2^{n^\epsilon}$.*

Later, we show that a stronger density theorem holds if $2^{n^\epsilon}$-secure one-way functions exist.

## 2.4. A Counter Conjecture and Relativizations

After Mahaney's result, there was not much progress on the conjecture although researchers believed it to be true. However, this changed in 1984 when Joseph and Young [40] argued that the conjecture is false. Their argument was as follows (paraphrased by Selman [53]). Let $f$ be any 1-1, size-increasing, $2^{n^\epsilon}$-secure one-way function. Consider the set $A = f(\mathsf{SAT})$. Set $A$ is clearly NP-complete. If it is p-isomorphic to SAT, there must exist a 1-1, honest p-reduction of SAT to $A$ which is also p-invertible. However, the set $A$ is, in a sense, a "coded" version of SAT such that on most of the strings of $A$, it is hard to "decode" it (because $f$ is not p-invertible on most of the strings). Thus, there is unlikely to be a 1-1, honest p-reduction of SAT to $A$ which is also p-invertible, and so $A$ is unlikely to be p-isomorphic to SAT. This led them to make a counter conjecture:

**Encrypted Complete Set Conjecture.** *There exists a 1-1, size-increasing, one-way function $f$ such that* SAT *and* $f(\mathsf{SAT})$ *are not p-isomorphic to each other.*

It is useful to observe here that this conjecture is false in computable setting: The inverse of any 1-1, size-increasing, computable function is also computable. The restriction to polynomial-time computability is what gives rise to the possible existence of one-way functions.

It is also useful to observe that this conjecture too implies $\mathsf{DP} \neq \mathsf{NP}$:

**Proposition 2.2.** *If the Encrypted Complete Set Conjecture is true then* $\mathsf{DP} \neq \mathsf{NP}$.

***Proof.*** If $\mathsf{DP} = \mathsf{NP}$ then every 1-1, size-increasing p-function is also p-invertible. Hence for every such function, SAT and $f(\mathsf{SAT})$ are p-isomorphic. □

The Encrypted Complete Set conjecture fails if one-way functions do not exist. Can it be shown to follow from the existence of strong one-way functions, such as $2^{n^\epsilon}$-secure one-way functions? This is not clear. (In fact, later we argue the opposite.) Therefore, to investigate the two conjectures further, the focus moved to relativized worlds. Building on a result of Kurtz [42], Hartmanis and Hemachandra [33] showed that there is an oracle relative to which DP = UP and the Isomorphism Conjecture is false. This shows that *both* the conjectures fail in a relativized world since DP = UP implies that no one-way functions exist.

Kurtz, Mahaney, and Royer [46] defined the notion of *scrambling functions*:

**Definition 2.10.** Function $f$ is *scrambling function* if $f$ is 1-1, size-increasing, p-computable, and there is no dense polynomial-time subset in range$(f)$.

Kurtz et al. observed that,

**Proposition 2.3.** *If scrambling functions exist then the Encrypted Complete Set Conjecture is true.*

***Proof.*** Let $f$ be a scrambling function, and consider $A = f(\mathsf{SAT})$. Set $A$ is NP-complete. Suppose it is p-isomorphic to SAT and let $p$ be the isomorphism between SAT and $A$. Since SAT has a dense polynomial-time subset, say $D$, $p(D)$ is a dense polynomial time subset of $A$. This contradicts the scrambling property of $f$. □

Kurtz et al., [46], then showed that,

**Theorem 2.6. (Kurtz, Mahaney, Royer)** *Relative to a random oracle, scrambling functions exist.*

***Proof Sketch.*** Let $O$ be an oracle. Define function $f$ as:

$$f(x) = O(x)O(x1)O(x11)\cdots O(x1^{2|x|})$$

where $O(z) = 1$ if $z \in O$, 0 otherwise. For a random choice of $O$, $f$ is 1-1 with probability 1. So, $f$ is a 1-1, size-increasing, $p^O$-computable function. Suppose a polynomial-time TM $M$ with oracle $O$ accepts a subset of range$(f)$. In order to distinguish a string in range of $f$ from those outside, $M$ needs to check the answer of oracle $O$ on several unique strings. And since $M$ can query only polynomially many strings from $O$, $M$ can accept only a sparse subset of range$(f)$. □

Therefore, relative to a random oracle, the Encrypted Complete Set Conjecture is true and the Isomorphism Conjecture is false. The question of existence of an oracle relative to which the Isomorphism Conjecture is true was resolved by Fenner, Fortnow, and Kurtz [27]:

**Theorem 2.7. (Fenner, Fortnow, Kurtz)** *There exists an oracle relative to which Isomorphism Conjecture is true.*

Thus, there are relativizations in which each of the three possible answers to the two conjectures is true. However, the balance of evidence provided by relativizations is towards the Encrypted Complete Set Conjecture since properties relative to a random oracle are believed to be true in unrelativized world too.[b]

### 2.5. The Conjectures for Other Classes

In search of more evidence for the two conjectures, researchers translated them to classes bigger than NP. The hope was that diagonalization arguments that do not work within NP can be used for these classes to prove stronger results about the structure of complete sets. This hope was realized, but not completely. In this section, we list the major results obtained for classes EXP and NEXP which were the two main classes considered.

Berman [20] showed that,

**Theorem 2.8. (Berman)** *Let $A$ be a p-complete set for* EXP. *Then for every $B \in$ EXP, $B \leq^p_{1,si} A$.*

***Proof Sketch.*** Let $M_1, M_2, \ldots$ be an enumeration of all polynomial-time TMs such that $M_i$ halts, on input $x$, within time $|x|^{|i|} + |i|$ steps. Let $B \in$ EXP and define $\hat{B}$ to be the set accepted by the following algorithm:

> Input $(i, x)$. Let $M_i(i, x) = y$. If $|y| \leq |x|$, accept iff $y \notin A$. If there exists a $z$, $z < x$ (in lexicographic order), such that $M_i(i, z) = y$, then accept iff $z \notin B$. Otherwise, accept iff $x \in B$.

The set $\hat{B}$ is clearly in EXP. Let $\hat{B} \leq^p_m A$ via $f$. Let the TM $M_j$ compute $f$. Define function $g$ as: $g(x) = f(j, x)$. It is easy to argue that $f$ is 1-1 and size-increasing on inputs of the form $(j, \star)$ using the definition of $\hat{B}$ and the fact that $f$ is a reduction. It follows that $g$ is a 1-1, size-increasing p-reduction of $B$ to $A$. □

**Remark 2.1.** A case can be made that the correct translation of the isomorphism result of [50] to the polynomial-time realm is to show that the complete sets are also complete under 1-1, size-increasing reductions. As observed earlier, the non-trivial part of the result in the setting of computability is to show the above implication. Inverting computable reductions is trivial. This translation will also avoid the conflict with Encrypted Complete Set Conjecture as it does not require p-invertibility. In fact, as

[b]There are notable counterexamples of this though. The most prominent one is the result IP = PSPACE [48, 54] which is false relative to a random oracle [24].

will be shown later, one-way functions help in proving an analog of the above theorem for the class NP! However, the present formulation has a nice symmetry to it (both the isomorphism and its inverse require the same amount of resources) and hence is the preferred one.

For the class NEXP, Ganesan and Homer [29] showed that,

**Theorem 2.9. (Ganesan–Homer)** *Let $A$ be a p-complete set for* NEXP. *Then for every $B \in$ NEXP, $B \leq_1^p A$.*

The proof of this uses ideas similar to the previous proof for EXP. The result obtained is not as strong since enforcing the size-increasing property of the reduction requires accepting the complement of a NEXP set which cannot be done in NEXP unless NEXP is closed under complement, a very unlikely possibility. Later, the author [5] proved the size-increasing property for reductions to complete sets for NEXP under a plausible hypothesis.

While the two conjectures could not be settled for the complete p-degree of EXP (and NEXP), answers have been found for p-degrees *close* to the complete p-degree of EXP. The first such result was shown by Ko, Long, and Du [41]. We need to define the notion of truth-table reductions to state this result.

**Definition 2.11.** Set $A$ *$k$-truth-table reduces* to set $B$ if there exists a p-function $f$, $f : \Sigma^* \mapsto \underbrace{\Sigma^* \times \Sigma^* \times \cdots \times \Sigma^*}_{k} \times \Sigma^{2^k}$ such that for every $x \in \Sigma^*$, if $f(x) = (y_1, y_2, \ldots, y_k, T)$ then $x \in A$ iff $T(B(y_1)B(y_2)\cdots B(y_k)) = 1$ where $B(y_i) = 1$ iff $y_i \in B$ and $T(s)$, $|s| = k$, is the $s$th bit of string $T$.

Set $B$ is *$k$-truth-table complete* for EXP if $B \in$ EXP and for every $A \in$ EXP, $A$ $k$-truth-table reduces to $B$.

The notion of truth-table reductions generalizes p-reductions. For both EXP and NEXP, it is known that complete sets under 1-truth-table reductions are also p-complete [23, 38], and not all complete sets under 2-truth-table reductions are p-complete [55]. Therefore, the class of 2-truth-table complete sets for EXP is the smallest class properly containing the complete p-degree of EXP.

Ko, Long, and Du [41] related the structure of certain p-degrees to the existence of worst-case one-way functions:

**Theorem 2.10. (Ko–Long–Du)** *If there exist worst-case one-way functions then there is a p-degree in* EXP *such that the sets in the degree are not* *all p-isomorphic to each other. Further, sets in this degree are 2-truth-table complete for* EXP.

Kurtz, Mahaney, and Royer [43] found a p-degree for which the sets are unconditionally not all p-isomorphic to each other:

**Theorem 2.11. (Kurtz–Mahaney–Royer)** *There exists a p-degree in* EXP *such that the sets in the degree are not all p-isomorphic to each other. Further, sets in this degree are 2-truth-table complete for* EXP.

Soon afterwards, Kurtz, Mahaney, and Royer [44] found another p-degree with the opposite structure:

**Theorem 2.12. (Kurtz–Mahaney–Royer)** *There exists a p-degree in* EXP *such that the sets in the degree are all p-isomorphic to each other. Further, this degree is located inside the 2-truth-table complete degree of* EXP.

The set of results above on the structure of complete (or nearly complete) p-degree of EXP and NEXP do not favor any of the two conjectures. However, they do suggest that the third possibility, viz., both the conjectures being false, is unlikely.

## 2.6. The Conjectures for Other Reducibilities

Another direction from which to approach the two conjectures is to weaken the power of reductions instead of the class NP, the hope being that for reductions substantially weaker than polynomial-time, one can prove unconditional results. For several weak reductions, this was proven correct and in this section we summarize the major results in this direction.

The two conjectures for $r$-reductions can be formulated as:

**$r$-Isomorphism Conjecture.** *All $r$-complete sets for* NP *are $r$-isomorphic to each other.*

**$r$-Encrypted Complete Set Conjecture.** *There is a 1-1, size-increasing, $r$-function $f$ such that* SAT *and $f$(*SAT*) are not $r$-isomorphic to each other.*

Weakening p-reductions to *logspace-reductions* (functions computable by TMs with read-only input tape and work tape space bounded by $O(\log n)$, $n$ is the input size) does not yield unconditional results as any such result will separate NP from L, another long-standing open problem. So we need to weaken it further. There are three major ways of doing this.

### 2.6.1. *Restricting the input head movement*

Allowing the input head movement in only one direction leads to the notion of 1-L-functions.

**Definition 2.12.** A *1-L-function* is computed by deterministic TMs with read-only input tape, the workspace bounded by $O(\log n)$ where $n$ is the input length, and the input head restricted to move in one direction only (left-to-right by convention). In other words, the TM is allowed only one scan of its input. To ensure the space bound, the first $O(\log n)$ cells on the work tape are marked at the beginning of the computation.

These functions were defined by Hartmanis, Immerman, and Mahaney [34] to study the complete sets for the class L. They also observed that the "natural" NP-complete sets are also complete under 1-L-reductions. Structure of complete sets under 1-L-reductions attracted a lot of attention, and the first result was obtained by Allender [14]:

**Theorem 2.13. (Allender)** *For the classes* PSPACE *and* EXP, *complete sets under 1-L-reductions are p-isomorphic to each other.*

While this shows a strong structure of complete sets of some classes under 1-L-reductions, it does not answer the 1-L-Isomorphism Conjecture. After a number of extensions and improvements [10, 29, 37], the author [1] showed that,

**Theorem 2.14. (Agrawal)** *Let $A$ be a 1-L-complete set for* NP. *Then for every $B \in \mathsf{NP}$, $B \leq_{1,si,i}^{1-L} A$.*

***Proof Sketch.*** We first show that $A$ is also complete under *forgetful* 1-L-reductions. Forgetful 1-L-reductions are computed by TMs that, immediately after reading a bit of the input, forget its value. This property is formalized by defining configurations: A *configuration* of a 1-L TM is a tuple $\langle q, j, w \rangle$ where $q$ is a state of the TM, $j$ its input head position, and $w$ the contents of its worktape including the position of the worktape head. A forgetful TM, after reading a bit of the input and before reading the next bit, reaches a configuration which is independent of the value of the bit that is read.

Let $B \in \mathsf{NP}$, and define $\hat{B}$ to be the set accepted by the following algorithm:

> Input $x$. Let $x = y10^b1^k$. Reject if $b$ is odd or $|y| \neq tb$ for some integer $t$. Otherwise, let $y = y_1y_2\cdots y_t$ with $|y_i| = b$. Let $v_i = 1$ if $y_i = uu$ for some $u$, $|u| = \frac{b}{2}$; $v_i = 0$ otherwise. Accept iff $v_1v_2\cdots v_t \in B$.

The set $\hat{B}$ is a "coded" version of set $B$ and reduces to $B$ via a p-reduction. Hence, $\hat{B} \in \mathsf{NP}$. Let $f$ be a 1-L-reduction of $\hat{B}$ to $A$ computed by TM $M$. Consider the workings of $M$ on inputs of size $n$. Since $M$ has $O(\log n)$ space, the number of configurations of $M$ will be bounded by a polynomial, say $q(\cdot)$, in $n$. Let $b = k\lceil \log n \rceil$ such that $2^{b/2} > q(n)$. Let $C_0$ be the initial configuration of $M$. By the Pigeon Hole Principle, it follows that there exist two distinct strings $u_1$ and $u_1'$, $|u_1| = |u_1'| = \frac{b}{2}$, such that $M$ reaches the same configuration, after reading either of $u_1$ and $u_1'$. Let $C_1$ be the configuration reached from this configuration after reading $u_1$. Repeat the same argument starting from $C_1$ to obtain strings $u_2$, $u_2'$, and configuration $C_2$. Continuing this way, we get triples $(u_i, u_i', C_i)$ for $1 \leq i \leq t = \lfloor \frac{n-b-1}{b} \rfloor$. Let $k = n - b - 1 - bt$. It follows that the TM $M$ will go through the configurations $C_0$, $C_1$, ..., $C_t$ on any input of the form $y_1y_2\ldots y_t10^b1^k$ with $y_i \in \{u_iu_i, u_i'u_i\}$. Also, that the pair $(u_i, u_i')$ can be computed in logspace without reading the input.

Define a reduction $g$ of $B$ to $\hat{B}$ as follows: On input $v$, $|v| = t$, compute $b$ such that $2^{b/2} > q(b+1+bt)$, and consider $M$ on inputs of size $b+1+bt$. For each $i$, $1 \leq i \leq t$, compute the pair $(u_i, u_i')$ and output $u_iu_i$ if the $i$th bit of $v$ is 1, output $u_iu_i'$ otherwise. It is easy to argue that the composition of $f$ and $g$ is a forgetful 1-L-reduction of $B$ to $A$.

Define another set $B'$ as accepted by the following algorithm:

> Input $x$. Reject if $|x|$ is odd. Otherwise, let $x = x_1x_2\cdots x_ns_1s_2\cdots s_n$. Accept if exactly one of $s_1$, $s_2$, ..., $s_n$, say $s_j$, is zero and $x_j = 1$. Accept if all of $s_1$, $s_2$, ..., $s_n$ are one and $x_1x_2\cdots x_n \in B$. Reject in all other cases.

Set $B' \in \mathsf{NP}$. As argued above, there exists a forgetful 1-L-reduction of $B'$ to $A$, say $h$. Define a reduction $g'$ of $B$ to $B'$ as: $g'(v) = v1^{|v|}$. It is easy to argue that $h \circ g'$ is a size-increasing, 1-L-invertible, 1-L-reduction of $B$ to $A$ and $h \circ g'$ is 1-1 on strings of size $n$ for all $n$. Modifying this to get a reduction that is 1-1 everywhere is straightforward. □

The above result strongly suggests that the 1-L-Isomorphism Conjecture is true. However, the author [1] showed that,

**Theorem 2.15. (Agrawal)** *1-L-complete sets for* NP *are all 2-L-isomorphic to each other but not 1-L-isomorphic.*

The *2-L-isomorphism* above is computed by logspace TMs that are allowed two left-to-right scans of their input. Thus, the 1-L-Isomorphism Conjecture fails and a little more work shows that the 1-L-Encrypted Complete Set Conjecture is true! However, the failure of the Isomorphism Conjecture here is for a very different reason: it is because 1-L-reductions are not powerful enough to carry out the isomorphism construction as in Theorem 2.2. For a slightly more powerful reducibility, 1-NL-reductions, this is not the case.

**Definition 2.13.** A *1-NL-function* is computed by TMs satisfying the requirements of definition 2.12, but allowed to be non-deterministic. The non-deterministic TM must output the same string on all paths on which it does not abort the computation.

For 1-NL-reductions, the author [1] showed, using proof ideas similar to the above one, that,

**Theorem 2.16. (Agrawal)** *1-NL-complete sets for* NP *are all 1-NL-isomorphic to each other.*

The author [1] also showed similar results for *c-L-reductions* for constant $c$ (functions that are allowed at most $c$ left-to-right scans of the input).

### 2.6.2. *Reducing space*

The second way of restricting logspace reductions is by allowing the TMs only *sublogarithmic* space, i.e., allowing the TM space $o(\log n)$ on input of size $n$; we call such reductions *sublog-reductions.* Under sublog-reductions, NP has no complete sets, and the reason is simple: Every sublog-reduction can be computed by deterministic TMs in time $O(n^2)$ and hence if there is a complete set for NP under sublog-reductions, then $\mathsf{NTIME}(n^{k+1}) = \mathsf{NTIME}(n^k)$ for some $k > 0$, which is impossible [26]. On the other hand, each of the classes $\mathsf{NTIME}(n^k)$, $k \geq 1$, has complete sets under sublog-reductions.

The most restricted form for sublog-reductions is *2-DFA-reductions*:

**Definition 2.14.** A *2-DFA-function* is computed by a TM with read-only input tape and no work tape.

2-DFA functions do not require any space for their computation, and therefore are very weak. Interestingly, the author [4] showed that sublog-reductions do not add any additional power for complete sets:

**Theorem 2.17. (Agrawal)** *For any $k \geq 1$, sublog-complete sets for* $\mathsf{NTIME}(n^k)$ *are also 2-DFA-complete.*

For 2-DFA-reductions, the author and Venkatesh [12] proved that,

**Theorem 2.18. (Agrawal-Venkatesh)** *Let $A$ be a 2-DFA-complete set for* $\mathsf{NTIME}(n^k)$ *for some $k \geq 1$. Then, for every $B \in \mathsf{NTIME}(n^k)$, $B \leq_{1,si}^{2DFA} A$ via a reduction that is* mu-DFA-invertible.

*muDFA-functions* are computed by TMs with no space and multiple heads, each moving in a single direction only. The proof of this is also via forgetful TMs. The reductions in the theorem above are not 2-DFA-invertible, and in fact, it was shown in [12] that,

**Theorem 2.19. (Agrawal-Venkatesh)** *Let $f(x) = xx$. Function $f$ is a 2-DFA-function and for any $k \geq 1$, there is a 2-DFA-complete set $A$ for* $\mathsf{NTIME}(n^k)$ *such that $A \not\leq_{1,si,i}^{2DFA} f(A)$.*

The above theorem implies that 2-DFA-Encrypted Complete Set Conjecture is true.

### 2.6.3. *Reducing depth*

Logspace reductions can be computed by (unbounded fan-in) *circuits* of logarithmic depth.[c] Therefore, another type of restricted reducibility is obtained by further reducing the depth of the circuit family computing the reduction. Before proceeding further, let us define the basic notions of a circuit model.

**Definition 2.15.** A *circuit family* is a set $\{C_n : n \in \mathbb{N}\}$ where each $C_n$ is an acyclic circuit with $n$ Boolean inputs $x_1, \ldots, x_n$ (as well as the constants 0 and 1 allowed as inputs) and some number of output gates $y_1, \ldots, y_r$. $\{C_n\}$ has *size* $s(n)$ if each circuit $C_n$ has at most $s(n)$ gates; it has *depth* $d(n)$ if the length of the longest path from input to output in $C_n$ is at most $d(n)$.

A circuit family has a notion of uniformity associated with it:

[c] For a detailed discussion on the circuit model of computation, see [52].

**Definition 2.16.** A family $\mathcal{C} = \{C_n\}$ is *uniform* if the function $n \mapsto C_n$ is easy to compute in some sense. This can also be defined using the complexity of the *connection set* of the family:

$$\text{conn}(\mathcal{C}) = \{(n, i, j, T_i, T_j) \mid \text{the output of gate } i \text{ of type } T_i \text{ is input to gate } j \text{ of type } T_j \text{ in } C_n\}.$$

Here, gate type $T_i$ can be Input, Output, or some Boolean operator.

Family $\mathcal{C}$ is *Dlogtime-uniform* [18] if conn($\mathcal{C}$) is accepted by a linear-time TM. It is *p-uniform* [15] if conn($\mathcal{C}$) is accepted by a exponential-time TM (equivalently, by a TM running in time bounded by a polynomial in the circuit size). If we assume nothing about the complexity of conn($\mathcal{C}$), then we say that the family is *non-uniform.*

An important restriction of logspace functions is to functions computed by *constant depth* circuits.

**Definition 2.17.** Function $f$ is a *u-uniform* $\mathsf{AC}^0$*-function* if there is a $u$-uniform circuit family $\{C_n\}$ of size $n^{O(1)}$ and depth $O(1)$ consisting of unbounded fan-in AND and OR and NOT gates such that for each input $x$ of length $n$, the output of $C_n$ on input $x$ is $f(x)$.

Note that with this definition, an $\mathsf{AC}^0$-function cannot map strings of equal size to strings of different sizes. To allow this freedom, we adopt the following convention: Each $C_n$ will have $n^k + k\log(n)$ output bits (for some $k$). The last $k \log n$ output bits will be viewed as a binary number $r$, and the output produced by the circuit will be the binary string contained in the first $r$ output bits.

It is worth noting that, with this definition, the class of Dlogtime-uniform $\mathsf{AC}^0$-functions admits many alternative characterizations, including *expressibility in first-order logic with* $\{+, \times, \leq\}$ [18, 47], the *logspace-rudimentary reductions* [17, 39], *logarithmic-time alternating Turing machines with* $O(1)$ *alternations* [18] etc. Moreover, almost all known NP-complete sets are also complete under Dlogtime-uniform $\mathsf{AC}^0$-reductions (an exception is provided by [7]). We will refer to Dlogtime-uniform $\mathsf{AC}^0$-functions also as *first-order-functions.*

$\mathsf{AC}^0$-reducibility is important for our purposes too, since the complete sets under the reductions of the previous two subsections are also complete under $\mathsf{AC}^0$-reductions (with uniformity being Dlogtime- or p-uniform). This follows from the fact that these sets are also complete under some appropriate notion of forgetful reductions. Therefore, the class of $\mathsf{AC}^0$-complete sets for NP is larger than all of the previous classes of this section.

The first result for depth-restricted functions was proved by Allender, Balcázar, and Immerman [16]:

**Theorem 2.20. (Allender–Balcázar–Immerman)** *Complete sets for* NP *under first-order projections are first-order-isomorphic to each other.*

*First-order projections* are computed by a very restricted kind of Dlogtime-uniform $\mathsf{AC}^0$ family in which no circuit has AND and OR gates. This result was generalized by the author and Allender [6] to $\mathsf{NC}^0$*-functions*, which are functions computed by $\mathsf{AC}^0$ family in which the fan-in of every gate of every circuit is at most two.

**Theorem 2.21. (Agrawal–Allender)** *Let $A$ be a non-uniform* $\mathsf{NC}^0$*-complete set for* NP*. Then for any $B \in$* NP*, $B$ non-uniform* $\mathsf{NC}^0$*-reduces to $A$ via a reduction that is 1-1, size-increasing, and non-uniform* $\mathsf{AC}^0$*-invertible. Further, all non-uniform* $\mathsf{NC}^0$*-complete sets for* NP *are non-uniform* $\mathsf{AC}^0$*-isomorphic to each other where these isomorphisms can be computed and inverted by depth three non-uniform* $\mathsf{AC}^0$ *circuits.*

***Proof Sketch.*** The proof we describe below is the one given in [3]. Let $B \in \mathsf{NP}$, and define $\hat{B}$ to be the set accepted by the following algorithm:

> On input $y$, let $y = 1^k0z$. If $k$ does not divide $|z|$, then reject. Otherwise, break $z$ into blocks of $k$ consecutive bits each. Let these be $u_1u_2u_3 \ldots u_p$. Accept if there is an $i$, $1 \le i \le p$, such that $u_i = 1^k$. Otherwise, reject if there is an $i$, $1 \le i \le p$, such that $u_i = 0^k$. Otherwise, for each $i$, $1 \le i \le p$, label $u_i$ as *null* if the number of ones in it is 2 modulo 3; as *zero* if the number of ones in it is 0 modulo 3; and as *one* otherwise. Let $v_i = \epsilon$ if $u_i$ is null, 0 if $u_i$ is zero, and 1 otherwise. Let $x = v_1v_2 \cdots v_p$, and accept iff $x \in B$.

Clearly, $\hat{B} \in \mathsf{NP}$. Let $\{C_n\}$ be the $\mathsf{NC}^0$ circuit family computing a reduction of $\hat{B}$ to $A$. Fix size $n$ and consider circuit $C_{k+1+n}$ for $k = 4\lceil \log n \rceil$. Let $C$ be the circuit that results from setting the first input $k+1$ bits of $C_{k+1+n}$ to $1^k0$. Randomly set each of the $n$ input bits of $C$ in the following way: With probability $\frac{1}{2}$, leave it unset; with probability $\frac{1}{4}$ each, set it to 0 and 1 respectively. The probability that any block of $k$ bits is completely set is at most $\frac{1}{n^4}$. Similarly, the probability that there is a block that has at most three unset bits is at most $\frac{1}{n}$, and therefore, with high probability, every block has at least four unset bits.

Say that an output bit is *good* if, after the random assignment to the input bits described above is completed, the value of the output bit depends on exactly one unset input bit. Consider an output bit. Since $C$ is an $\mathsf{NC}^0$

circuit, the value of this bit depends on at most a constant, say $c$, number of input bits. Therefore, the probability that this bit is good after the assignment is at least $\frac{1}{2} \cdot \frac{1}{4^{c-1}}$. Therefore, the expected number of good output bits is at least $\frac{m}{4^c}$, where $m$ is the number of output bits of $C$ whose value depends on some input bit. Using the definition of set $\hat{B}$, it can be argued that $\Omega(n)$ output bits depend on some input bit, and hence $\Omega(n)$ output bits are expected to be good after the assignment. Fix any assignment that does this, as well as leaves at least four unset bits in each block. Now set some more input bits so that each block that is completely set is null, each block that has exactly two unset bits has number of ones equal to 0 modulo 3, and there are no blocks with one, three, or more unset bits. Further, for at least one unset input bit in a block, there is a good output bit that depends on the bit, and there are $\Omega(\frac{n}{\log n})$ unset input bits. It is easy to see that all these conditions can be met.

Now define a reduction of $B$ to $\hat{B}$ as: On input $x$, $|x| = p$, consider $C_{k+1+n}$ such that the number of unset input bits in $C_{k+1+n}$ after doing the above process is at least $p$. Now map the $i$th bit of $x$ to the unset bit in a block that influences a good output bit and set the other unset input bit in the block to zero. This reduction can be computed by an $\mathsf{NC}^0$ circuit (in fact, the circuit does not need any AND or OR gate).

Define a reduction of $B$ to $A$ given by the composition of the above two reductions. This reduction is a *superprojection*: it is computed by circuit family $\{D_p\}$ with each $D_p$ being an $\mathsf{NC}^0$ circuit such that for every input bit to $D_p$, there is an output bit that depends exactly on this input bit. A superprojection has the input written in certain bit positions of the output. Therefore, it is 1-1 and size-increasing. Inverting the function is also easy: Given string $y$, identify the locations where the input is written, and check if the circuit $D_p$ ($p$ = number of locations) on this input outputs $y$. This checking can be done by a depth two $\mathsf{AC}^0$ circuit.

This gives a 1-1, size-increasing, $\mathsf{AC}^0$-invertible, $\mathsf{NC}^0$-reduction of $B$ to $A$. The circuit family is non-uniform because it is not clear how to deterministically compute the settings of the input bits. Exploiting the fact that the input is present in the output of the reductions, an $\mathsf{AC}^0$-isomorphism, computed by depth three circuits, can be constructed between two complete sets following [21] (see [8] for details). □

Soon after, the author, Allender, and Rudich [8] extended it to all $\mathsf{AC}^0$-functions, proving the Isomorphism Conjecture for non-uniform $\mathsf{AC}^0$-functions.

**Theorem 2.22. (Agrawal–Allender–Rudich)** *Non-uniform* $\mathsf{AC}^0$*-complete sets for* NP *are non-uniform* $\mathsf{AC}^0$*-isomorphic to each other. Further, these isomorphisms can be computed and inverted by depth three non-uniform* $\mathsf{AC}^0$ *circuits.*

***Proof Sketch.*** The proof shows that complete sets for NP under $\mathsf{AC}^0$-reductions are also complete under $\mathsf{NC}^0$-reductions and invokes the above theorem for the rest. Let $A$ be a complete set for NP under $\mathsf{AC}^0$-reductions. Let $B \in \mathsf{NP}$. Define set $\hat{B}$ exactly as in the previous proof. Fix an $\mathsf{AC}^0$-reduction of $\hat{B}$ to $A$ given by family $\{C_n\}$. Fix size $n$, and consider $C_{k+1+n}$ for $k = n^{1-\epsilon}$ for a suitable $\epsilon > 0$ to be fixed later. Let $D$ be the circuit that results from setting the first $k+1$ input bits of $C_{k+1+n}$ to $1^k0$.

Set each input bit of $D$ to 0 and 1 with probability $\frac{1}{2} - \frac{1}{2n^{1-2\epsilon}}$ each and leave it unset with probability $\frac{1}{n^{1-2\epsilon}}$. By the Switching Lemma of Furst, Saxe, and Sipser [28], the circuit $D$ will reduce, with high probability, to an $\mathsf{NC}^0$ circuit on the unset input bits for a suitable choice of $\epsilon > 0$. In each block of $k$ bits, the expected number of unset bits will be $n^\epsilon$, and therefore, with high probability, each block has at least three unset bits. Fix any settings satisfying both of the above.

Now define a reduction of $B$ to $\hat{B}$ that, on input $x$, $|x| = p$, identifies $n$ for which the circuit $D$ has at least $p$ blocks, and then maps $i$th bit of input $x$ to an unset bit of the $i$th block of the input to $D$, setting the remaining bits of the block so that the sum of ones in the block is 0 modulo 3. Unset bits in all remaining blocks are set so that the sum of ones in the block equals 2 modulo 3.

The composition of the reduction of $B$ to $\hat{B}$ and $\hat{B}$ to $A$ is an $\mathsf{NC}^0$-reduction of $B$ to $A$. Again, it is non-uniform due to the problem of finding the right settings of the input bits. □

The focus then turned towards removing the non-uniformity in the above two reductions. In the proof of Theorem 2.21 given in [6], the uniformity condition is p-uniform. In [7], the uniformity of 2.22 was improved to p-uniform by giving a polynomial-time algorithm that computes the correct settings of input bits. Both the conditions were further improved to logspace-uniform in [3] by constructing a more efficient derandomization of the random assignments. And finally, in [2], the author obtained very efficient derandomizations to prove that,

**Theorem 2.23. (Agrawal)** *First-order-complete sets for* NP *are first-order-isomorphic.*

The isomorphisms in the theorem above are no longer computable by depth three circuits; instead, their depth is a function of the depth of the circuits computing reductions between the two complete sets.

### 2.6.4. *Discussion*

At first glance, the results for the weak reducibilities above seem to provide equal support to both the conjectures: The Isomorphism Conjecture is true for 1-NL and $\mathsf{AC}^0$-reductions for any reasonable notion of uniformity, while the Encrypted Complete Set Conjecture is true for 1-L and 2-DFA reductions. However, on a closer look a pattern begins to emerge. First of all, we list a common feature of all the results above:

**Corollary 2.1.** *For $r \in$ {1-L, 1-NL, 2-DFA, $\mathsf{NC}^0$, $\mathsf{AC}^0$}, $r$-complete sets for* NP *are also complete under 1-1, size-increasing, $r$-reductions.*

The differences arise in the resources required to invert the reductions and to construct the isomorphism. Some of the classes of reductions that we consider are so weak, that for a given function $f$ in the class, there is no function in the class that can check, on input $x$ and $y$, whether $f(x) = y$. For example, suppose $f$ is an $\mathsf{NC}^0$-function and one needs to construct a circuit that, on input $x$ and $y$, outputs 1 if $y = f(x)$, and outputs 0 otherwise. Given $x$ and $y$, an $\mathsf{NC}^0$ circuit can compute $f(x)$, and can check if the bits of $f(x)$ are equal to the corresponding bits of $y$; however, it cannot output 1 if $f(x) = y$, since this requires taking an AND of $|y|$ bits. Similarly, some of the reductions are too weak to construct the isomorphism between two sets given two 1-1, size-increasing, and invertible reductions between them. Theorems 2.14 and 2.15 show this for 1-L-reductions, and the same can be shown for $\mathsf{NC}^0$-reductions too. Observe that p-reductions do not suffer from either of these two drawbacks. Hence we cannot read too much into the failure of the Isomorphism Conjecture for $r$-reductions. We now formulate another conjecture that seems better suited to getting around the above drawbacks of some of the weak reducibilities. This conjecture was made in [1].

Consider a 1-1, size-increasing $r$-function $f$ for a resource bound $r$. Consider the problem of accepting the set range($f$). A TM accepting this set will typically need to guess an $x$ and then verify whether $f(x) = y$. It is, therefore, a non-deterministic TM with resource bound at least $r$. Let $r^{range} \geq r$ be the resource bound required by this TM. For a circuit accepting range($f$), the non-determinism is provided as additional "guess bits" and its output is 1 if the circuit evaluates to 1 on some settings of the guess bits. We can similarly define $r^{range}$ to be the resource bound required by such a *non-deterministic* circuit to accept range($f$).

**$r$-Complete Degree Conjecture.** *$r$-Complete sets for* NP *are also complete under 1-1, size-increasing, $r$-reductions that are $r^{range}$-invertible.*

Notice that the invertibility condition in the conjecture *does not allow non-determinism.* For p-reductions,

**Proposition 2.4.** *The p-Complete Degree Conjecture is equivalent to the Isomorphism Conjecture.*

***Proof.*** Follows from the observation that $p^{range} = p$ as range of a p-function can be accepted in non-deterministic polynomial-time, and from Theorem 2.2. □

Moreover, for the weaker reducibilities that we have considered, one can show that,

**Theorem 2.24.** *For $r \in \{$1-L, 1-NL, 2-DFA, $\mathsf{NC}^0$, $\mathsf{AC}^0\}$, the $r$-Complete Degree Conjecture is true.*

***Proof.*** It is an easy observation that for $r \in \{$1-L, 1-NL, $\mathsf{AC}^0\}$, $r^{range} = r$. The conjecture follows from Theorems 2.14, 2.16, and 2.23.

Accepting range of a 2-DFA-function requires verifying the output of 2-DFA TM on each of its constant number of passes on the input. The minimum resources required for this are to have multiple heads stationed at the beginning of the output of each pass, guess the input bit-by-bit, and verify the outputs on this bit for each pass simultaneously. Thus, the TM is a non-deterministic TM with no space and multiple heads, each moving in one direction only. So Theorem 2.18 proves the conjecture.

Accepting range of an $\mathsf{NC}^0$-function requires a non-deterministic $\mathsf{AC}^0$ circuit. Therefore, Theorems 2.21 and 2.23 prove the conjecture for $r = \mathsf{NC}^0$. □

In addition to the reducibilities in the above theorem, the $r$-Complete Degree Conjecture was proven for some more reducibilities in [1].

These results provide evidence that $r$-Complete Degree Conjecture is true for all reasonable resource bounds; in fact, *there is no known example of a reasonable reducibility for which the conjecture is false.*

The results above also raise doubts about the intuition behind the Encrypted Complete Set Conjecture as we shall argue now. Consider $\mathsf{AC}^0$-reductions. There exist functions computable by depth $d$, Dlogtime-uniform $\mathsf{AC}^0$ circuits that cannot be inverted on most of the strings by depth three, non-uniform $\mathsf{AC}^0$ circuits [35]. However, by Theorem 2.22, $\mathsf{AC}^0$-complete sets are also complete under $\mathsf{AC}^0$-reductions that are invertible by depth two, non-uniform $\mathsf{AC}^0$ circuits and the isomorphisms between all such sets are computable and invertible by depth three, non-uniform $\mathsf{AC}^0$ circuits. So, for every 1-1, size-increasing, $\mathsf{AC}^0$-function, it is possible to efficiently find a dense subset on which the function is invertible by depth two $\mathsf{AC}^0$ circuits.

Therefore, the results for weak reducibilities provide evidence that the Isomorphism Conjecture is true.

## 2.7. A New Conjecture

In this section, we revert to the conjectures in their original form. The investigations for weak reducibilities provide some clues about the structure of NP-complete sets. They strongly suggest that all NP-complete sets should also be complete under 1-1, size-increasing p-reductions. Proving this, of course, is hard as it implies $\mathsf{DP} \neq \mathsf{NP}$ (Proposition 2.1). Can we prove this under a reasonable assumption? This question was addressed and partially answered by the author in [5], and subsequently improved by the author and Watanabe [13]:

**Theorem 2.25. (Agrawal–Watanabe)** *If there exists a 1-1, $2^{n^\epsilon}$-secure one-way function for some $\epsilon > 0$, then all* NP*-complete sets are also complete under 1-1, and size-increasing,* P/poly*-reductions.*

In the above theorem, P/poly-*functions* are those computed by polynomial-size, non-uniform circuit families.

***Proof Sketch.*** Let $A$ be an NP-complete set and let $B \in \mathsf{NP}$. Let $f_0$ be a 1-1, $2^{n^\epsilon}$-secure one-way function. Recall that we have assumed that $|f_0(y)|$ is determined by $|y|$ for all $y$. Håstad et al., [36], showed how to construct a *pseudorandom generator* using any one-way function. Pseudorandom generators are size-increasing functions whose output cannot be distinguished from random strings by polynomial-time probabilistic TMs. Let $G$ be the pseudorandom generator constructed from $f_0$. Without loss of generality, we can assume that $|G(y)| = 2|y| + 1$ for all $y$. We also modify $f_0$ to $f$ as: $f(y, r) = f_0(y)rb$ where $|r| = |y|$ and $b = y \cdot r$, the inner product of strings $y$ and $r$. It is known that the bit $b$ is a *hard-core bit*, i.e., it cannot be predicted by polynomial-time probabilistic TMs on input $f_0(y)r$ [32].

Define $B_1$ to be the set:

$$B_1 = \{(x, w) \mid x \in B \ \wedge \ |w| = |x|^{2/\epsilon}\} \cup \text{range}(G),$$

and $B_2$ to be the set:

$$B_2 = \{f(z) \mid z \in B_1\}.$$

Both the sets are in NP. Let $B_2$ reduce to $A$ via polynomial-time reduction $g$. Since $f$ is 1-1, $h = g \circ f$ is a reduction of $B_1$ to $A$. We now show that $h$ rarely maps a large number of strings to a single string. For an odd $n$, let

$$p_n = \Pr_{z,z' \in_U \Sigma^n}[h(z) = h(z')].$$

In other words, $p_n$ is the collision probability of the function $h$ for strings of length $n$. Define function $\bar{f}(y, r) = f_0(y)r\bar{b}$ where $\bar{b}$ is the complement of the inner product value $y \cdot r$. Since $f_0$ is 1-1, range($f$) and range($\bar{f}$) are disjoint and therefore, range($\bar{F}$) is a subset of $\bar{B}_2$. Let

$$\bar{p}_n = \Pr_{z,z' \in_U \Sigma^n}[h(z) = g(\bar{f}(z'))].$$

Define a probabilistic TM $M^+$ that on input $u$, $|u| = |f(z)|$ for $|z| = n$, randomly picks $z' \in \Sigma^n$ and accepts iff $g(u) = h(z')$. The probability, over random $z \in \Sigma^n$, that $M^+$ accepts $f(z)$ is exactly $p_n$. The probability, over random $y, r \in \Sigma^{\frac{n-1}{2}}$ and $b \in \Sigma$, that $M^+$ accepts $u = f_0(y)rb$ is exactly $\frac{1}{2}p_n + \frac{1}{2}\bar{p}_n$ (since $b$ is either $y \cdot r$ or its complement with probability $\frac{1}{2}$ each). Hence the gap between the two probabilities is exactly $|\frac{1}{2}p_n - \frac{1}{2}\bar{p}_n|$. If this is large, then $M^+$ can be used to predict the hard-core bit of $f$ with high probability, which is not possible. Therefore, the difference of $p_n$ and $\bar{p}_n$ is small.

To show that $\bar{p}_n$ is small, define another TM $M^-$ that on input $z$, $|z| = n$, randomly picks $z' \in \Sigma^n$ and accepts iff $h(z) = g(\bar{f}(z'))$. On a random $z \in \Sigma^n$, the probability that $M^-$ accepts is exactly $\bar{p}_n$. On input $G(x)$ when $x$ is randomly chosen from $\Sigma^{\frac{n-1}{2}}$, the probability that $M^-$ accepts is zero since range($G$) is contained in $B_1$ and range($\bar{f}$) is contained in $\bar{B}_2$. Hence the difference between the two probabilities is exactly $\bar{p}_n$. This cannot be large as otherwise it violates the pseudorandomness of $G$. Therefore, $p_n$ is small.

Now define function $t$ as follows. For every $n$, randomly choose a $w_n$, $|w_n| = n^{2/\epsilon}$; let $t(x) = (x, w_{|x|})$. Note that $t$ is a probabilistic function. It can be argued that with high probability (over the choices of $w_n$),

(1) range($t$) does not intersect with range($G$), and so $t$ is a reduction of $B$ to $B_1$, and (2) $h \circ t$ is 1-1, and size-increasing. Non-uniformly fixing a choice of $w_n$ for every $n$, we get that $h \circ t$ is a 1-1, size-increasing, non-uniform polynomial-time reduction of $B$ to $A$. □

In hindsight, the above theorem is not surprising since the analogous result for EXP was shown using diagonalization [20] and one-way functions provide a strong form of diagonalization that works within NP in contrast to standard diagonalization techniques. It is a little unsatisfactory though, since it only shows completeness under *non-uniform* 1-1, size-increasing reductions. It is, however, sufficient to conclude that,

**Corollary 2.2.** *If there exists a 1-1, $2^{n^\epsilon}$-secure one-way function for some $\epsilon > 0$, then all* NP*-complete sets are dense.*

***Proof.*** By the above theorem, all NP-complete sets are also complete under 1-1, size-increasing, P/poly-reductions. It is an easy observation that if $A$ is dense and reduces to $B$ via a 1-1 reduction then $B$ is also dense. The corollary follows from the fact that SAT is dense. □

Another suggestion from the previous section is that one-way functions may have easily identifiable dense subsets on which they are p-invertible. This was investigated in [13], where the *easy cylinder* property was defined.

**Definition 2.18.** Let $f$ be a 1-1, size-increasing, P/poly-function. The function $f$ has an *easy cylinder* if there exist

- polynomials $q(\cdot)$, $q'(\cdot)$, and $\ell(\cdot)$ with $\ell(n) \geq 2q(q'(n)+n+\lceil\log(q'(n)+n)\rceil)$, and
- a P/poly embedding function $e$, computable by circuits of size $\leq q(|e(y)|)$ on input $y$,

such that for every $n$ and for every string $u$ of length $\ell(n)$, there exists a polynomial size circuit $C_u$, and string $s_u$, $|s_u| \leq q'(n)$, such that $C_u(f(u, e(s_u, x))) = x$ for all $x \in \Sigma^n$.

Intuitively, a function $f$ has an easy cylinder if there exists a parametrized (on $u$) dense subset in its domain on which it is easy to invert, and the dense subset depends on the parameter in a simple way (via the string $s_u$). Note that the circuit $C_u$ can be chosen depending on $f$ as well as $u$ but the embedding function $e$ must be independent of $u$.

Define set $K$ as:

$$K = \{(p, y) \mid p \text{ is a code of an NTM } M_p \text{ such that } M_p \text{ accepts } y \text{ in at most } |py|^2 \text{ steps}\}.$$

$K$ is easily seen to be NP-complete. The author and Watanabe [13] showed that,

**Theorem 2.26. (Agrawal–Watanabe)** *Suppose $K$ reduces to $A$ via $f$ and $f$ is a 1-1, size-increasing, P/poly-reduction with an easy cylinder. Then $K$ is P/poly-isomorphic to $A$.*

***Proof Sketch.*** Suppose $f$ has an easy cylinder with embedding function $e$. We define a P/poly-reduction $h$ from $K$ to $K$ such that $f$ is easy to invert on the range of $h$. Fix any $n$, and consider a non-deterministic Turing machine $M$ that executes as follows:

> Input $(u, y)$. Guess $x$, $s$, $|x| = n$, $|s| \le q'(n)$, and check whether $e(s, x)$ equals $y$; if not, reject; if yes, accept if and only if $x$ is in $K$.

Here we note that the advice of size $q(q'(n) + n + \lceil \log(q'(n) + n) \rceil)$ for computing $e$ on $\Sigma^{q'(n)+n+\lceil \log(q'(n)+n) \rceil}$ is hardwired in $M$. Further, from the complexity of $e$, $M(y)$ halts within $2q(q'(n) + n + \lceil \log(q'(n) + n) \rceil)$ steps. Thus, by letting $p_n$ be a code of this machine $M$ that is (with some padding) of size $\ell(n) \ge 2q(q'(n) + n + \lceil \log(q'(n) + n) \rceil)$, we have that $M_{p_n}$ halts and accepts $(p_n, e(s, x))$ in $|p_n e(s, x)|^2$ steps iff $M$ accepts $(p_n, e(s, x))$ iff $x \in K$ for all $x \in \Sigma^n$.

With these machine codes $p_n$ for all $n$, the reduction $h$ of $K$ to itself is defined as follows for each $n$ and each $x \in \Sigma^n$:

$$h(x) = (p_n, e(s_{p_n}, x)).$$

It follows from the above argument that $h$ is a reduction of $K$ to $K$. Furthermore, $h$ is P/poly-function.

Let $g = f \circ h$. Function $g$ is clearly a 1-1, size-increasing P/poly-reduction of $K$ to $A$. We show that $g$ is also P/poly-invertible. This follows from the existence of circuit $C_{p_n}$ such that $x = C_{p_n}(f(p_n, e(s_{p_n}, x)))$ for all $x \in \Sigma^n$. □

Finally, [13] showed that many of the candidate one-way functions do have easy cylinders. For example, the function Mult defined above:

> Mult has two inputs numbers $x$ and $y$. Fix polynomials $q'(n) = 0$, $q(n) = n$, and $\ell(n) = 2(n + \lceil \log n \rceil)$. Fix $s_u = \epsilon$ and the embedding function $e(s_u, z) = (s_u, z) = zt$ where $|t| = \lceil \log |z| \rceil$ and $t$ equals the

number $|z|$ in binary. Therefore, $\text{Mult}(u, e(s_u, z)) = \text{Mult}(u, zt)$. Since $|u| \geq |zt|$, fixing $u$ fixes the first number $x$ and $z$ determines the second number $y$. Therefore, given $u$, it is trivial to invert $\text{Mult}(u, zt)$.

The function Mult also has an easy cylinder: use $u$ to fix all but the second string of the last pair. It is also proved in [13] that all 1-1, size-increasing, $\mathsf{AC}^0$-functions have easy cylinders. The notion of easy cylinders is a formalization of the property of $\mathsf{AC}^0$ functions identified at the end of the last section. As already observed, many well-known candidate one-way functions do have easy cylinders. Based on this, [13] conjectured that,

**Easy Cylinder Conjecture.** *All 1-1, size-increasing,* P/poly*-functions have an easy cylinder.*

The following corollary follows from the above two theorems.

**Corollary 2.3.** *If there exists a $2^{n^\epsilon}$-secure one-way function and the Easy Cylinder Conjecture is true, then all sets complete for* NP *under* P/poly*-reductions are* P/poly*-isomorphic to each other.*

It is not clear if the Easy Cylinder Conjecture is true. The only indication we have is that the conjecture is true when translated to $\mathsf{AC}^0$ settings, and that many well-known candidate one-way functions have easy cylinders. Goldreich [31] argued against the conjecture by defining a candidate one-way function of the form $f^n$ where $f$ is a candidate one-way function in $\mathsf{NC}^0$ based on expander graphs. He argued that it is not clear whether $f^n$ has an easy cylinder, and conjectured that it does not.

## 2.8. Future Directions

The results of the previous two sections suggest that the Isomorphism Conjecture is true. However, the evidence is far from overwhelming. Answers to the following questions should make the picture clearer:

- Can one prove the $r$-Complete Degree Conjecture for other reducibilities, for example, $\mathsf{AC}^0[2]$ (computed by constant depth circuits with AND and PARITY gates)?
- Does Goldreich's function have an easy cylinder? Can one prove it does not under a reasonable hypothesis?
- Even if the Easy Cylinder Conjecture is true and strong one-way functions exist, the Isomorphism Conjecture is true only for P/poly-

reductions. Can one define alternative and plausible conjecture(s) from which the Isomorphism Conjecture for p-reductions follows?

**Acknowledgement** The author wishes to thank the anonymous referee whose suggestions helped improve the paper substantially.

## References

[1] M. Agrawal, On the isomorphism problem for weak reducibilities, *J. Comput. System Sci.* **53**(2), 267–282, (1996).
[2] M. Agrawal. The first order isomorphism theorem. In *Proceedings of the FST&TCS*, vol. 2245, *LNCS*, pp. 70–82, (2001).
[3] M. Agrawal. Towards uniform $AC^0$ isomorphisms. In *Proceedings of the Conference on Computational Complexity*, pp. 13–20, (2001).
[4] M. Agrawal, For completeness, sublogarithmic space is no space, *Inform. Process. Lett.* **82**(6), 321–325, (2002).
[5] M. Agrawal. Pseudo-random generators and structure of complete degrees. In *Proceedings of the Conference on Computational Complexity*, pp. 139–147, (2002).
[6] M. Agrawal and E. Allender. An isomorphism theorem for circuit complexity. In *Proc. 11th Conference on Computational Complexity*, pp. 2–11, (1996).
[7] M. Agrawal, E. Allender, R. Impagliazzo, T. Pitassi, and S. Rudich, Reducing the complexity of reductions, *Comput. Complexity.* **10**(2), 117–138, (2001).
[8] M. Agrawal, E. Allender, and S. Rudich, Reductions in circuit complexity: An isomorphism theorem and a gap theorem, *J. Comput. System Sci.* **57**, 127–143, (1998).
[9] M. Agrawal and V. Arvind, Quasi-linear truth-table reductions to p-selective sets, *Theoret. Comput. Sci.* **158**, 361–370, (1996).
[10] M. Agrawal and S. Biswas, Polynomial-time isomorphism of 1-L-complete sets, *J. Comput. System Sci.* **53**(2), 155–160, (1996).
[11] M. Agrawal, N. Kayal, and N. Saxena, PRIMES is in P, *Ann. of Math.* **160** (2), 781–793, (2004).
[12] M. Agrawal and S. Venkatesh, The isomorphism conjecture for 2-DFA reductions, *Internat. J. Found. Comput. Sci.* **7**(4), 339–352, (1996).
[13] M. Agrawal and O. Watanabe. One-way functions and Berman–Hartmanis conjecture. In *Proceedings of the Conference on Computational Complexity*, pp. 194–202, (2009).
[14] E. Allender, Isomorphisms and 1-L reductions, *J. Comput. System Sci.* **36** (6), 336–350, (1988).
[15] E. Allender, P-uniform circuit complexity, *J. ACM.* **36**, 912–928, (1989).
[16] E. Allender, J. Balcázar, and N. Immerman, A first-order isomorphism theorem, *SIAM J. Comput.* **26**(2), 557–567, (1997).
[17] E. Allender and V. Gore, Rudimentary reductions revisited, *Inform. Process. Lett.* **40**, 89–95, (1991).

[18] D. Barrington, N. Immerman, and H. Straubing, On uniformity within $NC^1$, *J. Comput. System Sci.* **74**, 274–306, (1990).

[19] R. Beigel, M. Kummer, and F. Stephan, Approximable sets, *Inform. and Comput.* **120**(2), 73–90, (1995).

[20] L. Berman. *Polynomial Reducibilities and Complete Sets*. PhD thesis, Cornell University, (1977).

[21] L. Berman and J. Hartmanis, On isomorphism and density of NP and other complete sets, *SIAM J. Comput.* **1**, 305–322, (1977).

[22] H. Buhrman and J. Hitchcock. NP-hard sets are exponentially dense unless coNP $\subseteq$ NP/poly. In *Proceedings of the Conference on Computational Complexity*, pp. 1–7, (2008).

[23] H. Buhrman, S. Homer, and L. Torenvliet, Completeness for nondeterministic complexity classes, *Mathematical Systems Theory.* **24**(1), 179–200, (1991).

[24] R. Chang, B. Chor, O. Goldreich, J. Hartmanis, J. Håstad, D. Ranjan, and P. Rohatgi, The random oracle hypothesis is false, *J. Comput. System Sci.* **49**(1), 24–39, (1994).

[25] S. Cook. The complexity of theorem proving procedures. In *Proceedings of Annual ACM Symposium on the Theory of Computing*, pp. 151–158, (1971).

[26] S. Cook, A hierarchy for nondeterministic time hierarchy, *J. Comput. System Sci.* **7**(4), 343–353, (1973).

[27] S. Fenner, L. Fortnow, and S. Kurtz, The isomorphism conjecture holds relative to an oracle, *SIAM J. Comput.* **25**(1), 193–206, (1996).

[28] M. Furst, J. Saxe, and M. Sipser, Parity, circuits, and the polynomial hierarchy, *Mathematical Systems Theory.* **17**, 13–27, (1984).

[29] K. Ganesan and S. Homer, Complete problems and strong polynomial reducibilities, *SIAM J. Comput.* **21**, 733–742, (1992).

[30] O. Goldreich, *Foundation of Cryptography I: Basic Tools*. Cambridge University Press, (2001).

[31] O. Goldreich. A candidate counterexample for the easy cylinder conjecture. Technical report, TR09-028, Electronic Colloquium on Computational Complexity. Available at: http://www.eccc.uni-trier.de/eccc, (2009).

[32] O. Goldreich and L. A. Levin. A hardcore predicate for all one-way functions. In *Proceedings of Annual ACM Symposium on the Theory of Computing*, pp. 25–32, (1989).

[33] J. Hartmanis and L. Hemchandra, One-way functions and the nonisomorphism of NP-complete sets, *Theoret. Comput. Sci.* **81**(1), 155–163, (1991).

[34] J. Hartmanis, N. Immerman, and S. Mahaney. One-way log-tape reductions. In *Proceedings of Annual IEEE Symposium on Foundations of Computer Science*, pp. 65–72, (1978).

[35] J. Håstad. Almost optimal lower bounds for small depth circuits. In *Proceedings of Annual ACM Symposium on the Theory of Computing*, pp. 6–20, (1986).

[36] J. Håstad, R. Impagliazzo, L. Levin, and M. Luby, A pseudo-random gener-

ator from any one-way function, *SIAM J. Comput.* pp. 221–243, (1998).
[37] L. A. Hemchandra and A. Hoene, Collapsing degrees via strong computation, *J. Comput. System Sci.* **46**(3), 363–380, (1993).
[38] S. Homer, S. Kurtz, and J. Royer, On 1-truth-table hard languages, *Theoret. Comput. Sci.* **155**, 383–389, (1993).
[39] N. Jones, Space-bounded reducibility among combinatorial problems, *J. Comput. System Sci.* **11**, 68–85, (1975).
[40] D. Joseph and P. Young, Some remarks on witness functions for nonpolynomial and noncomplete sets in NP, *Theoret. Comput. Sci.* **39**, 225–237, (1985).
[41] K. Ko, T. Long, and D. Du, A note on one-way functions and polynomial-time isomorphisms, *Theoret. Comput. Sci.* **47**, 263–276, (1987).
[42] S. Kurtz. A relativized failure of Berman–Hartmanis conjecture. Unpublished Manuscript, (1983).
[43] S. Kurtz, S. Mahaney, and J. Royer. Noncollapsing degrees. Technical report 87-001, Department of Computer Science, University of Chicago, (1987).
[44] S. Kurtz, S. Mahaney, and J. Royer, Collapsing degrees, *J. Comput. System Sci.* **37**, 247–268, (1988).
[45] S. Kurtz, S. Mahaney, and J. Royer. The structure of complete degrees. In ed. A. Selman, *Complexity Theory Retrospective*, pp. 108–146. Springer-Verlag, (1988).
[46] S. Kurtz, S. Mahaney, and J. Royer, The isomorphism conjecture fails relative to a random oracle, *J. ACM.* **42**(2), 401–420, (1995).
[47] S. Lindell. A purely logical characterization of circuit complexity. In *Proceedings of the Structure in Complexity Theory Conference*, pp. 185–192, (1992).
[48] C. Lund, L. Fortnow, H. Karloff, and N. Nissan, Algebraic methods for interactive proof systems, *J. ACM.* **39**(4), 859–868, (1992).
[49] S. Mahaney, Sparse complete sets for NP: Solution of a conjecture of Berman and Hartmanis, *J. Comput. System Sci.* **25**(2), 130–143, (1982).
[50] J. Myhill, Creative sets, *Z. Math. Logik Grundlag. Math.* **1**, 97–108, (1955).
[51] M. Ogihara, Polynomial time membership comparable sets, *SIAM J. Comput.* **24**(5), 1068–1081, (1995).
[52] C. Papadimitriou, *Computational Complexity*. Addison-Wesley, (1995).
[53] A. L. Selman, A survey of one-way functions in complexity theory, *Mathematical Systems Theory.* **25**, 203–221, (1992).
[54] A. Shamir, IP = PSPACE, *J. ACM.* **39**(4), 869–877, (1992).
[55] O. Watanabe, A comparison of polynomial time completeness notions, *Theoret. Comput. Sci.* **54**, 249–265, (1987).

# Chapter 3

# The Ershov Hierarchy

Marat M. Arslanov*

*Department of Mathematics*
*Kazan State University*
*420008 Kazan, Russia*
*E-mail: Marat.Arslanov@ksu.ru*

In this chapter we investigate set-theoretic properties and the Turing degree structure of the hierarchy of $\Delta_2^0$-sets, which is well known in the literature as the Ershov hierarchy.

## Contents



*The author's research was partially supported by the Russian Foundation for Basic Research 05-01-00605. The author would like to thank the anonymous referee for many suggestions and improvements throughout the chapter.

## 3.1. The Hierarchy of Sets

The notion of a *computably enumerable (c.e.) set*, i.e. a set of integers whose members can be effectively listed, is a fundamental one. Another way of approaching this definition is via an approximating function $\{A_s\}_{s\in\omega}$ to the set $A$ in the following sense: We begin by guessing $x \notin A$ at stage 0 (i.e. $A_0(x) = 0$); when later $x$ enters $A$ at a stage $s+1$, we change our approximation from $A_s(x) = 0$ to $A_{s+1}(x) = 1$. Note that this approximation (for fixed) $x$ may change at most once as $s$ increases, namely when $x$ enters $A$. An obvious variation of this definition is to allow more than one change: A set $A$ is 2-*c.e.* (or *d-c.e.*) if for each $x$, $A_s(x)$ change at most twice as $s$ increases. This is equivalent to requiring the set $A$ to be the difference of two c.e. sets $A_1 - A_2$. Similarly, one can define $n$-c.e. sets by allowing $n$ changes for each $x$. A direct generalization of this reasoning leads to sets which are computably approximable in the following sense: For a set $A$ there is a set of uniformly computable sequences $\{f(0,x), f(1,x), \ldots, f(s,x), \ldots \mid x \in \omega\}$ consisting of 0 and 1 such that for any $x$ the limit of the sequence $f(0,x), f(1,x), \ldots$ exists and is equal to the value of the characteristic function $A(x)$ of $A$. The well-known Shoenfield Limit Lemma states that the class of such sets coincides with the class of all $\Delta^0_2$-sets. Thus, for a set $A$, $A \leqslant_T \emptyset'$ if and only if there is a computable function $f(s,x)$ such that $A(x) = \lim_s f(s,x)$.

The notion of $d$-c.e. and $n$-c.e. sets goes back to Putnam [51] and Gold [37], and was first investigated and generalized by Ershov [33–35]. The arising hierarchy of sets is now known as *the Ershov difference hierarchy.* The position of a set $A$ in this hierarchy is determined by the number of changes in the approximation of $A$ described above, i.e. by the number of different pairs of neighboring elements of the sequence.

The Ershov hierarchy consists of the finite and infinite levels. The finite levels of the hierarchy consist of the $n$-c.e. sets for $n \in \omega$. Otherwise a set belongs to one of the infinite levels of the hierarchy. The infinite levels of the hierarchy are defined using infinite constructive ordinals. As it turns out, the resulting hierarchy of sets exhausts the whole class of $\Delta^0_2$-sets. Each subsequent level of the hierarchy contains all previous ones but does not coincide with any of them. At the same time the levels of the hierarchy are arranged so uniformly, that even the following conjecture was stated: The semilattices of the Turing degrees of the sets from the finite levels of the hierarchy starting with the second level are indistinguishable in first order predicate logic. This conjecture became well known as *Downey's Conjecture* and involved a whole series of publications.

The Turing degrees of the sets from the finite levels of the Ershov hierarchy have been intensively studied since the 1970s. It turned out that they (partially ordered by Turing reducibility) have a sufficiently rich inner structure, in many respects repeating its paramount representative, the class of c.e. degrees.

Our notation and terminology are standard and generally follow Soare [56]. In particular, the standard enumerations of the c.e. sets and partial computable functions are denoted by $\{W_x\}_{x\in\omega}$ and $\{\Phi_x\}_{x\in\omega}$ respectively. As usual, we append $[s]$ to various functionals such as $\Phi_e^A(x)[s]$ to indicate the state of affairs at stage $s$. In particular if $A$ is c.e. (or otherwise being approximated) we mean by this notation the result of running the $e^{th}$ Turing machine for $s$ steps on input $x$ with oracle $A_s$, the subset of $A$ enumerated by stage $s$ (or the approximation to $A$ at stage $s$). We take the use of this computation to be the greatest number about which it queries the oracle and denote it by $\varphi_e(A;x)[s]$; so changing the oracle at $\varphi_e(A;x)[s]$ destroys the computation. We also use a modified version of the restriction notation for functions to mesh with this definition of the use: $f\lceil x$ means the restriction of the function $f$ to numbers $y \leq x$. Thus if $\Phi_e^A(x)$ is convergent, then the use is $A\lceil\varphi_e(A;x)$ and changing $A$ at $\varphi_e(A;x)$ destroys this computation (and similarly for computations and approximations at stage $s$ of a construction). For a set $A \subseteq \omega$ its complement $\omega - A$ is denoted by $\bar{A}$. The cardinality of a set $A$ is denoted by $|A|$.

The pairing function $\langle x,y\rangle$ is defined as $\langle x,y\rangle := \frac{(x+y)^2+3x+y}{2}$ and bijectively maps $\omega^2$ onto $\omega$. We denote by $l$ and $r$ the uniquely defined functions such that for all $x,y$, $l(\langle x,y\rangle) = x, r(\langle x,y\rangle) = y$ and $\langle l(x),r(x)\rangle = x$; the $n$-place function $\langle x_1,\dots x_n\rangle$ for $n > 2$ is defined as $\langle x_1,\dots x_n\rangle = \langle\langle\dots\langle x_1,x_2\rangle,x_3\rangle,\dots,x_n\rangle$. In this case the $s$-th component of $\langle x_1,\dots x_n\rangle$ is denoted as $c_{n,s}$. Thus, $\langle c_{n,1}(x),\dots c_{n,n}(x)\rangle = x$ and $c_{n,s}(\langle x_1,\dots x_n\rangle) = x_s$. If a function $f$ is defined at $x$, then we write $f(x)\downarrow$, otherwise $f(x)\uparrow$. The characteristic function of a set $A$ is denoted by the same letter: $A(x) = 1$, if $x \in A$, and otherwise $A(x) = 0$.

### 3.1.1. *The finite levels of the Ershov hierarchy*

We begin with the following characterization of the $\Delta_2^0$-sets (i.e. sets $A \leqslant_T \emptyset'$).

**Lemma 3.1.** (*Shoenfield Limit Lemma*) *A set $A$ is a $\Delta_2^0$-set if and only if* *there is a computable function of two variables $f$ such that $f(s,x) \in \{0,1\}$ for all $s, x$, $f(0,x) = 0$ and $\lim_s f(s,x)$ exists for each $x$ (i.e. $|\{s : f(s,x) \neq f(s+1,x)\}| < \infty$), and $\lim_s f(s,x) = A(x)$.*

It follows easily from the Limit Lemma that

**Theorem 3.1.** *A set $A$ is Turing reducible (T-reducible) to $\emptyset'$ if and only if there is a uniformly computably enumerable sequence of c.e. sets $\{R_x\}_{x\in\omega}$ such that*

$$R_0 \supseteq R_1 \supseteq \ldots, \quad \bigcap_{x=0}^{\infty} R_x = \emptyset, \quad \textit{and} \quad A = \bigcup_{x=0}^{\infty}(R_{2x} - R_{2x+1}). \tag{1}$$

***Proof.*** $(\rightarrow)$ Let $A \leqslant_T \emptyset'$. By the Limit Lemma there is a computable function $f$ such that $A = \lim_s f(s,x)$, and for all $x$, $f(0,x) = 0$. Define c.e. sets $R_n, n \in \omega$, as follows:
$R_0 = \{y : \exists s(f(s,y) = 1)\}$;
$R_1 = \{y : \exists s_0, s_1(s_0 < s_1, f(s_0,y) = 1, f(s_1,y) = 0)\}$, and in general for $n > 0$;
$R_n = \{y : \exists s_0 < s_1 < \ldots < s_n(f(s_0,y) = 1, f(s_1,y) = 0, \ldots, f(s_n,y) = n+1 \bmod 2\}$.

Obviously, all sets $R_n$ are c.e., the sequence $\{R_x\}_{x\in\omega}$ is uniformly c.e., and $R_0 \supseteq R_1 \supseteq \ldots$. It is also easy to check that $\bigcap_{x=0}^{\infty} R_x = \emptyset$ and $A = \bigcup_{x=0}^{\infty}(R_{2x} - R_{2x+1})$.

$(\leftarrow)$ For this direction the proof is straightforward. □

Note that if $A$ is an arbitrary $\Sigma_2^0$-set then it is easy to show that $A = \bigcup_{x=0}^{\infty}(R_{2x} - R_{2x+1})$ for a uniformly computably enumerable sequence of c.e. sets $\{R_x\}_{x\in\omega}$ such that $R_0 \supseteq R_1 \supseteq R_2 \supseteq \ldots$. Therefore, in Theorem 3.1 the condition $\bigcap_{x=0}^{\infty} R_x = \emptyset$ is necessary.

If in (1) starting from some $n$ all elements of the sequence $\{R_x\}_{x\in\omega}$ are empty, then we obtain sets from the finite levels of the Ershov hierarchy.

**Definition 3.1.** A set $A$ is *$n$-computably enumerable* (an *$n$-c.e.* set), if either $n = 0$ and $A = \emptyset$, or $n > 0$ and there are c.e. sets $R_0 \supseteq R_1 \supseteq R_2 \supseteq \ldots \supseteq R_{n-1}$ such that

$$A = \bigcup_{i=0}^{\left[\frac{n-1}{2}\right]} (R_{2i} - R_{2i+1}) \quad \text{(here if } n \text{ is an odd number then } R_n = \emptyset).$$

It follows from this definition that if $n > 1$ and $n$ is an even number $(n = 2m)$ then

$$A = \bigcup_{x=0}^{m-1} (R_{2x} - R_{2x+1}),$$

and if $n > 1$ and $n$ is an odd number $(n = 2m + 1)$ then

$$A = \{ \bigcup_{x=0}^{m-1} (R_{2x} - R_{2x+1}) \} \cup R_{2m}.$$

Therefore, the class of 1-c.e. sets coincides with the class of c.e. sets, 2-c.e. sets can be written as $R_1 - R_2$, where $R_1 \supseteq R_2$ c.e. sets, therefore they are also called *d*-c.e. (difference-c.e.) sets, 3-c.e. sets can be written as $(R_1 - R_2) \cup R_3$ etc.

The $n$-c.e. sets constitute the level $\Sigma_n^{-1}$ of the Ershov hierarchy. They are also called $\Sigma_n^{-1}$-sets. The complements of the $\Sigma_n^{-1}$-sets constitute the level $\Pi_n^{-1}$ of the hierarchy ($\Pi_n^{-1}$-sets). The intersection of these two classes is denoted by $\Delta_n^{-1}$:

$$\Delta_n^{-1} = \Sigma_n^{-1} \cap \Pi_n^{-1}.$$

The proof of the following statement is straightforward.

**Theorem 3.2.** *A set $A$ is an $n$-c.e. set for some $n \geqslant 0$ if and only if there is a computable function $g$ of two variables $s$ and $x$ such that $A(x) = \lim_s g(s, x)$ for every $x$, $g(0, x) = 0$, and*

$$|\{s | g(s + 1, x) \neq g(s, x)\}| \leqslant n. \tag{1}$$

The class of the $n$-c.e. sets is denoted by $\mathcal{R}_n$. It is clear that every $n$-c.e. set is also $(n + 1)$-c.e., therefore $\mathcal{R}_0 \subseteq \mathcal{R}_1 \subseteq \mathcal{R}_2 \subseteq \dots$. It is easy to see that the reverse inclusions do not hold and that for every $n > 0$ there is an $(n + 1)$-c.e. set with an $(n + 1)$-c.e. complement which is not $n$-c.e. and not even co-$n$-c.e.

Therefore, we have

**Theorem 3.3.** *(Hierarchy Theorem) For every $n > 0$,*

$$\Sigma_n^{-1} \cup \Pi_n^{-1} \subsetneq \Sigma_{n+1}^{-1} \cap \Pi_{n+1}^{-1}.$$

*Comment.* The Limit Lemma appeared for the first time in Shoenfield [54]. The finite levels of the Ershov hierarchy were defined and studied (under different names) also in Putnam [51] and Gold [37]. Addison [1] considered a general method of constructing "difference" hierarchies. In particular, his hierarchy, generated by c.e. sets, defines the same classes of $n$- and $\omega$-c.e. sets ($\omega$-c.e. sets will be defined later). In the same paper he also obtained several properties of $n$- and $\omega$-c.e. sets, for instance, the Hierarchy Theorem 3.3. The notations $\Sigma_n^{-1}, \Pi_n^{-1}$ and $\Delta_n^{-1}$ for the finite levels of the hierarchy, as well as analogous notations for further levels (see Theorem 3.14) were introduced by Ershov [33, 34].

### 3.1.2. *The properties of productiveness and creativeness on the n-c.e. sets*

On the class of $n$-c.e. sets Ershov [33] introduced the notion of creative sets which is similar to the appropriate definition on c.e. sets and preserves its main properties.

**Definition 3.2.** A set $P$ is $\Sigma_n^{-1}$-*productive*, $n \geqslant 2$, if there is an $n$-place computable function $f(x_1, \ldots, x_n)$ such that for any c.e. sets $W_{x_1} \supseteq W_{x_2} \supseteq \ldots \supseteq W_{x_n}$

$$\bigcup_{i=1}^{[\frac{n+1}{2}]} (W_{x_{2i-1}} - W_{x_{2i}}) \subseteq P \rightarrow f(x_1, \ldots, x_n) \in P - \bigcup_{i=1}^{[\frac{n+1}{2}]} (W_{x_{2i-1}} - W_{x_{2i}})$$

(for odd $n$, set $W_{x_{n+1}} = \emptyset$).

An $n$-c.e. set $A$ is $\Sigma_n^{-1}$-*creative* if its complement is $\Sigma_n^{-1}$-productive.

For simplicity we will consider only the case $n = 2$, the general case is similar.

For $d$-c.e. sets the definition of $\Sigma_2^{-1}$-productive sets can be reformulated as follows: A set $P$ is $\Sigma_2^{-1}$-productive, if there is a unary computable function $f$ such that for any $x$,

$$W_{l(x)} \supseteq W_{r(x)} \,\&\, (W_{l(x)} - W_{r(x)}) \subseteq P \rightarrow f(x) \in P - (W_{l(x)} - W_{r(x)}).$$

Similarly to the case of c.e. sets, $\Sigma_2^{-1}$-productive sets cannot be $d$-c.e. sets. Indeed, if $P = W_x - W_y$, $W_x \supseteq W_y$, then $f(\langle x, y\rangle) \in P - (W_x - W_y) = \emptyset$, a contradiction.

Define:

$$R_1 = \{x | x \in W_{l(x)} \cup W_{r(x)}\};$$
$$R_2 = \{x | x \in W_{r(x)}\}.$$

It is clear that $R_1 \supseteq R_2$ and $R_1 - R_2 = \{x | x \in W_{l(x)} \,\&\, x \notin W_{r(x)}\}$.

**Theorem 3.4.** *The set $R_1 - R_2$ is $\sum_2^{-1}$-creative.*

***Proof.*** We have to prove that the set $\omega - (R_1 - R_2) = (\omega - R_1) \cup R_2$ is $\sum_2^{-1}$-productive.

Let $W_x - W_y \subseteq (\omega - R_1) \cup R_2$. If $\langle x, y\rangle \in R_1 - R_2$, then $\langle x, y\rangle \in R_1 \,\&\, \langle x, y\rangle \notin R_2$ implies $\langle x, y\rangle \in W_x$ and $\langle x, y\rangle \notin W_y$, which implies $\langle x, y\rangle \in W_x - W_y$. But this is impossible, since $W_x - W_y \subseteq (\omega - R_1) \cup R_2$. Therefore $\langle x, y\rangle \in \omega - (R_1 - R_2)$. If $\langle x, y\rangle \in W_x - W_y$, then $\langle x, y\rangle \in W_x$ and $\langle x, y\rangle \notin W_y$. It follows that $\langle x, y\rangle \in R_1 - R_2$, a contradiction. Therefore, for all $x$ and $y$, $W_x - W_y \subseteq \omega - (R_1 - R_2)$, which implies $\langle x, y\rangle \in (\omega - (R_1 - R_2)) - (W_x - W_y)$. □

**Theorem 3.5.** *The set $R_1 - R_2$ is $\Sigma_2^{-1}$-complete in the sense that every d-c.e. set is m-reducible to $R_1 - R_2$.*

***Proof.*** We have $R_1 - R_2 = \{x | x \in W_{l(x)} \quad \& \quad x \notin W_{r(x)}\}$. It follows from the proof of Theorem 3.4 that the function $\langle x, y\rangle$ is a productive function for $\overline{R_1 - R_2}$, i.e. $W_x - W_y \subseteq \overline{R_1 - R_2} \to \langle x, y\rangle \in \overline{R_1 - R_2} - (W_x - W_y)$.

Let $A_1$ and $A_2$ be c.e. sets and $A_1 \supseteq A_2$. Now we define a computable function $h$ which $m$-reduces $A_1 - A_2$ to $R_1 - R_2$.

We first define computable functions $g_1$ and $g_2$ as follows:
$W_{g_1(x)} = \{t | \Phi_x(t, 0) \downarrow = 0\}, W_{g_2(x)} = \{t | \Phi_x(t, 0) \downarrow = 0 \,\&\, \Phi_x(t, 1) \downarrow = 1\}$.
Now define

$$q(y, z, t, n) = \begin{cases} 0, & \text{if } y \in A_1,\ t \in W_z,\ n = 0; \\ 1, & \text{if } y \in A_2,\ t \in W_z,\ n = 1; \\ \uparrow, & \text{in all other cases.} \end{cases}$$

By the s-m-n-theorem there is a computable function $\alpha$ such that $\Phi_{\alpha(y,z)}(t, n) = q(y, z, t, n)$. It follows that

$$\Phi_{\alpha(y,z)}(t, n) = \begin{cases} n, & \text{if } y \in A_{n+1},\ t \in W_z,\ n \leqslant 1; \\ \uparrow, & \text{otherwise.} \end{cases}$$

Define $p(x) = \langle g_1(x), g_2(x)\rangle$. Let $\beta$ be a computable function such that for all $y, z$, $W_{\beta(y,z)} = \{p(\alpha(y, z))\}$.

By the Recursion Theorem there is a computable function $f$ such that for each $y$,

$$W_{\beta(y,f(y))} = W_{f(y)}.$$

It follows from the definition of the function $\beta$ that $W_{f(y)} = \{p(\alpha(y, f(y)))\}$. Finally define $h(y) = p(\alpha(y, f(y)))$.

To prove that for any $y$, $y \in A_1 - A_2 \leftrightarrow h(y) \in R_1 - R_2$, suppose that $y \in A_1 - A_2$. We have $\Phi_{\alpha(y,f(y))}(t, 0) \downarrow= 0$ if and only if $t \in W_{f(y)}$ if and only if $t = p(\alpha(y, f(y)))$ if and only if $t = h(y)$. Therefore, $W_{g_1(\alpha(y,f(y)))} = \{h(y)\} = W_{f(y)}$. Since $y \notin A_2$, $W_{g_2(\alpha(y,f(y)))} = \emptyset$.

Let us denote $\langle g_1(\alpha(y, f(y))), g_2(\alpha(y, f(y)))\rangle$ by $x$. Then $x = p(\alpha(y, f(y)))$. If $W_{g_1(\alpha(y,f(y)))} - W_{g_2(\alpha(y,f(y)))} \subseteq \overline{R_1 - R_2}$, then $x \in \overline{R_1 - R_2} - W_{g_1(\alpha(y,f(y)))}$, and since $x = p(\alpha(y, f(y))) = h(y)$, $h(y) \in \overline{R_1 - R_2} - W_{g_1(\alpha(y,f(y)))}$, a contradiction.

Therefore, $W_{g_1(\alpha(y,f(y)))} - W_{g_2(\alpha(y,f(y)))} \subseteq R_1 - R_2$. But the set $W_{g_1(\alpha(y,f(y)))} - W_{g_2(\alpha(y,f(y)))}$ consists of a single element $h(y)$, therefore $h(y) \in R_1 - R_2$.

Now suppose that $y \notin A_1 - A_2$. In this case we have either a) $y \notin A_1$, or b) $y \in A_1 \cap A_2$.

Case a) If $y \notin A_1$, then the function $q$ is undefined at this $y$ and all $z, t, n$, therefore the function $\Phi_{\alpha(y,f(y))}$ is also undefined for all $t, n$. It follows that the sets $W_{g_1(\alpha(y,f(y)))}$ and $W_{g_2(\alpha(y,f(y)))}$ are empty and

$$h(y) = \langle g_1(\alpha(y, f(y))), g_2(\alpha(y, f(y)))\rangle \in \overline{R_1 - R_2},$$

since the set $\overline{R_1 - R_2}$ is productive.

Case b) If $y \in A_1 \cap A_2$, then it follows from their definitions that the sets $W_{g_1(\alpha(y,f(y)))}$ and $W_{g_2(\alpha(y,f(y)))}$ coincide. Therefore, since the set $\overline{R_1 - R_2}$ is productive, we have $h(y) \in \overline{R_1 - R_2}$.

It follows that the function $h(y)$ $m$-reduces the set $A_1 - A_2$ to the $\Sigma_2^{-1}$-creative set $R_1 - R_2$, as required. □

The proof of Theorem 3.5 can be reorganized to prove a more general claim: Any $\Sigma_n^{-1}$-creative set is $\Sigma_n^{-1}$-complete in the sense that any $n$-c.e. set is $m$-reducible to this set.

**Theorem 3.6.** *Let* $Q_n = \bigcup_{i=1}^{[\frac{n+1}{2}]} (R_{2i-1} - R_{2i})$ *(letting* $R_{n+1} = \emptyset$*), where c.e. sets* $R_1 \supseteq R_2 \supseteq \ldots \supseteq R_n$ *are defined as follows: for every* $i$, $1 \leqslant i \leqslant n$, $R_i = \{x | x \in \bigcup_{s=i}^{n} W_{c_{ns}(x)}\}$.

*a) The sets $Q_n$ are $\Sigma_n^{-1}$-creative sets for all $n$, $2 \leqslant n < \omega$;*
*b) The sets $Q_n$ are $\Sigma_n^{-1}$-complete.*

The proof is similar to the proof of Theorem 3.5. Now the functions $g_i$, $1 \leqslant i \leqslant n$, and $p$ are defined as follows: $W_{g_i(x)} = \{t | \Phi_x(t, 0) \downarrow = 0 \,\&\, \ldots \,\&\, \Phi_x(t, i-1) \downarrow = i - 1\}$, $p(x) = g(\langle g_1(x), \ldots, g_n(x)\rangle)$, where $g$ is the productive function for $\overline{Q}_n$.

*Comment.* Theorems 3.4, 3.5 and 3.6 are from Ershov [33].

### 3.1.3. *The class of the $\omega$-c.e. sets*

As we can see, the $n$-c.e. sets for $n < \omega$ does not exhaust the collection of $\Delta_2^0$-sets. Therefore, to obtain in this way a description of all $\Delta_2^0$-sets we need to consider infinite levels of the hierarchy.

In the definition of $n$-c.e. sets ($n < \omega$) we have used non-increasing sequences $R_0 \supseteq R_1 \supseteq \ldots \supseteq R_{n-1}$ of c.e. sets. The infinite levels of the Ershov hierarchy are defined using uniformly c.e. sequences of c.e. sets, such that the c.e. sets in these sequences satisfy the same $\subseteq$-relations which are consistent with the order type of the ordinal which defines the level of this set in the hierarchy.

**Definition 3.3.** Let $P(x, y)$ be a computable binary relation which partially orders the set of natural numbers (for convenience instead of $P(x, y)$ we will write $x \leqslant_P y$.) By definition, a uniformly c.e. sequence $\{R_x\}$ of c.e. sets is a $P$- (or $\leqslant_P$-)*sequence* if for all pairs $x, y$, $x \leqslant_P y$ implies that $R_x \subseteq R_y$.

Note that we can easily redefine the $n$-c.e. sets for $n < \omega$ according to this definition. Indeed, if, for instance, for some c.e. sets $A_1 \supseteq A_2 \supseteq \ldots A_n$ we have $A = (A_1 - A_2) \cup \ldots \cup (A_{n-1} - A_n)$ (where $n$ is an even number), then let $R_0 = A_n, R_1 = A_{n-1}, \ldots, R_{n-1} = A_1$. We have thus obtained an $n$-sequence ($n = \{0 < 1 < \ldots < n-1\}$) $R_0 \subseteq R_1 \subseteq \ldots \subseteq R_{n-1}$ such that

$$A = \bigcup_{i=0}^{\frac{n-1}{2}} (R_{2i+1} - R_{2i}).$$

The sets from the first infinite level of the Ershov hierarchy are the $\omega$-c.e. sets. They are defined using $\omega$-sequences of c.e. sets, i.e. sequences $\{R_x\}_{x \in \omega}$, in which the relation $R_x \subseteq R_y$ is consistent with the order type of $\omega = \{0 < 1 < \ldots\}$: $R_0 \subseteq R_1 \subseteq \ldots$.

**Definition 3.4.** A set $A \subseteq \omega$ belongs to level $\Sigma_\omega^{-1}$ of the Ershov hierarchy (or $A$ is a $\Sigma_\omega^{-1}$-set) if there is an $\omega$-sequence $\{R_x\}_{x\in\omega}$ such that $A = \bigcup_{n=0}^{\infty}(R_{2n+1} - R_{2n})$. $A$ belongs to level $\Pi_\omega^{-1}$ of the Ershov hierarchy (or $A$ is a $\Pi_\omega^{-1}$-set), if $\overline{A} \in \Sigma_\omega^{-1}$. Finally, $A$ belongs to level $\Delta_\omega^{-1}$ of the Ershov hierarchy ($A$ is a $\Delta_\omega^{-1}$-set), if $A$ and $\overline{A}$ both are $\Sigma_\omega^{-1}$-sets, i.e. $\Delta_\omega^{-1} = \Sigma_\omega^{-1} \cap \Pi_\omega^{-1}$. $\Delta_\omega^{-1}$-sets are also called $\omega$-c.e. sets.

**Theorem 3.7.** *(Epstein, Haas, and Kramer [32]) A set $A \subseteq \omega$ belongs to level $\Sigma_\omega^{-1}$ of the Ershov hierarchy if and only if there is a partial computable function $\psi$ such that for every $x$,*

*$x \in A$ implies $\exists s(\psi(s,x)\downarrow)$ and $A(x) = \psi(\mu s(\psi(s,x)\downarrow), x)$;*

*$x \notin A$ implies either $\forall s(\psi(s,x)\uparrow)$,*

*or $\exists s(\psi(s,x)\downarrow)$ & $A(x) = \psi(\mu s(\psi(s,x)\downarrow), x)$.*

*In other words, $A \subseteq \mathrm{dom}(\psi(\mu s(\psi(s,x)\downarrow), x))$, and for every $x$,*

*$x \in \mathrm{dom}(\psi(\mu s(\psi(s,x)\downarrow), x))$ implies $A(x) = \psi(\mu s(\psi(s,x)\downarrow), x)$.*

***Proof.*** ($\rightarrow$) Let $A = \bigcup_{n=0}^{\infty}(R_{2n+1} - R_{2n})$ for some $\omega$-sequence $\{R_x\}_{x\in\omega}$. Define the required partial computable function $\psi(s,x)$ as follows: For a given $x$, wait for a stage $s$ such that $x \in R_{2m+1,s}$ for some (least) $m$. (If this never happens then $\psi(s,x)\uparrow$ for all $s$.) Then define $\psi(2m+1,x) = 1$ and wait for a stage $s_1 > s$ and a number $n < 2m+1$ such that $x \in R_{n,s_1} - R_{n-1,s_1}$. Then define $\psi(n,x) = 1$, if $n$ is an odd number, and $\psi(n,x) = 0$, if $n$ is an even number, and so on. Obviously, the function $\psi$ is the required function.

($\leftarrow$) Define c.e. sets $R_i$, $i \geqslant 0$, as follows:

$$R_0 = \{x|\psi(0,x)\downarrow= 0\},$$
$$R_1 = R_0 \cup \{x|\psi(0,x)\downarrow= 1\},$$
$$\cdots\cdots\cdots\cdots\cdots$$
$$R_{2m} = R_{2m-1} \cup \{x|\psi(m,x)\downarrow= 0\},$$
$$R_{2m+1} = R_{2m} \cup \{x|\psi(m,x)\downarrow= 1\}.$$

Obviously, $\{R_n\}_{n\in\omega}$ is a uniformly c.e. sequence of c.e. sets $R_i, i \in \omega$, and $R_0 \subseteq R_1 \subseteq \ldots$.

Now suppose that $x \in A$. Then there is an integer $s$ such that $\psi(s,x)\downarrow$, $A(x) = \psi(s,x) = 1$, and if $s > 0$, then $\psi(s-1,x)\uparrow$. Therefore, $x \in R_{2n+1}$ for some (least) $n$, $x \notin \bigcup_{m<2n+1} R_m$ and $x \in \bigcup_{n=0}^{\infty}(R_{2n+1} - R_{2n})$.

Conversely, if $x \notin A$, then either $\psi(s,x)\uparrow$ for all $s$, or there is an integer $s$ such that $\psi(s,x)\downarrow = 0$, and if $s > 0$ then $\psi(s-1,x)\uparrow$. Therefore, either $x \notin R_i$ for all $i$, or $x \in R_{2n}$ for some (least) $n$ and $x \notin \bigcup_{m<2n} R_m$. This means that $x \notin \bigcup_{n=0}^{\infty}(R_{2n+1} - R_{2n})$. □

**Definition 3.5.** Let $f$ be a total unary function. A set $A$ is called *$f$-computably enumerable* (an $f$-c.e. set), if there is a computable function $g$ such that for all $s$ and $x$, $A(x) = \lim_s g(s,x)$, and

$$|\{s : g(s,x) \neq g(s+1,x)\}| \leqslant f(x).$$

**Theorem 3.8.** *a) There is an id-c.e. set (where id is the identity function) which is not $n$-c.e. for any $n \in \omega$;*

*b) Let $f$ and $g$ be computable functions such that $\exists^\infty x(f(x) < g(x))$. Then there is a $g$-c.e., but not $f$-c.e. set;*

*c) There is a $\Delta^0_2$-set which is not $f$-c.e. for any computable function $f$;*

*d) Let $A$ be an $f$-c.e. set for some computable function $f$, $A \neq \emptyset$, and let $g$ be a computable function such that $\forall y \exists x(g(x) \geqslant y)$. Then there exists a $g$-c.e. set $B$ such that $A \equiv_T B$.*

***Proof.*** For parts a)–c) use Cantor's diagonalization argument. For part d) let $h$ be the following computable function: $h(0) = 0$, and $h(x+1) = \mu y\{y > h(x) \,\&\, g(y) \geqslant f(x+1)\}$. Define $B = \{x : \exists z \in A(x = h(z))\}$. Then $B$ is $g$-c.e. and $B \equiv_T A$. □

**Theorem 3.9.** *Let $A \subseteq \omega$. The following are equivalent:*

*a) $A$ is $\omega$-c.e.*

*b) There is an $\omega$-sequence $\{R_x\}_{x\in\omega}$ such that $\bigcup_{x\in\omega} R_x = \omega$; and $A = \bigcup_{n=0}^{\infty}(R_{2n+1} - R_{2n})$.*

*c) $A$ is $f$-c.e. for some computable function $f$.*

*d) There is a partial computable function $\psi$ such that for all $x$,*

$$A(x) = \psi(k,x), \quad \textit{where} \quad k = \mu t(\psi(t,x)\downarrow). \qquad (1)$$

*(In this case we write $A(x) = \psi(\mu t(\psi(t,x)\downarrow), x)$.)*

*e) $A$ is tt-reducible to $\emptyset'$.*

***Proof.*** *c*) → *d*) Let $A$ be $\omega$-c.e. and

$$A(x) = \lim_s g(s,x), |\{s|g(s+1,x) \neq g(s,x)\}| \leqslant f(x)$$

for some computable functions $g$ and $f$. Define a partial computable function $\psi$ as follows: For any $x$

$$\psi(f(x), x) = g(0, x).$$

If $\exists s(g(s+1,x) \neq g(s,x))$, then let $s_1$ be the least such $s$. Define $\psi(f(x)-1, x) = g(s_1+1, x)$. Further we proceed by induction: let $\psi(f(x)-i, x) = g(s_i+1, x)$ be the last value of $\psi$ defined this way. If $\exists s > s_i(g(s+1,x) \neq g(s,x))$, then let $s_{i+1}$ be the least such $s$. Define $\psi(f(x)-(i+1), x) = g(s_{i+1}, x)$. It is clear that the function $\psi$ is partial computable and for all $x$, $A(x) = \psi(\mu s(\psi(s,x)\downarrow), x)$.

Part *d*) → *a*) immediately follows from Theorem 3.7.

*a*) → *b*) Let $\{P_x\}_{x\in\omega}$ and $\{Q_x\}_{x\in\omega}$ be $\omega$-sequences such that $A = \bigcup_{n=0}^{\infty}(P_{2n+1} - P_{2n})$ and $\bar{A} = \bigcup_{n=0}^{\infty}(Q_{2n+1} - Q_{2n})$. Define a new $\omega$-sequence $\{R_x\}_{x\in\omega}$ as follows: $R_0 = P_0$. For $x > 0$, $R_x = P_x \cup Q_{x-1}$. It is clear, that $A = \bigcup_{n=0}^{\infty}(R_{2n+1} - R_{2n})$ and $\bigcup_{x\in\omega} R_x = \omega$;.

*b*) → *c*) Let $A = \bigcup_{n=0}^{\infty}(R_{2n+1} - R_{2n})$ for an $\omega$-sequence $\{R_x\}_{x\in\omega}$ such that $\bigcup_{x\in\omega} R_x = \omega$;. Define computable functions $g(s,x)$ and $f(x)$ as follows: For a given $x \in \omega$, first find the first stage $t$ such that either $x \in R_{0,t}$ or $x \in R_{m,t} - R_{m-1,t}$ for some $m > 0$. If $x \in R_{0,t}$ then $f(x) = 0$ and $g(s,x) = 0$ for all $s \in \omega$. Otherwise define $f(x) = m$, $g(0,x) = 1$, if $m$ is an odd number, and $g(0,x) = 0$, otherwise. Further, for $s > 0$ define $g(s,x) = g(s-1,x)$, if for any $m$, $x \in R_{m,s}$ implies $x \in R_{m,s-1}$. Otherwise, let $n = \mu m(x \in R_{m,s})$. Define $g(0,x) = 1$, if $n$ is an odd number, and $g(0,x) = 0$ otherwise.

Obviously, the functions $g$ and $f$ are computable, $A(x) = \lim_s g(s,x)$ for all $x$, and $|\{s : g(s,x) \neq g(s+1,x)\}| \leqslant f(x)$.

*c*) → *e*) Let $A(x) = \lim_s g(s,x)$ and $|\{s : g(s,x) \neq g(s+1,x)\}| \leq f(x)$ for some computable functions $g$ and $f$. Define

$$M = \{\langle i, x, a\rangle : \exists t(|\{s \leq t : g(s,x) \neq g(s+1,x)\}| = i \,\&\, g(t,x) = a)\}.$$

Obviously, $M$ is c.e. and $x \in A$ if and only if $((\langle 0,x,1\rangle \in M \,\&\, \langle 1,x,1\rangle \notin M \,\&\, \langle 1,x,0\rangle \notin M) \vee ((\langle 1,x,1\rangle \in M \,\&\, \langle 2,x,1\rangle \notin M \,\&\, \langle 2,x,0\rangle \notin M) \vee \ldots \vee \langle f(x),x,1\rangle \in M$.

The last condition can be written as a *tt*-condition and, therefore, $A \leq_{tt} M$.

*e)* $\to$ *c)* For a given $x$ we can effectively find an integer $n$, a Boolean function $\alpha : \{0,1\}^n \to \{0,1\}$, and a finite set $\{t_1, \ldots t_n\}$ such that $x \in A$ if and only if $\alpha(K(t_1), \ldots K(t_n)) = 1$. Define

$$g(s,x) = \alpha(K_s(t_1), \ldots K_s(t_n)).$$

(Here $K$ is the creative set $\{e : e \in W_e\}$ and $\{K_s\}_{s\in\omega}$ is an effective enumeration of $K$.) It is clear, that $A(x) = \lim_s g(s,x)$, and $|\{s : g(s,x) \neq g(s+1,x)\}| \leq n$. □

If we replace in part d) of Theorem 3.9 $\mu t$ by a bounded search opearator $\mu t \leqslant n$, then we obtain a similar description of the $n$-c.e. sets (more precisely, the weakly $n$-c.e. sets) for $1 \leqslant n < \omega$.

**Definition 3.6.** (Epstein, Haas, and Kramer [32]) A set $A$ is *weakly $n$-c.e.* for some $n \geqslant 0$, if there is a computable function $g$ of two variables $s$ and $x$ such that $A(x) = \lim_s g(s,x)$ and

$$|\{s | g(s+1,x) \neq g(s,x)\}| \leqslant n$$

(in the definition of $n$-c.e. sets the condition "$g(0,x) = 0$ for every $x$" is omitted).

The following properties of the weakly $n$-c.e. sets are straightforward.

a) A set is weakly 0-c.e. if and only if it is computable; b) Every $n$-c.e. set also is weakly $n$-c.e.; c) A set $A$ is weakly $n$-c.e. for an arbitrary $n \geqslant 0$ if and only if its complement $\bar{A}$ is also weakly $n$-c.e.; d) The sets $A$ and $\bar{A}$ are both $(n+1)$-c.e. (i.e. $A \in \Delta^{-1}_{n+1}$) if and only if they are both weakly $n$-c.e.; e) For any $n > 0$ there is a weakly $n$-c.e. set $A$ such that neither $A$ nor $\bar{A}$ is $n$-c.e.

**Theorem 3.10.** *(Epstein, Haas, and Kramer [32], Carstens [15]) Let $A \subseteq \omega$ and $n > 0$. The following are equivalent:*

*a) $A$ is weakly $n$-c.e.;*

*b) There is a partial computable function $\psi$ such that for every $x$,*

$$A(x) = \psi(\mu t \leqslant n(\psi(t,x)\downarrow), x); \qquad (1)$$

*c) A is bounded truth-table reducible to $\emptyset'$ with norm n.*

***Proof.*** *a)* $\to$ *b)* Let $A(x) = \lim_s g(s,x)$ for some computable function $g$ and $|\{s|g(s+1,x) \neq g(s,x)\}| \leqslant n$ for every $x$.

The required function $\psi$ is defined as follows: for every $x$, $\psi(n,x) = g(0,x)$. If $\exists s(g(s+1,x) \neq g(s,x))$, then let $s_1$ be the least such $s$. Define $\psi(n-1,x) = g(s_1+1,x)$. Further proceed by induction: let $\psi(n-i,x) = g(s_i+1,x)$ be the last value of $\psi$ which was defined. If $\exists s > s_i(g(s+1,x) \neq g(s,x))$, then let $s_{i+1}$ be the least such $s$. Define $\psi(n-(i+1),x) = g(s_{i+1},x)$.

It is clear, that $\psi$ is partial computable and (1) holds.

*b)* $\to$ *a)* In this direction the proof is straightforward.

*c)* $\to$ *a)* Let $A$ be *btt*-reducible to the creative set $K$ with norm $n$. This means that for any $x$ we can effectively find an $n$-place Boolean function $\alpha_x$ and a finite set $F_x = \{t_1, t_2, \ldots, t_n\}$ such that $x \in A$ if and only if $\alpha_x(K(t_1), \ldots, K(t_n))$.
Define a computable function $g$ as follows:

$$g(s,x) = \alpha(K_s(x_1), \ldots, K_s(x_n)).$$

Obviously, $A(x) = \lim_s g(s,x)$, and $|\{s|g(s+1,x) \neq g(s,x)\}| \leq n$.

*a)* $\to$ *c)* The proof of this part is similar to the proof of part *c)* $\to$ *e)* of Theorem 3.9. □

### 3.1.4. *A description of the $\Delta^0_2$-sets using constructive ordinals*

The $\omega$-c.e. sets are the first examples of sets from infinite levels of the Ershov hierarchy. Later we will consider sets from other infinite levels of the hierarchy exhausting all $\Delta^0_2$-sets.

In what follows we use Kleene's system of ordinal notations $(O, <_O)$ (Kleene [43], see also Rogers [53]). Recall that if $a \in O$ then $|a|_O$ denotes the ordinal $\alpha$, which has $O$-notation $a$.

On $O$ a computable function $+_O$ is defined which for all $x, y$ and $z$, has the following properties:

a) $x, y \in O \to x +_O y \in O$;
b) $x, y \in O \,\&\, y \neq 1 \to x <_O x +_O y$;
c) $x, y \in O \to |x +_O y|_O = |x|_O + |y|_O$;

d) $x \in O \,\&\, y <_O z \leftrightarrow x +_O y <_O x +_O z$.

**Remark 3.1.** In general, the relation $y <_O x +_O y$ does not necessarily hold for all $x, y \in O$. But it follows from part c), that for all $x, y \in O$ such that $1 \leq_O x$, we have the following: $|y|_O \leq |x +_O y|_O$.

**Definition 3.7.** Let $a, b \in O$ and $|a|_O = \alpha, |b|_O = \beta$. We say that $a$ is *monotonically reducible* to $b$ (written $a \preceq_O b$), if there is a partial computable function $h$ such that $\{x : x <_O a\} \subseteq \mathrm{dom} h$, $\forall x <_O a(h(x) <_O b)$, and

1) $(\forall c, d <_O a)(c <_O d \leftrightarrow h(c) <_O h(d))$, and
2) $(\forall d <_O a)(k_O(d) = k_O(h(d)))$,

where $k_O$ is the partial computable function which is used in the definition of $O$ as a system of notations: then, for $x \in O$, $k_O(x) = 0$, if $|x|_O = 0$; $k_O(x) = 1$, if $|x|_O$ is a successor; and $k_O(x) = 2$, if $|x|_O$ is a limit ordinal.

It is clear that the relation $\preceq_O$ is reflexive and transitive, and for all $a, b \in O$, $a \preceq_O b$ implies $\Sigma_a^{-1} \subseteq \Sigma_b^{-1}$. Now the properties c) and d) of $+_O$ stated above imply the following useful property of notations from $O$ which will be used in Theorem 3.19.

**Proposition 3.1.** *For all $x, y \in O$, $y \preceq_O x +_O y$.*

**Definition 3.8.** Let $S$ be a univalent system of notations for constructive ordinals, let $\alpha$ be an ordinal which has an $S$-notation, and let $\Psi$ be a partial computable function and $f$ a unary function. We write $\Psi \rightarrow_{\{\alpha,S\}} f$, if for all $x \in \mathrm{dom} f$ we have $f(x) = \Psi(n, x)$, where $n$ is a notation for the least ordinal $\lambda < \alpha$ such that $\Psi(n, x) \downarrow$.

For simplicity in this case we will also write (cf. Theorems 3.9 and 3.10) $f(x) = \Psi((\mu\lambda < \alpha)_S(\Psi((\lambda)_S, x)\downarrow), x)$.

Let $\alpha$ be an ordinal. The *parity function* on ordinals is defined as follows: $\alpha$ is an even ordinal if either it is 0 or a limit ordinal, or it is the successor of an odd ordinal. Otherwise $\alpha$ is an odd ordinal. Therefore, if $\alpha$ is even then $\alpha'$ (the successor of $\alpha$) is odd and vice versa.

In the system of notations $S$ the parity function $e(x)$ is defined as follows: Let $n \in D_S$. Then $e(n) = 1$, if $|n|_S$ is an odd ordinal, and $e(n) = 0$, if $|n|_S$ is an even ordinal.

Let $\alpha$ be an ordinal which has a notation $a$ in a notation system $S$, i.e. $|a|_S = \alpha$. Suppose that for a set $A$, a partial computable function $\Psi$ and for every $x$ we have

$$A(x) = \Psi((\mu\lambda< \alpha)_S(\Psi((\lambda)_S, x)\downarrow, x)) \tag{1}$$

(in symbols $\Psi \to_{\{\alpha,S\}} A$).

We define an $a$-sequence of c.e. sets $\{R_x\}$ as follows: For every $x <_S a$,

$$R_x = \bigcup_{y<_S x} R_y \cup \begin{cases} \{z| \quad \exists t \leqslant_S x(\Psi(t,z)\downarrow= 1)\}, \text{ if } e(x) \neq e(a); \\ \{z| \quad \exists t \leqslant_S x(\Psi(t,z)\downarrow= 0)\}, \text{ if } e(x) = e(a). \end{cases}$$

Clearly,

$$A = \{z|\exists x <_S a(z \in R_x \,\&\, e(x) \neq e(a) \,\&\, \forall y <_S x(z \notin R_y)\}. \tag{2}$$

In particular, if $\alpha = \omega$ this agrees with our previous description of $\omega$-c.e. sets via $\omega$-sequences, and if $\alpha = n < \omega$ ($\alpha$ is a natural number) with our description of $n$-c.e. sets.

If a set $A$ is defined as in (2) using some $\alpha$-sequence $\{R_x\}$ such that $\bigcup_{x<_S a} R_x = \omega$, then the converse claim also holds: The set $A$ can be defined as in (1) for some partial computable function $\Psi$.

Indeed, let $\Psi$ be the following function:

$$\Psi(x,z) = \begin{cases} 1, & \text{if } z \in R_x,\ e(x) \neq e(a); \\ 0, & \text{if } z \in R_x,\ e(x) = e(a); \\ \uparrow, & \text{otherwise.} \end{cases}$$

Since $\bigcup_{x<_S a} R_x = \omega$ we have $\forall z \exists x(\Psi(x,z)\downarrow)$. Now it is easy to see that $A(x) = \Psi((\mu\lambda< \alpha)_S(\Psi((\lambda)_S, x)\downarrow, x))$.

**Remark 3.2.** Here the condition $\bigcup_{x<_S a} R_x = \omega$ is necessary, otherwise the condition $\forall z \exists x(\Psi(x,z)\downarrow)$, which we need for (1), does not hold. It is easy to see that in (2) the $a$-sequence $\{R_x\}$ has this property.

We have proved the following:

**Theorem 3.11.** *Let $S$ be a univalent system of notations for constructive ordinals, $A \subseteq \omega$ and $\alpha$ an ordinal which has $S$-notation $a$. Then the following are equivalent:*

*a) There is a partial computable function $\Psi$ such that for every $x$, $A(x) = \Psi((\mu\lambda< \alpha)_S(\Psi((\lambda)_s, x)\downarrow, x))$;*

*b) There is an a-sequence* $\{R_x\}_{x<_s a}$ *such that* $\bigcup_{x<_s a} R_x = \omega$, *and*

$$A = \{z|\ \exists x <_S a(z \in R_x \,\&\, e(x) \neq e(a) \,\&\, \forall y <_S x(z \notin R_y))\}.$$

Theorem 3.11 generalizes the previously obtained descriptions of $\omega$-c.e. and $n$-c.e. sets for $n < \omega$ using $\omega$-sequences and $n$-sequences of c.e. sets respectively. Now we will show that any $\Delta_2^0$-set has such a description for some $a$-sequence $\{R_x\}$, where $a$ is a notation for some ordinal $\alpha$ in the notation system $S$. Moreover, we will have that $\alpha \leqslant \omega^2$.

We first prove that any set which can be so defined using an $a$-sequence $\{R_x\}$ for some $a \in S$, is a $\Delta_2^0$-set, i.e. these definitions do not take us out of the class of $\Delta_2^0$-sets.

**Theorem 3.12.** *Let $S$ be a univalent and recursively related system of notations, $\alpha$ a constructive ordinal which has a notation in $S$, $\Psi$ a partial computable function, and let $f$ be a function such that $\Psi \rightarrow_{\{\alpha,S\}} f$. Then $f \leqslant_T \emptyset'$.*

***Proof.*** Let $x$ be a given integer. To $\emptyset'$-compute $f(x)$ find the first (if any) integer $n$ such that $\Psi(n,x)\downarrow$. (If there is no such $n$, then $f(x)\uparrow$.)

Let $\Psi(n,x)\downarrow$ for some $n$. Using the oracle $\emptyset'$ find (if it exists) an integer $m$ such that $\nu_S(m) < \nu_S(n)$ and $\Psi(m,x)\downarrow$. This is possible since $S$ is recursively related. Now repeat the same, replacing $n$ by $m$ and so on. Since the set $\alpha$ is well-ordered we will repeat this process only finitely many times. Now let $m$ be an integer such that $\nu_s(m)$ is the least ordinal such that $\Psi(m,x)\downarrow$. We have $f(x) = \Psi(m,x)$. □

**Theorem 3.13.** *Let $f \leqslant_T \emptyset'$ be a total function. There is a partial computable function $\Psi$ such that for every $x$,*

$$f(x) = \Psi(|\mu\lambda{<}\, \omega^2|_O(\Psi(|\lambda|_O, x)\downarrow), x).$$

***Proof.*** Since $O$ is a universal system of notations, it is enough to construct a univalent and recursively related system of notations $S$ and a partial computable function $\Psi$ such that for every $x$,

$$f(x) = \Psi((\mu\lambda{<}\, \omega^2)_S(\Psi((\lambda)_S, x)\downarrow), x).$$

Let $f(x) = \lim_s g(s,x)$ for all $x$ and some computable function $g$. Let $0 = s_0^x < s_1^x < \ldots < s_{k_x}^x$ be all integers $s$, for which $g(s,x) \neq g(s+1,x)$. Therefore $k_x$ is the number of different values of the function $g$ on the set of pairs $\{(s,x)|s \in \omega\}$.

Arrange all pairs $(y, x)$ (or rather, the indices $\langle y, x\rangle$ of these pairs) into an $\omega^2$-sequence as follows:

$$\text{Block } 0 \begin{cases} \langle s^0_{k_0}, 0\rangle, \langle s^0_{k_0}+1, 0\rangle, \langle s^0_{k_0}+2, 0\rangle, \dots, \langle s^0_{k_0}+j, 0\rangle, \dots \\ \langle s^0_{k_0-1}, 0\rangle, \langle s^0_{k_0-1}+1, 0\rangle, \langle s^0_{k_0-1}+2, 0\rangle, \dots, \langle s^0_{k_0}-1, 0\rangle \\ \dots\dots\dots\dots\dots\dots\dots\dots\dots\dots\dots\dots\dots\dots \\ \langle s^0_1, 0\rangle, \langle s^0_1+1, 0\rangle, \langle s^0_1+2, 0\rangle, \dots, \langle s^0_2-1, 0\rangle \\ \langle 0, 0\rangle, \langle 1, 0\rangle, \langle 2, 0\rangle, \dots, \langle s^0_1-1, 0\rangle \end{cases}$$

$$\text{Block } 1 \begin{cases} \langle s^1_{k_1}, 1\rangle, \langle s^1_{k_1}+1, 1\rangle, \langle s^1_{k_1}+2, 1\rangle, \dots, \langle s^1_{k_1}+j, 1\rangle, \dots \\ \langle s^1_{k_1-1}, 1\rangle, \langle s^1_{k_1-1}+1, 1\rangle, \langle s^1_{k_1-1}+2, 1\rangle, \dots, \langle s^1_{k_1}-1, 1\rangle \\ \dots\dots\dots\dots\dots\dots\dots\dots\dots\dots\dots\dots\dots\dots \\ \langle s^1_1, 1\rangle, \langle s^1_1+1, 1\rangle, \langle s^1_1+2, 1\rangle, \dots, \langle s^1_2-1, 1\rangle \\ \langle 0, 1\rangle, \langle 1, 1\rangle, \langle 2, 1\rangle, \dots, \langle s^1_1-1, 1\rangle \end{cases}$$

$$\dots\dots\dots\dots\dots\dots\dots\dots\dots\dots\dots\dots\dots\dots\dots .$$

Each $i$-th row (except the 0th) of the $x$-th block ($0 \leqslant x < \infty$) is filled with numbers $\langle s^x_i, x\rangle, \langle s^x_i+1, x\rangle, \langle s^x_i+2, x\rangle, \dots, \langle s^x_{i+1}-1, x\rangle$, and the 0th row consists of the infinite sequence of numbers $\langle s^x_{k_x}+j, x\rangle$ , $j \geqslant 0$. It is clear that $x$-th block of this matrix contains all numbers $\langle j, x\rangle, j \geqslant 0$, without repetition.

Now, for each $x$, we transform rows of the $x$-th block so that its $i$-th row for each $i \geqslant 0$ (not only for $i = 0$) contains infinitely many integers, but nevertheless we still have the following conditions:

1) The first element of the $i$-th row is the number $\langle s^x_{k_x-i}, x\rangle$,
2) Each block contains all natural numbers $\langle j, x\rangle, j \geqslant 0$ without repetition.

For this we fill the rows of the $x$-th block as follows: Sequentially compute $g(0, x), g(1, x), \dots$ and simultaneously fill with numbers $\langle 0, x\rangle, \langle 1, x\rangle, \langle 2, x\rangle, \dots$ the positions of the last row from left to right until we reach the number $s^x_1$, for which we have $g(s^x_1-1, x) \neq g(s^x_1, x)$. After that we begin to fill with numbers $\langle s^x_1, x\rangle, \langle s^x_1+1, x\rangle, \dots$ simultaneously from left to right positions of the last two rows until we reach the number $s^x_2$, for which we have $g(s^x_2-1, x) \neq g(s^x_2, x)$. Then we fill with numbers $\langle s^x_2, x\rangle, \langle s^x_2+1, x\rangle, \dots$ simultaneously from left to right positions of the last three rows: the third to last, second to last and last (in this order), until we reach the number $s^x_3$ and so on. Let us denote this process of enumerating elements of the constructed matrix by $\mathcal{M}$, and by $a_{i,j}$ the element of the matrix which is in the $j$-th place of its $i$-th row.

Thus, inside each block we have finitely many rows of order type $\omega$. It is clear that for each pair $(x, y)$ the number $\langle x, y\rangle$ belongs to exactly one row of the matrix. Define a linear ordering $<_\varphi$ on the elements of the matrix as follows: $a_{i,j} <_\varphi a_{k,l}$, if either $i < k$, or $i = k$, but $j < l$. Therefore, each of the $\omega$ blocks has order type $\omega \cdot n$ for some $n \geqslant 1$, and all the numbers in the matrix give order type $\omega^2$.

Now we define a univalent system of notations $S$ for ordinals $< \omega^2$ as follows: We map, in an order-preserving way (and denoting this map as $\nu_S$) the integer $a_{i,j}$ to the ordinal $\omega \cdot i + j$, $0 \leqslant i < \omega$: $\alpha < \beta$ if and only if $(\alpha)_S <_\varphi (\beta)_S$.

To verify that $S$ is a univalent system of notations, define computable functions $k_S$, $p_S$ and a partial computable function $q_S$ as follows:

$$k_S(\langle s, x\rangle) = \begin{cases} 0, & \text{if } s = s^0_{k_0}, x = 0; \\ 2, & \text{if } s = 0 \vee (s > 0 \ \&\ g(s, x) \neq g(s-1, x)); \\ 1, & \text{otherwise.} \end{cases}$$

Obviously $k_S$ is a computable function, and if $k_S(x) = 0$, then $\nu_S(x) = 0$; if $\nu_S(x)$ a successor, then $k_S(x) = 1$, and if $\nu_S(x)$ a limit ordinal then $k_S(x) = 2$.

We define the function $p_S(x)$ as follows: If $l(x) \neq s_i^{r(x)}$ for some $i \leqslant k_{r(x)}$, then $p_S(x) = \langle l(x) - 1, r(x)\rangle$.

It is clear that $p_S$ is a partial computable function, and if $\nu_S(x)$ is a successor then $p_S(x)$ is defined and $\nu_S(x) = \nu_S(p_S(x)) + 1$.

To define the function $q_S$ consider the following two cases.

Case 1. $x = \langle n, 0\rangle$. (The number $x$ belongs to the 0-th block of the table.) Define $q_S(x)$ as an index of the following partial computable function $f$: If $n < s^0_{k_0}$, then we sequentially compute values of $g(n, 0), g(n+1, 0), \ldots$ until we obtain a number $s > n$ such that $g(s-1, 0) \neq g(s, 0)$. (It is clear that if $n = s^0_i$ for some $i < k_0$ then $s = s^0_{i+1}$.) Define $f(t) = g(s+t, 0)$ for all $t \geqslant 0$.

Case 2. $x = \langle n, m\rangle, m > 0$. (The number $x$ belongs to the $m$-th nonzero block.) In this case $q_S(x)$ is an index of the following partial computable function $f$: We define $f(0), f(1), \ldots$ sequentially as $\langle 0, m-1\rangle, \langle 1, m-1\rangle, \langle 2, m-1\rangle, \ldots$ (the subsequent elements of the last row of the $(m-1)$th block) and simultaneously compute $g(n, m), g(n+1, m), g(n+2, m), \ldots$ until we again obtain a number $s > n$ such that $g(s-1, 0) \neq g(s, 0)$. (It is clear that if $n = s^m_i$ for some $i < k_m$, then again $s = s^m_{i+1}$.) After that, the remaining values of the function $f$ are defined as the subsequent values of the function $g(s+t,m), t \geqslant 0$.

It is easy to see that if $\nu_S(x)$ is a limit ordinal then $q_S(x)$ is defined and $\{\nu_S(\Phi_{q_S(x)}(n))\}_{n\in\omega}$ is an increasing sequence whose limit is $\nu_S(x)$.

Obviously $\nu_S(x) < \nu_S(y)$ if and only if either $r(x) < r(y)$ or $r(x) = r(y)$, but then $l(x) < l(y)$. Therefore $S$ is also recursively related system of notations for ordinals.

Now let

$$\Psi(\langle s,y\rangle,x) = \begin{cases} g(s,x), & \text{if } x = y; \\ \uparrow, & \text{otherwise.} \end{cases}$$

Obviously, $\Psi$ is a partial computable function and for every $x$, $f(x) = \Psi((\mu\lambda < \omega^2)_S(\Psi((\lambda)_S, x)\downarrow), x)$. □

Since $O$ is a universal system of notations, Theorems 3.11, 3.12, and 3.13 imply the following:

**Corollary 3.1.** *For any set $A \subseteq \omega$, $A \leqslant_T \emptyset'$ if and only if there is an $a$-sequence $\{R_x\}_{x<_O a}$, $|a|_O \leqslant \omega^2$, such that $\bigcup_{x<_O a} R_x = \omega$, and*

$$A = \{z| \quad \exists x <_O a(z \in R_x \,\&\, e(x) \neq e(a) \,\&\, \forall y <_O x(z \notin R_y)\}.$$

*Comment.* Theorems 3.12 and 3.13 are due to Ershov [34]. In the proofs, we have used an approach suggested by Epstein, Haas, and Kramer [32]. Theorem 3.11 is also from this work.

### 3.1.5. *The infinite levels of the Ershov hierarchy*

Since $|a|_O$ has order-type $\langle\{x : x <_O a\}, <_O\rangle$, the sentence "$a$-sequence of c.e. sets $\{R_x\}$" for $a \in O$ has to be understood in the sense of Definition 3.3. Define for $a \in O$ the operations $S_a$ and $P_a$, which map $a$-sequences $\{R_x\}_{x<_O a}$ into subsets of $\omega$ as follows:

$$S_a(R) = \{z|\exists x <_O a(z \in R_x \,\&\, e(x) \neq e(a) \,\&\, \forall y <_O x(z \notin R_y))\}.$$
$$P_a(R) = \{z|\exists x <_O a(z \in R_x \,\&\, e(x) = e(a) \,\&\, \forall y <_O x(z \notin R_y))\} \cup \{\omega - \bigcup_{x<_O a} R_x\}.$$

It follows from these definitions that $P_a(R) = \overline{S_a(R)}$ for all $a \in O$ and all $a$-sequences $R$.

By definition the class $\Sigma_a^{-1}$ ($\Pi_a^{-1}$) for $a \in O$ is the class of sets $S_a(R)$ ($P_a(R)$, respectively), where $R = \{R_x\}_{x<_O a}$ runs through all $a$-sequences of c.e. sets. Let $\Delta_a^{-1} = \Sigma_a^{-1} \cap \Pi_a^{-1}$.

It is easy to see that for natural numbers $n > 0$ and for $a \in O$ such that $|a|_O = \omega$ these definitions coincide with the previous ones. (The finite levels of the Ershov hierarchy are denoted by ordinals, not by their $O$-notations.)

**Theorem 3.14.** *(Hierarchy Theorem) Let $a, b \in O$ and $a <_O b$. Then $\Sigma_a^{-1} \cup \Pi_a^{-1} \subsetneq \Sigma_b^{-1} \cap \Pi_b^{-1}$.*

***Proof.*** It follows immediately from the definitions of the classes of $\Sigma_a^{-1}$- and $\Pi_a^{-1}$- sets that if $a <_O b$ then $\Sigma_a^{-1} \cup \Pi_a^{-1} \subseteq \Sigma_b^{-1} \cap \Pi_b^{-1}$. It is easy to see that here all the inclusions are proper. □

**Corollary 3.2.** *For every $a \in O$, $\Sigma_a^{-1} \subsetneq \Sigma_2^0 \cap \Pi_2^0$.*

***Proof.*** Suppose, for the sake of contradiction, that for some $a \in O$ we have $\Sigma_a^{-1} = \Sigma_2^0 \cap \Pi_2^0$. Let $b \in O$ be a notation such that $a <_O b$. Then, by Theorem 3.14, $\Sigma_a^{-1} \subset \Sigma_b^{-1}$. Therefore, $\Sigma_2^0 \cap \Pi_2^0 = \Sigma_a^{-1} \subset \Sigma_b^{-1}$, a contradiction. □

**Theorem 3.15.** *Let $|a|_O$ be a limit ordinal. The set $A$ belongs to the class $\Delta_a^{-1}$ if and only if there is an $a$-sequence $\mathcal{R}$ such that $A = S_a(\mathcal{R})$ and $\bigcup_{b<_O a} R_b = \omega$.*

***Proof.*** ($\rightarrow$) Let $A \in \Delta_a^{-1}$. Then $A = S_a(\mathcal{R}_0)$ and $\omega - A = S_a(\mathcal{R}_1)$ for $a$-sequences of c. e. sets $\mathcal{R}_0 = \{R_{0,x}\}_{x<_O a}$ and $\mathcal{R}_1 = \{R_{1,x}\}_{x<_O a}$.

We define a new $a$-sequence $\mathcal{P} = \{P_x\}_{x<_O a}$ as follows: If in $\{x \mid x <_O a\}$ $x$ is a notation for a limit ordinal, then we define $P_x = R_{0,x}$, otherwise $|x|_O$ is the successor of an ordinal $|y|_O$ such that $y <_O x$. Define $P_x = R_{0,x} \cup R_{1,y}$.

Since $A \subseteq \bigcup\limits_{x<_O a} R_{0,x}$ and $\omega - A \subseteq \bigcup\limits_{x<_O a} R_{1,x}$, and for all $y <_O a$ we have the inclusions $R_{0,y} \subseteq P_y$ and $R_{1,y} \subseteq P_x$, where $y <_O x <_O a$, we concluded that $\bigcup\limits_{x<_O a} P_x = \omega$. The verification of the condition $A = S_a(\mathcal{P})$ is straightforward.

($\leftarrow$) Now suppose that $A = S_a(\mathcal{P})$ for some $a$-sequence $\mathcal{P} = \{P_x\}_{x<_O a}$, suppose also that $\bigcup\limits_{x<_O a} P_x = \omega$. Define a new $a$-sequence $\mathcal{R} = \{R_x\}_{x<_O a}$

as follows: $R_1 = \emptyset$. Further, for an arbitrary $x \in O$, $1 <_O x <_O a$, we set $R_x = \cup_{y<_O x} R_y$, if $x$ is a notation of a limit ordinal in $\{x \mid x <_O a\}$. Otherwise, we set $R_x = P_y$ for some $y <_O x$ such that $|x|_O$ is a successor of $|y|_O$. Again it is easy to check that $\omega - A = S_a(\mathcal{R})$. □

Theorem 3.11 now immediately implies the following:

**Corollary 3.3.** *Let $|a|_O$ be a limit ordinal. The set $A$ belongs to the class $\Delta_a^{-1}$ if and only if there is a partial computable function $\Psi$ such that for every $x$,*
$A(x) = \Psi(|\mu\lambda < \alpha|_O(\Psi(|\lambda|_O, x)\downarrow, x))$.

The proof of the following theorem is similar to the proofs of Theorems 3.11 and 3.15:

**Theorem 3.16.** *Let $A \subseteq \omega$ and $a \in O$. The following are equivalent:*
*a) $A$ belongs to the class $\Sigma_a^{-1}$;*
*b) There is a partial computable function $\Psi$ such that for every $x$, $x \in A$ if and only if $\Psi(|\mu\lambda < \alpha|_O(\Psi(|\lambda|_O, x)\downarrow, x))$.*

Generalizing Definition 3.4 of the $\omega$-c.e. sets to infinite ordinals we introduce the following definition:

**Definition 3.9.** Let $|a|_O$ be a limit ordinal. If $A \in \Delta_a^{-1}$, then the set $A$ is called an $|a|_O$-*c.e.* set (or an $\alpha$-c.e. set, if $|a|_O = \alpha$).

It is clear that if $A \in \Sigma_a^{-1}$ for some $a \in O$, and $B \leqslant_m A$, then $B \in \Sigma_a^{-1}$, and if $A$ is $|a|_O$-c.e. for some limit ordinal $|a|_O$, $a \in O$, and $B \leqslant_m A$, then $B$ is also $|a|_O$-c.e. set.

The following theorem is a direct corollary of Theorems 3.11, 3.12 and 3.13.

**Theorem 3.17.** $\displaystyle\bigcup_{a\in O} \Sigma_a^{-1} = \bigcup_{a\in O, |a|_O=\omega^2} \Sigma_a^{-1} = \Sigma_2^0 \cap \Pi_2^0.$

Theorem 3.17 cannot be strengthened:

**Theorem 3.18.** $\displaystyle\bigcup_{a\in O, |a|_O<\omega^2} \Sigma_a^{-1} \neq \Sigma_2^0 \cap \Pi_2^0.$

***Proof.*** Let $a, b \in O$ be notations such that $|a|_O = \omega^2$, $|b|_O < \omega^2$. It is easy to see that $b \preceq_O a$, which implies $\Sigma_b^{-1} \subseteq \Sigma_a^{-1}$. Therefore, for each $a \in O$ such that $|a|_O \geqslant \omega^2$, we have $\displaystyle\bigcup_{b\in O, |b|_O<\omega^2} \Sigma_b^{-1} \subseteq \Sigma_a^{-1} \subset \Sigma_2^0 \cap \Pi_2^0.$ □

**Theorem 3.19.** *a) For any $a \in O$ there is a path $T_0$ in $O$ through $a$ such that* $\bigcup_{b \in T_0} \Sigma_b^{-1} = \Sigma_2^0 \cap \Pi_2^0$;

*b) There is a path $T$ in $O$ such that $|T|_0 = \omega^3$ and* $\bigcup_{a \in T} \Sigma_a^{-1} = \Sigma_2^0 \cap \Pi_2^0$.

***Proof.*** a) Let $a \in O$, and let $\{b_0, b_1, \ldots\}$ be a listing of all $b \in O$ such that $|b|_0 = \omega^2$. We define $T_0$ as a path through $c_0, c_1, c_2, \ldots$, where $c_0 = a$, $c_1 = a +_0 b_0$, $c_2 = (a +_0 b_0) +_0 b_1, \ldots, c_n = (\ldots (a +_0 b_0) +_0 \ldots) +_0 b_n$. Obviously $c_0 <_0 c_1 <_0 c_2 <_0 \ldots$, and the order type of $T_0$ is $|a|_0 + \omega^3$. Since for each $n < \omega$ we have $c_n = d +_0 b_n$ for some $d \in O$, and for all $x, y$, $y \preceq_0 x +_0 y$ (see Proposition 3.1), we have, for every $n$, that $b_n \preceq_0 c_n$. Now it follows from Theorem 3.17 that $\bigcup_{b \in T_0} \Sigma_b^{-1} = \Sigma_2^0 \cap \Pi_2^0$.

b) Immediate by the preceding proof for $a = 1$. □

The following claim shows that Theorem 3.19 b) cannot be strengthened:

**Proposition 3.2.** *If a path $T$ in $O$ is such that $|T|_0 < \omega^3$, then* $\bigcup_{a \in T} \Sigma_a^{-1} \neq \Sigma_2^0 \cap \Pi_2^0$.

***Proof.*** We first prove the following:

**Lemma 3.2.** *For any $a \in O$,*

$$\bigcup_{a \leqslant_0 b, |b|_0 - |a|_0 < \omega^2} \Sigma_b^{-1} \neq \Sigma_2^0 \cap \Pi_2^0.$$

***Proof of Lemma.*** Let $d \in O$ be a notation such that $a \leqslant_0 d$ and $|d|_0 = |a|_0 + \omega^2$. It is not difficult to see that for every $b \in O$ such that $a \leqslant_0 b$ and $|b|_0 - |a|_0 < \omega^2$ we have $b \preceq_0 d$. Therefore, $\bigcup_{a \leqslant_0 b, |b|_0 - |a|_0 < \omega^2} \Sigma_b^{-1} \subseteq \Sigma_d^{-1} \neq \Sigma_2^0 \cap \Pi_2^0$. □ (of Lemma)

(***Proof of Proposition 3.2 continued.***) Since $|T|_0 < \omega^3$, in $T$ there is an element $a$ such that for some ordinal $\rho < \omega^2$ we have $|T|_0 = |a|_0 + \rho$. Hence, if $b \in T$ and $a \leqslant_0 b$, then $|b|_0 - |a|_0 < \omega^2$. Therefore,

$$\bigcup_{b \in T} \Sigma_b^{-1} \subseteq \bigcup_{a \leqslant_0 b, |b|_0 - |a|_0 < \omega^2} \Sigma_b^{-1}.$$

Now it remains to apply the preceding lemma. □

*Comment.* All results of this section are due to Ershov [34].

### 3.1.6. *Levels of the Ershov hierarchy containing Turing jumps*

M. C. Faizrahmanov [36] has investigated the levels of the Ershov hierarchy containing Turing jumps. Not every level of the hierarchy contains the Turing jump of a set. For instance, its finite levels contain Turing jumps only of computable sets. Indeed, if $A'$ is $n$-c.e. and $A$ is non-computable, then there is a non-$n$-c.e. set $B <_T A$. Therefore $B' \leqslant_1 A'$. It follows that $B <_1 A'$ and, hence, $B$ is an $n$-c.e. set, a contradiction.

**Theorem 3.20.** *(M. C. Faizrahmanov) If $A' \in \Pi_a^{-1}$ for a set $A$ and a notation $a \in O$, then $A' \in \Delta_a^{-1}$.*

***Proof.*** As usual, we denote $\mathrm{dom}\Phi_e^A$ by $W_e^A$ for every $e \in \omega$. Here $\{\Phi_e^A\}_{e\in\omega}$ is the standard enumeration of all unary functions partial computable in $A$.

Since $\omega - A' \in \Sigma_a^{-1}$, then it follows from Theorem 3.16 that there is a partial computable function $\Psi$ such that $x \in \overline{A'}$ if and only if $\Psi(|\mu\lambda <\alpha|_O(\Psi(|\lambda|_O, x)\downarrow, x)) = 1$. (In this case we also say that $\omega - A' \in \Sigma_a^{-1}$ *with function* $\Psi$.)

Let $B = \{x : \exists t \in O(\Psi(t,x)\downarrow \ \ \& \ t <_O a)\}$. Obviously, $B$ is c.e. and $\overline{A'} \subseteq B$. Let $\{B_s\}_{s\in\omega}$ be an effective enumeration of $B$. Since $A$ is a $\Delta_2^0$-set, there is a uniformly computable sequence $\{A_s\}_{s\in\omega}$ such that $A = \lim_s A_s$. Let $e$ be an integer that $A'(x) = \lim_s W_{e,s}^{A_s}(x)$ for all $x$.

Now we define a set $U$ c.e. in $A$. Using the Recursion Theorem we initially fix an index of $U$ (in the enumeration $\{W_e^A\}_{e\in\omega}$) and, therefore, we can fix a computable function $f$ such that

$$(\forall x)\{x \in U \leftrightarrow f(x) \in A'\}.$$

Stage $s = 0$. Set $U_0 \ = \ \emptyset$.

Stage $s > 0$. Let $(s)_0 = i$. If $f(i) \in B_s$ and $\Phi_{e,s}^A(i)\downarrow$, then enumerate $i$ into $U_s$.

Let $U = \bigcup_s U_s$.

For every $i$ we have $f(i) \in B$. Indeed, suppose $f(i) \notin B$ for some $i$. Then $i \notin U$ and, therefore, $f(i) \in \overline{A'}$. It follows that $f(i) \in B$, a contradiction.

Now we define a partial computable function $\Theta$ as follows:

$$\Theta(t,x) = \begin{cases} 0, & \text{if } \Psi(t, f(x)) \downarrow = 1; \\ 1, & \text{if } \Psi(t, f(x)) \downarrow \neq 1; \\ \uparrow, & \text{if } \Psi(t, f(x)) \uparrow. \end{cases}$$

It is clear, that the function $\Theta$ defines $A'$ as a $\Delta_a^{-1}$-set. □

Theorem 3.21 is proved for the following *natural system of notations* $\langle D_C, |.|_C\rangle$ for ordinals below $\omega^\omega$ with its domain $D_C$ and the map $|.|_C : D_C \to \omega^\omega$.

$$D_C = \{x : \exists m, k_0, \ldots, k_m(x = \langle m, k_0, \ldots, k_m\rangle \ \& \ m \neq 0 \to k_0 \neq 0)\},$$

$$|\langle m, k_0, \ldots, k_m\rangle|_C = \omega^m k_0 + \omega^{m-1} k_1 + \cdots + k_m.$$

In this theorem the levels of the Ershov hierarchy $\Sigma_a^{-1}, \Pi_a^{-1}$ and $\Delta_a^{-1}$ are also defined for $a \in C$. It is clear that $C$ is a univalent and recursively related system and for simplicity, in what follows we identify ordinals with their notations.

Let $\alpha$ and $\beta$ be ordinals $< \omega^\omega$ and

$$\alpha = \omega^m p_0 + \omega^{m-1} p_1 + \ \cdots \ p_m,$$

$$\beta = \omega^m q_0 + \omega^{m-1} q_1 + \ \cdots \ q_m,$$

for some $m, p_0, \ldots, p_m, q_0, \ldots, q_m$.

The ordinal $\alpha(+)\beta$ defined as

$$\alpha(+)\beta = \omega^m(p_0 + q_0) + \omega^{m-1}(p_1 + q_1) + \ \cdots \ (p_m + q_m)$$

is called the *natural sum* of $\alpha$ and $\beta$.

**Theorem 3.21.** *(M. C. Faizrahmanov) Let $A \subset \omega$.*

*a) If $n > 0$ and $A' \in \Sigma_{\omega^n}^{-1}$, then $A' \in \Delta_{\omega^n}^{-1}$;*

*b) If $m, n > 0$ and $A' \in \Sigma_{\omega^n m}^{-1}$, then $A' \in \Delta_{\omega^n}^{-1}$;*

*c) For every $n > 0$ there is a set $A$ such that $A' \in \Delta_{\omega^{n+1}}^{-1} - \Delta_{\omega^n}^{-1}$.*

***Proof.*** We present only part $a)$ of the theorem. The proof of part $b)$ is based on part $a)$ and uses induction on $m$. The proof of part $c)$ is achieved by means of a direct construction using a finite injury priority argument.

Let $n > 0$ and $A' \in \Sigma_{\omega^n}^{-1}$. Define

$$S = \{2^x(2y+1) : x \in A' \ \& \ y \in \omega\}.$$

It is clear that $S$ recursively isomorphic to $A'$ and, therefore, there is a partial computable function $\Psi$ such that $x \in S$ if and only if $(\exists \beta < \omega^n)(\Psi(\beta, x)\downarrow = 1 \ \& \ (\forall \gamma < \beta)(\Psi(\gamma, x)\uparrow)$ (i.e. $\Psi$ defines $S$ as a $\Sigma^{-1}_{\omega^n}$-set).

Define a c.e. set $B$ as follows: $B = \{x : \exists \beta < \omega^n(\Psi(\beta, x)\downarrow)\}$. Let $\{B_s\}_{s\in\omega}$ be a computable enumeration of $B$. Since $A$ is a $\Delta^0_2$-set, there is a uniformly computable sequence $\{A_s\}_{s\in\omega}$ such that $A = \lim_s A_s$. Let $S = \lim_s W^A_e[s]$ for some integer $e$.

Define for all $x$, $i$ and $s$

$$r(x, i) = 2^x 3^i,$$

$$q(x, i, s) = |\{t < s : \Phi^A_e(r(x, i))[t] \neq \Phi^A_e(r(x, i))[t + 1]\}|,$$

$$p(x, i, s) = 3^x 5^{i+1} 7^{q(x,i,s)}.$$

For each partial computable function $\Phi_n$ we define partial computable functions $h^n_0$ and $h^n_1$ as follows:

Let $h^n_{0,0} = h^n_{1,0} = \emptyset$. Suppose that $h^n_{0,s}$ and $h^n_{1,s}$ are already defined and let $i = min\{k : h^n_{0,s}(k)\uparrow\}$. Let $x \leqslant s$ be the least (if any) integer such that $\Phi_{n,s}(r(x, i))\downarrow \in B_s$, $\Phi_{n,s}(p(x, i, s))\downarrow \in B_s$, and $\Phi^A_e(r(x, i))[s]\uparrow$. If there is such $x$, then define $h^n_{0,s+1} = h^n_{0,s} \cup \{(i, r(x, i))\}$, $h^n_{1,s+1} = h^n_{1,s} \cup \{(i, p(x, i, s))\}$. Otherwise, define $h^n_{0,s+1} = h^n_{0,s}$, $h^n_{1,s+1} = h^n_{1,s}$. Let $h^n_0 = \bigcup_s h^n_{0,s}$, $h^n_1 = \bigcup_s h^n_{1,s}$. It follows from the definitions of $h^n_0$ and $h^n_1$ that $h^n_{0,s}(i)\downarrow$ if and only if $h^n_{1,s}(i)\downarrow$. Since the ranges of $r$ and $p$ are disjoint, the ranges of values of $h^n_0$ and $h^n_1$ are also disjoint.

Now we construct set $U$ c.e. in $A$. By the Recursion Theorem we can initially fix an index of $U$ and, therefore, fix a computable function $f$ such that $x \in U \leftrightarrow f(x) \in S$. Let $n$ be an integer such that $f = \Phi_n$ and denote $h_k = h^n_k$, $h_{k,s} = h^n_{k,s}$ for $k = 0, 1$.

Stage $s = 0$. Let $U_0 = \emptyset$.

Stage $s + 1$ consists of two steps.

Step 1. Let $i = \mu j(j \notin \text{dom } h_{0,s})$. For each $x \leqslant s$,

(a) If $\Phi^A_e(r(x, i))[s]\downarrow$ and $A_s \upharpoonright \varphi_e(A, r(x, i))[s] = A \upharpoonright \varphi_e(A, r(x, i))[s]$, then enumerate $r(x, i)$ into $U_{s+1}$;

(b) Let $\hat{r} = max\{\varphi_e(A, r(x, i))[t] : t \leqslant s\}$. If $A_s \upharpoonright \hat{r} = A \upharpoonright \hat{r}$, then enumerate $p(x, i, s)$ into $U_{s+1}$.

Step 2. For each $j \in \text{dom } h_{0,s}$ such that $\Phi^A_{e,s+1}(j)\downarrow$, enumerate $h_{0,s}(j)$, $h_{1,s}(j)$ into $U_{s+1}$.

Now let $U = \bigcup_s U_s$. There are two possibilities:

Case 1. The function $h_0$ is not total.

Let $i = min\{k : h_0(k)\uparrow\}$. In this case for all $x$ we have that

$$f(r(x,i)) \in B \rightarrow r(x,i) \in S.$$

Indeed, suppose that $f(r(x,i)) \in B$. Since $S = \lim_s W_e^A[s]$, there is a stage $s$ such that $A_s \restriction \hat{r} = A \restriction \hat{r}$, where $\hat{r}$ is defined as above. Moreover, the stage $s$ can be chosen so that this equality will be preserved at all subsequent stages. Then for all $t \geqslant s$ we have $p(x,i,t) \in U$ and, therefore, $f(p(x,i,t)) \in B$. Since $h_0(i)$ is undefined, $\Phi_e^A(r(x,i))$ is defined, which means that $r(x,i) \in S$.

It follows from part $(a)$ of the construction that

$$\{r(x,i) : r(x,i) \in U\} = \{r(x,i) : r(x,i) \in S\}.$$

Therefore, if $r(x,i) \in S$, then $f(r(x,i)) \in B$ and we have $A' \leqslant_1 B$ via the reduction function $g(x) = f(r(x,i))$. This means that $A'$ is c.e. and, therefore, $A' \in \Delta^{-1}_{\omega^n}$.

Case 2. The function $h_0$ is total.

It follows that $h_1$ is also total. Let $i \in \omega$ be an arbitrary integer and $s$ be the least stage such that $h_{0,s+1}(i)\downarrow$. By definition of $h_0$ we have $\Phi_e^A(h_0(i))[s]\uparrow$. Let $s_0 = min\{t \leqslant s : \forall s' \in [t,s](\Phi_e^A(h_0(i))[s']\uparrow)\}$. Obviously, for all $t \in [s_0, s)$ we have $q(x,i,t) = q(x,i,t+1)$. Hence $h_1(i) = p(x,i,s_0)$. Also it is clear that $p(x,i,s_0) \notin U_{s_0}$.

Now either $h_0(i) \notin U_{s+1}$, or $h_1(i) \notin U_{s+1}$. Indeed, suppose that $h_0(i) \in U_{s+1}$. Then there is a stage $t < s_0$ such that $\Phi_e^A(h_0(i))[t]\downarrow$ and $A_t \restriction \varphi_e(A, h_0(i))[t] = A \restriction \varphi_e(A, h_0(i))[t]$. Since for all $u \in [s_0, s]$ we have $\Phi_e^A(h_0(i))[u]\uparrow$, we also have for all $u \in [s_0, s]$, $A_u \restriction \hat{h} \neq A \restriction \hat{h}$, where $\hat{h} = max\{\varphi_e(A, h_0(i))[v] : v \leqslant s_0\}$. Therefore, $p(x,i,s_0) \notin U_{s+1}$. Now step 2 of the construction ensures that

$$\forall i(i \in S \leftrightarrow (f(h_0(i)) \in S \ \& \ f(h_1(i)) \in S)). \tag{1}$$

Now we define a function $\Theta$, which defines $S$ as a $\Delta^{-1}_{\omega^n}$-set. Let $\Psi_s$ denote the part of the function $\Psi$ defined at the end of stage $s$.

For a given $i$ find a stage $v$ and ordinals $\beta_0, \beta_1 < \omega^n$ such that $\Psi_v(\beta_0, f(h_0(i)))\downarrow$ and $\Psi_v(\beta_1, f(h_1(i)))\downarrow$. (Such ordinals $\beta_0$ and $\beta_1$ exist, since by construction for all $j$ we have $\{f(h_0(j)),\ f(h_1(j))\} \subset B$.)

Define a partial computable function $\Theta_0$ so that

$$\Theta_0(\beta_0(+)\beta_1, i) = \Psi_v(\beta_0, f(h_0(i))) \cdot \Psi_v(\beta_1, f(h_1(i))).$$

Now suppose that a partial computable function $\Theta_s$ is already defined and for $k \in \{0,1\}$ let $\gamma_k = \mu\delta(\Psi_{v+s}(\delta, f(h_k(i)) \downarrow)$. Define a partial computable function $\Theta_{s+1}$ so that

$$\Theta_{s+1}(\gamma_0(+)\gamma_1, i) = \Psi_{v+s}(\gamma_0, f(h_0(i))) \cdot \Psi_{v+s}(\gamma_1, f(h_1(i))).$$

Let $\Theta = \bigcup_s \Theta_s$. To show that $\Theta$ defines $S$ as a $\Delta^{-1}_{\omega^n}$-set, take an arbitrary integer $x$ and let $\alpha_k = \mu\beta(\Psi(\beta, f(h_k(x))) \downarrow)$, $k = 0, 1$. Then $\alpha_0(+)\alpha_1 = \mu\gamma(\Theta(\gamma, x) \downarrow)$. Since $\alpha_0$ and $\alpha_1$ are below $\omega^n$, $\alpha_0(+)\alpha_1 < \omega^n$. Now it follows from (1) that

$$S(x) = \Psi(\alpha_0, f(h_0(x))) \cdot \Psi(\alpha_1, f(h_1(x))) = \Theta(\alpha_0(+)\alpha_1, x).$$

This means that $A' \in \Delta^{-1}_{\omega^n}$. □

## 3.2. The Turing Degrees of the $n$-c.e. Sets

### 3.2.1. *The class of the $n$-c.e. degrees*

The first results on the Turing degrees of the sets from different levels of the Ershov hierarchy were obtained in 1970's of the last century when S.B. Cooper in his dissertation (Cooper [16]) proved the existence of a Turing degree which contains a 2-c.e. set, but does not contain c.e. sets (below such degrees are called *properly* 2-*c.e.* degrees), and A.H. Lachlan (unpublished) proved that for any $n > 1$ below any properly $n$-c.e. degree there is a non-computable c.e. degree. These two results show that the class of $n$-c.e. degrees for $n > 1$ differs from the class of c.e. degrees as well as from the class of degrees below $\mathbf{0}'$: By Lachlan's above-mentioned result no $n$-c.e. degree can be minimal while there are minimal degrees below $< \mathbf{0}'$.

These results provoked a certain interest among mathematicians and became the starting point for the investigation of properties of the $n$-c.e. degrees. Generalizing Cooper's theorem, M.Lerman and L. Hay established that for any $n > 1$ there are $(n+1)$-c.e. degrees $\mathbf{c}$ and $\mathbf{d}$ such that the interval $\{\mathbf{b} | \ \mathbf{c} \leqslant \mathbf{b} \leqslant \mathbf{d}\}$ does not contain $n$-c.e. degrees. They also noted that combining Cooper's proof with the permitting method, one can construct below any non-computable c.e. degree a properly 2-c.e. degree. Further, R.A. Shore and L. Hay combined Cooper's method with Sacks's coding technique to construct a properly 2-c.e. degree above any given T-incomplete c.e. degree. (These results are not published, they are mentioned in Epstein, Haas, and Kramer [32].)

More active investigations toward the development of the structural theory of the $n$-c.e. (mainly the 2-c.e.) degrees began after publications by Arlsanov [5, 6], and Downey [26]. In these papers the authors prove that the elementary theories of the semilattices of c.e. degrees and $n$-c.e. degrees are different. M.M. Arslanov proved that for any $n \geqslant 1$ and for any $n$-c.e. degree $\mathbf{a} > \mathbf{0}$ there exists a 2-c.e. degree $\mathbf{d} < \mathbf{0}'$ such that $\mathbf{a} \cup \mathbf{d} = \mathbf{0}'$. Earlier S.B. Cooper and C.E.M. Yates (unpublished, see Miller [50]) independently proved that this result fails in the c.e. degrees. This shows that these theories are different at the $\Sigma_3^0$-level. R.G. Downey proved that the four-element lattice $\diamondsuit$ which is also called *the diamond lattice*, is embeddable into the 2-c.e. degrees preserving $\mathbf{0}$ and $\mathbf{0}'$ (earlier Lachlan [44] had proved that this is impossible in the c.e. degrees). Therefore, these theories are different also at the $\Sigma_2^0$-level (at the $\Sigma_1^0$-level they coincide, which easily follows from Lachlan's above-mentioned result on the $n$-c.e. degrees). In his paper Downey also stated his famous conjecture on the elementarily equivalence of the semilattices of $n$- and $m$- c.e. degrees for $n \neq m, n, m > 1$.

At present the structural theory of the $n$-c.e. degrees is worked out fairly well. Most important results obtained in this area of research in the past forty years are (in addition to the above-mentioned results of Arslanov and Downey) the proof of the non-density of the ordering of the $n$-c.e. degrees for any $n > 1$ (Cooper, Harrington, Lachlan, Lempp, and Soare [21]), the recent work of Arslanov, Kalimullin, and Lempp [11] on the non-elementary equivalence of the semilattices of 2-c.e. and 3-c.e. degrees, the work of Yang and Yu [59], where it is proved that in the signature $\{\leqslant\}$ the c.e. degrees do not form a $\Sigma_1$-substructure of the $n$-c.e. degrees for any $n \geqslant 2$, and a series of papers by Cooper, Li, Yi, and Ishmukhametov, in which the authors investigated the splitting properties of the $n$-c.e. degrees for the different $n \geqslant 1$.

But a whole number of natural and important questions still remain open. First of all there is the problem of definability of the c.e. degrees in the ordering of the $n$-c.e. degrees for $n > 1$ (in a more general setting the question on definability of the $m$-c.e. degrees in the orderings of $n$-c.e. degrees for $1 \leqslant m < n$), the problem on the elementary equivalence of the structures of $n$-c.e. degrees for different $n > 2$, the decidability of the restricted fragments of theories of these structures, in particular the problem of the decidability of the $\exists\forall$-theory of the 2-c.e. degrees.

**Definition 3.10.** A Turing degree $\mathbf{a}$ is *$n$-computably enumerable* (an *$n$-c.e.* degree), if it contains some $n$-c.e. set; an $n$-c.e. degree $\mathbf{a}$ is properly $n$-c.e. degree, if it contains no $m$-c.e. sets for any $m < n$.

The set of all $n$-c.e. degrees we denote by $\mathcal{D}_n$, the class of all Turing degrees by $\mathcal{D}$, and the set of all Turing degrees below $\mathbf{0}'$ by $\mathcal{D}(\leqslant \mathbf{0}')$ . $\mathcal{D}_\omega$ denotes the set of all $\omega$-c.e. degrees. We have

$$\mathcal{D}_0 \subsetneq \mathcal{D}_1 \subsetneq \mathcal{D}_2 \subsetneq \ldots \subsetneq \mathcal{D}_\omega \subsetneq \mathcal{D}(\leqslant \mathbf{0}').$$

**Theorem 3.22.** *(Lachlan, unpublished) Let* $\mathbf{a}$ *be a properly n-c.e. degree for some* $n > 1$. *There are degrees* $\mathbf{a}_1, \mathbf{a}_2 \ldots, \mathbf{a}_n$ *such that* $\mathbf{0} < \mathbf{a}_1 < \ldots < \mathbf{a}_n = \mathbf{a}$ *and for every* $m, 1 < m \leqslant n$, $\mathbf{a}_m$ *is c.e. in* $\mathbf{a}_{m-1}$, *and* $\mathbf{a}_1$ *is a properly c.e. degree. In particular, below any n-c.e. degree* $\mathbf{a} > \mathbf{0}$ *there is a non-computable c.e. degree.*

Note that in this theorem we don't require that every $\mathbf{a}_m$, $1 < m < n$, must be a *properly* c.e. degree. It will follow from Corollary 3.5 that if $\mathbf{a}_3$ is a properly 3-c.e. degree then $\mathbf{a}_2$ also must be a properly $d$-c.e. degree. We don't know whether this is true for any $n > 3$. Probably, not. An indirect argument toward this conjecture is Theorem 3.29.

Theorem 3.22 allows us to transfer some properties of the c.e. degrees to the case of the $n$-c.e. degrees for $n > 1$. We demonstrate this in the following two examples.

It follows from Lachlan's non-diamond theorem, Lachlan [44], that there are no c.e. degrees, except $\mathbf{0}$ and $\mathbf{0}'$, which have complements in the c.e. degrees. (By *a complement* of a c.e. degree $\mathbf{a}$ we mean a degree $\mathbf{b}$ such that $\mathbf{a} \cup \mathbf{b} = \mathbf{0}'$ and $\mathbf{a} \cap \mathbf{b} = \mathbf{0}$. It is clear that $\mathbf{0}$ and $\mathbf{0}'$ are complements to each other.) Later in Theorem 3.44 we show that this result does not hold in the $n$-c.e. degrees for any $n > 1$, but it easily follows from Theorem 3.22 that a similar result holds in case of the $n$-c.e. degrees in the following weaker formulation.

**Theorem 3.23.** *For every* $n > 1$ *there is a n-c.e. degree which has no complement.*

***Proof.*** Let $\mathbf{a} > \mathbf{0}$ be a c.e. degree such that $\mathbf{a} \cap \mathbf{b} \neq \mathbf{0}$ for any c.e. degree $\mathbf{b} > \mathbf{0}$ (Yates [60]). If there is an $n$-c.e. degree $\mathbf{b} > \mathbf{0}$ such that $\mathbf{a} \cap \mathbf{b} = \mathbf{0}$, then by Theorem 3.22 we have $\mathbf{a} \cap \mathbf{c} = \mathbf{0}$ for some c.e. degree $\mathbf{c} > \mathbf{0}$, a contradiction. □

Further, it follows from Theorem 3.22 that if a pair $(\mathbf{d}_0, \mathbf{d}_1)$ is a minimal pair of degrees in $\mathcal{D}_2$ then there is a pair $(\mathbf{a}_0, \mathbf{a}_1)$ of c.e. degrees minimal in $\mathcal{R}$ such that $\mathbf{a}_0 < \mathbf{d}_0$ and $\mathbf{a}_1 < \mathbf{d}_1$. Therefore, Lachlan's theorem

(Lachlan [46]) on the existence of a c.e. degree $\mathbf{a} > \mathbf{0}$ such that there is no minimal pair of c. e. degrees below $\mathbf{a}$ immediately gives the following:

**Theorem 3.24.** *There is a non-computable c.e. degree such that below it there is no minimal pair of d-c.e. degrees.*

In Epstein [31] by a permitting argument below any given c.e. degree $\mathbf{a} > \mathbf{0}$ a minimal degree is constructed. Obviously, any such construction produces an $\omega$-c.e. set. Therefore, we have the following:

**Theorem 3.25.** *For every c.e. degree $\mathbf{a} > \mathbf{0}$ there is a minimal $\omega$-c.e. degree $\mathbf{m} < \mathbf{a}$.*

### 3.2.2. *The degrees of the n-c.e. sets in the n-CEA hierarchy*

It follows from Theorem 3.22 that the hierarchy of the $n$-c.e. sets is closely connected with the hierarchy of $n$-CEA ($n$-computably enumerable and above) sets, which was first defined and studied in Arslanov [2, 4], and Jockusch and Shore [40, 41].

**Definition 3.11.** The c.e. sets are 1-*CEA sets*. Further by induction, a set $A$ is an $(n+1)$-*CEA* set for some $n \geq 1$ if it is c.e. in an $n$-CEA set $B \leqslant_T A$. Furthermore, if a set $A$ c.e. in a set $B \leqslant_T A$, then $A$ is called a $B$-*CEA* set. A degree $\mathbf{a}$ is an $n$-*CEA* degree for some $n \geqslant 1$, if it contains an $n$-CEA set.

By Theorem 3.22 every $n$-c.e. set is also an $n$-CEA set. The converse, obviously, does not hold: For instance, the $n$-th jump of any c.e. set is also an $n$-CEA set. Moreover, the hierarchy of the $n$-c.e. degrees does not coincide with the hierarchy of the $n$-CEA degrees even among the degrees below $\mathbf{0}'$:

**Theorem 3.26.** *There is a 2-CEA degree $\mathbf{a} < \mathbf{0}'$, which is not an $\omega$-c.e. degree.*

***Proof.*** Let $\mathbf{d} < \mathbf{0}'$ be a $d$-c.e. degree such that the interval $(\mathbf{d}, \mathbf{0}')$ does not contain $\omega$-c.e. degrees (see Theorem 3.48 below). By Theorem 3.22 $\mathbf{d}$ is $CEA(\mathbf{a})$ for some c.e. degree $\mathbf{a} \leqslant \mathbf{d}$, and by the Sacks Density Theorem (relativized to $\mathbf{a}$) there is a degree $\mathbf{b}$ c.e. in $\mathbf{a}$ such that $\mathbf{d} < \mathbf{b} < \mathbf{0}'$. By choice of $\mathbf{d}$, the degree $\mathbf{b}$ is not $\omega$-c.e. □

The following theorem asserts that, conversely, for any $n > 1$ there are $n$-c.e. degrees, which are not $(n-1)$-CEA degrees.

**Theorem 3.27.** *(Arslanov [4], Jockusch and Shore [41]) Let $n > 1$. There is an $n$-c.e. set $D$ such that the degree of $D$ does not contain $(n-1)$-CEA sets.*

**Theorem 3.28.** *(Arslanov, LaForte, and Slaman [12]) Let $C$ be an $\omega$-c.e. set and let $A$ be a c.e. set. If $C \leqslant_T A \oplus W^A$, then there is a d-c.e. set $D$ such that $C \leqslant_T D \leqslant_T A \oplus W^A$.*

Notice that the *d*-c.e. set $D$ constructed in the above theorem is itself c.e. in $A$ as a set, rather than merely being of $A$-c.e. degree.

**Corollary 3.4.** *If $C$ is $\omega$-c.e., $A$ is c.e., and the degree of $C$ is $A$-CEA, then there exists a d-c.e. set $D$ which is itself c.e. in $A$ as a set such that $C \equiv_T D$.*

***Proof.*** Take $C \equiv_T A \oplus W^A$ in the previous theorem. Then $D \leqslant_T C$, so $C \equiv_T D$. □

Theorem 3.28 immediately yields the following:

**Corollary 3.5.** *Any $\omega$-c.e. degree which is 2-CEA is also d-c.e.*

It is natural to assume that a similar result holds for all $n$ in the sense that the $n$-c.e. and the $n$-CEA degrees agree on the $\omega$-c.e. degrees. But this is not true:

**Theorem 3.29.** *(Arslanov, LaForte, and Slaman [12]) There exists a d-c.e. set $D$ such that, for every $n \geqslant 3$, there exists a set $A_n$ which is simultaneously $D$-CEA and $(n+1)$-c.e., yet fails to be of $n$-c.e. degree.*

Now we turn to the discussion of the following question which has a long history: Let $\mathbf{a} < \mathbf{0}'$ be a non-computable c.e. degree. Is there a degree $\mathbf{b} < \mathbf{0}'$ CEA in $\mathbf{a}$ such that $\mathbf{b}$ is not c.e.? The following result is due to Soare and Stob [57], and it is the first result in this direction.

**Theorem 3.30.** *Let $\mathbf{a}$ be a non-computable c.e. degree such that $\mathbf{a}' = \mathbf{0}'$. Then there is a non c.e. degree $\mathbf{b} > \mathbf{a}$ c. e. in $\mathbf{a}$.*

In Arslanov, Lempp, and Shore [14] we answer this question negatively in the following very strong form:

**Theorem 3.31.** *There is an incomplete non-computable c.e. set $A$ such that every set CEA in $A$ and computable in $0'$ is of c.e. degree.*

On the other hand, in this paper, we also obtain the following result in the positive direction:

**Theorem 3.32.** *Let* $\mathbf{c} < \mathbf{h}$ *be c.e. degrees such that* $\mathbf{c}$ *is low and* $\mathbf{h}$ *is high. Then there is a degree* $\mathbf{a} < \mathbf{h}$ *such that* $\mathbf{a}$ *is CEA in* $\mathbf{c}$.

Soare and Stob [57] also claimed that a modification of their strategy for a low $\mathbf{a}$ would make $\mathbf{b}$ 2-c.e. They have since withdrawn this claim (personal communication) but Theorems 3.30 and 3.32 suggest the following conjecture:

**Conjecture 3.1.** *For every low c.e. degree* $\mathbf{a} > \mathbf{0}$ *there is a d-c.e. degree* $\mathbf{b}$ *CEA in* $\mathbf{a}$ *which is not c.e.*

Unfortunately, we did not succeed in answering this question. The only results we obtained in this direction are Theorems 3.33, 3.34 and 3.35.

**Theorem 3.33.** *(Arslanov, Lempp, and Shore [14]) For all high c.e. degrees* $\mathbf{h} < \mathbf{g}$ *there is a properly d-c.e. degree* $\mathbf{a}$ *such that* $\mathbf{h} < \mathbf{a} < \mathbf{g}$ *and* $\mathbf{a}$ *c.e. in* $\mathbf{h}$.

**Theorem 3.34.** *(Arslanov, Lempp, and Shore [14]) There is a c.e. degree* $\mathbf{a}, \mathbf{0} < \mathbf{a} < \mathbf{0}'$, *such that for any degree* $\mathbf{b} > \mathbf{a}$ *c.e. in* $\mathbf{a}$, *if* $\mathbf{b} \leqslant \mathbf{0}'$ *then* $\mathbf{b}$ *is c.e.*

Now suppose $\mathbf{c}$ is a low, non-computable c.e. degree and $\mathbf{a}$ the degree CEA in $\mathbf{c}$ constructed by Soare and Stob [57].

Let $C \in \mathbf{c}$ be a c.e. set and a set $A \in \mathbf{a}$ c.e. in $C$, $A \geqslant_T C$. Let $\Phi$ be a p.c. functional such that $A = \mathrm{dom}\ \Phi^C$. Since $\mathbf{c}$ is low there is a computable function $g$ such that $\Phi^C(x)\downarrow$ if and only if $\lim_s g(s,x) = 1$, and $\Phi^C(x)\uparrow$ if and only if $\lim_s g(s,x) = 0$.

Let us construct a $d$-c.e. set $V$ which is c.e. in $C$:

For each $x$, wait for a stage $s$ such that $\Phi^C(x)[s]\downarrow$ and $g(s,x) = 1$. Enumerate $< x, 0 >$ into $V$ and wait for a stage $t > s$ such that $C_t \restriction \varphi(C,x)[s] \neq C_s \restriction \varphi(C,x)[s]$ and $g(t,x) = 0$, then remove $\langle x, 0\rangle$ from $V$. Wait for a stage $s'$ such that again $\Phi^C(x)[s']\downarrow$ with a new value of $\varphi(C,x)[s']$ and $g(s',x) = 1$, then put $\langle x, 1\rangle$ into $V$, and so on.

Obviously, $V$ is a $d$-c.e. set c.e. in $C$ such that $V \leqslant_T C \oplus A$.

Now, if in addition $C'$ is $\omega$-c.e. (and, therefore, by Theorem 3.9 $C' \leq_{tt} \emptyset'$) then there are computable functions $f$ and $g$ such that for all $x$, $C'(x) = \lim_s g(s,x)$ and

$$|\{s : g(s+1,x) \neq g(s,x)\}| \leqslant f(x).$$

In this case we have $A \leqslant_T C \oplus V$: To compute $A(x)$ find some $i \leqslant f(x)$ such that $< x, i > \in V$. If there is no such $i$ then $x \notin A$, if there is some such $i$ then $x \in A$.

Therefore, if $C'$ is an $\omega$-c.e. set then $C \oplus A \equiv_T C \oplus V$, and the degree **a** CEA in **c** from Soare and Stob [57] is itself *d*-c.e., without an additional construction. Recall that a set $A$ is called *superlow* if $A' \equiv_{tt} \emptyset'$. A degree is superlow if it contains a superlow set. Therefore, we have the following:

**Theorem 3.35.** *Let* $\mathbf{a} > \mathbf{0}$ *be a superlow degree. Then there is a properly d-c.e. degree* $\mathbf{d} > \mathbf{a}$ *such that* $\mathbf{d}$ *is c.e. in* $\mathbf{a}$.

### 3.2.3. *The relative arrangement of the n-c.e. degrees*

In this section we study the relative arrangement of degrees from finite levels of the Ershov hierarchy. We begin with the following theorem which generalizes an unpublished result of R. Shore and L. Hay and can be proved similarly to Cooper's proof of the existence of a properly $n$-c.e. degree.

**Theorem 3.36.** *For all* $n > 1$ *there are n-c.e. sets* $V <_T U$ *such that between degrees* $V$ *and* $U$ *there are no* $(n-1)$*-c.e. degrees.*

The assertion of the following theorem is wrong if $n = 1$ (see Theorem 3.55 below).

**Theorem 3.37.** *For all* $n > 1$, *if* $\mathbf{a}$ *is a properly* $(n+1)$*-c.e. degree, then there is an n-c.e. degree* $\mathbf{b} < \mathbf{a}$ *such that between* $\mathbf{b}$ *and* $\mathbf{a}$ *there are no c.e. degrees.*

***Proof.*** Let $n > 1$ and let $A$ be an $(n+1)$-c.e. set of properly $(n+1)$-c.e. degree. By Theorem 3.22 there is an $n$-c.e. set $\tilde{A}$ such that $A$ is an $\tilde{A}$-CEA set. Obviously, $\tilde{A} <_T A$. Suppose that $\tilde{A} \leqslant_T W <_T A$ for some c.e. set $W$. Then $A$ is a $W$-CEA set and, therefore, 2-CEA. But then by Corollary 3.5 the degree of $A$ is $d$-c.e., a contradiction. □

Suppose that in Theorem 3.36 one of the sets $U >_T \emptyset$ or $V <_T \emptyset'$ is fixed. It is natural to ask the following question: Is there another set such that the claim of Theorem 3.36 still holds? In general the answer to this question is unknown. But:

**Theorem 3.38.** *(R. Shore and L. Hay, unpublished) There is, for instance, a d-c.e. set* $V$ *of low degree such that between* $V$ *and* $\emptyset'$ *in Turing reducibility there are no c.e. sets.*

The properly $n$-c.e. degrees are situated dense enough among the degrees $\leq \mathbf{0}'$, in particular, between any two c.e. degrees $\mathbf{a} < \mathbf{b}$ there is a properly $n$-c.e. degree, for any $n > 1$. For the case $n = 2$ this is proved in Cooper, Lempp, and Watson [22], the proof for $n > 2$ is similar.

**Theorem 3.39.** *Let $V <_T U$ be c.e. sets. There exists a d-c.e. set $D$ such that $V <_T D <_T U$ and $\forall x(W_x \not\equiv_T D)$.*

Is it possible in Theorem 3.39 to make the degree of $D$ c.e. in $V$? This question has been extensively studied. It follows from Theorem 3.34 that in general it is impossible even if $U = \emptyset'$. Can we do that if $V' \equiv_T U = \emptyset'$? This is an open question (see Conjecture 3.1 above).

**Theorem 3.40.** *(Cooper and Yi [25] for $n = 2$; Arslanov, LaForte, and Slaman [12] for $n > 2$) For any c.e. degree $\mathbf{x}$ and any $n$-c.e. degree $\mathbf{y}$, if $\mathbf{x} < \mathbf{y}$ then $\mathbf{x} < \mathbf{z} < \mathbf{y}$ for some d-c.e. degree $\mathbf{z}$.*

***Proof.*** For $n = 1$ this is the Sacks Density Theorem, and for $n = 2$ this is Theorem 3.55, part *(iii)*. For $n > 2$ use an induction argument: Assume, that the theorem is proved for $m$-c.e. sets for all $m \leqslant n$ and let $B$ be an $(n+1)$-c.e. set, $n > 1$, and let $A <_T B$ be a c.e. set. There is an $n$-c.e. set $\tilde{B} \leqslant_T B$, in which $B$ is c.e. Then the set $A \oplus \tilde{B}$ is also $n$-c.e. If $A <_T A \oplus \tilde{B}$, then, by assumption, there is a $d$-c.e. set $C$ such that $A <_T C <_T A \oplus \tilde{B} \leqslant_T B$. If $A \equiv_T A \oplus \tilde{B}$, then the $(n+1)$-c.e. set $B$ is c.e. in $A$ and, therefore, by Corollary 3.5 the degree of $B$ is $d$-c.e. Now the claim follows from the case $n = 2$. □

### 3.2.4. *The cupping, capping and density properties*

We begin with the following

**Theorem 3.41.** *(Arslanov [5, 6]) Let $\mathbf{a} > \mathbf{0}$ be an $n$-c.e. degree for some $n > 2$. Then there is a d-c.e. degree $\mathbf{d} < \mathbf{0}'$ such that $\mathbf{a} \cup \mathbf{d} = \mathbf{0}'$.*

Since there is a non-computable c.e. set $A$ such that $A \oplus U <_T \emptyset'$ for every c.e. set $U <_T \emptyset'$ (Cooper, Yates, unpublished, see Miller [50]), it follows from Theorem 3.41 that for every $n \geqslant 2$, the structures $\mathcal{D}_n$ and $\mathcal{R}$ are not elementarily equivalent at the $\Sigma_3$-level.

Generalizing Theorem 3.41 Cooper, Lempp, and Watson [22], proved the following,

**Theorem 3.42.** *If $\mathbf{a} > \mathbf{0}$ is a c.e. degree and $\mathbf{h} > \mathbf{a}$ is a high c.e. degree then there is a d-c.e. degree $\mathbf{b} < \mathbf{h}$ such that $\mathbf{a} \cup \mathbf{b} = \mathbf{h}$.*

In turn Harrington (see Miller [50]) strengthened the above mentioned result of Cooper and Yates replacing $\mathbf{0}'$ by an arbitrary high c.e. degree. Therefore, for every $n \geqslant 2$ and any high c.e. degree $\mathbf{h}$, the structures $\mathcal{D}_n(\leqslant \mathbf{h})$ and $\mathcal{R}(\leqslant \mathbf{h})$ are also non-elementarily equivalent at the $\Sigma_3^0$-level. Further, Arslanov and Cooper (unpublished, see Arslanov [8] for the case $\mathbf{h} = \mathbf{0}'$) generalized Theorem 3.42 in the following way:

**Theorem 3.43.** *Let* $\mathbf{h}$ *be a high c.e. degree,* $\mathbf{a} < \mathbf{h}$ *and* $\mathbf{b} < \mathbf{h}$ *arbitrary non-computable c.e. degrees. Then there is a d-c.e. degree* $\mathbf{d} < \mathbf{h}$ *such that* $\mathbf{h} = \mathbf{a} \cup \mathbf{d} = \mathbf{b} \cup \mathbf{d}$.

As we already mentioned, the diamond lattice is not embeddable into the c.e. degrees preserving $\mathbf{0}$ and $\mathbf{0}'$. Downey [26], proved that in the $n$-c.e. degrees such an embedding is possible for any $n, n \geqslant 2$.

**Theorem 3.44.** *There are incomparable d-c.e. degrees* $\mathbf{a}$ *and* $\mathbf{b}$ *such that* $\mathbf{a} \cup \mathbf{b} = \mathbf{0}'$ *and* $\mathbf{a} \cap \mathbf{b} = \mathbf{0}$.

Therefore, for every $n \geqslant 2$, the structures $\mathcal{D}_n$ and $\mathcal{R}$ are not elementarily equivalent at the $\Sigma_2$-level.

One can try to strengthen the Diamond Theorem 3.44 in several directions. First of all the following natural question arises: Is it possible in this theorem to replace the degrees $\mathbf{0}$ and $\mathbf{0}'$ by arbitrary c.e. degrees $\mathbf{a}$ and $\mathbf{b}$, $\mathbf{a} < \mathbf{b}$, respectively? Further, it follows from Lachlan's non-diamond theorem, (Lachlan [44]), that in Theorem 3.44 at least one of the *d*-c.e. degrees $\mathbf{a}$ or $\mathbf{b}$ cannot be c.e. Can we make one of these degrees c.e.? Finally, Theorem 3.44 states that there is a non-trivial *d*-c.e. degree $\mathbf{d}$ which has a complement in $\mathcal{D}_2$. (A degree $\mathbf{c}$ is *a complement* for $\mathbf{d}$ if $\mathbf{d} \cup \mathbf{c} = \mathbf{0}'$ and $\mathbf{d} \cap \mathbf{c} = \mathbf{0}$.) A natural question asks: Which degrees in $\mathcal{D}_2$ have complements?

The first question is connected with a general question on the decomposability of a c.e. degree $\mathbf{a}$ over a given c.e. degree $\mathbf{b} \leqslant \mathbf{a}$ into two incomparable *d*-c.e. degrees, i.e. on the existence of incomparable *d*-c.e. degrees $\mathbf{c}_0$ and $\mathbf{c}_1$ such that $\mathbf{c}_0 > \mathbf{b}$, $\mathbf{c}_1 > \mathbf{b}$ and $\mathbf{a} = \mathbf{c}_0 \cup \mathbf{c}_1$. (It follows from Lachlan's non-splitting theorem (Lachlan [45]) that in $\mathcal{R}$ such an assertion does not hold for $\mathbf{a} = \mathbf{0}'$ and some c.e. degree $\mathbf{b} > \mathbf{0}$.)

A negative answer to the first question was obtained by Kaddah [42]:

**Theorem 3.45.** *In* $\mathcal{D}_2$ *below any c.e. degree* $\mathbf{a} > \mathbf{0}$ *there is a c.e. degree* $\mathbf{b} \leqslant \mathbf{a}$ *non-branching in d-c.e. degrees. (A degree* $\mathbf{a}$ *is branching, if there are degrees* $\mathbf{b} > \mathbf{a}$ *and* $\mathbf{c} > \mathbf{a}$ *such that* $\mathbf{b} \cap \mathbf{c} = \mathbf{a}$.*)*

Therefore, there is, for instance, a low c.e. degree $\mathbf{l} > \mathbf{0}$ such that the diamond lattice is not embeddable between degrees $\mathbf{0}'$ and $\mathbf{l}$, preserving $\mathbf{l}$ as its least element.

The answer to the second question turned out to be positive. It follows from the next theorem:

**Theorem 3.46.** *(Li and Yi [49]) There are incomparable d-c.e. degrees $\mathbf{a}_0$ and $\mathbf{a}_1$ such that for every n-c.e. degree $\mathbf{x} > \mathbf{0}$, either $\mathbf{a}_0 \cup \mathbf{x} = \mathbf{0}'$, or $\mathbf{a}_1 \cup \mathbf{x} = \mathbf{0}'$.*

**Corollary 3.6.** *There are a c.e. degree $\mathbf{a} > \mathbf{0}$ and a d-c.e. degree $\mathbf{b} > \mathbf{0}$ such that $\mathbf{a} \cup \mathbf{b} = \mathbf{0}'$ and $\mathbf{a} \cap \mathbf{b} = \mathbf{0}$.*

***Proof.*** Let $\mathbf{a}_0$ and $\mathbf{a}_1$ be as in Theorem 3.46. It is clear that $\mathbf{a}_0 \cap \mathbf{a}_1 = \mathbf{0}$. Let $\mathbf{d} > \mathbf{0}$ be an arbitrary *d*-c.e. degree. Suppose for definiteness that $\mathbf{d} \cup \mathbf{a}_0 = \mathbf{0}'$. If $\mathbf{d} \cap \mathbf{a}_0 \neq \mathbf{0}$, then there exists a c.e. degree $\mathbf{b} > \mathbf{0}$ such that $\mathbf{b} \leqslant \mathbf{d}$ and $\mathbf{b} \leqslant \mathbf{a}_0$. Since $\mathbf{a}_0 \cap \mathbf{a}_1 = \mathbf{0}$, $\mathbf{b} \cap \mathbf{a}_1 = \mathbf{0}$. Since $\mathbf{b} \cup \mathbf{a}_0 = \mathbf{a}_0$, we have $\mathbf{b} \cup \mathbf{a}_1 = \mathbf{0}'$ by Theorem 3.46. Therefore, the desired pair of degrees is either $\mathbf{d}$ and $\mathbf{a}_0$, or $\mathbf{b}$ and $\mathbf{a}_1$. □

**Theorem 3.47.** *(Jiang [39]) For any high degree $\mathbf{h}$ there is a c.e. set $H \in \mathbf{h}$ such that in Theorem 3.44 the set $\emptyset'$ can be replaced by $H$.*

This result can be also obtained using Cooper's proof in Cooper [17] where below any high c.e. degree a minimal pair of c.e. degrees is constructed. On the other hand, it follows from Theorem 3.24 that not every c.e. degree $\mathbf{a} > \mathbf{0}$ in $\mathcal{D}_2$ is the top of a diamond lattice.

The ordering of the *n*-c.e. degrees is not dense for any $n > 1$:

**Theorem 3.48.** *(Cooper, Harrington, Lachlan, Lempp, and Soare [21]) There is a d-c.e. degree $\mathbf{d} < \mathbf{0}'$ such that there are no $\omega$-c.e. degrees $\mathbf{b}$ such that $\mathbf{d} < \mathbf{b} < \mathbf{0}'$.*

For the class of 2-low *n*-c.e. degrees with $n > 1$ we have another picture:

**Theorem 3.49.** *(Cooper [18]) For every $n > 1$ the partial ordering of the 2-low n-c.e. degrees is dense. Moreover, if $\mathbf{b} < \mathbf{a}$ are 2-low n-c.e. degrees, then there are n-c. e. degrees $\mathbf{a}_0$ and $\mathbf{a}_1$ such that $\mathbf{a} = \mathbf{a}_0 \cup \mathbf{a}_1$ and $\mathbf{b} < \mathbf{a}_0, \mathbf{a}_1$.*

Theorem 3.48 states that there is a maximal *d*-c.e. degree. But there are no maximal low *d*-c.e. degrees (Arslanov, Cooper and Li [9, 10]). Jiang [39] strengthened Theorem 3.48 establishing:

**Theorem 3.50.** *For any $n \geq 1$, above any low $n$-c.e. degree there is a maximal d-c.e. degree.*

On the other hand,

**Theorem 3.51.** *(Yi, unpublished, see Cooper [20]) There is a high c.e. degree $\mathbf{h} < \mathbf{0}'$ such that below $\mathbf{h}$ there are no maximal d-c.e. degrees.*

It follows from this theorem that the semilattices $\mathcal{D}_2$ and $\mathcal{D}_2(\leqslant \mathbf{h})$ are not elementarily equivalent.

Theorem 3.49 leaves open the question on the elementary equivalence of the semilattices of the low$_2$ c.e. and the low$_2$ *d*-c.e. degrees. So far we have no example which would distinguish these two semilattices.

### 3.2.5. *Splitting properties*

Let $\mathbf{a} > \mathbf{0}$ be a properly $n$-c.e. degree for some $n > 1$, and let $\mathbf{b}$ be a c. e. degree such that $\mathbf{b} < \mathbf{a}$. Since $\mathbf{a}$ is c.e. in some $(n-1)$-c.e. degree $\mathbf{a}_0 < \mathbf{a}$ (Theorem 3.22), it follows from the Sacks Splitting Theorem, relativized to $\mathbf{a}_0 \cup \mathbf{b} < \mathbf{a}$, that $\mathbf{a}$ is splittable into two $\Delta^0_2$-degrees which are above $\mathbf{b}$, i.e. there are $\Delta^0_2$-degrees $\mathbf{c}_0$ and $\mathbf{c}_1$ such that $\mathbf{c}_0 \cup \mathbf{c}_1 = \mathbf{a}$ and $\mathbf{b} < \mathbf{c}_0 < \mathbf{a}, \mathbf{b} < \mathbf{c}_1 < \mathbf{a}$. It turns out that such a splitting is possible also in the *d*-c.e. degrees.

**Theorem 3.52.** *(Cooper and Li [23]) Any d-c.e. degree $\mathbf{a} > \mathbf{0}$ is non-trivially splittable in $\mathcal{D}_2$ over any c.e. degree $\mathbf{b} < \mathbf{a}$.*

Since the ordering of the *d*-c.e. degrees is non-dense, it follows that in Theorem 3.52 we cannot replace the c.e. degree $\mathbf{b}$ by a *d*-c.e. degree. Moreover, it follows from Theorem 3.48 that in general, this is impossible even if $\mathbf{a}$ is a c.e. degree. However,

**Theorem 3.53.** *(Arslanov, Cooper, and Li [9, 10]) Any c.e. degree is splittable in the d-c.e. degrees over any low d-c.e. degree.*

It follows from Theorem 3.49 that the properties of density and splitting can be combined in the low$_2$ $n$-c.e. degrees. In the class of the low$_2$ c.e. degrees this result also holds (Shore and Slaman [55]). These and some other similarities between the low$_2$ c.e. and the low$_2$ $n$-c.e. degrees for $n > 1$ suggest the following conjecture (Downey and Stob [28]):

**Conjecture 3.2.** *The ordering of the low$_2$ $n$-c.e. degrees is elementarily equivalent to the ordering of the low$_2$ c. e. degrees.*

For the low$_3$ $n$-c.e. degrees Cooper and Li [23] proved the following:

**Theorem 3.54.** *For any $n > 1$, there is a low$_3$ $n$-c.e. degree $\mathbf{a}$ and a c.e. degree $\mathbf{b}$, $\mathbf{0} < \mathbf{b} < \mathbf{a}$, such that for any splitting of $\mathbf{a}$ into $n$-c.e. degrees $\mathbf{a_0}$ and $\mathbf{a_1}$, at least one of the degrees $\mathbf{a_0}$ or $\mathbf{a_1}$ is above $\mathbf{b}$.*

(In this case we say that $\mathbf{a}$ *is not splittable avoiding the upper cone of degrees above* $\mathbf{b}$.)

Since in $\mathcal{R}$ such a splitting of the low$_3$ c.e. degrees is possible, it follows that elementary theories of these two semilattices are different.

### 3.2.6. *Isolated d-c.e. degrees*

Cooper and Yi [25] defined the notion of an *isolated d-c.e. degree.* A $d$-c.e. degree $\mathbf{d}$ is isolated by a c.e. degree $\mathbf{a} < \mathbf{d}$ (we also say "$\mathbf{a}$ *isolates* $\mathbf{d}$"), if for any c.e. degree $\mathbf{b}$, $\mathbf{b} \leqslant \mathbf{d}$ implies $\mathbf{b} \leqslant \mathbf{a}$. Cooper and Yi [25] established the following results about such degrees:

**Theorem 3.55.** *(i) There exists an isolated d-c.e. degree;*

*(ii) There exists a non-isolated properly d-c.e. degree;*

*(iii) Given any c.e. degree $\mathbf{a}$ and any d-c.e. degree $\mathbf{d} > \mathbf{a}$, there is a d-c.e. degree $\mathbf{e}$ between $\mathbf{a}$ and $\mathbf{d}$.*

**Theorem 3.56.** *a) (LaForte [47], and Arslanov, Lempp, and Shore [13]) Given any two comparable c.e. degrees $\mathbf{v} < \mathbf{u}$, there exist an isolated d-c.e. degree $\mathbf{c}$ and a non-isolated d-c.e. degree $\mathbf{d}$ between them.*

*b) (Arslanov, Lempp, and Shore [13]) There is a non-computable c.e. degree $\mathbf{a}$ such that $\mathbf{a}$ does not isolate any degree $\mathbf{b} > \mathbf{a}$ which is c.e. in $\mathbf{a}$.*

The following two results show that the c.e. degrees $\mathbf{a}$ not isolating any $d$-c.e. degree $\mathbf{d}$ which is CEA in $\mathbf{a}$ are widely distributed in the c.e. degrees.

**Theorem 3.57.** *(Arslanov, Lempp, and Shore [13]) a) For every non-computable c.e. degree $\mathbf{c}$, there is a non-computable c.e. degree $\mathbf{a} \leqslant \mathbf{c}$ which isolates no degree CEA in it;*

*b) If $\mathbf{c}$ is a degree c.e. in $\mathbf{0}'$, then there is a c. e. degree $\mathbf{a}, \mathbf{a}' = \mathbf{c}$, which isolates no degree CEA in it.*

Suppose that a c.e. degree $\mathbf{a}$ isolates a $d$-c.e. degree $\mathbf{d} > \mathbf{a}$. Since between $\mathbf{a}$ and $\mathbf{d}$ there are no c.e. degrees except $\mathbf{a}$, then one might think that the degrees $\mathbf{a}$ and $\mathbf{d}$ are situated "close enough" to each other. But it follows from Theorem 3.58 that this is not true (if we agree that the high degrees are situated "close" to $\mathbf{0}'$, and the low degrees are situated "close" to $\mathbf{0}$).

**Theorem 3.58.** *(Ishmukhametov and Wu [38]) There are a high d-c.e. degree* $\mathbf{d}$ *and a low c.e. degree* $\mathbf{a} < \mathbf{d}$ *such that* $\mathbf{a}$ *isolates* $\mathbf{d}$.

The following result is due to Wu [58]. It can easily be derived from known results and is an interesting generalization of the idea of isolated degrees.

**Theorem 3.59.** *There are d-c.e. degrees* $\mathbf{a} < \mathbf{b}$ *such that there is exactly one c.e. degree* $\mathbf{c}$ *between them. Moreover, the degree* $\mathbf{b}$ *can be included into any given interval of high c.e. degree* $\mathbf{u}$ *and* $\mathbf{v}$, $\mathbf{u} < \mathbf{v}$.

***Proof.*** Let $\mathbf{u}$ and $\mathbf{v}$, $\mathbf{u} < \mathbf{v}$, be high c.e. degrees. Between $\mathbf{u}$ and $\mathbf{v}$ there is an isolated *d*-c.e. degree $\mathbf{b}$ (LaForte [47]). Let $\mathbf{c} < \mathbf{b}$ be a c.e. degree which isolates $\mathbf{b}$. It is easy to see that $\mathbf{u} \leqslant \mathbf{c}$, otherwise the c.e. degree $\mathbf{u} \cup \mathbf{c}$ contradicts the choice of $\mathbf{b}$. Therefore, since the degree $\mathbf{u}$ is high, the degree $\mathbf{c}$ is also high. It is known (Cooper [20]) that for any high c.e. degree, in particular for the degree $\mathbf{c}$, there exists a c.e. degree $\mathbf{d} < \mathbf{c}$, such that for every c.e. degree $\mathbf{x} < \mathbf{c}$ we have $\mathbf{x} \cup \mathbf{d} < \mathbf{c}$. Also, in Cooper, Lempp, and Watson [22] it is proved that for any high c.e. degree, in particular for the degree $\mathbf{c}$, and for any nonzero c.e. degree below it, in particular for the degree $\mathbf{d}$, there exists a *d*-c.e. degree $\mathbf{a} < \mathbf{c}$ such that $\mathbf{a} \cup \mathbf{d} = \mathbf{c}$. It follows that between $\mathbf{a}$ and $\mathbf{c}$ there are no c.e. degrees. (Since for any such c.e. degree $\mathbf{x}$ we would have $\mathbf{x} \cup \mathbf{d} = \mathbf{c}$, which contradicts the choice of $\mathbf{d}$.) Therefore, the c.e. degree $\mathbf{c}$ is the unique c.e. degree between $\mathbf{a}$ and $\mathbf{b}$. □

### 3.2.7. *A generalization*

There are several ways to generalize the notion of isolated *d*-c.e. degrees. Some of them can be found in Efremov [29, 30] and Wu [58]. Here we consider the following common generalization.

**Definition 3.12.** Let $\mathcal{A}$ and $\mathcal{B}$ be classes of sets such that $\mathcal{A} \subseteq \mathcal{B}$. By definition, a set $A \in \mathcal{A}$ *isolates* a set $B \in \mathcal{B}$, if $A <_T B$ and for any set $W \in \mathcal{A}$, $W \leqslant_T B \to W \leqslant_T A$. In this case we also say that the set $B$ is $\mathcal{A}$*-isolated by* the set $A$. A Turing degree $\mathbf{b}$ is $\mathcal{A}$*-isolated*, if it contains an $\mathcal{A}$-isolated set.

In particular, if $\mathcal{A}$ is the class of $m$-c.e. degrees, and $\mathcal{B}$ is the class of $n$-c.e. degrees, $m < n$, then we obtain the notion of a $n$-c.e. degree which is isolated by some $m$-c.e. degree, and of an $m$-c.e. degree, which isolates some $n$-c.e. degree. (It is clear that if $n = 2$ and $m = 1$ this definition coincides with the Cooper/Yi definition of isolated degrees.)

**Definition 3.13.** Let $\mathcal{A}$ and $\mathcal{B}$ be two classes of sets such that $\mathcal{A} \subseteq \mathcal{B}$, and let $A, B \in \mathcal{B}$. We define a relation $A \leqslant^{\{\mathcal{A},\mathcal{B}\}} B$ on $\mathcal{B}$ by

$A \leqslant^{\{\mathcal{A},\mathcal{B}\}} B$ if and only if for any set $W \in \mathcal{A}$ we have $W \leqslant_T A$ implies $W \leqslant_T B$.

If here $\mathcal{A}$ is the class of all $m$-c.e. sets and $\mathcal{B}$ is the class of all $n$-c.e. sets for some $1 \leqslant m \leqslant n$, then instead of $A \leqslant^{\{\mathcal{A},\mathcal{B}\}} B$ we write $A \leqslant^{\{m,n\}} B$.

If in this definition $\mathcal{A} = \mathcal{B}$, then we obtain the usual notion of Turing reducibility. In particular, for any $n \geqslant 1$, $A \leqslant^{\{n,n\}} B$ if and only if $A \leqslant_T B$. Also, it is obvious that for all $\mathcal{A} \subseteq \mathcal{B}$, $A, B \in \mathcal{B}$,

1) $A \leqslant_T B \rightarrow A \leqslant^{\{\mathcal{A},\mathcal{B}\}} B$;
2) $A \leqslant^{\{\mathcal{A},\mathcal{B}\}} B, B \leqslant^{\{\mathcal{A},\mathcal{B}\}} C \rightarrow A \leqslant^{\{\mathcal{A},\mathcal{B}\}} C$.

We call the corresponding equivalency classes the $^{\{\mathcal{A},\mathcal{B}\}}$-degrees. It follows from 1), that every $^{\{\mathcal{A},\mathcal{B}\}}$-degree is a collection of possibly several Turing degrees.

These definitions are naturally connected with the notion of isolated degrees. For instance, if a c.e. degree $\mathbf{a}$ isolates a $d$-c.e. degree $\mathbf{b}$, then this means that $\mathbf{b} \leqslant^{\{1,2\}} \mathbf{a}$. Therefore, $\mathbf{a} =^{\{1,2\}} \mathbf{b}$, i.e. all isolated $d$-c.e. degrees and their isolating c.e. degrees belong to the same $^{\{1,2\}}$-degree. Cooper and Yi's theorem on the existence of an isolated $d$-c.e. degree $\mathbf{d}$ now means that there exists a $^{\{1,2\}}$-degree, which contains a c.e. degree and a non c.e. $d$-c.e. degree. On the other hand, the existence of a non-isolated $d$-c.e. degree means that there exists a $^{\{1,2\}}$-degree, which contains a $d$-c.e. degree and does not contain c.e. degrees. Theorem 3.66 states that there exists a $^{\{1,2\}}$-degree which consists of a single c.e. degree. Theorem 3.55, part (*iii*) states that no c.e. degree $d$-c.e.-isolates a $d$-c.e. degree (on the class of all $d$-c.e. degrees). Similarly, Theorem 3.40 states that no c.e. degree $d$-c.e.-isolates any $n$-c.e. degree, for any $n > 1$.

Below, we will deal with the classes of $d$-c.e. sets and $^{\{1,2\}}$-degrees.

It is clear that each $^{\{1,2\}}$-degree contains at most one c. e. degree and, in general, may contain several $d$-c.e. degrees. As usual, we call a $^{\{1,2\}}$-degree as a c.e. $^{\{1,2\}}$-degree if it contains a c.e. degree.

The following theorem states that each c.e. $^{\{1,2\}}$-degree either does not contain any non c.e. *d*-c.e. degree or contains infinitely many such degrees.

**Theorem 3.60.** *Any c.e. $^{\{1,2\}}$-degree either consists of a single c. e. degree or contains an infinite descending chain of non c.e. d-c.e. degrees.*

***Proof.*** Let a c.e. $^{\{1,2\}}$-degree contain a c.e. set $A$ and a *d*-c.e. set $D$, which is not T-equivalent to any c.e. sets. Since $A \equiv^{\{1,2\}} D$, $A <_T D$ and $A$ isolates $D$. By Theorem 3.55 there is a *d*-c.e. set $C$ such that $A <_T C <_T D$. It is easy to see that the set $A$ also isolates $C$, therefore $A \equiv^{\{1,2\}} C \equiv^{\{1,2\}} D$. Now we repeat the same argument with $A$ and $C$ instead of $A$ and $D$ and so on. □

It is easy to see that the c.e. $^{\{1,2\}}$-degrees form an upper semilattice where the least upper bound for the $^{\{1,2\}}$-degrees of c.e. sets $A$ and $B$ is the degree of the set $A \oplus B$. Indeed, if $A \leqslant^{\{1,2\}} C$ and $B \leqslant^{\{1,2\}} C$ for some set $C$ then we have $A \leqslant_T C$ and $B \leqslant_T C$, otherwise the c.e. sets $A$ and $B$ refute the $^{\{1,2\}}$-reducibility of $A$ and $B$ to $C$, accordingly. Therefore, $A \oplus B \leqslant_T C$ and, hence, $A \oplus B \leqslant^{\{1,2\}} C$. We don't know, whether the $^{\{1,2\}}$-degrees of the *d*-c.e. sets form an upper semilattice. In general, the join operator $A \oplus B$ does not give the least upper bound for the $^{\{1,2\}}$-degrees of sets $A$ and $B$. (This can be easily proved by a routine finite injury priority argument.)

**Theorem 3.61.** *For each $n \geqslant 2$ there exists a $^{\{1,2\}}$-degree, which contains at least $n$ incomparable Turing degrees.*

***Proof.*** The proof is a direct generalization of the proof of Theorem 3.55, part (*i*). □

Theorem 3.62 states that there are no maximal $^{\{1,2\}}$-degrees among the low$_2$ degrees.

**Theorem 3.62.** *Let $D$ be a d-c.e. set such that $D'' \equiv_T \emptyset''$. Then there is a c.e. set $A$ such that $D <^{\{1,2\}} A <^{\{1,2\}} \emptyset'$.*

***Proof.*** It is enough to consider only the case when the degree of $D$ is not computably enumerable. Otherwise, since $\mathbf{a} \leqslant \mathbf{b}$ implies $\mathbf{a} \leqslant^{\{1,2\}} \mathbf{b}$, the theorem follows from the Sacks Density Theorem.

The following lemma is proved in Arslanov [3] (see also Soare [56, Theorem XII.5.1]).

**Lemma 3.3.** *For any function $\psi$ which is computable in $\emptyset''$, there is a computable function $g$ such that $W_{g(e)} \equiv_T W_{\psi(e)}$ for all $e$.*

Let $S = \{\langle i,j\rangle \mid W_i = \Phi_j^D\}$. It is easy to see that $S \in \Pi_2^D$, therefore $S$ is computable in $D'' \equiv_T \emptyset''$, i.e. $S \leqslant_T \emptyset''$. Now we define a function $g \leqslant_T \emptyset''$:

$$W_{g(\langle i,j\rangle)} = \begin{cases} W_i, & \text{if}\langle i,j\rangle \in S; \\ \emptyset, & \text{otherwise.} \end{cases}$$

Since $g$ is a total function and $g \leqslant_T \emptyset''$, by Lemma 3.3 there exists a computable function $f$ such that $W_{f(\langle i,j\rangle)} \equiv_T W_{g(\langle i,j\rangle)}$. Let $E = \oplus_{1\leqslant k<\infty} W_{f(k)}$. Since for any $k \in \omega$ we have $W_{f(k)} \leqslant_T D$ and the degree of $D$ does not contain c.e. sets, the set $E$ is computably enumerable and $\forall n\{D \not\leqslant_T E^{[<n]} \equiv_T \oplus_{1\leqslant k\leqslant n} W_{f(k)}\}$.

By Shoenfield's Thickness Lemma (see Soare [56, Lemma VIII.1.1]) there is a c.e. set $A \leqslant_T E$ such that $A$ is a *thick subset* of $E$ (i. e. $A \subseteq E$, and $A^{[e]} =^* E^{[e]}$), $D \not\leqslant_T A$. (We denote by $X =^* Y$ that $(X-Y)\cup(Y-X)$ is finite, and let $X^{[e]} = \{\langle x,e\rangle : \langle x,e\rangle \in X\}$ be the $e$-th section of $X$.)

We have $W_i = \Phi_j^D$ implies $W_i \equiv_T E^{[\langle i,j\rangle]} =^* A^{[\langle i,j\rangle]}$, i.e. $W_i \leqslant_T D$ implies $W_i \leqslant_T A$.

There are the following two possibilities:

1) $A \leqslant_T D$. Then $A$ isolates $D$. Let $B$ be an arbitrary c.e. set such that $A <_T B <_T \emptyset'$. We have $D <^{\{1,2\}} B <^{\{1,2\}} \emptyset'$.

2) $A \not\leqslant_T D$. Then, obviously, $D <^{\{1,2\}} A <^{\{1,2\}} \emptyset'$. □

**Remark 3.3.** Analyzing this proof we can see that we have proved a slightly stronger result. For instance, let $D$ be a *d*-c.e. set such that its degree is not computably enumerable and there is a computable function $f$ with the following properties:

a) $W_{f(e)} \leqslant_T D$ for any $e \in \omega$;
b) $(\forall e \in \omega)(W_e \leqslant_T D \to (\exists x \in \omega)[W_e \leqslant_T W_{f(x)}])$.

Then, defining the set $E$ again as $\oplus_{1\leqslant k<\infty} W_{f(k)}$ and repeating the construction of the set $A$, we obtain that $D <^{\{1,2\}} A <^{\{1,2\}} \emptyset'$. Moreover, if we have *d*-c.e. sets $D_1$ and $D_2$ such that $D_1 <^{\{1,2\}} D_2$, and if for the set $D_1$ there is a computable function $f$ with properties *a*), *b*), and the additional property $E \leqslant_T D_2$, then, again repeating the previous argument we obtain that $D_1 <^{\{1,2\}} A <^{\{1,2\}} D_2$ for some c.e. set $A$.

### 3.2.8. *Further results and open questions*

The following questions are the main open questions on the arrangement of the $n$-c.e. degrees for various $n \geqslant 1$:

- Is the relation "$\mathbf{x}$ is c.e." definable in $\mathcal{D}_n$ for each $n \geqslant 2$? Are there non-trivial *finite* sets of c.e. degrees definable in $\mathcal{D}_n$? (For an infinite definable set of c.e. degrees see Corollary 3.11 below.)
- Is the relation "$\mathbf{x}$ is $m$-c.e." definable in $\mathcal{D}_n$ for each pair $n, m$, $n > m \geqslant 2$?
- Are $\{\mathcal{D}_m, <\}$ and $\{\mathcal{D}_n, <\}$ elementarily equivalent for each $n \neq m, m, n \geqslant 2$ ? (For the case $m = 2, n = 3$ see Corollary 3.10 below.)
- Are there $n \neq m, m, n \geqslant 1$, such that $\mathcal{D}_m(\mathbf{a}, \mathbf{b})$ is elementarily equivalent to $\mathcal{D}_n(\mathbf{a}, \mathbf{b})$ for some c.e. degrees $\mathbf{a} < \mathbf{b}$?
- Are there numbers $n > m \geqslant 1$ such that $\{\mathcal{D}_m, <\}$ is a $\Sigma_1$-substructure of $\{\mathcal{D}_n, <\}$?

An investigation of the problems listed above is driven by the need to better understand the level of the structural similarity of the classes of c.e. and $n$-c.e. degrees for different $n > 1$, as well as of the level of the homogeneity for the notion of c.e. with respect to $n$-c.e. degrees in the sense of the level of the similarity of orderings of the c.e. degrees and of the $n$-c.e. degrees which are CEA in some $\mathbf{d}$.

**a)** *Elementary equivalence.*

We first consider questions on the elementary equivalence.

**Theorem 3.63.** *(Arslanov, Kalimullin, and Lempp [11]) There are* 2*-c.e. degrees* $\mathbf{d}$ *and* $\mathbf{e}$ *such that* $\mathbf{0} < \mathbf{d} < \mathbf{e}$ *and for any* 2*-c.e. degree* $\mathbf{u} < \mathbf{e}$ *either* $\mathbf{u} \leqslant \mathbf{d}$ *or* $\mathbf{d} \leqslant \mathbf{u}$.

**Theorem 3.64.** *(Arslanov, Kalimullin, and Lempp [11]) For all c.e. degrees* $\mathbf{x}$ *and* 2*-c.e. degrees* $\mathbf{d}$ *and* $\mathbf{e}$ *such that both* $\mathbf{d}, \mathbf{e}$ *are c.e. in* $\mathbf{x}$ *and* $\mathbf{0} < \mathbf{x} < \mathbf{d} < \mathbf{e}$, *there is a* 2*-c.e. degree* $\mathbf{u}$ *c.e. in* $\mathbf{x}$ *such that* $\mathbf{x} < \mathbf{u} < \mathbf{e}$ *and* $\mathbf{d}|\mathbf{u}$.

The following theorem is a refinement of Theorem 3.63.

**Theorem 3.65.** *a) In Theorem 3.63 the degree* $\mathbf{d}$ *is necessarily c.e. and b) for each* 2*-c.e. degree* $\mathbf{e}$ *there is at most one c.e. degree* $\mathbf{d} < \mathbf{e}$ *with this property.*

***Proof.*** By Theorem 3.22 the degree $\mathbf{e}$ is c.e. in a c.e. degree $\mathbf{b} < \mathbf{e}$. If $\mathbf{b} > \mathbf{d}$, then by the Sacks Splitting Theorem we split $\mathbf{b}$ into two c.e. degrees $\mathbf{b}_0$ and $\mathbf{b}_1$ avoiding the upper cone of $\mathbf{d}$ (avoiding $\mathbf{d}$, for short). At least one of these degrees must be incomparable with $\mathbf{d}$, a contradiction.

If $\mathbf{b} < \mathbf{d}$, then consider the c.e. degree $\mathbf{c} = \mathbf{b} \cup \mathbf{a}$, where $\mathbf{a} < \mathbf{d}$ is a c.e. degree such that $\mathbf{d}$ is c.e. in $\mathbf{a}$. Obviously, $\mathbf{c} \leqslant \mathbf{d}$. If $\mathbf{c} < \mathbf{d}$ then we obtain a contradiction with Theorem 3.64, since both the 2-c.e. degrees $\mathbf{e}$ and $\mathbf{d}$ are c.e. in $\mathbf{c}$. Therefore, $\mathbf{d} = \mathbf{c}$. Similar arguments prove also the second part of the theorem. □

**Corollary 3.7.** *(of Theorem 3.65). There are no strong minimal covers in the 2-c.e. degrees.*

***Proof.*** Indeed, if $\mathbf{b}$ is a strong minimal cover for $\mathbf{a}$, then by Theorem 3.65, $\mathbf{a}$ is c.e. and, therefore, by Theorem 3.55 there is a $d$-c.e. degree strictly between $\mathbf{a}$ and $\mathbf{b}$. □

**Corollary 3.8.** *(of Theorem 3.65). There are no 2-c.e. degrees* $\mathbf{f} > \mathbf{e} > \mathbf{d} > \mathbf{0}$ *such that for any $u$,*

*(i) if* $\mathbf{u} \leqslant \mathbf{f}$ *then either* $\mathbf{e} \leqslant \mathbf{u}$ *or* $\mathbf{u} \leqslant \mathbf{e}$*, and*
*(ii) if* $\mathbf{u} \leqslant \mathbf{e}$ *then either* $\mathbf{d} \leqslant \mathbf{u}$ *or* $\mathbf{u} \leqslant \mathbf{d}$*.*

***Proof.*** If there are such degrees $\mathbf{f} > \mathbf{e} > \mathbf{d} > \mathbf{0}$ then by Theorem 3.65 the degree $\mathbf{e}$ is c.e. and by the Sacks Splitting Theorem is splittable avoiding $\mathbf{d}$ which is a contradiction. □

**Question 3.1.** *Are there 3-c.e. degrees* $\mathbf{f} > \mathbf{e} > \mathbf{d} > \mathbf{0}$ *with this property?*

Obviously, an affirmative answer to this question refutes the elementary equivalence of $\mathcal{D}_2$ and $\mathcal{D}_3$.

Though this question still remains open, we can weaken a little this property of degrees $(\mathbf{d}, \mathbf{e}, \mathbf{f})$ to carry out the mission imposed to these degrees to refute the Downey's Conjecture. We consider triples of non-computable $n$-c.e. degrees $\{(\mathbf{d}, \mathbf{e}, \mathbf{f}) \mid \mathbf{0} < \mathbf{d} < \mathbf{e} < \mathbf{f}\}$ with the following (weaker) property: For any $n$-c.e. degree $\mathbf{u}$,

(i) if $\mathbf{u} \leqslant \mathbf{f}$ then either $\mathbf{u} \leqslant \mathbf{e}$ or $\mathbf{e} \leqslant \mathbf{d} \cup \mathbf{u}$, and
(ii) if $\mathbf{u} \leqslant \mathbf{e}$ then either $\mathbf{d} \leqslant \mathbf{u}$ or $\mathbf{u} \leqslant \mathbf{d}$.

(In the first line the former condition $\mathbf{e} \leqslant \mathbf{u}$ was replaced by a weaker condition $\mathbf{e} \leqslant \mathbf{d} \cup \mathbf{u}$.)

We still have the following corollary from Theorems 3.63 and 3.64:

**Corollary 3.9.** *There are no* 2*-c.e. degrees* $\mathbf{f} > \mathbf{e} > \mathbf{d} > \mathbf{0}$ *such that for any* 2*-c.e. degree u,*

*(i) if* $\mathbf{u} \leqslant \mathbf{f}$ *then either* $\mathbf{u} \leqslant \mathbf{e}$ *or* $\mathbf{e} \leqslant \mathbf{d} \cup \mathbf{u}$*, and*
*(ii) if* $\mathbf{u} \leqslant \mathbf{e}$ *then either* $\mathbf{d} \leqslant \mathbf{u}$ *or* $\mathbf{u} \leqslant \mathbf{d}$*.*

***Proof.*** Suppose that there are such degrees $\mathbf{f} > \mathbf{e} > \mathbf{d} > \mathbf{0}$. Let $\mathbf{f_1} \leqslant \mathbf{f}$ and $\mathbf{e_1} \leqslant \mathbf{e}$ be c.e. degrees such that $\mathbf{f}$ and $\mathbf{e}$ are c.e. in $\mathbf{f_1}$ and $\mathbf{e_1}$, respectively. Consider the degree $\mathbf{x} = \mathbf{d} \cup \mathbf{e_1} \cup \mathbf{f_1}$. Obviously, $\mathbf{d} \leqslant \mathbf{x} \leqslant \mathbf{f}$.

By Theorem 3.65 the degree $\mathbf{x}$ is c.e. and $\mathbf{e} \not< \mathbf{x}$, otherwise $\mathbf{x}$ is splittable in the c.e. degrees avoiding $\mathbf{e}$, which is a contradiction. Also $\mathbf{x} \neq \mathbf{e}$, since in this case we can split $\mathbf{x}$ avoiding $\mathbf{d}$, which is again a contradiction. Finally, if $\mathbf{x} \not\leqslant \mathbf{e}$ then it follows from condition $(i)$ that $\mathbf{e} \leqslant \mathbf{d} \cup \mathbf{x} = \mathbf{x}$, a contradiction. Therefore, $\mathbf{x} < \mathbf{e}$. Since $\mathbf{f}$ and $\mathbf{e}$ are both c.e. in $\mathbf{x}$, it follows now from Theorem 3.64 that there is a 2-c.e. degree $\mathbf{u}$ such that $\mathbf{x} < \mathbf{u} < \mathbf{f}$ and $\mathbf{u} | \mathbf{e}$, a contradiction. □

**Theorem 3.66.** *(Arslanov, Kalimullin, and Lempp [11]) There are a c.e. degree* $\mathbf{d} > \mathbf{0}$*, a* 2*-c.e. degree* $\mathbf{e} > \mathbf{d}$*, and a* 3*-c.e. degree* $\mathbf{f} > \mathbf{e}$ *such that for any* 3*-c.e. degree* $\mathbf{u}$*,*

*(i) if* $\mathbf{u} \leqslant \mathbf{f}$ *then either* $\mathbf{u} \leqslant \mathbf{e}$ *or* $\mathbf{e} \leqslant \mathbf{d} \cup \mathbf{u}$*, and*
*(ii) if* $\mathbf{u} \leqslant \mathbf{e}$ *then either* $\mathbf{d} \leqslant \mathbf{u}$ *or* $\mathbf{u} \leqslant \mathbf{d}$*.*

**Corollary 3.10.** $\mathcal{D}_2 \not\equiv \mathcal{D}_3$ *at the* $\Sigma_2$*- level.*

Theorems 3.63 and 3.66 raise a whole series of new questions, study of which could lead to the better understanding of the inner structure of the ordering of the $n$-c.e. degrees. Below we consider some of these questions.

**Definition 3.14.** Let $n > 1$. An $(n+1)$-tuple of degrees $\mathbf{a}_0, \mathbf{a}_1, \ldots \mathbf{a}_{n-1}, \mathbf{a}_n$ forms an *n-bubble* in $\mathcal{D}_m$ for some $m \geqslant 1$, if $\mathbf{0} = \mathbf{a}_0 < \mathbf{a}_1 < \mathbf{a}_2 < \ldots < \mathbf{a}_{n-1} < \mathbf{a}_n$, $\mathbf{a}_k$ is $k$-c.e. for each $k, 1 \leqslant k \leqslant n$, and for any $m$-c.e. degree $\mathbf{u}$, if $\mathbf{u} \leqslant \mathbf{a}_k$ then either $\mathbf{u} \leqslant \mathbf{a}_{k-1}$ or $\mathbf{a}_{k-1} \leqslant \mathbf{u}$.

An $(n+1)$-tuple of degrees $\mathbf{a}_0, \mathbf{a}_1, \mathbf{a}_2, \ldots \mathbf{a}_{n-1}, \mathbf{a}_n$ forms a *weak n-bubble* in $\mathcal{D}_m$ for some $m \geqslant 1$, if $\mathbf{0} = \mathbf{a}_0 < \mathbf{a}_1 < \mathbf{a}_2 < \ldots < \mathbf{a}_{n-1} < \mathbf{a}_n$, $\mathbf{a}_k$ is $k$-c.e. for each $k, 1 \leqslant k \leqslant n$, and for any $m$-c.e. degree $\mathbf{u}$, if $\mathbf{u} \leqslant \mathbf{a}_k$ then either $\mathbf{u} \leqslant \mathbf{a}_{k-1}$ or $\mathbf{a}_{k-1} \leqslant \mathbf{u} \cup \mathbf{a}_{k-2}$.

Obviously, every $n$-bubble is also an $n$-weak bubble for every $n > 1$, but we don't know if the reverse holds. Theorem 3.63 and Corollary 3.8 state that in the 2-c.e. degrees there are 2-bubbles and there are no $n$-bubbles (and even that there are no $n$-weak bubbles), for every $n > 2$. Theorem 3.66 states that in the 3-c.e. degrees there are 3-weak bubbles. Questions on the existence of $n$-bubbles (and even on $n$-week bubbles) in the $n$-c.e. degrees for $n > 3$, and on the existence of the $n$-bubbles in the $m$-c.e. degrees for $2 < m < n$ are open.

**Conjecture 3.3.** *For every $n, 1 < n < \omega$, $\mathcal{D}_n$ contains an $n$-bubble, but does not contain $m$-bubbles for any $m > n$. (As we already saw this is true for $n = 2$.)*

Obviously, if this conjecture holds for some $n > 1$ then this means that $\mathcal{D}_n$ is not elementarily equivalent to $\mathcal{D}_m, m > n$.

**b)** *Definability.*

**Definition 3.15.** (Cooper and Li [23]). A *Turing approximation to the class of the c.e. degrees $\mathcal{R}$ in the $n$-c.e. degrees* is a Turing definable class $\mathcal{S}_n$ of $n$-c.e. degrees such that

(i) either $\mathcal{R} \subseteq \mathcal{S}_n$ (in this case we say that $\mathcal{S}_n$ is an approximation to $\mathcal{R}$ from above), or
(ii) $\mathcal{S}_n \subseteq \mathcal{R}$ ($\mathcal{S}_n$ is an approximation to $\mathcal{R}$ from below).

Obviously, $\mathcal{R}$ is definable in the $n$-c.e. degrees if and only if there is a Turing definable class $\mathcal{S}_n$ of $n$-c.e. degrees which is a Turing approximation to the class $\mathcal{R}$ in the $n$-c.e. degrees simultaneously from above and from below.

There are a number of known nontrivial Turing approximations from above to the class of the c.e. degrees in the $n$-c.e. degrees. For instance, such an approximation can be obtained from Theorem 3.55 (iii).

A nontrivial Turing approximation from below can be obtained from Theorems 3.1 and 3.65. Consider the following set of c.e. degrees: $\mathcal{S}_2 = \{\mathbf{0}\} \bigcup \{\mathbf{x} > \mathbf{0} | (\exists \mathbf{y} > \mathbf{x})(\forall \mathbf{z})(\mathbf{z} \leqslant \mathbf{y} \rightarrow \mathbf{z} \leqslant \mathbf{x} \vee \mathbf{x} \leqslant \mathbf{z})\}$. It follows from Theorem 3.65 that

**Corollary 3.11.** *$\mathcal{S}_2 \subseteq \mathcal{R}$ and $\mathcal{S}_2 \neq \{\mathbf{0}\}$.*

Therefore, $\mathcal{S}_2$ is a nontrivial approximation from below to the class of the c.e. degrees $\mathcal{R}$ in the class of the $d$-c.e. degrees. A small additional construction in Theorem 3.63 allows to achieve that $\mathcal{S}_2$ contains infinitely many c.e. degrees.

Since each non-computable c.e. degree $\mathbf{d}$ from $\mathcal{S}_2$ isolates some $d$-c.e. degree $\mathbf{e}$, it follows from Theorem 3.57 that $\mathcal{S}_2$ does not coincide with the class of all c.e. degrees.

**Open Question.** Is there for every pair of c.e. degrees $\mathbf{a} < \mathbf{b}$ a degree $\mathbf{c} \in \mathcal{S}_2$ such that $\mathbf{a} < \mathbf{c} < \mathbf{b}$ (i.e. $\mathcal{S}_2$ is dense in $\mathcal{R}$)?

An affirmative answer to this question implies definability of $\mathcal{R}$ in $\mathcal{D}_2$ as follows: Given a c.e. degree $\mathbf{a} > \mathbf{0}$ we first split $\mathbf{a}$ into two incomparable c.e. degrees $\mathbf{a}_0$ and $\mathbf{a}_1$, then using the density of $\mathcal{S}_2$ in $\mathcal{R}$ find between $\mathbf{a}$ and $\mathbf{a}_i, i \leqslant 1$, a c.e. degree $\mathbf{c}_i, i \leqslant 1$, such that $\mathbf{a} = \mathbf{c}_0 \cup \mathbf{c}_1$. This shows that in this case a nonzero 2-c.e. degree is c.e. if and only if it is the least upper bound of two incomparable 2-c.e. degrees from $\mathcal{S}_2$.

**Conjecture 3.4.** *Each c.e. degree $\mathbf{a} > \mathbf{0}$ is the least upper bound of two incomparable degrees from $\mathcal{S}_2$ and, therefore, the class of the c.e. degrees is definable in $\mathcal{D}_2$.*

**Question 3.2.** *Is $\mathcal{R}$ definable in $\mathcal{D}_2$? Is $\mathcal{D}_m$ definable in $\mathcal{D}_n$ for some $1 < m < n$?*

**c)** $\Sigma_1$*-substructures.*

There are only a few known results in this direction.

(T. Slaman, unpublished) The partial ordering of the $n$-c.e. degrees is not a $\Sigma_1$-substructure of $\{\mathcal{D}(\leqslant \mathbf{0}'), <\}$.

(Yang and Yu [59]) The structure $\{\mathcal{R}, <\}$ is not a $\Sigma_1$-substructure of $\{\mathcal{D}_2, <\}$.

In Theorem 3.66 we have a c.e. degree $\mathbf{d} > \mathbf{0}$ and a 2-c.e. degree $\mathbf{e} > \mathbf{d}$ such that every 3-c.e. degree $\mathbf{u} \leqslant \mathbf{e}$ is comparable with $\mathbf{d}$. Can this condition be strengthened in the following sense: there are a c.e. degree $\mathbf{d} > \mathbf{0}$ and a 2-c.e. degree $\mathbf{e} > \mathbf{d}$ such that every $n$-c.e. degree $\leqslant \mathbf{e}$ for every $n < \omega$ is comparable with $\mathbf{d}$?

**Question 3.3.** *Are there a c.e. degree $\mathbf{d} > \mathbf{0}$ and a 2-c.e. degree $\mathbf{e} > \mathbf{d}$ such that for any $n < \omega$ and any $n$-c.e. degree $\mathbf{u} \leqslant \mathbf{e}$ either $\mathbf{u} \leqslant \mathbf{d}$ or $\mathbf{d} \leqslant \mathbf{u}$?*

An affirmative answer to this question would reveal an interesting property of the finite levels of the Ershov difference hierarchy with far-reaching consequences. From other side, if the question has a negative answer, then let $\mathbf{d} > \mathbf{0}$ and $\mathbf{e} > \mathbf{d}$ be a c.e. degree and a 2-c.e. degree, respectively, and let $n \geqslant 3$ be the greatest natural number such that every $n$-c.e. degree $\mathbf{u} \leqslant \mathbf{e}$ is comparable with $\mathbf{d}$ and there is an $(n+1)$-c.e. degree $\mathbf{v} \leqslant \mathbf{e}$ which is incomparable with $\mathbf{d}$. Now consider the following $\Sigma_1$-formula:

$$\varphi(x, y, z) \equiv \exists u(x < y < z \,\&\, u \leqslant z \,\&\, u \not\leqslant y \,\&\, y \not\leqslant u).$$

Let $\mathbf{d}$ and $\mathbf{e}$ be degrees and $n$ be the integer whose existence is assumed by the negative answer to the previous question. Then we have $\mathcal{D}_{n+1} \models \varphi(\mathbf{0}, \mathbf{d}, \mathbf{e})$, and $\mathcal{D}_n \models \neg\varphi(\mathbf{0}, \mathbf{d}, \mathbf{e})$, which means that in this case $\mathcal{D}_n$ is not a $\Sigma_1$-substructure of $\mathcal{D}_{n+1}$. This is a well-known open question.

We see that an answer to this question in either direction leads to very interesting consequences.

All sentences known so far in the language of partial ordering, which are true in the $n$-c.e. degrees and false in the $(n+1)$-c.e. degrees for some $n \geqslant 1$, belong to the level $\forall\exists$ or to a higher level of the arithmetic hierarchy. This and some other observations allow us to state the following plausible conjecture:

**Conjecture 3.5.** *For any $n \geqslant 1$ and for any $\exists\forall$-sentence $\varphi$, $\mathcal{D}_n \models \varphi \rightarrow \mathcal{D}_{n+1} \models \varphi$. (The $\exists\forall$-theory of the $n$-c.e. degrees is a subtheory of the $\exists\forall$-theory of the $(n+1)$-c.e. degrees.)*

How many parameters are needed in formulas which are witnesses in the proof that $\mathcal{D}_1$ is not a $\Sigma_1$-substructure of $\mathcal{D}(\leqslant \mathbf{0}')$ and $\mathcal{D}_2$?

- Slaman's result ($\mathcal{R} \not\preceq_{\Sigma_1} \mathcal{D}(\mathbf{0}')$): 3 parameters;
- Yang and Yu ($\mathcal{R} \not\preceq_{\Sigma_1} \mathcal{D}_2$): 4 parameters.

**Question 3.4.** *Can these numbers be reduced?*

## References

[1] J. W. Addison. The method of alternating chains. In *Theory of Models*, pp. 1–16, North–Holland, Amsterdam, (1965).

[2] M. M. Arslanov, Weakly recursively enumerable sets and limiting computability, *Ver. Metodi i Kibernetika.* **15**, 3–9, (1979).

[3] M. M. Arslanov, On some generalizations of the fixed-point theorem, *Sov. Math.* **228**, 9–16, (1981).

[4] M. M. Arslanov, On a hierarchy of degrees of unsolvability, *Ver. Metodi i Kibernetika.* **18**, 10–18, (1982). (In Russian).

[5] M. M. Arslanov, Structural properties of the degrees below $\mathbf{0}'$, *Dokl. Nauk. SSSR.* pp. 270–273, (1985).
[6] M. M. Arslanov, On the upper semilattice of Turing degrees below $\mathbf{0}'$, *Sov. Math.* **7**, 27–33, (1988).
[7] M. M. Arslanov, Completeness in the arithmetical hierarchy and fixed-points, *Algebra and Logic.* **28**, 3–17, (1989).
[8] M. M. Arslanov. Degree structures in the local degree theory. In ed. A. Sorbi, *Complexity, Logic, and Recursion Theory 187*, Lecture Notes in Pure and Applied Mathematics, pp. 49–74. Marcel Dekker, New York, (1997).
[9] M. M. Arslanov, S. B. Cooper, and A. Li, There is no low maximal d.c.e. degree, *Math. Logic Quart.* **46**, 409–416, (2000).
[10] M. M. Arslanov, S. B. Cooper, and A. Li, There is no low maximal d.c.e. degree – corrigendum, *Math. Logic Quart.* **50**, 628–636, (2004).
[11] M. M. Arslanov, I. Sh. Kalimullin, and S. Lempp, On Downey's conjecture, *J. Symbolic Logic.* **75**, 401–441, (2010).
[12] M. M. Arslanov, G. L. LaForte, and T. A. Slaman, Relative recursive enumerability in the difference hierarchy, *J. Symbolic Logic.* **63**, 411–420, (1998).
[13] M. M. Arslanov, S. Lempp, and R. A. Shore. On isolating r.e. and isolated $d$-r.e. degrees. In *London Math. Soc. Lect. Note Series*, vol. 224, pp. 41–80. Cambridge University Press, (1996).
[14] M. M. Arslanov, S. Lempp, and R. A. Shore, Interpolating $d$-r.e. and REA degrees between r.e. degrees, *Ann. Pure Appl. Logic.* **78**, 29–56, (1996).
[15] H. G. Carstens, $\Delta_2^0$-mengen, *Arch. Math. Log. Grundlag.* **18**, 55–65, (1978).
[16] S. B. Cooper, *Degrees of Unsolvability*, Ph. D. Thesis, Leicester University, Leicester, England, (1971).
[17] S. B. Cooper, Minimal pairs and high recursively enumerable degrees, *J. Symbolic Logic.* **39**, 655–660, (1974).
[18] S. B. Cooper, The density of the low$_2$ $n$-r. e. degrees, *Arch. Math. Logic.* **30** (1), 19–24, (1991).
[19] S. B. Cooper, A splitting theorem for the $n$-recursively enumerable degrees, *Proc. Amer. Math. Soc.* **115**, 461–472, (1992).
[20] S. B. Cooper. Local degree theory. In ed. E. R. Griffor, *Handbook of Computability Theory*, pp. 121–153. Elsevier, Amsterdam, New York, Tokyo, (1999).
[21] S. B. Cooper, L. A. Harrington, A. H. Lachlan, S. Lempp, and R. I. Soare, The d.r.e. degrees are not dense, *Ann. Pure Appl. Logic.* **55**, 125–151, (1991).
[22] S. B. Cooper, S. Lempp, and P. Watson, Weak density and cupping in the $d$-r.e. degrees, *Israel J. Math.* **67**, 137–152, (1989).
[23] S. B. Cooper and A. Li, Turing definability in the Ershov hierarchy, *Journal of London Math. Soc.* **66**(2), 513–526, (2002).
[24] S. B. Cooper and A. Li, Splitting and cone avoidance in the d.c.e. degrees, *Science in China (Series A).* **45**, 1135–1146, (2002).
[25] S. B. Cooper and X. Yi. Isolated $d$-r.e. degrees. Preprint Series 17, University of Leeds, Dept. of Pure Math., (1995).
[26] R. G. Downey, D-r.e. degrees and the nondiamond theorem, *Bull. London Math. Soc.* **21**, 43–50, (1989).

[27] R. G. Downey, A. Li, and G. Wu, Complementary cappable degrees in the difference hierarchy, *Ann. Pure Appl. Logic.* **125**, 101–118, (2004).
[28] R. G. Downey and M. Stob, Splitting theorems in recursion theory, *Ann. Pure Appl. Logic.* **65**, 1–106, (1993).
[29] A. A. Efremov, Isolated from above $d$-r.e. degrees, I, *Sov. Math.* **2**, 20–28, (1998).
[30] A. A. Efremov, Isolated from above $d$-r.e. degrees, II, *Sov. Math.* **7**, 18–25, (1998).
[31] R. L. Epstein, Minimal Degrees of Unsolvability and the Full Approximation Construction, *162*, *Memoirs Amer. Math. Soc.*, Amer. Math. Soc., Providence, R.I, (1975).
[32] R. L. Epstein, R. Haas, and R. L. Kramer. Hierarchies of sets and degrees below $\mathbf{0}'$. In eds. M. Lerman, J. H. Shmerl, and R. I. Soare, *Logic Year 1979–80*, *859*, Lecture Notes in Mathematics, pp. 32–48, Springer–Verlag, Heidelberg, Tokyo, New York, (1981).
[33] Yu. L. Ershov, A hierarchy of sets, I, *Algebra and Logic.* **7**, 47–73, (1968).
[34] Yu. L. Ershov, A hierarchy of sets, II, *Algebra and Logic.* **7**, 15–47, (1968).
[35] Yu. L. Ershov, A hierarchy of sets, III, *Algebra and Logic.* **9**, 34–51, (1970).
[36] M. C. Faizrahmanov. Turing jumps in the Ershov hierarchy. to appear in Algebra and Logic (2010).
[37] E. M. Gold, Limiting recursion, *J. Symbolic Logic.* **30**(1), 28–48, (1965).
[38] S. Ishmukhametov and G. Wu, Isolation and the high/low hierarchy, *Arch. Math. Logic.* **41**, 259–266, (2002).
[39] Z. Jiang, Diamond lattice embedded into d.r.e. degrees, *Science in China (Series A).* **36**, 803–811, (1993).
[40] C. G. Jockusch, Jr. and R. A. Shore, Pseudo-jump operators I: The r.e. case, *Trans. Amer. Math. Soc.* **275**, 599–609, (1983).
[41] C. G. Jockusch, Jr. and R. A. Shore, Pseudo-jump operators II: Transfinite iterations, hierarchies, and minimal covers, *J. Symbolic Logic.* **49**, 1205–1236, (1984).
[42] D. Kaddah, Infima in the d.r.e. degrees, *Ann. Pure Appl. Logic.* **62**, 207–263, (1993).
[43] S. C. Kleene, On notation for ordinal numbers, *J. Symbolic Logic.* **3**, 150–155, (1938).
[44] A. H. Lachlan, Lower bounds for pairs of recursively enumerable degrees, *Proc. London Math. Soc.* **16**, 537–569, (1966).
[45] A. H. Lachlan, A recursively enumerable degree which will not split over all lesser ones, *Ann. Math. Logic.* **9**, 307–365, (1975).
[46] A. H. Lachlan, Bounding minimal pairs, *J. Symbolic Logic.* **44**, 626–642, (1979).
[47] G. LaForte, The isolated d.r.e degrees are dense in the r.e. degrees, *Math. Log. Quarterly.* **42**, 83–103, (1996).
[48] A. Li, G. Wu, and Y. Yang, Bounding computable enumerable degrees in the Ershov hierarchy, *Ann. Pure Appl. Logic.* **141**, 79–88, (2006).
[49] A. Li and X. Yi, Cupping the recursively enumerable degrees by d.r.e. degrees, *Proc. London Math. Soc.* **78**(3), 1–21, (1999).

[50] D. Miller. High recursively enumerable degrees and the anticupping property. In eds. M. Lerman, J. H. Schmerl, and S. R. I., *Logic Year 1979–80: University of Connecticut, 859*, Lecture Notes in Mathematics, pp. 230–245, Springer–Verlag, Berlin, Heidelberg, Tokyo, New York, (1981).

[51] H. Putnam, Trial and error predicates and the solution to a problem of Mostowski, *J. Symbolic Logic.* **30**(1), 49–57, (1965).

[52] R. W. Robinson, Interpolation and embedding in the recursively enumerable degrees, *Ann. of Math.* **93**, 285–314, (1971).

[53] H. Rogers, Jr., *Theory of Recursive Functions and Effective Computability.* McGraw-Hill, New York, (1967).

[54] J. R. Shoenfield, On degrees of unsolvability, *Ann. of Math.* **69**, 644–653, (1959).

[55] R. A. Shore and T. A. Slaman, Working below a $\text{low}_2$ recursively enumerable degree, *Ann. Pure Appl. Logic.* **52**, 1–25, (1990).

[56] R. I. Soare, *Recursively Enumerable Sets and Degrees.* Perspectives in Mathematical Logic, Omega Series, Springer–Verlag, Berlin, Heidelberg, (1987).

[57] R. I. Soare and M. Stob. Relative recursive enumerability. In ed. J. Stern, *Proceedings of the Herbrand Symposium, Logic Colloquium 19881*, pp. 299–324, North–Holland, Amsterdam, New York, Oxford, (1982).

[58] G. Wu, Bi-isolation in the d.c.e. degrees, *J. Symbolic Logic.* **69**, 409–420, (2004).

[59] Y. Yang and L. Yu, On $\Sigma_1$-structural differences among finite levels of the Ershov hierarchy, *J. Symbolic Logic.* **71**, 1223–1236, (2006).

[60] C. E. M. Yates, A minimal pair of recursively enumerable degrees, *J. Symbolic Logic.* **31**, 159–168, (1966).

# Chapter 4

## Complexity and Approximation in Reoptimization

Giorgio Ausiello, Vincenzo Bonifaci* and Bruno Escoffier

*Sapienza University of Rome,*
*Department of Computer and Systems Science,*
*00185 Rome, Italy*
*E-mail: ausiello@dis.uniroma1.it*

*Sapienza University of Rome,*
*Department of Computer and Systems Science,*
*00185 Rome, Italy, and*
*University of L'Aquila,*
*Department of Electrical and Information Engineering,*
*67040 L'Aquila, Italy*
*E-mail: bonifaci@dis.uniroma1.it*

*LAMSADE,*
*Université Paris Dauphine and CNRS,*
*75775 Paris Cedex 16, France*
*E-mail: escoffier@lamsade.dauphine.fr*

In this chapter the following model is considered: We assume that an instance $I$ of a computationally hard optimization problem has been solved and that we know the optimum solution of such an instance. Then a new instance $I'$ is proposed, obtained by means of a slight perturbation of instance $I$. How can we exploit the knowledge we have on the solution of instance $I$ to compute an (approximate) solution of instance $I'$ in an efficient way? This computation model is called *reoptimization* and is of practical interest in various circumstances. In this chapter we first discuss what kind of performance we can expect for specific classes of problems and then we present some classical optimization problems (i.e. Max Knapsack, Min Steiner Tree, Scheduling) in which this approach has been fruitfully applied. Subsequently, we address vehicle routing

*This work was partially supported by the Future and Emerging Technologies Unit of EC (IST priority - 6th FP), under contract no. FP6-021235-2 (project ARRIVAL).

problems and we show how the reoptimization approach can be used to obtain good approximate solutions in an efficient way for some of these problems.

## Contents



## 4.1. Introduction

In this chapter we illustrate the role that a new computational paradigm called *reoptimization* plays in the solution of NP-*hard problems* in various practical circumstances. As it is well known a great variety of relevant optimization problems are intrinsically difficult and no solution algorithms running in polynomial time are known for such problems. Although the existence of efficient algorithms cannot be ruled out at the present state of knowledge, it is widely believed that this is indeed the case. The most renowned approach to the solution of NP-hard problems consists in resorting to *approximation algorithms* which, in polynomial time, provide a suboptimal solution whose quality (measured as the ratio between the values of the optimum and approximate solution) is somehow guaranteed. In the last twenty years the definition of better and better approximation algorithms and the classification of problems based on the quality of approximation that can be achieved in polynomial time have been among the most important research directions in theoretical computer science and have produced a huge flow of literature [4, 36].

More recently a new computational approach to the solution of NP-hard problems has been proposed [1]. This approach can be meaningfully adopted when the following situation arises: Given a problem $\Pi$, the instances of $\Pi$ that we need to solve are indeed all obtained by means of a slight perturbation of a given reference instance $I$. In such a case we can devote enough time to the exact solution of the reference instance $I$ and then, any time that the solution for a new instance $I'$ is required, we can apply a simple heuristic that efficiently provides a good approximate solution to $I'$. Let us imagine, for example, that we know that a traveling salesman has to visit a set $S$ of, say, one thousand cities plus a few more cities that may change from time to time. In such case it is quite reasonable to devote a conspicuous amount of time to the exact solution of the traveling salesman problem on the set $S$ and then to *reoptimize* the solution whenever the modified instance is known, with a (hopefully) very small computational effort.

To make the concept more precise let us consider the following simple example (Max Weighted Sat): Let $\phi$ be a Boolean formula in conjunctive normal form, consisting of $m$ weighted clauses over $n$ variables, and let us suppose we know a truth assignment $\tau$ such that the weight of the clauses satisfied by $\tau$ is maximum; let this weight be $W$. Suppose that now a new clause $c$ with weight $w$ over the same set of variables is provided and that we have to find a "good" although possibly not optimum truth assignment $\tau'$ for the new formula $\phi' = \phi \wedge c$. A very simple heuristic can always guarantee a 1/2 approximate truth assignment in constant time. The heuristic is the following: If $W \geq w$ then put $\tau' = \tau$, otherwise take as $\tau'$ any truth assignment that satisfies $c$. It is easy to see that, in any case, the weight provided by this heuristic will be at least 1/2 of the optimum.

Actually the reoptimization concept is not new. A similar approach has been applied since the early 1980s to some polynomial time solvable optimization problems such as minimum spanning tree [16] and shortest path [14, 32] with the aim to maintain the optimum solution of the given problem under input modification (say elimination or insertion of an edge or update of an edge weight). A big research effort devoted to the study of efficient algorithms for the dynamic maintenance of the optimum solution of polynomial time solvable optimization problems followed the first results. A typical example of this successful line of research has been the design of algorithms for the partially or fully dynamic maintenance of a minimum spanning tree in a graph under edge insertion and/or edge elimination [12, 22] where at any update, the computation of the new optimum solution requires at most $O(n^{1/3} \log n)$ amortized time per operation, much less than recomputing the optimum solution from scratch.

A completely different picture arises when we apply the concept of reoptimization to NP-hard optimization problems. In fact, reoptimization provides very different results when applied to polynomial time optimization problems with respect to what happens in the case of NP-hard problems.

In the case of NP-hard optimization problems, unless P=NP polynomial time reoptimization algorithms can only help us to obtain approximate solutions, since if we knew how to maintain an optimum solution under input updates, we could solve the problem optimally in polynomial time (see Section 4.3.1).

The application of the reoptimization computation paradigm to NP-hard optimization problems is hence aimed at two possible directions: either at achieving an approximate solution of better quality than we would have obtained without knowing the optimum solution of the base instance, or achieving an approximate solution of the same quality but at a lower computational cost (as is the case in our previous example).

In the first place the reoptimization model has been applied to classical NP-hard optimization problems such as scheduling (see Bartusch et al. [6], Schäffter [34], or Bartusch et al. [7] for practical applications). More recently it has been applied to various other NP-hard problems such as Steiner Tree [9, 13] or the Traveling Salesman Problem [1, 5, 8]. In this chapter we will discuss some general issues concerning reoptimization of NP-hard optimization problems and we will review some of the most interesting applications.

The chapter is organized as follows. First, in Section 4.2 we provide basic definitions concerning complexity and approximability of optimization problems and we show simple preliminary results. Then in Section 4.3 the computational power of reoptimization is discussed and results concerning the reoptimization of various NP-hard optimization problems are shown. Finally Section 4.4 is devoted to the application of the reoptimization concept to a variety of vehicle routing problems. While most of the results contained in Section 4.3 and Section 4.4 derive from the literature, it is worth noting that a few of the presented results – those for which no reference is given – appear in this paper for the first time.

## 4.2. Basic Definitions and Results

In order to characterize the performance of reoptimization algorithms and analyze their application to specific problems we have to provide first a basic introduction to the class of NP optimization problems (NPO problems) and to the notion of approximation algorithms and approximation classes. For a more extensive presentation of the theory of approximation the reader can refer to [4].

**Definition 4.1.** An NP *optimization (*NPO*) problem* $\Pi$ is defined as a four-tuple $(\mathcal{I}, \mathrm{Sol}, m, \mathrm{opt})$ such that:

- $\mathcal{I}$ is the set of *instances* of $\Pi$ and it can be recognized in polynomial time;
- given $I \in \mathcal{I}$, $\mathrm{Sol}(I)$ denotes the set of *feasible solutions* of $I$; for every $S \in \mathrm{Sol}(I)$, $|S|$ (the size of $S$) is polynomial in $|I|$ (the size of $I$); given any $I$ and any $S$ polynomial in $|I|$, one can decide in polynomial time if $S \in \mathrm{Sol}(I)$;
- given $I \in \mathcal{I}$ and $S \in \mathrm{Sol}(I)$, $m(I, S)$ denotes the value of $S$; $m$ is polynomially computable and is commonly called *objective function*;
- $\mathrm{opt} \in \{\min, \max\}$ indicates the *type* of optimization problem.

As it is well known, several relevant optimization problems, known as NP-hard problems, are intrinsically difficult and no solution algorithms running in polynomial time are known for such problems. For the solution of NP-hard problems we have to resort to *approximation algorithms*, which in polynomial time provide a suboptimal solution of guaranteed quality.

Let us briefly recall the basic definitions regarding approximation algorithms and the most important approximation classes of NPO problems.

Given an NPO problem $\Pi = (\mathcal{I}, \mathrm{Sol}, m, \mathrm{opt})$, an optimum solution of an instance $I$ of $\Pi$ is denoted $S^*(I)$ and its measure $m(I, S^*(I))$ is denoted $\mathrm{opt}(I)$.

**Definition 4.2.** Given an NPO problem $\Pi = (\mathcal{I}, \mathrm{Sol}, m, \mathrm{opt})$, an *approximation algorithm* $A$ is an algorithm that, given an instance $I$ of $\Pi$, returns a feasible solution $S \in \mathrm{Sol}(I)$.

If $A$ runs in polynomial time with respect to $|I|$, $A$ is called a *polynomial time approximation algorithm* for $\Pi$.

The quality of an approximation algorithm is usually measured as the ratio $\rho_A(I)$, *approximation ratio*, between the value of the approximate solution, $m(I, A(I))$, and the value of the optimum solution $\mathrm{opt}(I)$. For minimization problems, therefore, the approximation ratio is in $[1, \infty)$, while for maximization problems it is in $[0, 1]$. According to the quality of approximation algorithms that can be designed for their solution, NPO problems can be classified as follows:

**Definition 4.3.** An NPO problem $\Pi$ belongs to the class APX if there exists a polynomial time approximation algorithm $A$ and a rational value $r$ such that, given any instance $I$ of $\Pi$, $\rho_A(I) \leqslant r$ (resp. $\rho_A(I) \geqslant r$) if $\Pi$ is a minimization problem (resp. a maximization problem). In such case $A$ is called an $r$-approximation algorithm.

Examples of combinatorial optimization problems belonging to the class APX are Max Weighted Sat, Min Vertex Cover, and Min Metric TSP.

For particular problems in APX a stronger form of approximability can indeed be shown. For such problems, given any rational $r > 1$ (or $r \in (0, 1)$ for a maximization problem), there exists an algorithm $A_r$ and a suitable polynomial $p_r$ such that $A_r$ is an $r$-approximation algorithm whose running time is bounded by $p_r$ as a function of $|I|$. The family of algorithms $A_r$ parametrized by $r$ is called a *polynomial time approximation scheme* (PTAS).

**Definition 4.4.** An NPO problem $\Pi$ belongs to the class PTAS if it admits a polynomial time approximation scheme $A_r$.

Examples of combinatorial optimization problems belonging to the class PTAS are Min Partitioning, Max Independent Set on Planar Graphs, and Min Euclidean TSP.

Notice that in the definition of PTAS, the running time of $A_r$ is polynomial in the size of the input, but it may be exponential (or worse) in the inverse of $|r - 1|$. A better situation arises when the running time is polynomial in both the input size and the inverse of $|r - 1|$. In the favorable case when this happens, the algorithm is called a *fully polynomial time approximation scheme* (FPTAS).

**Definition 4.5.** An NPO problem $\Pi$ belongs to the class FPTAS if it admits a fully polynomial time approximation scheme.

It is important to observe that, under the (reasonable) hypothesis that $\mathsf{P} \neq \mathsf{NP}$, it is possible to prove that $\mathsf{FPTAS} \subsetneq \mathsf{PTAS} \subsetneq \mathsf{APX} \subsetneq \mathsf{NPO}$.

### 4.3. Reoptimization of NP-hard Optimization Problem

As explained in the introduction, the reoptimization setting leads to interesting optimization problems for which the complexity properties and the existence of good approximation algorithms have to be investigated. This section deals with this question, and is divided into two parts: In Subsection 4.3.1, we give some general considerations on these reoptimization problems, both on the positive side (obtaining good approximate solutions) and on the negative side (hardness of reoptimization). In Subsection 4.3.2, we survey some results achieved on reoptimizing three well-known problems (the Min Steiner Tree problem, a scheduling problem, and the Max Knapsack problem).

### 4.3.1. *General properties*

As mentioned previously, if one wishes to get an approximate solution on the perturbed instance, she/he can compute it by applying directly, from scratch, a known approximation algorithm for the problem dealt (on the modified instance). In other words, reoptimizing is at least as easy as approximating. The goal of reoptimization is to determine if it is possible to fruitfully use our knowledge on the initial instance in order to:

- either achieve better approximation ratios;
- or devise much faster algorithms;
- or both!

In this section, we present some general results dealing with reoptimization properties of some NPO problems. We first focus on a class of hereditary problems, then we discuss the differences between weighted and unweighted versions of classical problems, and finally present some ways to achieve hardness results in reoptimization.

Of course, many types of problems can be considered, and for each of them many ways to modify the instances might be investigated. We mainly focus here on graph problems where a modification consists of adding a new vertex on the instance, but show with various examples that the approaches we present are also valid in many other cases.

#### 4.3.1.1. *Hereditary problems*

We say that a property on graphs is hereditary if the following holds: If $G = (V, E)$ satisfies this property, then for any $V' \subseteq V$, the subgraph $G[V']$ induced by $V'$ verifies the property. Following this definition, for instance, being independent [a], being bipartite, or being planar are three hereditary properties. Now, let us define problems based on hereditary properties.

**Definition 4.6.** We call Hered the class of problems consisting, given a vertex-weighted graph $G = (V, E, w)$, of finding a subset of vertices $S$ (i) such that $G[S]$ satisfies a given hereditary property (ii) that maximizes $w(S) = \sum_{v \in S} w(v)$.

[a] i.e. having no edge.

Hereditary problems have been studied before as a natural generalization of important combinatorial problems [27]. For instance, Max Weighted Independent Set, Max Weighted Bipartite Subgraph, Max Weighted Planar Subgraph are three famous problems in Hered that correspond to the three hereditary properties given above.

For all these problems, we have a simple reoptimization strategy that achieves a ratio 1/2, based on the same idea used in the introduction. Note that this is a huge improvement for some problems respect to their approximability properties; for instance, it is well known that Max Weighted Independent Set is not approximable within any constant ratio, if P $\neq$ NP[b].

**Theorem 4.1.** *Let* $\Pi$ *be a problem in* Hered. *Under a vertex insertion, reoptimizing* $\Pi$ *is approximable within ratio 1/2 (in constant time).*

***Proof.*** Let $I = (G, w)$ be the initial instance of $\Pi$, $I' = (G', w')$ be the final instance (a new vertex $v$ has been inserted), $S^*$ be an optimum solution on $I$, and $S^*_{I'}$ be an optimum solution on $I'$. Notice that $w'(u) = w(u)$ for all $u \neq v$.

Getting a 1/2-approximate solution is very easy: just consider the best solution among $S^*$ and (if feasible) $S_1 := \{v\}$. Solution $S^*$ is feasible by heritability. We can also assume $S_1$ feasible, as otherwise by heritability no feasible solution can include $v$, and $S^*$ must be optimal. Finally, by heritability, $S^*_{I'} \setminus \{v\}$ is a feasible solution on the initial instance. Then, $w'(S^*_{I'}) \leq w'(S^*) + w'(v) = w'(S^*) + w'(S_1) \leq 2\max(w'(S^*), w'(S_1))$. □

Now, let us try to outperform this trivial ratio 1/2. A first idea that comes to mind is to improve the solution $S_1$ of the previous proof since it only contains one vertex. In particular, one can think of applying an approximation algorithm on the "remaining instance after taking $v$". Consider for instance Max Weighted Independent Set, and revisit the proof of the previous property. If $S^*_{I'}$ does not take the new vertex $v$, then our initial solution $S^*$ is optimum. If $S^*_{I'}$ takes $v$, then consider the remaining instance $I_{\overline{v}}$ after having removed $v$ and its neighbors. Suppose that we have a $\rho$-approximate solution $S_2$ on this instance $I_{\overline{v}}$. Then $S_2 \cup \{v\}$ is a feasible solution of weight:

$$w(S_2 \cup \{v\}) \geq \rho(w(S^*_{I'}) - w(v)) + w(v) = \rho w(S^*_{I'}) + (1 - \rho)w(v). \quad (4.1)$$

On the other hand, of course :

$$w(S^*) \geq w(S^*_{I'}) - w(v). \quad (4.2)$$

[b]And not even within ratio $n^{1-\varepsilon}$ for any $\varepsilon > 0$, under the same hypothesis [37].

If we output the best solution $S$ among $S^*$ and $S_2 \cup \{v\}$, then, by adding equations (4.1) and (4.2) with coefficients 1 and $(1-\rho)$, we get:

$$w(S) \geq \frac{1}{2-\rho} w(S^*_{I'}).$$

Note that this ratio is always better than $\rho$.

This technique is actually quite general and applies to many problems (not only graph problems and maximization problems). We illustrate this on two well-known problems: Max Weighted Sat (Theorem 4.2) and Min Vertex Cover (Theorem 4.3). We will also use it for Max Knapsack in Section 4.3.2.

**Theorem 4.2.** *Under the insertion of a clause, reoptimizing* Max Weighted Sat *is approximable within ratio 0.81.*

***Proof.*** Let $\phi$ be a conjunction of clauses over a set of binary variables, each clause being given with a weight, and let $\tau^*(\phi)$ be an initial optimum solution. Let $\phi' := \phi \wedge c$ be the final formula, where the new clause $c = l_1 \vee l_2 \vee \ldots \vee l_k$ (where $l_i$ is either a variable or its negation) has weight $w(c)$.

We consider $k$ solutions $\tau_i$, $i = 1, \ldots, k$. Each $\tau_i$ is built as follows:

- We set $l_i$ to true;
- We replace in $\phi$ each occurrence of $l_i$ and $\overline{l_i}$ with its value;
- We apply a $\rho$-approximation algorithm on the remaining instance (note that the clause $c$ is already satisfied); together with $l_i$, this is a particular solution $\tau_i$.

Then, our reoptimization algorithm outputs the best solution $\tau$ among $\tau^*(\phi)$ and the $\tau_i$s.

As previously, if the optimum solution $\tau^*(\phi')$ on the final instance does not satisfy $c$, then $\tau^*(\phi)$ is optimum. Otherwise, at least one literal in $c$, say $l_i$, is true in $\tau^*(\phi')$. Then, it is easy to see that

$$w(\tau_i) \geq \rho(w(\tau^*(\phi')) - w(c)) + w(c) = \rho w(\tau^*(\phi')) + (1-\rho)w(c).$$

On the other hand, $w(\tau^*(\phi)) \geq w(\tau^*(\phi')) - w(c)$, and the following result follows:

$$w(\tau) \geq \frac{1}{2-\rho} w(\tau^*(\phi')).$$

The fact that Max Weighted Sat is approximable within ratio $\rho = 0.77$ [3] concludes the proof. $\square$

It is worth noticing that the same ratio $(1/(2-\rho))$ is achievable for other satisfiability or constraint satisfaction problems. For instance, using the result of Johnson [24], reoptimizing Max Weighted E3SAT[c] when a new clause is inserted is approximable within ratio 8/9.

Let us now focus on a minimization problem, namely Min Vertex Cover. Given a vertex-weighted graph $G=(V,E,w)$, the goal in this problem is to find a subset $V' \subseteq V$ such that (i) every edge $e \in E$ is incident to at least one vertex in $V'$, and (ii) the global weight of $V'$, that is, $\sum_{v \in V'} w(v)$ is minimized.

**Theorem 4.3.** *Under a vertex insertion, reoptimizing* Min Vertex Cover *is approximable within ratio 3/2.*

***Proof.*** Let $v$ denote the new vertex and $S^*$ the initial given solution. Then, $S^* \cup \{v\}$ is a vertex cover on the final instance. If $S^*_{I'}$ takes $v$, then $S^* \cup \{v\}$ is optimum.

From now on, suppose that $S^*_{I'}$ does not take $v$. Then it has to take all its neighbors $N(v)$. $S^* \cup N(v)$ is a feasible solution on the final instance. Since $w(S^*) \le w(S^*_{I'})$, we get:

$$w(S^* \cup N(v)) \le w(S^*_{I'}) + w(N(v)). \tag{4.3}$$

Then, as for Max Weighted Independent Set, consider the following feasible solution $S_1$:

- Take all the neighbors $N(v)$ of $v$ in $S_1$;
- Remove $v$ and its neighbors from the graph;
- Apply a $\rho$-approximation algorithm on the remaining graph and add these vertices to $S_1$.

Since we are in the case where $S^*_{I'}$ does not take $v$, it has to take all its neighbors, and finally:

$$w(S_1) \le \rho(w(S^*_{I'}) - w(N(v))) + w(N(v)) = \rho w(S^*_{I'}) - (\rho-1)w(N(v)). \tag{4.4}$$

Of course, we take the best solution $S$ among $S^* \cup N(v)$ and $S_1$. Then, a convex combination of equations (4.3) and (4.4) leads to:

$$w(S) \le \frac{2\rho - 1}{\rho} w(S^*_{I'}).$$

The results follows since Min Vertex Cover is well known to be approximable within ratio 2. □

[c]Restriction of Max Weighted Sat when all clauses contain exactly three literals.

To conclude this section, we point out that these results can be generalized when several vertices are inserted. Indeed, if a constant number $k > 1$ of vertices are added, one can reach the same ratio with similar arguments by considering all the $2^k$ possible subsets of new vertices in order to find the ones that will belong to the new optimum solution. This brute force algorithm is still very fast for small constant $k$, which is the case in the reoptimization setting with slight modifications of the instance.

4.3.1.2. *Unweighted problems*

In the previous subsection, we considered the general cases where vertices (or clauses) have a weight. It is well known that all the problems we focused on are already NP-hard in the unweighted case, i.e. when all vertices/clauses receive weight 1. In this (very common) case, the previous approximation results on reoptimization can be easily improved. Indeed, since only one vertex is inserted, the initial optimum solution has an *absolute error* of at most one on the final instance, i.e.:

$$|S^*| \geq |S^*_{I'}| - 1.$$

Then, in some sense we don't really need to reoptimize since $S^*$ is already a very good solution on the final instance (note also that since the reoptimization problem is NP-hard, we cannot get rid of the constant $-1$). Dealing with approximation ratio, we derive from this remark, with a standard technique, the following result:

**Theorem 4.4.** *Under a vertex insertion, reoptimizing any unweighted problem in* Hered *admits a PTAS.*

***Proof.*** Let $\varepsilon > 0$, and set $k = \lceil 1/\varepsilon \rceil$. We consider the following algorithm:

(1) Test all the subsets of $V$ of size at most $k$, and let $S_1$ be the largest one such that $G[S_1]$ satisfies the hereditary property;
(2) Output the largest solution $S$ between $S_1$ and $S^*$.

Then, if $S^*_{I'}$ has size at most $1/\varepsilon$, we found it in step 1. Otherwise, $|S^*_{I'}| \geq 1/\varepsilon$ and:

$$\frac{|S^*|}{|S^*_{I'}|} \geq \frac{|S^*_{I'}| - 1}{|S^*_{I'}|} \geq 1 - \varepsilon.$$

Of course, the algorithm is polynomial as long as $\varepsilon$ is a constant. □

In other words, the PTAS is derived from two properties: the absolute error of 1, and the fact that problems considered are *simple.* Following [30], a problem is called *simple* if, given any fixed constant $k$, it is polynomial to determine whether the optimum solution has value at most $k$ (maximization) or not.

This result easily extends to other simple problems, such as Min Vertex Cover, for instance. It also generalizes when several (a constant number of) vertices are inserted, instead of only 1.

However, it is interesting to notice that, for some other (unweighted) problems, while the absolute error 1 still holds, we cannot derive a PTAS as in Theorem 4.4 because they are not simple. One of the most famous such problems is the Min Coloring problem. In this problem, given a graph $G = (V, E)$, one wishes to partition $V$ into a minimum number of independent sets (called colors) $V_1, \ldots, V_k$. When a new vertex is inserted, an absolute error 1 can be easily achieved while reoptimizing. Indeed, consider the initial coloring and add a new color which contains only the newly inserted vertex. Then this coloring has an absolute error of 1 since a coloring on the final graph cannot use fewer colors than an optimum coloring on the initial instance.

However, deciding whether a graph can be colored with 3 colors is an NP-hard problem. In other words, Min Coloring is not simple. We will discuss the consequence of this fact in the section on hardness of reoptimization.

To conclude this section, we stress the fact that there exist, obviously, many problems that do not involve weights and for which the initial optimum solution cannot be directly transformed into a solution on the final instance with absolute error 1. Finding the longest cycle in a graph is such a problem: adding a new vertex may change considerably the size of an optimum solution.

#### 4.3.1.3. *Hardness of reoptimization*

As mentioned earlier, the fact that we are interested in slight modifications of an instance on which we have an optimum solution makes the problem somehow simpler, but unfortunately does not generally allow a jump in complexity. In other words, reoptimizing is generally NP-hard when the underlying problem is NP-hard.

In some cases, the proof of NP-hardness is immediate. For instance, consider a graph problem where modifications consists of inserting a new vertex. Suppose that we had an optimum reoptimization algorithm for this problem. Then, starting from the empty graph, and adding the vertices one by one, we could find an optimum solution on any graph on $n$ vertices by using iteratively $n$ times the reoptimization algorithm. Hence, the underlying problem would be polynomial. In conclusion, the reoptimization version is also NP-hard when the underlying problem is NP-hard. This argument is also valid for other problems under other kinds of modifications. Actually, it is valid as soon as, *for any instance $I$, there is a polynomial-time solvable instance $I'$ (the empty graph in our example) that can be generated in polynomial time and such that a polynomial number of modifications transform $I'$ into $I$.*

In other cases, the hardness does not directly follow from this argument, and a usual polynomial time reduction has to be provided. This situation occurs, for instance, in graph problems where the modification consists of deleting a vertex. As we will see later, such hardness proofs have been given, for instance, for some vehicle routing problems (in short, VRP).

Let us now focus on the hardness of approximation in the reoptimization setting. As we have seen in particular in Theorem 4.4, the knowledge of the initial optimum solution may help considerably in finding an approximate solution on the final instance. In other words, it seems quite hard to prove a lower bound on reoptimization. And in fact, few results have been obtained so far.

One method is to transform the reduction used in the proof of NP-hardness to get an inapproximability bound. Though more difficult than in the usual setting, such proofs have been provided for reoptimization problems, in particular for VRP problems, mainly by introducing very large distances (see Section 4.4).

Let us now go back to Min Coloring. As we have said, it is NP-hard to determine whether a graph is colorable with 3 colors or not. In the usual setting, this leads to an inapproximability bound of $4/3 - \varepsilon$ for any $\varepsilon > 0$. Indeed, an approximation algorithm within ratio $\rho = 4/3 - \varepsilon$ would allow us to distinguish between 3-colorable graphs and graphs for which we need at least 4 colors. Now, we can show that this result *remains true for the reoptimization of the problem*:

**Theorem 4.5.** *Under a vertex insertion, reoptimizing* Min Coloring *cannot be approximated within a ratio $4/3 - \varepsilon$, for any $\varepsilon > 0$.*

***Proof.*** The proof is actually quite straightforward. Assume you have such a reoptimization algorithm A within a ratio $\rho = 4/3-\varepsilon$. Let $G = (V, E)$ be a graph with $V = \{v_1, \cdots, v_n\}$. We consider the subgraphs $G_i$ of $G$ induced by $V_i = \{v_1, v_2, \cdots, v_i\}$ (in particular $G_n = G$). Suppose that you have a 3-coloring of $G_i$, and insert $v_{i+1}$. If $G_{i+1}$ is 3-colorable, then A outputs a 3-coloring. Moreover, if $G_i$ is not 3-colorable, then neither is $G_{i+1}$. Hence, starting from the empty graph, and iteratively applying A, we get a 3-coloring of $G_i$ if and only if $G_i$ is 3-colorable. Eventually, we are able to determine whether $G$ is 3-colorable or not. □

This proof is based on the fact that Min Coloring is not simple (according to the definition previously given). A similar argument, leading to inapproximability results in reoptimization, can be applied to other non simple problems (under other modifications). It has been in particular applied to a scheduling problem (see Section 4.3.2).

For other optimization problems however, such as MinTSP in the metric case, finding a lower bound in approximability (if any!) seems a challenging task.

Let us finally mention another kind of negative result. In the reoptimization setting, we look somehow for a possible stability when slight modifications occur on the instance. We try to measure how much the knowledge of a solution on the initial instance helps to solve the final one. Hence, it is natural to wonder whether one can find a good solution in the "neighborhood" of the initial optimum solution, or if one has to change almost everything. Do neighboring instances have neighboring optimum/good solutions? As an answer to these questions, several results show that, for several problems, approximation algorithms that only "slightly" modify the initial optimum solution cannot lead to good approximation ratios. For instance, for reoptimizing MinTSP in the metric case, if you want a ratio better than 3/2 (guaranteed by a simple heuristic), then you have to change (on some instances) a significant part of your initial solution [5]. This kind of result, weaker than an inapproximability bound, provides information on the stability under modifications and lower bounds on classes of algorithms.

### 4.3.2. *Results on some particular problems*

In the previous section, we gave some general considerations on the reoptimization of NP-hard optimization problems. The results that have been presented follow, using simple methods, from the structural properties of the problem dealt with and/or from known approximation results. We now focus on particular problems for which specific methods have been devised, and briefly mention, without proofs, the main results obtained so far. We concentrate on the Min Steiner Tree problem, on a scheduling problem, and on the Max Knapsack problem. Vehicle routing problems, which concentrated a large attention in reoptimization, deserve, in our opinion, a full section (Section 4.4), in which we also provide some of the most interesting proofs in the literature together with a few new results.

#### 4.3.2.1. *Min Steiner Tree*

The Min Steiner Tree problem is a generalization of the Min Spanning Tree problem where only a subset of vertices (called terminal vertices) have to be spanned. Formally, we are given a graph $G = (V, E)$, a non-negative distance $d(e)$ for any $e \in E$, and a subset $R \subseteq V$ of *terminal* vertices. The goal is to connect the terminal vertices with a minimum global distance, i.e. to find a tree $T \subseteq E$ that spans all vertices in $R$ and minimizes $d(T) = \sum_{e \in T} d(e)$. It is generally assumed that the graph is complete, and the distance function is metric (i.e. $d(x, y) + d(y, z) \geq d(x, z)$ for any vertices $x, y, z$): indeed, the general problem reduces to this case by initially computing shortest paths between pairs of vertices.

Min Steiner Tree is one of the most famous network design optimization problems. It is NP-hard, and has been studied intensively from an approximation viewpoint (see [18] for a survey on these results). The best known ratio obtained so far is $1 + \ln(3)/2 \simeq 1.55$ [31].

Reoptimization versions of this problem have been studied with modifications on the vertex set [9, 13]. In Escoffier et al. [13], the modification consists of the insertion of a new vertex. The authors study the cases where the new vertex is terminal or non-terminal.

**Theorem 4.6 ([13]).** *When a new vertex is inserted (either terminal or not), then reoptimizing the* Min Steiner Tree *problem can be approximated within ratio 3/2.*

Moreover, the result has been generalized to the case in which several vertices are inserted. Interestingly, when $p$ non-terminal vertices are inserted, then reoptimizing the problem is still 3/2-approximable (but the running time grows very fast with $p$). On the other hand, when terminal vertices are added, the obtained ratio decreases (but the running time remains very low). The strategies consist, roughly speaking, of merging the initial optimum solution with Steiner trees computed on the set of new vertices and/or terminal vertices. The authors tackle also the case where a vertex is removed from the vertex set, and provide a lower bound for a particular class of algorithms.

Böckenhauer et al. [9] consider a different instance modification. Rather than inserting/deleting a vertex, the authors consider the case where the status of a vertex changes: either a terminal vertex becomes non-terminal, or vice versa. The obtained ratio is also 3/2.

**Theorem 4.7 ([9]).** *When the status (terminal / non-terminal) of a vertex changes, then reoptimizing the* Min Steiner Tree *problem can be approximated within ratio 3/2.*

Moreover, they exhibit a case where this ratio can be improved. When all the distances between vertices are in $\{1, 2, \cdots, r\}$, for a fixed constant $r$, then reoptimizing Min Steiner Tree (when changing the status of one vertex) is still NP-hard but admits a PTAS.

Note that in both cases (changing the status of a vertex or adding a new vertex), no inapproximability results have been achieved, and this is an interesting open question.

4.3.2.2. *Scheduling*

Due to practical motivations, it is not surprising that scheduling problems received attention dealing with the reconstruction of a solution (often called *rescheduling*) after an instance modification, such as a machine breakdown, an increase of a job processing time, etc. Several works have been proposed to provide a sensitivity analysis of these problems under such modifications. A typical question is to determine under which modifications and/or conditions the initial schedule remains optimal. We refer the reader to the comprehensive article [20] where the main results achieved in this field are presented.

Dealing with the reoptimization setting we develop in this chapter, Schäffter [34] proposes interesting results on a problem of *scheduling with forbidden sets*. In this problem, we have a set of jobs $V = \{v_1, \cdots, v_n\}$, each job having a processing time. The jobs can be scheduled in parallel (the number of machines is unbounded), but there is a set of constraints on these parallel schedules: A constraint is a set $F \subseteq V$ of jobs that cannot be scheduled in parallel (all of them at the same time). Then, given a set $\mathcal{F} = \{F_1, \cdots, F_k\}$ of constraints, the goal is to find a schedule that respects each constraint and that minimizes the latest completion time (makespan). Many situations can be modeled this way, such as the $m$-Machine Problem (for fixed $m$), hence the problem is NP-complete (and even hard to approximate).

Schäffter considers reoptimization when either a new constraint $F$ is added to $\mathcal{F}$, or a constraint $F_i \in \mathcal{F}$ disappears. Using reductions from the Set Splitting problem and from the Min Coloring problem, he achieves the following inapproximability results:

**Theorem 4.8 ([34]).** *If* P $\neq$ NP, *for any* $\varepsilon > 0$, *reoptimizing the scheduling with forbidden sets problem is inapproximable within ratio* $3/2-\varepsilon$ *under a constraint insertion, and inapproximable within ratio* $4/3-\varepsilon$ *under a constraint deletion.*

Under a constraint insertion Schäffter also provides a reoptimization strategy that achieves approximation ratio 3/2, thus matching the lower bound of Theorem 4.8. It consists of a simple local modification of the initial scheduling, by shifting one task (at the end of the schedule) in order to ensure that the new constraint is satisfied.

4.3.2.3. *Max Knapsack*

In the Max Knapsack problem, we are given a set of $n$ objects $O = \{o_1, \ldots, o_n\}$, and a capacity $B$. Each object has a weight $w_i$ and a value $v_i$. The goal is to choose a subset $O'$ of objects that maximizes the global value $\sum_{o_i \in O'} v_i$ but that respects the capacity constraint $\sum_{o_i \in O'} w_i \leq B$.

This problem is (weakly) NP-hard, but admits an FPTAS [23]. Obviously, the reoptimization version admits an FPTAS too. Thus, Archetti et al. [2] are interested in using classical approximation algorithms for Max Knapsack to derive reoptimization algorithms with better approximation ratios but with the same running time. The modifications considered consist of the insertion of a new object in the instance.

Though not being a graph problem, it is easy to see that the Max Knapsack problem satisfies the required properties of heritability given in Section 4.3.1 (paragraph on hereditary problems). Hence, the reoptimization version is 1/2-approximable in constant time; moreover, if we have a $\rho$-approximation algorithm, then the reoptimization strategy presented in Section 4.3.1 has ratio $\frac{1}{2-\rho}$ [2]. Besides, Archetti et al. [2] show that this bound is tight for several classical approximation algorithms for Max Knapsack.

Finally, studying the issue of sensitivity presented earlier, they show that any reoptimization algorithm that does not consider objects discarded by the initial optimum solution cannot have ratio better than 1/2.

## 4.4. Reoptimization of Vehicle Routing Problems

In this section we survey several results concerning the reoptimization of vehicle routing problems under different kinds of perturbations. In particular, we focus on several variants of the Traveling Salesman Problem (TSP), which we define below.

The TSP is a well-known combinatorial optimization problem that has been the subject of extensive studies – here we only refer the interested reader to the monographs by Lawler et al. [26] and Gutin and Punnen [19]. The TSP has been used since the inception of combinatorial optimization as a testbed for experimenting a whole array of algorithmic paradigms and techniques, so it is just natural to also consider it from the point of view of reoptimization.

**Definition 4.7.** An instance $I_n$ of the Traveling Salesman Problem is given by the distance between every pair of $n$ nodes in the form of an $n \times n$ matrix $d$, where $d(i,j) \in \mathbb{Z}_+$ for all $1 \leq i,j \leq n$. A feasible solution for $I_n$ is a *tour*, that is, a directed cycle spanning the node set $N := \{1, 2, \ldots, n\}$.

Notice that we have not defined an objective function yet; so far we have only specified the structure of instances and feasible solutions. There are several possibilities for the objective function and each of them gives rise to a different optimization problem. We need a few definitions. The *weight* of a tour $T$ is the quantity $w(T) := \sum_{(i,j) \in T} d(i,j)$. The *latency of a node* $i \in N$ with respect to a given tour $T$ is the total distance along the cycle $T$ from node 1 to node $i$. The *latency* of $T$, denoted by $\ell(T)$, is the sum of the latencies of the nodes of $T$.

Table 4.1. Best known results on the approximability of the standard and reoptimization versions of vehicle routing problems (AR = approximation ratio, $\Pi+$ = vertex insertion, $\Pi-$ = vertex deletion, $\Pi\pm$ = distance variation).

| **Problem** $\Pi$ | AR($\Pi$) | **Ref.** | AR($\Pi+$) | AR($\Pi-$) | AR($\Pi\pm$) | **Ref.** |
|---|---|---|---|---|---|---|
| Min TSP | unbounded | [33] | unb. | unb. | unb. | [5, 8] |
| Min MTSP | 1.5 | [11] | 1.34 | - | 1.4 | [1, 9] |
| Min ATSP | $O(\log n)$ | [15] | 2 | 2 | - | this work |
| Max TSP | 0.6 | [25] | $0.66 - O(n^{-1})$ | - | - | this work |
| Max MTSP | 0.875 | [21] | $1 - O(n^{-1/2})$ | - | - | [5] |
| MLP | 3.59 | [10] | 3 | - | - | this work |

The matrix $d$ obeys the *triangle inequality* if for all $i, j, k \in N$ we have $d(i,j) \leq d(i,k) + d(k,j)$. The matrix $d$ is said to be a *metric* if it obeys the triangle inequality and $d(i,j) = d(j,i)$ for all $i, j \in N$.

In the rest of the section we will consider the following problems:

(1) Minimum Traveling Salesman Problem (Min TSP): find a tour of minimum weight;
(2) Minimum Metric TSP (Min MTSP): restriction of Min TSP to the case when $d$ is a metric;
(3) Minimum Asymmetric TSP (Min ATSP): restriction of Min TSP to the case when $d$ obeys the triangle inequality;
(4) Maximum TSP (Max TSP): find a tour of maximum weight;
(5) Maximum Metric TSP (Max MTSP): restriction of Max TSP to the case when $d$ is a metric;
(6) Minimum Latency Problem (MLP): find a tour of minimum latency; $d$ is assumed to be a metric.

TSP-like problems other than those above have also been considered in the literature from the point of view of reoptimization; in particular, see Böckenhauer et al. [8] for a hardness result on the TSP with deadlines.

Given a vehicle routing problem $\Pi$ from the above list, we will consider the following reoptimization variants, each corresponding to a different type of perturbation of the instance: insertion of a node ($\Pi+$), deletion of a node ($\Pi-$), and variation of a single entry of the matrix $d$ ($\Pi\pm$).

In the following, we will sometimes refer to the initial problem $\Pi$ as the *static* problem. In Table 4.1 we summarize the approximability results known for the static and reoptimization versions of the problems above under these types of perturbations.

Some simple solution methods are common to several of the problems we study in this section. We define here two such methods; they will be used in the remainder of the section.

**Algorithm 1 (Nearest Insertion).** Given an instance $I_{n+1}$ and a tour $T$ on the set $\{1, \ldots, n\}$, find a node $i^* \in \text{argmin}_{1 \leq i \leq n} d(i, n+1)$. Obtain the solution by inserting node $n+1$ either immediately before or immediately after $i^*$ in the tour (depending on which of these two solutions is best).

**Algorithm 2 (Best Insertion).** Given an instance $I_{n+1}$ and a tour $T$ on the set $\{1, \ldots, n\}$, find a pair $(i^*, j^*) \in \text{argmin}_{(i,j) \in T}\, d(i, n+1) + d(n+1, j) - d(i, j)$. Obtain the solution by inserting node $n+1$ between $i^*$ and $j^*$ in the tour.

### 4.4.1. *The Minimum Traveling Salesman Problem*

#### 4.4.1.1. *The general case*

We start by considering the Min TSP. It is well known that in the standard setting the problem is very hard to approximate in the sense that it cannot be approximated within any factor that is polynomial in the number of nodes [33]. It turns out that the same result also holds for the reoptimization versions of the problem, which shows that in this particular case the extra information available through the optimal solution to the original instance does not help at all.

**Theorem 4.9 ([5, 8]).** *Let $p$ be a polynomial. Then each of* Min TSP+, Min TSP−, *and* Min TSP± *is not $2^{p(n)}$-approximable, unless* P=NP.

***Proof.*** We only give the proof for Min TSP−; the other proofs follow a similar approach. We use the so-called *gap technique* from Sahni and Gonzales [33]. Consider the following problem, Restricted Hamiltonian Cycle (RHC): Given an undirected graph $G = (V, E)$ and a Hamiltonian path $P$ between two nodes $a$ and $b$ in $G$, determine whether there exists a Hamiltonian cycle in $G$. This problem is known to be NP-complete [28]. We prove the claim of the theorem by showing that any approximation algorithm for Min TSP− with ratio $2^{p(n)}$ can be used to solve RHC in polynomial time.

Consider an instance of RHC, that is, a graph $G = (V, E)$ on $n$ nodes, two nodes $a, b \in V$ and a Hamiltonian path $P$ from $a$ to $b$. Without loss of generality we can assume that $V = \{1, \ldots, n\}$. We can construct in polynomial time the following TSP instance $I_{n+1}$ on node set $\{1, \ldots, n, n+1\}$:

- $d(i, j) = 1$ if $(i, j) \in E$;
- $d(n+1, a) = d(b, n+1) = 1$;
- all other entries of the matrix $d$ have value $2^{p(n)} \cdot n + 1$.

Since all entries are at least 1, the tour $T^*_{n+1} := P \cup \{(b, n+1), (n+1, a)\}$ is an optimum solution of $I_{n+1}$, with weight $w(T^*_{n+1}) = n + 1$. Thus, $(I_{n+1}, T^*_{n+1})$ is an instance of Min TSP−. Let $T^*_n$ be an optimum solution of instance $I_n$. Then $w(T^*_n) = n$ if and only if $G$ has a Hamiltonian cycle. Finally, a $2^{p(n)}$-approximation algorithm for Min TSP− allows us to decide whether $w(T^*_n) = n$. □

4.4.1.2. *Minimum Metric TSP*

In the previous section we have seen that no constant-factor approximation algorithm exists for reoptimizing the Minimum TSP in its full generality. To obtain such a result, we are forced to restrict the problem somehow. A very interesting case for many applications is when the matrix $d$ is a metric, that is, the Min MTSP. This problem admits a 3/2-approximation algorithm, due to Christofides [11], and it is currently open whether this factor can be improved. Interestingly, it turns out that the reoptimization version Min MTSP+ is (at least if one considers the currently best known algorithms) easier than the static problem: It allows a 4/3-approximation – although, again, we do not know whether even this factor may be improved via a more sophisticated approach.

**Theorem 4.10 ([5]).** Min MTSP+ *is approximable within ratio* 4/3.

***Proof.*** The algorithm used to prove the upper bound is a simple combination of Nearest Insertion and of the well-known algorithm by Christofides [11]; namely, both algorithms are executed and the solution returned is the one having the lower weight.

Consider an optimum solution $T^*_{n+1}$ of the final instance $I_{n+1}$, and the solution $T^*_n$ available for the initial instance $I_n$. Let $i$ and $j$ be the two neighbors of vertex $n+1$ in $T^*_{n+1}$, and let $T_1$ be the tour obtained from $T^*_n$ with the Nearest Insertion rule. Furthermore, let $v^*$ be the vertex in $\{1, \ldots, n\}$ whose distance to $n+1$ is the smallest.

Using the triangle inequality, we easily get $w(T_1) \leq w(T^*_{n+1}) + 2d(v^*, n+1)$ where, by definition of $v^*$, $d(v^*, n+1) = \min\{d(k, n+1) : k = 1, \ldots, n\}$. Thus

$$w(T_1) \leq w(T^*_{n+1}) + 2\max(d(i, n+1), d(j, n+1)). \tag{4.5}$$

Now consider the algorithm of Christofides applied on $I_{n+1}$. This gives a tour $T_2$ of length at most $(1/2)w(T^*_{n+1}) + \text{MST}(I_{n+1})$, where $\text{MST}(I_{n+1})$ is the weight of a minimum spanning tree on $I_{n+1}$. Note that $\text{MST}(I_{n+1}) \leq w(T^*_{n+1}) - \max(d(i,n+1), d(j,n+1))$. Hence

$$w(T_2) \leq \frac{3}{2}w(T^*_{n+1}) - \max(d(i,n+1), d(j,n+1)). \tag{4.6}$$

The result now follows by combining equations (4.5) and (4.6), because the weight of the solution given by the algorithm is $\min(w(T_1), w(T_2)) \leq (1/3)w(T_1) + (2/3)w(T_2) \leq (4/3)w(T^*_{n+1})$. □

The above result can be generalized to the case when more than a single vertex is added in the perturbed instance. Let Min MTSP+$k$ be the corresponding problem when $k$ vertices are added. Then it is possible to give the following result, which gives a trade-off between the number of added vertices and the quality of the approximation guarantee.

**Theorem 4.11 ([5]).** *For any* $k \geq 1$, Min MTSP+$k$ *is approximable within ratio* $3/2 - 1/(4k+2)$.

Reoptimization under variation of a single entry of the distance matrix (that is, problem Min MTSP±) has been considered by Böckenhauer et al. [9].

**Theorem 4.12 ([9]).** Min MTSP± *is approximable within ratio* 7/5.

#### 4.4.1.3. *Minimum Asymmetric TSP*

The Minimum Asymmetric Traveling Salesman Problem is another variant of the TSP that is of interest for applications, as it generalizes the Metric TSP. Unfortunately, in the static case there seems to be a qualitative difference with respect to the approximability of Minimum Metric TSP: While in the latter case a constant approximation is possible, for Min ATSP the best known algorithms give an approximation ratio of $\Theta(\log n)$. The first such algorithm was described by Frieze et al. [17] and has an approximation guarantee of $\log_2 n$. The currently best algorithm is due to Feige and Singh [15] and gives approximation $(2/3)\log_2 n$. The existence of a constant approximation for Min ATSP is an important open problem.

Turning now to reoptimization, there exists a non-negligible gap between the approximability of the static version and of the reoptimization version. In fact, reoptimization drastically simplifies the picture: Min ATSP+ is approximable within ratio 2, as we proceed to show.

**Theorem 4.13.** Min ATSP+ *is approximable within ratio* 2.

***Proof.*** The algorithm used to establish the upper bound is extremely simple: just add the new vertex between an arbitrarily chosen pair of consecutive vertices in the old optimal tour. Let $T$ be the tour obtained by inserting node $n+1$ between two consecutive nodes $i$ and $j$ in $T_n^*$. We have:

$$w(T) = w(T_n^*) + d(i, n+1) + d(n+1, j) - d(i, j).$$

By triangle inequality, $d(n+1, j) \leq d(n+1, i) + d(i, j)$. Hence

$$w(T) \leq w(T_n^*) + d(i, n+1) + d(n+1, i).$$

Again by triangle inequality, $w(T_n^*) \leq w(T_{n+1}^*)$, and $d(i, n+1) + d(n+1, i) \leq w(T_{n+1}^*)$, which concludes the proof. □

We remark that the above upper bound of 2 on the approximation ratio is tight, even if we use Best Insertion instead of inserting the new vertex between an arbitrarily chosen pair of consecutive vertices.

**Theorem 4.14.** Min ATSP− *is approximable within ratio 2.*

***Proof.*** The obvious idea is to skip the deleted node in the new tour, while visiting the remaining nodes in the same order. Thus, if $i$ and $j$ are respectively the nodes preceding and following $n+1$ in the tour $T_{n+1}^*$, we obtain a tour $T$ such that

$$w(T) = w(T_{n+1}^*) + d(i, j) - d(i, n+1) - d(n+1, j). \tag{4.7}$$

Consider an optimum solution $T_n^*$ of the modified instance $I_n$, and the node $l$ that is consecutive to $i$ in this solution. Since inserting $n+1$ between $i$ and $l$ would yield a feasible solution to $I_{n+1}$, we get, using triangle inequality:

$$\begin{aligned} w(T_{n+1}^*) &\leq w(T_n^*) + d(i, n+1) + d(n+1, l) - d(i, l) \\ &\leq w(T_n^*) + d(i, n+1) + d(n+1, i). \end{aligned}$$

By substituting in (4.7) and using triangle inequality again,

$$w(T) \leq w(T_n^*) + d(i, j) + d(j, i).$$

Hence, $w(T) \leq 2w(T_n^*)$. □

### 4.4.2. *The Maximum Traveling Salesman Problem*

#### 4.4.2.1. *Maximum TSP*

While the typical applications of the Minimum TSP are in vehicle routing and transportation problems, the Maximum TSP has applications to DNA sequencing and data compression [25]. Like the Minimum TSP, the Maximum TSP is also NP-hard, but differently from what happens for the Minimum TSP, it is approximable within a constant factor even when the distance matrix can be completely arbitrary. In the static setting, the best known result for Max TSP is a 0.6-approximation algorithm due to Kosaraju et al. [25]. Once again, the knowledge of an optimum solution to the initial instance is useful, as the reoptimization problem under insertion of a vertex can be approximated within a ratio of 0.66 (for large enough $n$), as we show next.

**Theorem 4.15.** Max TSP+ *is approximable within ratio* $(2/3) \cdot (1 - 1/n)$.

***Proof.*** Let $i$ and $j$ be such that $(i, n+1)$ and $(n+1, j)$ belong to $T^*_{n+1}$. The algorithm is the following:

(1) Apply Best Insertion to $T^*_n$ to get a tour $T_1$;
(2) Find a maximum cycle cover $\mathcal{C} = (C_0, \dots, C_l)$ on $I_{n+1}$ such that:
   (a) $(i, n+1)$ and $(n+1, j)$ belong to $C_0$;
   (b) $|C_0| \geq 4$;
(3) Remove the minimum-weight arc of each cycle of $\mathcal{C}$ and patch the paths obtained to get a tour $T_2$;
(4) Select the best solution between $T_1$ and $T_2$.

Note that Step 2 can be implemented in polynomial time as follows: We replace $d(i, n+1)$ and $d(n+1, j)$ by a large weight $M$, and $d(j, i)$ by $-M$ (we do not know $i$ and $j$, but we can try each possible pair of vertices and return the best tour constructed by the algorithm). Hence, this cycle cover will contain $(i, n+1)$ and $(n+1, j)$ but not $(j, i)$, meaning that the cycle containing $n+1$ will have at least 4 vertices.

Let $a := d(i, n+1) + d(n+1, j)$. Clearly, $w(T^*_{n+1}) \leq w(T^*_n) + a$. Now, by inserting $n+1$ in each possible position, we get

$$w(T_1) \geq (1 - 1/n) w(T^*_n) \geq (1 - 1/n)(w(T^*_{n+1}) - a).$$

Since $C_0$ has size at least 4, the minimum-weight arc of $C_0$ has cost at most $(w(C_0) - a)/2$. Since each cycle has size at least 2, we get a tour $T_2$

of value:

$$w(T_2) \geq w(\mathcal{C}) - \frac{w(C_0) - a}{2} - \frac{w(\mathcal{C}) - w(C_0)}{2}$$
$$= \frac{w(\mathcal{C}) + a}{2} \geq \frac{w(T^*_{n+1}) + a}{2}.$$

Combining the two bounds for $T_1$ and $T_2$, we get a solution which is $(2/3) \cdot (1 - 1/n)$-approximate. □

The above upper bound can be improved to 0.8 when the distance matrix is known to be symmetric [5].

#### 4.4.2.2. *Maximum Metric TSP*

The usual Maximum TSP problem does not admit a polynomial-time approximation scheme, that is, there exists a constant $c$ such that it is NP-hard to approximate the problem within a factor better than $c$. This result extends also to the Maximum Metric TSP [29]. The best known approximation for the Maximum Metric TSP is a randomized algorithm with an approximation guarantee of 7/8 [21].

By contrast, in the reoptimization of Max MTSP under insertion of a vertex, the Best Insertion algorithm turns out to be a very good strategy: It is asymptotically optimum. In particular, the following holds:

**Theorem 4.16** ([5]). Max MTSP+ *is approximable within ratio* $1 - O(n^{-1/2})$.

Using the above result one can easily prove that Max MTSP+ admits a polynomial-time approximation scheme: If the desired approximation guarantee is $1 - \epsilon$, for some $\epsilon > 0$, just solve by enumeration the instances with $O(1/\epsilon^2)$ nodes, and use the result above for the other instances.

### 4.4.3. *The Minimum Latency Problem*

Although superficially similar to the Minimum Metric TSP, the Minimum Latency Problem appears to be more difficult to solve. For example, in the special case when the metric is induced by a weighted tree, the MLP is NP-hard [35] while the Metric TSP is trivial. One of the difficulties in the MLP is that local changes in the input can influence the global shape of the optimum solution. Thus, it is interesting to notice that despite this fact, reoptimization still helps. In fact, the best known approximation so far for the static version of the MLP gives a factor of 3.59 and is achieved via a sophisticated algorithm due to Chaudhuri et al. [10], while it is possible to give a very simple 3-approximation for MLP+, as we show in the next theorem.

**Theorem 4.17.** MLP+ *is approximable within ratio* 3.

***Proof.*** We consider the *Insert Last* algorithm that inserts the new node $n+1$ at the "end" of the tour, that is, just before node 1. Without loss of generality, let $T_n^* = \{(1,2),(2,3),\ldots,(n-1,n)\}$ be the optimal tour for the initial instance $I_n$ (that is, the $k$th node to be visited is $k$). Let $T_{n+1}^*$ be the optimal tour for the modified instance $I_{n+1}$. Clearly $\ell(T_{n+1}^*) \geq \ell(T_n^*)$ since relaxing the condition that node $n+1$ must be visited cannot raise the overall latency.

The quantity $\ell(T_n^*)$ can be expressed as $\sum_{i=1}^n t_i$, where for $i = 1,\ldots,n$, $t_i = \sum_{j=1}^{i-1} d(j, j+1)$ can be interpreted as the "time" at which node $i$ is first visited in the tour $T_n^*$.

In the solution constructed by Insert Last, the time at which each node $i \neq n+1$ is visited is the same as in the original tour ($t_i$), while $t_{n+1} = t_n + d(n, n+1)$. The latency of the solution is thus $\sum_{i=1}^{n+1} t_i = \sum_{i=1}^n t_i + t_n + d(n, n+1) \leq 2\ell(T_n^*) + \ell(T_{n+1}^*) \leq 3\ell(T_{n+1}^*)$, where we have used $\ell(T_{n+1}^*) \geq d(n, n+1)$ (any feasible tour must include a subpath from $n$ to $n+1$ or vice versa). □

## 4.5. Concluding Remarks

In this chapter we have seen how the reoptimization model can often be applied to NP-hard combinatorial problems in order to obtain algorithms with approximation guarantees that improve upon the trivial approach of computing an approximate solution from scratch.

Apart from designing algorithms with good approximation guarantees for reoptimization problems – and from obtaining sharper negative results – there are some general open directions in the area. One is to investigate the more general issue of maintaining an approximate solution under input modifications. In our model we assumed that an optimal solution was available for the instance prior to the modification, but it is natural to relax this constraint by assuming only an approximate solution instead. In some cases the analysis of the reoptimization algorithm can be carried out in a similar way even with such a relaxed assumption, but this needs not be always true.

Another general question is that of studying the interplay between running time, approximation guarantee, and amount of data perturbation. If we devote enough running time (for example, exponential time for problems in NPO) to the solution of an instance, we can find an optimal solution independently of the amount of perturbation. On the other hand we saw that for many problems it is possible to find in polynomial time an almost optimal solution for any slightly perturbed instance. One could expect that there might be a general trade-off between the amount of data perturbation and the running time needed the reconstruct a solution of a given quality. It would be interesting to identify problems for which such trade-offs are possible.

## References

[1] C. Archetti, L. Bertazzi, and M. G. Speranza, Reoptimizing the traveling salesman problem, *Networks.* **42**(3), 154–159, (2003).

[2] C. Archetti, L. Bertazzi, and M. G. Speranza. Reoptimizing the 0–1 knapsack problem. Technical Report 267, Department of Quantitative Methods, University of Brescia, (2006).

[3] T. Asano, K. Hori, T. Ono, and T. Hirata. A theoretical framework of hybrid approaches to MAX SAT. In *Proc. 8th Int. Symp. on Algorithms and Computation*, pp. 153–162, (1997).

[4] G. Ausiello, P. Crescenzi, G. Gambosi, V. Kann, A. Marchetti-Spaccamela, and M. Protasi, *Complexity and approximation – Combinatorial optimization problems and their approximability properties*, Springer, Berlin, (1999).

[5] G. Ausiello, B. Escoffier, J. Monnot, and V. T. Paschos. Reoptimization of minimum and maximum traveling salesman's tours. In *Proc. 10th Scandinavian Workshop on Algorithm Theory*, pp. 196–207, (2006).

[6] M. Bartusch, R. Möhring, and F. J. Radermacher, Scheduling project networks with resource constraints and time windows, *Ann. Oper. Res..* **16**, 201–240, (1988).

[7] M. Bartusch, R. Möhring, and F. J. Radermacher, A conceptional outline of a DSS for scheduling problems in the building industry, *Decision Support Systems.* **5**, 321–344, (1989).

[8] H.-J. Böckenhauer, L. Forlizzi, J. Hromkovic, J. Kneis, J. Kupke, G. Proietti, and P. Widmayer. Reusing optimal TSP solutions for locally modified input instances. In *Proc. 4th IFIP Int. Conf. on Theoretical Computer Science*, pp. 251–270, (2006).

[9] H.-J. Böckenhauer, J. Hromkovic, T. Mömke, and P. Widmayer. On the hardness of reoptimization. In *Proc. 34th Conf. on Current Trends in Theory and Practice of Computer Science*, pp. 50–65, (2008).

[10] K. Chaudhuri, B. Godfrey, S. Rao, and K. Talwar. Paths, trees, and minimum latency tours. In *Proc. 44th Symp. on Foundations of Computer Sci-*

*ence*, pp. 36–45, (2003).

[11] N. Christofides. Worst-case analysis of a new heuristic for the travelling salesman problem. Technical Report 388, Graduate School of Industrial Administration, Carnegie-Mellon University, Pittsburgh, PA, (1976).

[12] D. Eppstein, Z. Galil, G. F. Italiano, and A. Nissenzweig, Sparsification-a technique for speeding up dynamic graph algorithms, *J. ACM.* **44**(5), 669–696, (1997).

[13] B. Escoffier, M. Milanic, and V. T. Paschos, *Simple and fast reoptimizations for the Steiner tree problem*, Cahier du LAMSADE 245, LAMSADE, Université Paris-Dauphine, (2007).

[14] S. Even and H. Gazit, Updating distances in dynamic graphs, *Methods Oper. Res.* **49**, 371–387, (1985).

[15] U. Feige and M. Singh. Improved approximation ratios for traveling salesperson tours and paths in directed graphs. In *Proc. 10th Int. Workshop on Approximation, Randomization, and Combinatorial Optimization*, pp. 104–118, (2007).

[16] G. N. Frederickson, Data structures for on-line updating of minimum spanning trees, with applications, *SIAM J. Comput.*. **14**(4), 781–798, (1985).

[17] A. M. Frieze, G. Galbiati, and F. Maffioli, On the worst-case performance of some algorithms for the asymmetric traveling salesman problem, *Networks.* **12**(1), 23–39, (1982).

[18] C. Gröpl, S. Hougardy, T. Nierhof, and H. Prömel. Approximation algorithms for the Steiner tree problem in graphs. In eds. D.-Z. Du and X. Cheng, *Steiner Trees in Industry*, pp. 235–279. Kluwer Academic Publishers, Dordrecht, (2000).

[19] G. Gutin and A. P. Punnen, Eds., *The Traveling Salesman Problem and its Variations.* Kluwer, Dordrecht, (2002).

[20] N. G. Hall and M. E. Posner, Sensitivity analysis for scheduling problems, *J. Sched.*. **7**(1), 49–83, (2004).

[21] R. Hassin and S. Rubinstein, A 7/8-approximation algorithm for metric Max TSP, *Inform. Process. Lett.*. **81**(5), 247–251, (2002).

[22] M. R. Henzinger and V. King, Maintaining minimum spanning forests in dynamic graphs, *SIAM J. Comput.*. **31**(2), 367–374, (2001).

[23] O. H. Ibarra and C. E. Kim, Fast approximation algorithms for the knapsack and sum of subset problems, *J. ACM.* **22**(4), 463–468, (1975).

[24] D. S. Johnson, Approximation algorithms for combinatorial problems, *J. Comput. Systems Sci.*. **9**, 256–278, (1974).

[25] S. R. Kosaraju, J. K. Park, and C. Stein. Long tours and short superstrings. In *Proc. 35th Symp. on Foundations of Computer Science*, pp. 166–177, (1994).

[26] E. L. Lawler, J. K. Lenstra, A. Rinnoy Kan, and D. B. Shymois, Eds., *The Traveling Salesman Problem: A Guided Tour of Combinatorial Optimization.* Wiley, Chichester, (1985).

[27] J. M. Lewis and M. Yannakakis, The node-deletion problem for hereditary properties is NP-complete, *J. Comput. Systems Sci.*. **20**(2), 219–230, (1980).

[28] C. H. Papadimitriou and K. Steiglitz, On the complexity of local search for

the traveling salesman problem, *SIAM J. Comput.*. **6**(1), 76–83, (1977).
[29] C. H. Papadimitriou and M. Yannakakis, The traveling salesman problem with distances one and two, *Math. Oper. Res.*. **18**(1), 1–11, (1993).
[30] A. Paz and S. Moran, Non-deterministic polynomial optimization problems and their approximations, *Theoret. Comput. Sci.*. **15**, 251–277, (1981).
[31] G. Robins and A. Zelikovsky, Tighter bounds for graph Steiner tree approximation, *SIAM J. Discrete Math.*. **19**(1), 122–134, (2005).
[32] H. Rohnert. A dynamization of the all-pairs least cost problem. In *Proc. 2nd Symp. on Theoretical Aspects of Computer Science*, pp. 279–286, (1985).
[33] S. Sahni and T. F. Gonzalez, P-complete approximation problems, *J. ACM*. **23**(3), 555–565, (1976).
[34] M. W. Schäffter, Scheduling with forbidden sets, *Discrete Appl. Math.*. **72** (1-2), 155–166, (1997).
[35] R. Sitters. The minimum latency problem is NP-hard for weighted trees. In *Proc. 9th Integer Programming and Combinatorial Optimization Conf.*, pp. 230–239, (2002).
[36] V. V. Vazirani, *Approximation Algorithms*. Springer, Berlin, (2001).
[37] D. Zuckerman, Linear degree extractors and the inapproximability of max clique and chromatic number, *Theory Comput.*. **3**(1), 103–128, (2007).

# Chapter 5

# Definability in the Real Universe

S. Barry Cooper *

*Department of Pure Mathematics*
*University of Leeds*
*Leeds LS2 9JT, UK*
*E-mail: S.B.Cooper@leeds.ac.uk*

Logic has its origins in basic questions about the nature of the real world and how we describe it. This chapter seeks to bring out the physical and epistemological relevance of some of the more recent technical work in logic and computability theory.

## Contents



---

*Chapter based on an invited talk at Logic Colloquium 2009, Sofia, Bulgaria, August, 2009. Research supported by U.K. EPSRC grant No. GR /S28730/01, and by a Royal Society International Collaborative Grant.

## 5.1. Introduction

Logic has an impressive history of addressing very basic questions about the nature of the world we live in. At the same time, it has clarified concepts and informal ideas about the world, and gone on to develop sophisticated technical frameworks within which these can be discussed. Much of this work is little known or understood by non-specialists, and the significance of it largely ignored. While notions such as set, proof and consistency have become part of our culture, other very natural abstractions such as that of definability are unfamiliar and disconcerting, even to working mathematicians. The widespread interest in Gödel's [46, 47] incompleteness results and their frequent application, often in questionable ways, shows both the potential for logicians to say something important about the world, while at the same time illustrating the limitations of what has been achieved so far. This article seeks to bring out the relevance of some of the more recent technical work in logic and computability theory. Basic questions addressed include: How do scientists represent and establish control over information about the universe? How does the universe itself exercise control over its own development? And more feasibly: How can we reflect that control via our scientific and mathematical representations?

Definability – what we can describe in terms of what we are given in a particular language – is a key notion. As Hans Reichenbach (Hilary Putnam is perhaps his best-known student) found in the 1920s onwards, formalising definability in the real world comes into its own when we need to clarify and better understand the content of a hard-to-grasp description of reality, such as Einstein's theory of general relativity. Reichenbach's seminal work [78] on axiomatising relativity has become an ongoing project, carried forward today by Istvan Nemeti, Hajnal Andreka and their co-workers (see, for example, Andréka, Madarász, Németi and Székely [2]). One can think of such work as paralleling the positive developments that models of computation enabled during the early days of computer science, bringing a surer grip on practical computation. But computability theory also gave an overview of what can be computed in principle, with corresponding technical developments apparently unrelated to applications. The real-world relevance of most of this theory remains conjectural.

The capture of natural notions of describability and real-world robustness via the precisely formulated ones of definability and invariance also brings a corresponding development of theory, which can be applied in different mathematical contexts. Such an application does not just bring interesting theorems, which one just adds to the existing body of theory with conjectural relevance. It fills out the explanatory framework to a point where it can be better assessed for power and validity. And it is this which is further sketched out below. The basic ingredients are the notions of definability and invariance, and a mathematical context which best describes the scientific description of familiar causal structure.

## 5.2. Computability versus Descriptions

In the modern world, *scientists* look for theories that enable predictions, and, if possible, predictions of a computational character. Everyone else lives with less constrained descriptions of what is happening, and is likely to happen. Albert Einstein [38] might have expressed the view in 1950 that:

> When we say that we understand a group of natural phenomena, we mean that we have found a constructive theory which embraces them.

But in everyday life people commonly use informal language to describe expectations of the real world from which constructive or computational content is not even attempted. And there is a definite mismatch between the scientist's drive to extend the reach of his or her methodology, and the widespread sense of an intrusion of algorithmic thinking into areas where it is not valid. A recent example is the controversy around Richard Dawkins' book [32], *The God Delusion.* This dichotomy has some basis in theorems from logic (such as Gödel's incompleteness theorems): but the basis is more one for argument and confusion than anything more consensual. Things were not always so.

If one goes back before the time of Isaac Newton, before the scientific era, informal *descriptions* of the nature of reality were the common currency of those trying to reason about the world. This might even impinge on mathematics – as when the Pythagoreans wrestled with the ontology of irrational numbers. Calculation had a quite specific and limited role in society.

## 5.3. Turing's Model and Incomputability

In 1936, Turing [100] modelled what he understood of how a then human "computer" (generally a young woman) might perform calculations – laying down rules that were very restrictive in a practical sense, but which enabled, as he plausibly argued, all that might be achieved with apparently more powerful computational actions. Just as the Turing machine's primitive actions (observing, moving, writing) were the key to modelling complex computations, so the Turing machine itself provided a route to the modelling of complex natural processes within structures which are discretely (or at least countably) presented. In this sense, it seemed we now had a way of making concrete the Laplacian model of science which had been with us in some form or other ever since the significance of what Newton had done became clear.

But the techniques for presenting a comprehensive range of computing machines gave us the *universal* Turing machine, so detaching computations from their material embodiments: and – a more uncomfortable surprise – by adding a quantifier to the perfectly down-to-earth description of the universal machine we get (and Turing [100] proved it) an incomputable object, the halting set of the machine. In retrospect, this becomes a vivid indication of how natural language has both an important real-world role, and quickly outstrips our computational reach. The need then becomes to track down material counterpart to the simple mathematical schema which give rise to incomputability. Success provides a link to a rich body of theory and opens a Pandora's box of new perceptions about the failings of science and the nature of the real universe.

## 5.4. The Real Universe as Discipline Problem

The Laplacian model has a deeply ingrained hold on the rational mind. For a bromeliad-like late flowering of the paradigm we tend to think of Hilbert and his assertion of very general expectations for axiomatic mathematics. Or of the state of physics before quantum mechanics. The problem is that modelling the universe is definitely not an algorithmic process, and that is why intelligent, educated people can believe very different things, even in science. Even in mathematics. So for many, the mathematical phase-transition from computability to incomputability, which a quantifier provides, is banned from the real world (see for example Cotogno [24]). However simple the *mathematical* route to incomputability, when looking out at the natural world, the trick is to hold the eyeglass to an unseeing eye. The global aspect of causality so familiar in mathematical structures is denied a connection with reality, in any shape or form. For a whole community, the discovery of incomputability made the real universe a real discipline problem. When Martin Davis [30] says:

> The great success of modern computers as all-purpose algorithm-executing engines embodying Turing's universal computer in physical form, makes it extremely plausible that the abstract theory of computability gives the correct answer to the question, 'What is a computation?', and, by itself, makes the existence of any more general form of computation extremely doubtful.

we have been in the habit of agreeing, in a mathematical setting. But in the context of a general examination of hypercomputational propositions (whatever the validity of the selected examples) it gives the definite impression of a defensive response to an uncompleted paradigm change. For convenience, we call this response [31] – that 'there is no such discipline as hypercomputation' - Davis' Thesis.

The universal Turing machine freed us from the need actually *embody* the machines needed to host different computational tasks. The importance of this for building programmable computers was immediately recognised by John von Neuman, and played a key role in the early history of the computer (see Davis [29]). The notion of a *virtual machine* is a logical extension of this tradition, which has found widespread favour amongst computer scientists and philosophers of a functionalist turn of mind – for instance, there is the Sloman and Chrisley [89] proposition for releasing consciousness from the philosophical inconvenience of embodiment (see also Torrance, Clowes and Chrisley [99]). Such attempts to tame nature are protected by a dominant paradigm, but there is plenty of dissatisfaction with them based on respect for the complex physicality of what we see.

## 5.5. A Dissenting Voice ...

Back in 1970, Georg Kreisel considered one of the simplest physical situations presenting mathematical predictive problems. Contained within the mathematics one detects uncompleted infinities of the kind necessary for incomputability to have any significance for the real world. In a footnote to Kreisel [56] he proposed a collision problem related to the 3-body problem, which might result in 'an analog computation of a non-recursive function'.

Even though Kreisel's view was built on many hours of deep thought about extensions of the Church–Turing thesis to the material universe – much of this embodied in Odifreddi's 20-page discussion of the Church–Turing thesis in his book [69] on Classical Recursion Theory – it is not backed up by any proof of of the inadequacy of the Turing model built on a precise description of the collision problem.

This failure has become a familiar one, what has been described as a failure to find 'natural' examples of incomputability other than those computably equivalent to the halting problem for a universal Turing machine – with even that not considered very natural by the mainstream mathematician. One requirement of a 'natural' incomputable set is that it be computably enumerable, like the set of solutions of a diophantine equation, or the set of natural numbers $n$ such that there exists a block of precisely $n$ 7s in the decimal expansion of the real number $\pi$ – or like the halting set of a given Turing machine. The problem is that given a computably enumerable set of numbers, there are essentially two ways of knowing its incomputability. One way is to have designed the set oneself to have complement different to any other set on a standard list of computably enumerable sets. Without working relative to some other incomputable set, one just gets canonical sets computably equivalent to the halting set of the universal Turing machine. Otherwise the set one built has no known robustness, no definable character one can recognise it by once it is built. The other way of knowing a particular computably enumerable set to be incomputable is to be able to compute one of the sets built via way one from the given set. But only the canonical sets have been found so far to work in this way. So it is known that there is a whole rich universe of computably inequivalent computably enumerable sets – but the only individual ones recognisably so are computably equivalent to the halting problem. Kreisel's failure is not so significant when one accepts that an arbitrary set picked from nature in some way is very unlikely to be a mathematically canonical object. It seems quite feasible that there is a mathematical theorem waiting to be proved, explaining why there is no accessible procedure for verifying incomputability in nature.

Since Kreisel's example, there have been other striking instances of infinities in nature with the potential for hosting incomputability. In *Off to Infinity in Finite Time* Donald Saari and Jeff Xia [86] describe how one can even derive singularities arising from the behaviour of five bodies moving under the influence of the familiar Newtonian inverse square law.

There is a range of more complex examples which are hard to fit into the standard Turing model, ones with more real-world relevance. There is the persistence of problems of predictability in a number of contexts. There is quantum uncertainty, constrained by computable probabilities, but hosting what looks very much like randomness; there are apparently emergent phenomena in many environments; and chaotic causal environments giving rise to strange attractors; and one has relativity and singularities (black holes), whose singular aspects can host incomputability. Specially interesting is the renewed interest in analog and hybrid computing machines, leading Jan van Leeuwen and Jiri Wiedermann [105] to observe that '... the classical Turing paradigm may no longer be fully appropriate to capture all features of present-day computing.' And – see later – there is mentality, consciousness and the observed shortcomings of the mathematical models of these.

The disinterested observer of Martin Davis' efforts to keep nature contained within the Turing/Laplacian model might keep in mind the well-known comment of Arthur C. Clarke [16] (Clarke's First Law) that:

> When a distinguished but elderly scientist states that something is possible, he is almost certainly right. When he states that something is impossible, he is very probably wrong.

In what follows we look in more detail at three key challenges to the attachment of Davis, and of a whole community, to the Turing model in the form of Davis' thesis.

There is a reason for this. At first sight, it may seem unimportant to know whether we have computational or predictive difficulties due to mere complexity of a real-world computational task, or because of its actual incomputability. And if there is no distinguishable difference between the two possibilities, surely it cannot matter which pertains. Well, no. Attached to two different mathematical characterisations one would expect different mathematical theories. And there is a rich and well-developed theory of incomputability. This mathematics may well constrain and give global form to the real world which it underlies. And these constraints and structurings may be very significant for our experience and understanding of the universe and our place in it.

## 5.6. The Quantum Challenge

In the early days of quantum computing, there was some good news for Davis' thesis from one of its most prominent supporters. David Deutsch was one of the originators of the standard model of quantum computation. In his seminal 1985 article [33] 'Quantum Theory, the Church-Turing Principle and the Universal Quantum Computer' in the *Proceedings of the Royal Society of London*, he introduced the notion of a 'universal quantum computer', and described how it might exploit quantum parallelism to compute more efficiently than a classical Turing machine. But Deutsch is quite clear that real computers based on this model would not compute anything not computable classically by a Turing machine. And, of course, there are many other instances of successful reductions of 'natural examples' of nature-based computational procedures to the Turing model.

But like Martin Davis, Deutsch [35] is keen to take things further – a lot further – attempting a reduction of human mentality to the Turing model in a way even Turing in his most constructive frame of mind might have had misgivings about:

> I am sure we will have [conscious computers], I expect they will be purely classical and I expect that it will be a long time in the future. Significant advances in our philosophical understanding of what consciousness is, will be needed.

Be this as it may, there are aspects of the underlying physics which are not fully used in setting up the standard model for quantum computing. It is true that *measurements* do play a role in a quantum computation, but in a tamed guise. This is how Andrew Hodges explains it, in his article *What would Alan Turing have done after 1954?* in the Teuscher volume [98]:

> Von Neumann's axioms distinguished the **U** (unitary evolution) and **R** (reduction) rules of quantum mechanics. Now, quantum computing so far (in the work of Feynman, Deutsch, Shor, etc.) is based on the **U** process and so computable. It has not made serious use of the **R** process: the unpredictable element that comes in with reduction, measurement or collapse of the wave function.

The point being that measurements in the quantum context are intrusive, with outcomes governed by computable probabilities, but with the mapping out of what goes on within those probabilities giving the appearance of randomness. There are well-established formalisations of the intuitive notion of randomness, largely coincident and a large body of mathematical theory built on these (see, for example, Chaitin [15], Downey and Hirschfeldt [36], Nies [67]). A basic feature of the theory is the fact that randomness implies incomputability (but not the converse). Calude and Svozil [14] have extracted a suitable mathematical model of quantum randomness, built upon assumptions generally acceptable to the physicists. Analysing the computability-theoretic properties of the model, they are able to show that quantum randomness does exhibit incomputability. But, interestingly, they are unable as yet to confirm that quantum randomness is mathematically random.

But quantum mechanics does not just present one of the toughest challenges to Davis' thesis. It also presents the observer with a long-standing challenge to its own *realism.* Interpretations of the theory generally fail to satisfy everyone, and the currently most widely accepted interpretations contain what must be considered metaphysical assumptions. When we have assembled the key ingredients, we will be in a position to argue that the sort of fundamental thinking needed to rescue the theory from such assumptions is based on some very basic mathematics.

## 5.7. Schrödinger's Lost States, and the Many-Worlds Interpretation

One way of describing the quantum world is via the Schrödinger wave equation. What Hodges refers to above are the processes for change of the wave equation describing the quantum state of a physical system. On the one hand, one has deterministic continuous evolution via Schrödinger's equation, involving superpositions of basis states. On the other, one has probabilistic non-local discontinuous change due to measurement. With this, one observes a jump to a single basis state. The interpretive question then is: *Where do the other states go?*

Writing with hindsight: If the physicists knew enough logic, they would have been able to make a good guess. And if the logicians had been focused enough on the foundations of quantum mechanics they might have been able to tell them.

As it is, physics became a little weirder around 1956. The backdrop to this is the sad and strange life-story of Hugh Everett III and his family, through which strode the formidable John Wheeler, Everett's final thesis advisor, and Bryce DeWitt, who in 1970 coined the term 'Many-Worlds' for Everett's neglected and belittled idea: an idea whose day came too late to help the Everett family, now only survived by the son Mark who relives parts of the tragic story via an autobiography [42] and appropriately left field confessional creations as leader of the Eels rock band.

Many-Worlds, with a little reworking, did away with the need to explain the transition from many superposed quantum states to the 'quasi-classical' uniqueness we see around us. The multiplicity survives and permeates micro- to macro-reality, via a decohering bushy branching of alternative histories, with us relegated to our own self-contained branch. Max Tegmark has organised the multiplying variations on the Many-Worlds theme into hierarchical levels of 'multiverses', from modest to more radical proposals, with even the underlying mathematics and the consequent laws of physics individuating at Level IV. Of course, if one does not bother any more to explain why our universe works so interestingly, one needs the 'anthropic principle' on which to base our experience of the world – 'We're here because we're here because we're here because we're here ...', as they sang during the Great War, marching towards the trenches. The attraction of this picture derives from the drive for a coherent overview, and the lack of a better one. As David Deutsch put it in *The Fabric of Reality* [34, p.48]:

> ... understanding the multiverse is a precondition for understanding reality as best we can. Nor is this said in a spirit of grim determination to seek the truth no matter how unpalatable it may be ... It is, on the contrary, because the resulting world-view is so much more integrated, and makes more sense in so many ways, than any previous world-view, and certainly more than the cynical pragmatism which too often nowadays serves as surrogate for a world-view amongst scientists.

Here is a very different view of the multiverse from the distinguished South African mathematician George Ellis [40, p.198], one-time collaborator of Stephen Hawking:

> The issue of what is to be regarded as an ensemble of 'all possible' universes is unclear, it can be manipulated to produce any result you want ... The argument that this infinite ensemble actually exists can be claimed to have a certain explanatory economy (Tegmark 1993), although others would claim that Occam's razor has been completely abandoned in favour of a profligate excess of existential multiplicity, extravagantly hypothesized in order to explain the one universe that we do know exists.

The way out of this foundational crisis, as with previous ones in mathematics and science, is to adopt a more constructive approach. In this way, one can combine the attractions of Tegmark's [96] *Mathematical Universe Hypothesis* (MUH) with the discipline one gets from the mathematics of what can be built from very small beginnings.

### 5.8. Back in the One World ...

A constructive approach is not only a key to clarifying the interpretive problem. Eliminating the redundancy of parallel universes, and the reliance on the anthropic principle, also entails the tackling of the unsatisfactory arbitrariness of various aspects of the standard model. The exact values of the constants of nature, subatomic structure, the geometry of space –

all confront the standard model of particle physics with a foundational problem. Alan Guth, inventor of the 'cosmic inflation' needed to make sense of our picture of the early universe, asks [48]:

> If the creation of the universe can be described as a quantum process, we would be left with one deep mystery of existence: What is it that determined the laws of physics?

And Peter Woit, in his recent book [106] *Not Even Wrong – The Failure of String Theory and the Continuing Challenge to Unify the Laws of Physics*, comments on the arbitrary constants one needs to give the right values to get the standard model to behave properly:

> One way of thinking about what is unsatisfactory about the standard model is that it leaves seventeen non-trivial numbers still to be explained, ...

Even though the exact number of constants undetermined by theory, but needing special fine-tuning to make the standard model fit with observation, does vary, even one is too many. This dissatisfaction with aspects of the standard model goes back to Einstein. Quoting from Einstein's *Autobiographical Notes* [39, p.63]:

> ... I would like to state a theorem which at present can not be based upon anything more than upon a faith in the simplicity, i.e. intelligibility, of nature ... nature is so constituted that it is possible logically to lay down such strongly determined laws that within these laws only rationally completely determined constants occur (not constants, therefore, whose numerical value could be changed without destroying the theory) ...

What is needed is mathematics which does more than express mechanistic relationships between basic entities. One needs theory expressed in language strong enough to encapsulate not just relations on the material world, but relations on such relations – relations which entail qualifications sophisticated enough to determine all aspects of the our universe, including the laws of nature themselves. Or, as Roger Penrose terms it [70, pp. 106–107], we need to capture *Strong Determinism*, whereby:

> ... all the complication, variety and apparent randomness that we see all about us, as well as the precise physical laws, are all exact and unambiguous consequences of one single coherent mathematical structure.

The article [13] of Calude, Campbell, Svozil and Stefanescu on *Strong determinism vs. computability* contains a useful discussion of the computability-theoretic ramifications of strong determinism.

In the next section we examine some more approachable phenomena than those at the quantum level. Even though the challenge these present to Davis' Thesis is less obvious than that of quantum uncertainty, they do point us in the direction of the mathematics needed to make sense of strong determinism.

## 5.9. The Challenge from Emergence

The waves on the seashore, the clouds scudding across the sky, the complexity of the Mandelbrot set – observing these, one is made aware of limits on what we can practically compute. The underlying rules governing them are known, but that is not enough. When we talk about the problem of 'seeing the wood for the trees' we are approaching the gap between micro and macro events from another direction. Either way, there are commonly encountered situations in which either reduction, or seeing the 'big picture', entails more than a computation.

Although an interest in such things goes back to Poincaré – we already mentioned the 3-body problem – it was the second half of the twentieth century saw the growth of chaos theory, and a greater of awareness of the generation of informational complexity via simple rules, accompanied by the *emergence* of new regularities. The most mundane and apparently uncomplicated situations could provide examples, such as Robert Shaw's [87] strange attractor arising from an appropriately paced dripping tap. And inhospitable as turbulent fluids might appear, there too higher order formations might emerge and be subject to mathematical description, as demonstrated by David Ruelle (see Ruelle [85]) another early pioneer in the area. Schematic metaphors for such examples are provided by the cellular automaton (CA) model, and famously by John Conway's Game of Life. Here is the musician Brian Eno [41] talking in relation to how his creative work on 'generative music' was influenced by 'Life':

> These are terribly simple rules and you would think it probably couldn't produce anything very interesting. Conway spent apparently about a year finessing these simple rules. ...He found that those were all the rules you needed to produce something that appeared life-like.
>
> What I have over here, if you can now go to this Mac computer, please. I have a little group of live squares up there. When I hit go I hope they are going to start behaving according to those rules. There they go. I'm sure a lot of you have seen this before. What's interesting about this is that so much happens. The rules are very, very simple, but this little population here will reconfigure itself, form beautiful patterns, collapse,

> open up again, do all sorts of things. It will have little pieces that wander around, like this one over here. Little things that never stop blinking, like these ones. What is very interesting is that this is extremely sensitive to the conditions in which you started. If I had drawn it one dot different it would have had a totally different history. This is I think counter-intuitive. One's intuition doesn't lead you to believe that something like this would happen.

Margaret Boden and Ernest Edmonds [7] make a case for generative art, emergent from automata-like computer environments, really qualifying as art. While computer pioneer Konrad Zuse was impressed enough by the potentialities of cellular automata to suggest [107] that the physics of the universe might be CA computable.

An especially useful key to a general mathematical understanding of such phenomena is the well-known link between emergent structures in nature, and familiar mathematical objects, such as the Mandelbrot and Julia sets. These mathematical metaphors for real-world complexity and associated patterns have caught the attention of many – such as Stephen Smale [6] and Roger Penrose – as a way of getting a better grip on the computability/complexity of emergent phenomena. Here is Penrose [71] describing his fascination with the Mandelbrot set:

> Now we witnessed ... a certain extraordinarily complicated looking set, namely the Mandelbrot set. Although the rules which provide its definition are surprisingly simple, the set itself exhibits an endless variety of highly elaborate structures.

As a mathematical analogue of emergence in nature, what are the distinctive mathematical characteristics of the Mandelbrot set? It is derived from a simple polynomial formula over the complex numbers, via the addition of a couple of quantifiers. In fact, with a little extra work, the quantifiers can be reduced to just one. This gives the definition the aspect of a familiar object from classical computability theory – namely, a $\Pi^0_1$ set. Which is just the level at which we might not be surprised to encounter incomputability. But we have the added complication of working with real (via complex) numbers rather than just the natural numbers. This creates room for a certain amount of controversy around the use of the BSS model of real computation (see Blum, Cucker, Shub and Smale [6]) to show the incomputability of the Mandelbrot set and most Julia sets. The 2009 book by Mark Braverman and Michael Yampolsky [9] on *Computability of Julia Sets* is a reliable guide to recent results in the area, including those using the more mainstream computable analysis model of real computation. The situation is not simple, and the computability of the Mandelbrot set, as of now, is still an open question.

What is useful, in this context, is that these examples both connect with emergence in nature, and share logical form with well-known objects which transcend the standard Turing model. As such, they point to the role of extended language in a real context taking us beyond models which are purely mechanistic. And hence give us a route to mathematically capturing the origins of emergence in nature, and to extending our understanding of how nature computes. We can now view the halting set of a universal Turing machine as an emergent phenomenon, despite it not being as pretty visually as our Mandelbrot and Julia examples.

One might object that there is no evidence that quantifiers and other globally defined operations have any existence in nature beyond the minds of logicians. But how does nature know anything about any logical construct? The basic logical operations derive their basic status from their association with elementary algorithmic relationships over information. Conjunction signifies an appropriate and very simple merging of information, of the kind commonly occurring in nature. Existential quantification expresses projection, analogous to a natural object throwing a shadow on a bright sunny day. And if a determined supporter of Davis' Thesis plays at God, and isolates a computational environment with the aim of bringing it within the Turing model, then the result is the delivery of an identity to that environment, the creating a natural entity – like a human being, perhaps – with undeniable naturally emergent global attributes.

There are earlier, less schematic approaches to the mathematics of emergence, ones which fit well with the picture so far.

It often happens that when one gets interested in a particular aspect of computability, one finds Alan Turing was there before us. Back in the 1950s, Turing proposed a simple reaction–diffusion system describing chemical reactions and diffusion to account for morphogenesis, i.e. the development of form and shape in biological systems. One can find a full account of the background to Turing's seminal intervention in the field at Jonathan Swinton's well-documented webpage [95] on *Alan Turing and morphogenesis*. One of Turing's main achievements was to come up with mathematical descriptions – differential equations – governing such phenomena as Fibonacci phyllotaxis: the surprising showing of Fibonacci progressions in such things as the criss-crossing spirals of a sunflower head. As Jonathan Swinton describes:

> In his reaction-diffusion system [Turing] had the first and one of the most compelling models mathematical biology has devised for the creation process. In his formulation of the Hypothesis of Geometrical Phyllotaxis he expressed simple rules adequate for the appearance of Fibonacci pattern. In his last, unfinished work he was searching for plausible reasons why those rules might hold, and it seems only in this that he did not succeed. It would take many decades before others, unaware of his full progress, would retrace his steps and finally pass them in pursuit of a rather beautiful theory.

Most of Turing's work in this area was unpublished in his lifetime, only appearing in 1992 in the *Collected Works* [103]. Later work, coming to fruition just after Turing died, was carried forward by his student Bernard Richards, appearing in his thesis [79]. See Richards [80] for an interesting account of Richards' time working with Turing.

The field of *synergetics*, founded by the German physicist Hermann Haken, provides another mathematical approach to emergence. Synergetics is a multi-disciplinary approach to the study of the origins and evolution of macroscopic patterns and spacio-temporal structures in interactive systems. An important feature of synergetics for our purposes is its focus on *self-organisational* processes in science and the humanities, particularly that of autopoiesis. An instance of an autopoietic system is a biological cell, and is distinguished by being sufficiently autonomous and operationally closed, to recognisably self-reproduce.

A particularly celebrated example of the technical effectiveness of the theory is Ilya Prigogine's achievement of the Nobel Prize for Chemistry in 1977 for his development of dissipative structure theory and its application to thermodynamic systems far from equilibrium, with subsequent consequences for self-organising systems. Nonlinearity and irreversibility are associated key aspects of the processes modelled in this context.

See Michael Bushev's comprehensive review of the field in his book [12] *Synergetics – Chaos, Order, Self-Organization.* Klaus Mainzer's book [61] on *Thinking in Complexity: The Computational Dynamics of Matter, Mind, and Mankind* puts synergetics in a wider context, and mentions such things as synergetic computers.

The emphasis of the synergetists on *self-organisation* in relation to the emergence of order from chaos is important in switching attention from the *surprise* highlighted by so many accounts of emergence, to the *autonomy* and *internal organisation* intrinsic to the phenomenon. People like Prigogine found within synergetics, as did Turing for morphogenesis, precise descriptions of previously mysteriously emergent order.

## 5.10. A Test for Emergence

There is a problem with the big claims made for emergence in many different contexts. Which is that, like with 'life', nobody has a good definition of it. Sometimes, this does matter. Apart from which, history is littered with instances of vague concepts clarified by science, with huge benefits to our understanding of the world and to the progress of science and technology. The clarification of what we mean by a *computation*, and the subsequent development of the computer and computer science is a specially relevant example here. Ronald C. Arkin, in his book [3, p.105] *Behaviour-Based Robotics*, summarises the problem as it relates to emergence:

> Emergence is often invoked in an almost mystical sense regarding the capabilities of behavior-based systems. Emergent behavior implies a holistic capability where the sum is considerably greater than its parts. It is true that what occurs in a behavior-based system is often a surprise to the system's designer, but does the surprise come because of a shortcoming of the analysis of the constituent behavioral building blocks and their coordination, or because of something else?

There is a salutary warning from the history of British Emergentists, who had their heyday in the early 1920s – Brian McLaughlin's book [64]. The notion of emergence has been found to be a useful concept from at least the time of John Stuart Mill, back in the nineteenth century. The emergentists of the 1920s used the concept to explain the irreducibility of the 'special sciences', postulating a hierarchy with physics at the bottom, followed by chemistry, biology, social science etc. The emergence was seen, anticipating modern thinking, as being irreversible, imposing the irreducibility of say biology to quantum theory. Of course the British emergentists experienced their heyday before the great quantum discoveries of the late 1920s, and as described in McLaughlin [64], this was in a sense their undoing. One of the leading figures of the movement was the Cambridge philosopher C. D. Broad, described by Graham Farmelo in his biography of Paul Dirac [43, p.39] as being, in 1920, 'one of the most talented young philosophers working in Britain'. In many ways a precursor of the current philosophers arguing for the explanatory role of emergence in the philosophy of mind, Charlie Broad was alive to the latest scientific developments, lecturing to the young Paul Dirac on Einstein's new theory of relativity while they were both at Bristol. But here is Broad writing in 1925 [10, p.59] about the 'emergence' of salt crystals:

> ... the characteristic behaviour of the whole ... could not, even in theory, be deduced from the most complete knowledge of the behaviour of its components ... This ... is what I understand by the 'Theory of Emergence'. I cannot give a conclusive example of it, since it is a matter of controversy whether it actually applies to anything ... I will merely remark that, so far as I know at present, the characteristic behaviour of Common Salt cannot be deduced from the most complete knowledge of the properties of Sodium in isolation; or of Chlorine in isolation; or of other compounds of Sodium, ...

The date 1925 is significant of course. It was in the years following that Dirac and others developed the quantum mechanics which would explain much of chemistry in terms of locally described interactions between subatomic particles. The reputation of the emergentists, for whom such examples had been basic to their argument for the far-reaching relevance of emergence, never quite recovered.

For Ronald, Sipper and Capcarrère in 1999, Turing's approach to pinning down intelligence in machines suggested a test for emergence. Part of the thinking would have been that emergence, like intelligence, is something we as observers think we can recognise; while the complexity of what we are looking for resists observer-independent analysis. The lesson is to police the observer's evaluation process, laying down some optimal rules for a human observer. Of course, the Turing Test is specially appropriate to its task, our own experience of human intelligence making us well qualified to evaluate the putative machine version. Anyway, the Emergence Test of Ronald, Sipper and Capcarrère [83] for emergence being present in a system, modelled on the Turing Test, had the following three ingredients:

(1) **Design:** The system has been constructed by the designer, by describing local elementary interactions between components (e.g. artificial creatures and elements of the environment) in a language $\mathfrak{L}_1$.
(2) **Observation:** The observer is fully aware of the design, but describes global behaviours and properties of the running system, over a period of time, using a language $\mathfrak{L}_2$.
(3) **Surprise:** The language of design $\mathfrak{L}_1$ and the language of observation $\mathfrak{L}_2$ are distinct, and the causal link between the elementary interactions programmed in $\mathfrak{L}_1$ and the behaviours observed in $\mathfrak{L}_2$ is non-obvious to the observer – who therefore experiences surprise. In other words, there is a cognitive dissonance between the observer's mental image of

the system's design stated in $\mathfrak{L}_1$ and his contemporaneous observation of the system's behaviour stated in $\mathfrak{L}_2$.

Much of what we have here is what one would expect, extracting the basic elements of the previous discussion, and expressing it from the point of view of the assumed observer. But an ingredient which should be noted is the formal distinction between the language $\mathfrak{L}_1$ of the design and that of the observer, namely $\mathfrak{L}_2$. This fits in with our earlier mathematical examples: the halting set of a universal Turing machine, and the Mandelbrot set, where the new language is got by adding a quantifier – far from a minor augmentation of the language, as any logician knows. And it points to the importance of the language used to describe the phenomena, an emphasis underying the next section.

## 5.11. Definability the Key Concept

We have noticed that it is often possible to get descriptions of emergent properties in terms of the elementary actions from which they arise. For example, this is what Turing did for the role of Fibonacci numbers in relation to the sunflower etc. This is not unexpected, it is characteristic of what science does. And in mathematics, it is well known that complicated descriptions may take us beyond what is computable. This could be seen as a potential source of surprise in emergence.

But one can turn this viewpoint around, and get something more basic. There is an intuition that entities do not just generate descriptions of the rules governing them: they actually *exist* because of, and according to mathematical laws. And that for entities that we can be aware of, these will be mathematical laws which are susceptible to description. That it is the describability that is key to their observability. But that the existence of such descriptions is not enough to ensure we can access them, even though they have algorithmic content which provides the stuff of observation.

It is hard to for one to say anything new. In this case Leibniz was there before us, essentially with his Principle of Sufficient Reason. According to Leibniz [60] in 1714:

> ... there can be found no fact that is true or existent, or any true proposition, without there being a sufficient reason for its being so and not otherwise, although we cannot know these reasons in most cases.

Taking this a little further – natural phenomena not only generate descriptions, but arise and derive form from them. And this connects with a useful abstraction – that of mathematical definability, or, more generally, invariance under the automorphisms of the appropriate structure. So giving precision to our experience of emergence as a potentially non-algorithmic determinant of events.

This is a familiar idea in the mathematical context. The relevance of definability for the real world is implicitly present in Hans Reichenbach's work [78] on the axiomatisation of relativity. It was, of course, Alfred Tarski who gave a precise logical form to the notion of definability. Since then logicians have worked within many different mathematical structures, succeeding in showing that different operations and relations are non-trivially definable, or in some cases undefinable, in terms of given features of the structure. Another familiar feature of mathematical structures is the relationship between definability within the structure and the decidability of its theory (see Marker [62]), giving substance to the intuition that knowledge of the world is so hard to capture, because so much can be observed and described. Tarski's proof of decidability of the real numbers, contrasting with the undecidability of arithmetic, fits with the fact that one cannot even define the integers in the structure of the real numbers.

Unfortunately, outside of logic, and certainly outside of mathematics, the usefulness of definability remains little understood. And the idea that features of the real world may actually be undefinable is, like that of incomputability, a recent and unassimilated addition to our way of looking at things.

At times, definability or its breakdown comes disguised within quite familiar phenomena. In science, particularly in basic physics, symmetries play an important role. One might be surprised at this, wondering where all these often beautiful and surprising symmetries come from. Maybe designed by some higher power? In the context of a mathematics in which undefinability and nontrivial automorphisms of mathematical structures is a common feature, such symmetries lose their unexpectedness. When Murray Gell-Mann demonstrated the relevance of SU(3) group symmetries to the quark model for classifying of elementary particles, it was based on lapses in definability of the strong nuclear force in relation to quarks of differing flavour. The automorphisms of which such symmetries are an expression give a clear route from fundamental mathematical structures and their automorphism groups to far-reaching macro-symmetries in nature. If one accepts that such basic attributes as *position* can be subject to failures of definability, one is close to restoring realism to various basic sub-atomic phenomena.

One further observation: identifying emergent phenomena with material expressions of definable relations suggests an accompanying *robustness* of such phenomena. One would expect the mathematical characterisation to strip away much of the mystery which has made emergence so attractive to theologically inclined philosophers of mind, such as Samuel Alexander [1, p.14]:

> The argument is that mind has certain specific characters to which there is or even can be no neural counterpart ... Mind is, according to our interpretation of the facts, an 'emergent' from life, and life an emergent from a lower physico-chemical level of existence.

And further [1, p.428]:

> In the hierarchy of qualities the next higher quality to the highest attained is deity. God is the whole universe engaged in process towards the emergence of this new quality, and religion is the sentiment in us that we are drawn towards him, and caught in the movement of the world to a higher level of existence.

In contrast, here is Martin Nowak, Director of the Program for Evolutionary Dynamics at Harvard University, writing in the collection [11] *What We Believe But Cannot Prove*, describing the sort of robustness we would expect:

> I believe the following aspects of evolution to be true, without knowing how to turn them into (respectable) research topics.
>
> Important steps in evolution are robust. Multicellularity evolved at least ten times. There are several independent origins of eusociality. There were a number of lineages leading from primates to humans. If our ancestors had not evolved language, somebody else would have.

What is meant by robustness here is that there is mathematical content which enables the process to be captured and moved between different platforms; though it says nothing about the relevance of embodiment or the viability of virtual machines hostable by canonical machines. We return to this later. On the other hand, it gives us a handle on representability of emergent phenomena, a key aspect of intelligent computation.

## 5.12. The Challenge of Modelling Mentality

Probably the toughest environment in which to road-test the general mathematical framework we have associated with emergence is that of human mental activity. What about the surprise ingredient of the Emergence Test?

Mathematical thinking provides an environment in which major ingredients – Turing called them intuition and ingenuity, others might call them creativity and reason – are easier to clearly separate. A classical source of information and analysis of such thinking is the French mathematician Jacques Hadamard's *The Psychology of Invention in the Mathematical Field* [49], based on personal accounts supplied by distinguished informants such as Poincaré, Einstein and Polya. Hadamard was particularly struck by Poincaré's thinking, including a 1908 address of his to the French Psychological Society in Paris on the topic of *Mathematical Creation.* Hadamard followed Poincaré and Einstein in giving an important role to unconscious thought processes, and their independence of the role of language and mechanical reasoning. This is Hadamard's account, built on that of Poincaré [73], of Poincaré's experience of struggling with a problem:

> At first Poincaré attacked [a problem] vainly for a fortnight, attempting to prove there could not be any such function ... [quoting Poincaré]:
>
> "Having reached Coutances, we entered an omnibus to go some place or other. At the moment when I put my foot on the step, the idea came to me, without anything in my former thoughts seeming to have paved the way for it ... I did not verify the idea ... I went on with a conversation already commenced, but I felt a perfect certainty. On my return to Caen, for conscience sake, I verified the result at my leisure."

This experience will be familiar to most research mathematicians – the period of incubation, the failure of systematic reasoning and the surprise element in the final discovery of the solution: a surprise that may, over a lifetime, lose some of its bite with repetition and familiarity, but which one is still compelled to recognise as being mysterious and worthy of surprise. Anyway, the important third part of the Emergence Test is satisfied here.

Perhaps even more striking is the fact that Poincaré's solution had that robustness we looked for earlier: the solution came packaged and mentally represented in a form which enabled it to be carried home and unpacked intact when back home. Poincaré just carried on with his conversation on the bus, his friend presumably unaware of the remarkable thoughts coursing through the mathematician's mind.

Another such incident emphasises the lack of uniqueness and the special character of such incidents – Jacques Hadamard [49] quoting Poincaré again:

> "Then I turned my attention to the study of some arithmetical questions apparently without much success ... Disgusted with my failure, I went to spend a few days at the seaside and thought of something else. One morning, walking on the bluff, the idea came to me, with just the same characteristics of brevity, suddenness and immediate certainty, that the arithmetic transformations of indefinite ternary quadratic forms were identical with those of non-Euclidian geometry."

What about the *design*, and the observer's *awareness* of the design? Here we have a large body of work , most notably from neuro-scientists and philosophers, and an increasingly detailed knowledge of the workings of the brain. What remains in question – even accepting the brain as the design (not as simple as we would like!) – is the exact nature of the connection between the design and the emergent level of mental activity. This is an area where the philosophers pay an important role in clarifying problems and solutions, while working through consequences and consistencies.

The key notion, providing a kind of workspace for working through alternatives, is that of *supervenience.* According to Jaegwon Kim [53, pp.14–15], supervenience:

> ... represents the idea that mentality is at bottom physically based, and that there is no free-floating mentality unanchored in the physical nature of objects and events in which it is manifested.

There are various formulations. This one is from the online *Stanford Encyclopedia of Philosophy*:

> A set of properties **A** supervenes upon another set **B** just in case no two things can differ with respect to **A**-properties without also differing with respect to their **B**-properties.

So in this context, it is the mental properties which are thought to supervene on the neuro-physical properties. All we need to know is are the details of how this supervenience takes place. And what throws up difficulties is our own intimate experience of the outcomes of this supervenience.

One of the main problems relating to supervenience is the so-called 'problem of mental causation', the old problem which undermined the Cartesian conception of mind–body dualism. The persistent question is: *How can mentality have a causal role in a world that is fundamentally physical?* Another unavoidable problem is that of 'overdetermination' – the problem of phenomena having both mental and physical causes. For a pithy expression of the problem, here is Kim [54] again:

> . . . the problem of mental causation is solvable only if mentality is physically reducible; however, phenomenal consciousness resists physical reduction, putting its causal efficacy in peril.

It is not possible here, and not even useful, to go into the intricacies of the philosophical debates which rage on. But it is important to take on board the lesson that a crude mechanical connection between mental activity and the workings of the brain will not do the job. Mathematical modelling is needed to clarify the mess, but has to meet very tough demands.

## 5.13. Connectionist Models to the Rescue?

Synaptic interactions are basic to the workings of the brain, and connectionist models based on these are the first hope. And there is optimism about such models from such leading figures in the field as Paul Smolensky [90], recipient of the 2005 David E. Rumelhart Prize:

> There is a reasonable chance that connectionist models will lead to the development of new somewhat-general-purpose self-programming, massively parallel analog computers, and a new theory of analog parallel computation: they may possibly even challenge the strong construal of Church's Thesis as the claim that the class of well-defined computations is exhausted by those of Turing machines.

And it is true that connectionist models have come a long way since Turing's 1948 discussion [102] of 'unorganised machines', and McCulloch and Pitts' 1943 early paper [65] on neural nets. (Once again, Turing was there at the beginning, see Teuscher's book [97] on *Turing's Connectionism.*)

But is that all there is? For Steven Pinker [72] '. . . neural networks alone cannot do the job'. And focusing on our elusive higher functionality, and the way in which mental images are recycled and incorporated in new mental processes, he points to a 'kind of mental fecundity called recursion':

> We humans can take an entire proposition and give it a role in some larger proposition. Then we can take the larger proposition and embed it in a still-larger one. Not only did the baby eat the slug, but the father saw the baby eat the slug, and I wonder whether the father saw the baby eat the slug, the father knows that I wonder whether he saw the baby eat the slug, and I can guess that the father knows that I wonder whether he saw the baby eat the slug, and so on.

Is this really something new? Neural nets can handle recursions of various kinds. They can exhibit imaging and representational capabilities. They can learn. The problem seems to be with modelling the holistic aspects of brain functionalism. It is hard to envisage a model at the level of neural networks which successfully represent and communicate its own global informational structures. Neural nets do have many of the basic ingredients of what one observes in brain functionality, but the level of developed synergy of the ingredients one finds in the brain does seem to occupy a different world. There seems to be a dependency on an evolved embodiment which goes against the classical universal machine paradigm. We develop these comments in more detail later in this section.

For the mathematician, definability is the key to representation. As previously mentioned, the language functions by representing basic modes of using the informational content of the structure over which the language is being interpreted. Very basic language corresponds to classical computational relationships, and is local in import. If we extend the language, for instance, by allowing quantification, it still conveys information about an algorithmic procedure for accessing information. The new element is that the information accessed may now be emergent, spread across a range of regions of the organism, its representation very much dependent on the material embodiment and with the information accessed via finitary computational procedures which also depend on the particular embodiment. One can observe this preoccupation with the details of the embodiment in the work of the neuro-scientist Antonio Damasio. One sees this in the following description from Damasio's book, *The Feeling Of What Happens*, of the kind of mental recursions Steven Pinker was referring to above [25, p.170]:

> As the brain forms images of an object – such as a face, a melody, a toothache, the memory of an event – and as the images of the object affect the state of the organism, yet another level of brain structure creates a swift nonverbal account of the events that are taking place in the varied brain regions activated as a consequence of the object-organism interaction. The mapping of the object-related consequences occurs in first-order neural maps representing the proto-self and object; the account of the causal relationship between object and organism can only be captured in second-order neural maps. ...one might say that the swift, second-order nonverbal account narrates a story: that of the organism caught in the act of representing its own changing state as it goes about representing something else.

Here we see the pointers to the elements working against the classical independence of the computational content from its material host. We may have a mathematical precision to the presentation of the process. But the presentation of the basic information has to deal with emergence of a possibly incomputable mathematical character, and so has to be dependent on the material instantiation. And the classical computation relative to such information, implicit in the quotations from Pinker and Damasio, will need to work relative to these material instantiations. The mathematics sets up a precise and enabling filing system, telling the brain how to work hierarchically through emergent informational levels, within an architecture evolved over millions of years.

There is some recognition of this scenario in the current interest in the evolution of hardware – see, for example, Hornby, Sekanina and Haddow [52]. We tend to agree with Steven Rose [84]:

> Computers are designed, minds have evolved. Deep Blue could beat Kasparov at a game demanding cognitive strategies, but ask it to escape from a predator, find food or a mate, and negotiate the complex interactions of social life outside the chessboard or express emotion when it lost a game, and it couldn't even leave the launchpad. Yet these are the skills that human survival depends on, the products of 3bn years of trial-and-error evolution.

From a computer scientist's perspective, we are grappling with the design of a *cyber-physical system* (CPS). And as Edward Lee from Berkeley describes [59]:

> To realize the full potential of CPS, we will have to rebuild computing and networking abstractions. These abstractions will have to embrace physical dynamics and computation in a unified way.

In Lee [58], he argues for 'a new systems science that is jointly physical and computational'.

Within such a context, connectionist models with their close relationship to synaptic interactions, and availability for ad hoc experimentation, do seem to have a useful role. But there are good reasons for looking for a more fundamental mathematical model with which to express the 'design' on which to base a definable emergence. The chief reason is the need for a general enough mathematical framework, capable of housing different computationally complex frameworks. Although the human brain is an important example, it is but one part of a rich and heterogeneous computational universe, reflecting in its workings many elements of that larger context. The history of mathematics has led us to look for abstractions which capture a range of related structures, and which are capable of theoretical development informed by intuitions from different sources, which become applicable in many different situations. And which provide basic understanding to take us beyond the particularities of individual examples.

## 5.14. Definability in What Structure?

In looking for the mathematics to express the design, we need to take account of the needs of physics as well as those of mentality or biology. In his *The Trouble With Physics* [91], Lee Smolin points to a number of deficiencies of the standard model, and also of popular proposals such as those of string theory for filling its gaps. And in successfully modelling the physical universe, Smolin declares [91, p.241]:

> ... *causality itself is fundamental.*

Referring to 'early champions of the role of causality' such as Roger Penrose, Rafael Sorkin (the inventor of causal sets), Fay Dowker and Fotini Markopoulou, Smolin goes on to explain [91, p.242]:

> It is not only the case that the spacetime geometry determines what the causal relations are. This can be turned around: Causal relations can determine the spacetime geometry ...
>
> It's easy to talk about space or spacetime emerging from something more fundamental, but those who have tried to develop the idea have found it difficult to realize in practice. ... We now believe they failed because they ignored the role that causality plays in spacetime. These days, many of us working on quantum gravity believe that causality itself is fundamental – and is thus meaningful even at a level where the notion of space has disappeared.

So, when we have translated 'causality' into something meaningful, and the model based on it has been put in place – the hoped-for prize is a theory in which even the background character of the universe is determined by its own basic structure. In such a scenario, not only would one be able to do away with the need for exotic multiverse proposals, patched with inflationary theories and anthropic metaphysics. But, for instance, one can describe a proper basis for the variation of natural laws near a mathematical singularity, and so provide a mathematical foundation for the reinstatement of the philosophically more satisfying cyclical universe as an alternative to the inflationary Big Bang hypothesis – see Paul Steinhardt and Neil Turok's book [94] for a well worked out proposal based on superstring theory.

## 5.15. The Turing Landscape, Causality and Emergence ...

If there is one field in which 'causality' can be said to be fundamental, it is that of computability. Although the sooner we can translate the term into something more precise, the better. 'Causality', despite its everyday usefulness, on closer inspection is fraught with difficulties, as John Earman [37, p.5] nicely points out:

> ...the most venerable of all the philosophical definitions [of determinism] holds that the world is deterministic just in case every event has a cause. The most immediate objection to this approach is that it seeks to explain a vague concept – determinism – in terms of a truly obscure one – causation.

Historically, one recognised the presence of a causal relationship when a clear mechanical interaction was observed. But Earman's book makes us aware of the subtleties beyond this at all stages of history. The success of science in revealing such interactions underlying mathematically signalled causality – even for Newton's gravitational 'action at a distance' – has encouraged us to think in terms of mathematical relationships being the essence of causality. Philosophically problematic as this may be in general, there are enough mathematical accompaniments to basic laws of nature to enable us to extract a suitably general mathematical model of physical causality, and to use this to improve our understanding of more complicated (apparent) causal relationships. The classical paradigm is still Isaac Newton's formulation of a mathematically complete formulation of his laws of motion, sufficient to predict an impressive range of planetary motions.

Schematically, logicians at least have no problem representing Newtonian transitions between mathematically well-defined states of a pair of particles at different times as the Turing reduction of one real to another, via a partial computable (p.c.) functional describing what Newton said would happen to the pair of particles. The functional expresses the computational and continuous nature of the transition. One can successfully use the functional to approximate, to any degree of accuracy, a particular transition.

This type of model, using partial computable functionals extracted from Turing's [101] notion of oracle Turing machine, is very generally applicable to basic laws of nature. However, it is well known that instances of a basic law can be composed so as to get much more problematic mathematical relationships, relationships which have a claim to be causal. We have mentioned cases above – for instance those related to the 3-body problem, or strange attractors emergent from complex confluences of applications of basic laws. See recent work Beggs, Costa, Loff and Tucker [4], Beggs and Tucker [5] concerning the modelling of physical interactions as computation relative to oracles, and incomputability from mathematical thought experiments based on Newtonian laws.

The technical details of the extended Turing model, providing a model of computable content of structures based on p.c. functionals over the reals, can be found in Cooper [19]. One can also find there details of how Emil Post [75] used this model to define the *degrees of unsolvability* – now known as the *Turing degrees* – as a classification of reals in terms of their relative computability. The resulting structure has turned out to be a very rich one, with a high degree of structural pathology. At a time when primarily mathematical motivations dominated the field – known for many years as a branch of mathematical logic called *recursive function theory* – this pathology was something of a disappointment. Subsequently, as we see below, this pathology became the basis of a powerful expressive language, delivering a the sort of richness of definable relations which qualify the structure for an important real-world modelling role.

Dominant as this Turing model is, and widely accepted to have a canonical role, there are more general types of relative computation. Classically, allowing non-deterministic Turing computations relative to well-behaved oracles gives one nothing new. But in the real world one often has to cope with data which is imperfect, or provided in real time, with delivery of computations required in real time. There is an argument that the corresponding generalisation is the 'real' relative computability. There are equivalent formalisations – in terms of *enumeration reducibility* between sets of data, due to Friedberg and Rogers [45], or (see Myhill [66]), in terms of *relative computability of partial functions* (extending earlier notions of Kleene and Davis). The corresponding extended structure provides an interesting and informative context for the better known Turing degrees – see, for example, Soskova and Cooper [93]. The Bulgarian research school, including D. Skordev, I. Soskov, A. Soskova, A. Ditchev, H. Ganchev, M. Soskova and others has played a special role in the development of the research area.

The universe we would like to model is one in which we can describe global relations in terms of local structure – so capturing the emergence of large-scale formations, and giving formal content to the intuition that such emergent higher structures 'supervene' on the computationally more basic local relationships.

Mathematically, there appears to be strong explanatory power in the formal modelling of this scenario as definability over a structure based on reducibilities closely allied to Turing functionals: or more generally, freeing the model from an explicit dependence on language, as invariance under automorphisms of the Turing structure. In the next section, we focus on the standard Turing model, although the evidence is that similar outcomes would be provided by the related models we have mentioned.

## 5.16. An Informational Universe, and Hartley Rogers' Programme

Back in 1967, the same year that Hartley Rogers' influential book *Theory of Recursive Functions and Effective Computability* appeared, a paper [81], based on an earlier talk of Rogers, appeared in the proceedings volume of the 1965 Logic Colloquium in Leicester. This short article initiated a research agenda which has held and increased its interest over a more than 40 year period. Essentially, *Hartley Rogers' Programme* concerns the fundamental problem of *characterising the Turing invariant relations.*

The intuition is that these invariant relations are key to pinning down how basic laws and entities emerge as mathematical constraints on causal structure: where the richness of the Turing structure discovered so far becomes the raw material for a multitude of non-trivially definable relations. There is an interesting relationship here between the mathematics and the use of the anthropic principle in physics to explain why the universe is as it is. It is well known that the development of the complex development we see around us is dependent on a subtle balance of natural laws and associated constants. One would like the mathematics to explain why this balance is more than an accidental feature of one of a multitude, perhaps infinitely many, randomly occurring universes. What the Turing universe delivers is a rich infrastructure of invariant relations, providing a basis for a correspondingly rich material instantiation, complete with emergent laws and constants, a provision of strong determinism, and a globally originating causality equipped with non-localism – though all in a very schematic framework. Of course, echoing Smolin, it is the underlying scheme that is currently missing. We have a lot of detailed information, but the skeleton holding it all together is absent.

However, the computability theorists have their own 'skeleton in the cupboard'. The modelling potential of the extended Turing model depends on it giving some explanation of such well-established features as quantum uncertainty, and certain experimentally verified uncertainties relating to human mentality. And there is a widely believed mathematical conjecture which would rob the Turing model of basic credentials for modelling observable uncertainty.

The *Bi-Interpretability Conjecture*, arising from Leo Harrington's familiarity with the model theoretic notion of bi-interpretability, can be roughly described as asserting that:

> *The Turing definable relations are exactly those with information content describable in second-order arithmetic.*

Moreover, given any description of information content in second-order arithmetic, one has a way of reading off the computability-theoretic definition in the Turing universe. Actually, a full statement of the conjecture would be in terms of 'interpreting' one structure in another, a kind of poor-man's isomorphism. Seminal work on formalising the global version of the conjecture, and proving partial versions of it complete with key consequences and equivalences, were due to Theodore Slaman and Hugh Woodin. See Slaman's 1990 International Congress of Mathematicians article [88] for a still-useful introduction to the conjecture and its associated research project.

An unfortunate consequence of the conjecture being confirmed would be the well-known rigidity of the structure second-order arithmetic being carried over to the Turing universe. The breakdown of definability we see in the real world would lose its model. However, work over the years makes this increasingly unlikely.

See Nies, Shore and Slaman [68] for further development of the requisite coding techniques in the local context, with the establishment of a number of local definability results. See Cooper [18, 22] for work in the other direction, both at the global and local levels. What is so promising here is the likelihood of the final establishment of a subtle balance between invariance and non-invariance, with the sort of non-trivial automorphisms needed to deliver a credible basis for the various symmetries, and uncertainties peculiar to mentality and basic physics: along with the provision via partial versions of bi-interpretability of an appropriate model for the emergence of the more reassuring 'quasi-classical' world from out of quantum uncertainty, and of other far-reaching consequences bringing such philosophical concepts as epistemological relativism under a better level of control.

**To summarise:** What we propose is that this most cartesian of research areas, classical computability theory, regain the real-world significance it was born out of in the 1930s. And that it structure the informational world of science in a radical and revealing way. The main features of this informational world, and its modelling of the basic causal structure of the universe would be:

- A universe described in terms of reals ...
- With basic natural laws modelled by computable relations between reals.
- Emergence described in terms of definability/invariance over the resulting structure ...
- With failures of definable information content modelling mental phenomena, quantum ambiguity, etc. ...
- Which gives rise to new levels of computable structure ...
- And a familiarly fragmented scientific enterprise.

As an illustration of the explanatory power of the model, we return to the problem of mental causation. Here is William Hasker, writing in *The Emergent Self* [50, p. 175], and trying to reconcile the automomy of the different levels:

> The "levels" involved are levels of organisation and integration, and the downward influence means that the behaviour of "lower" levels – that is, of the components of which the "higher-level" structure consists – is different than it would otherwise be, because of the influence of the new property that emerges in consequence of the higher-level organization.

The mathematical model, making perfect sense of this, treats the brain and its emergent mentality as an organic whole. In so doing, it replaces the simple everyday picture of what a causal relationship is with a more subtle confluence of mathematical relationships. Within this confluence, one may for different purposes or necessities adopt different assessment of what the relevant causal relationships are. For us, thinking about this article, we regard the mentality hosting our thoughts to provide the significant causal structure. Though we know full well that all this mental activity is emergent from an autonomous brain, modelled with some validity via a neural network.

So one might regard causality as a misleading concept in this context. Recognisable 'causality' occurs at different levels of the model, connected by relative definability. And the causality at different levels in the form of relations with identifiable algorithmic content, this content at higher levels being emergent. The diverse levels form a unity, with the 'causal' structure observed at one level reflected at other levels – with the possibility of non-algorithmic 'feedback' between levels. The incomputability involved in the transition between levels makes the supervenience involved have a non-reductive character.

## References

[1] S. Alexander, *Space, Time, and Deity: Gifford Lectures at Glasgow, 1916–1918.* vol. 2, Macmillan, London, (1920).
[2] H. Andréka, J. X. Madarász, I. Németi and G. Székely, Axiomatizing relativistic dynamics without conservation postulates, *Studia Logica.* **89**, 163–186, (2008).
[3] R. C. Arkin, *Behaviour-Based Robotics.* MIT Press, (1998).
[4] E. J. Beggs, J. F. Costa, B. Loff and J. V. Tucker, Computational complexity with experiments as oracles, *Proc. R. Soc. Ser. A.* pp. 2777–2801, (2008).
[5] E. J. Beggs and J. V. Tucker, Experimental computation of real numbers by newtonian machines, *Proc. R. Soc. Ser. A.* pp. 1541–1561, (2007).
[6] L. Blum, F. Cucker, M. Shub and S. Smale, *Complexity and Real Computation.* Springer, (1997).
[7] M. A. Boden and E. Edmonds, What is generative art?, *Digital Creativity.* **20**, 21–46, (2009).
[8] L. Bombelli, J. Lee, D. Meyer and R. D. Sorkin, Spacetime as a causal set, *Phys. Rev. Lett.* **59**, 521–524, (1987).
[9] M. Braverman and M. Yampolsky, *Computability of Julia Sets.* vol. 23, *Algorithms and Computation in Mathematics*, Springer, (2009).
[10] C. D. Broad, *The Mind and its Place in Nature.* Kegan-Paul, London, New York, (1925).
[11] J. Brockman, Ed., *What We Believe but Cannot Prove: Today's Leading Thinkers on Science in the Age of Certainty.* Harper Perennial, New York, (2006).
[12] M. Bushev, *Synergetics – Chaos, Order, Self-Organization.* World Scientific, Singapore, (1994).
[13] C. Calude, D. I. Campbell, K. Svozil, and D. Stefanescu. Strong determinism vs. computability. In eds. W. Depauli-Schimanovich, E. Koehler and F. Stadler, *The Foundational Debate, Complexity and Constructivity in Mathematics and Physics*, pp. 115–131. Kluwer, Dordrecht, (1995).
[14] C. S. Calude and K. Svozil, Quantum randomness and value indefiniteness, *Advanced Science Letters.* **1**, 165–168, (2008).
[15] G. J. Chaitin, *Algorithmic Information Theory.* Cambridge University Press, Cambridge, (1987).

[16] A. C. Clarke. Hazards of prophecy: The failure of imagination. In *Profiles of the Future.* Gollancz, London, (1962).
[17] S. B. Cooper. Clockwork or Turing U/universe? – Remarks on causal determinism and computability. In eds. S. B. Cooper and J. K. Truss, *Models and Computability,* vol. 259, *London Mathematical Society Lecture Notes Series,* pp. 63–116. Cambridge University Press, Cambridge, New York, Melbourne, (1999).
[18] S. B. Cooper, Upper cones as automorphism bases, *Siberian Adv. Math.* **9**, 1–61, (1999).
[19] S. B. Cooper, *Computability Theory.* Chapman & Hall/CRC, Boca Raton, London, New York, Washington, (2004).
[20] S. B. Cooper, Definability as hypercomputational effect, *Appl. Math. Comput.* pp. 72–82, (2006).
[21] S. B. Cooper. Computability and emergence. In eds. D. M. Gabbay, S. Goncharov and M. Zakharyaschev, *Mathematical Problems from Applied Logic I. Logics for the XXIst Century,* vol. 4, *Springer International Mathematical Series,* pp. 193–231. Springer, New York, (2006).
[22] S. B. Cooper. The limits of local Turing definability. In preparation.
[23] S. B. Cooper and P. Odifreddi. Incomputability in nature. In eds. S. Cooper and S. Goncharov, *Computability and Models,* pp. 137–160. Kluwer Academic/Plenum, New York, Boston, Dordrecht, London, Moscow, (2003).
[24] P. Cotogno, A brief critique of pure hypercomputation, *Mind and Machines.* **19**, 391–405, (2009).
[25] A. R. Damasio, *The Feeling of What Happens: Body and Emotion in the Making of Consciousness.* Harcourt Brace, New York, (1999).
[26] M. Davis, Arithmetical problems and recursively enumerable predicates. (abstract), *J. Symbolic Logic.* **15**(1), 77–78, (1950).
[27] M. Davis, Arithmetical problems and recursively enumerable predicates, *J. Symbolic Logic.* **18**(1), 33–41, (1953).
[28] M. Davis, Ed., *Solvability, Provability, Definability: The Collected Works of Emil L. Post.* Birkhäuser, Boston, Basel, Berlin, (1994).
[29] M. Davis, *The Universal Computer: The Road from Leibniz to Turing.* W. W. Norton, New York, London, (2000).
[30] M. Davis. The myth of hypercomputation. In ed. C. Teuscher, *Alan Turing: Life and Legacy of a Great Thinker,* pp. 195–211. Springer-Verlag, Berlin, Heidelberg, (2004).
[31] M. Davis, Why there is no such discipline as hypercomputation, *Appl. Math. Comput.* **178**, 4–7, (2006).
[32] R. Dawkins, *The God Delusion.* Bantam Press, London, (2006).
[33] D. Deutsch, Quantum Theory, the Church-Turing Principle and the Universal Quantum Computer, *Proc. R. Soc. Lond. Ser. A.* **400**, 97–117, (1985).
[34] D. Deutsch, *The Fabric of Reality.* Allen Lane, London, (1997).
[35] D. Deutsch. *Questions and answers with David Deutsch.* Available at `http:www.newscientist.com/article/dn10691-readers-q-amp-with-david-deutsch-.html`. [Accessed 21 December 2006].

[36] R. Downey and D. Hirschfeldt, *Algorithmic Randomness and Complexity.* Springer, New York, (2010).
[37] J. Earman, *A Primer On Determinism.* D. Reidel/Kluwer, Dordrecht, (1986).
[38] A. Einstein, *Out of My Later Years.* Philosophical Library, New York, (1950).
[39] A. Einstein. Autobiographical notes. In ed. P. Schilpp, *Albert Einstein: Philosopher-Scientist.* Open Court Publishing, Chicago, (1969).
[40] G. Ellis. The unique nature of cosmology. In eds. A. Ashtekar, R. S. Cohen, D. Howard, J. Renn and A. Shimony, *Revisiting the Foundations of Relativistic Physics: Festschrift in Honor of John Stachel.* Kluwer, Dordrecht, (1996).
[41] B. Eno, Generative music: evolving metaphors, in my opinion, is what artists do, *Motion Magazine* (7 July 1996).
[42] M. O. Everett, *Things the Grandchildren Should Know.* Little, Brown & Company, London, (2008).
[43] G. Farmelo, *The Strangest Man – The Hidden Life of Paul Dirac, Quantum Genius.* Faber & Faber, London, (2009).
[44] D. Friedan, A tentative theory of large distance physics, *J. High Energy Phys.* **10**, 063, (2003).
[45] R. M. Friedberg and H. Rogers, Jr., reducibility and completeness for sets of integers, *Z. Math. Logik Grundlag. Math.* **5**, 117–125, (1959).
[46] K. Gödel, Über formal unentscheidbare Sätze der Principia Mathematica und verwandter Systeme. I, *Monatsch Math. Phys.* **38**, 173–178, (1931). (English trans. in Davis [1965, 4–38], and in van Heijenoort [1967, 592–616]).
[47] K. Gödel, On undecidable propositions of formal mathematical systems. Notes by S. C. Kleene and J. B. Rosser on lectures at the Institute for Advanced Study, Princeton, New Jersey, pp. 30. (1934). (Reprinted in Davis [1965, pp. 39–74]).
[48] A. H. Guth, *The Inflationary Universe – The Quest for a New Theory of Cosmic Origins.* Addison-Wesley, Reading, (1997).
[49] J. Hadamard, *The Psychology of Invention in the Mathematical Field.* Princeton University Press, Princeton, (1945).
[50] W. Hasker, *The Emergent Self.* Cornell University Press, Ithaca, (1999).
[51] A. Hodges, *Alan Turing: The Enigma.* Vintage, London, Melbourne, Johannesburg, (1992).
[52] G. S. Hornby, L. Sekanina and P. C. Haddow. Evolvable systems: From biology to hardware. In *Proc. 8th International Conference on Evolvable Systems, ICES 2008, Prague, Czech Republic, September 21-24, 2008,* Springer, Berlin, (2008).
[53] J. Kim, *Mind in a Physical World.* MIT Press, Boston, (1998).
[54] J. Kim, *Physicalism, or Something Near Enough.* Princeton University Press, Princeton, (2005).
[55] G. Kreisel. Mathematical logic: What has it done for the philosophy of mathematics? In ed. R. Schoenman, *Bertrand Russell, Philosopher of the*

*Century.* Allen and Unwin, London, (1967).
[56] G. Kreisel. Church's Thesis: a kind of reducibility axiom for constructive mathematics. In eds. A. Kino, J. Myhill and R. E. Vesley, *Intuitionism and proof theory: Proceedings of the Summer Conference at Buffalo N.Y. 1968*, pp. 121–150. North-Holland, Amsterdam, London, (1970).
[57] T. S. Kuhn, *The Structure of Scientific Revolutions.* University of Chicago Press, Chicago, London, (1996), 3rd edition.
[58] E. A. Lee. Cyber-physical systems – are computing foundations adequate? Position Paper for NSF Workshop on *Cyber-Physical Systems: Research Motivation, Techniques and Roadmap, October 16–17, 2006, Austin, Texas.*
[59] E. A. Lee. Cyber physical systems: Design challenges. In *11th IEEE Symposium on Object Oriented Real-Time Distributed Computing (ISORC)*, pp. 363–369, (2008).
[60] G. W. Leibniz, *La Monadologie*, 1714. (English translation by G. M. Ross, Cambridge University Press, Cambridge, (1999)).
[61] K. Mainzer, *Thinking in Complexity: The Computational Dynamics of Matter, Mind, and Mankind.* Springer, (1994). 5th revised edn. 2007.
[62] D. Marker, Model theory and exponentiation, *Notices Amer. Math. Soc.* pp. 753–759, (1966).
[63] Y. Matiyasevich, *Hilbert's tenth problem.* MIT Press, Boston, (1993).
[64] B. P. McLaughlin. The Rise and Fall of British Emergentism. In eds. A. Beckermann, H. Flohr and J. Kim, *Emergence or Reduction? – Essays on the Prospects of Nonreductive Physicalism*, pp. 49–93. de Gruyter, Berlin, (1992). (Reprinted in *Emergence: Contemporary Readings in Philosophy and Science* (M. A. Bedau and P. Humphreys, eds.), MIT Press, Boston, London, 2008, pp. 19–59).
[65] W. McCulloch and W. Pitts, A logical calculus of the ideas immanent in nervous activity, *Bull. Math. Biophys.* **5**, 115–133, (1943).
[66] J. Myhill, A note on degrees of partial functions, *Proc. Amer. Math. Soc.* **12**, 519–521, (1961).
[67] A. Nies, *Computability and Randomness.* Oxford University Press, Oxford (2009).
[68] A. Nies, R. A. Shore and T. A. Slaman, Interpretability and definability in the recursively enumerable degrees, *Proc. London Math. Soc.* **77**(3), 241–291, (1998).
[69] P. Odifreddi, *Classical Recursion Theory.* North-Holland/Elsevier, Amsterdam, New York, Oxford, Tokyo, (1989).
[70] R. Penrose. Quantum physics and conscious thought. In eds. B. J. Hiley and F. D. Peat, *Quantum Implications: Essays in honour of David Bohm*, pp. 105–120. Routledge & Kegan Paul, London, New York, (1987).
[71] R. Penrose, *The Emperor's New Mind.* Oxford University Press, Oxford, (1994).
[72] S. Pinker, *How the Mind Works.* W.W. Norton, New York, (1997).
[73] H. Poincaré, *Science and method.* Dover Publications, New York, (1952).
[74] E. L. Post, Recursively enumerable sets of positive integers and their decision problems, *Bull. Amer. Math. Soc.* **50**, 461–494, (1944). (Reprinted

in E. L. Post, *Solvability, Provability, Definability: The Collected Works of Emil L. Post*, pp. 284–316).

[75] E. L. Post, Degrees of recursive unsolvability: preliminary report. (abstract), *Bull. Amer. Math. Soc.* **54**, 641–642, (1948).

[76] E. L. Post. Absolutely unsolvable problems and relatively undecidable propositions – account of an anticipation. In ed. Davis, *The Undecidable*, pp. 340–433. Raven Press, New York, (1965). (Reprinted in E. L. Post, *Solvability, Provability, Definability: The Collected Works of Emil L. Post*, pp. 375–441).

[77] H. Putnam, Trial and error predicates and the solution to a problem of mostowski, *J. Symbolic Logic.* **30**, 49–57, (1965).

[78] H. Reichenbach, *Axiomatik der relativistischen Raum-Zeit-Lehre.* 1924. (English Translation: *Axiomatization of the Theory of Relativity.* University of California Press, California, (1969).

[79] B. Richards. Morphogenesis of Radiolaria. MSc Thesis Manchester University, (1954).

[80] B. Richards, Turing, Richards and morphogenesis, *The Rutherford Journal.* **1**. Available at `http://www.rutherfordjournal.org/article010109.html` [Accessed December 2005].

[81] J. H. Rogers. Some problems of definability in recursive function theory. In ed. J. N. Crossley, *Sets, Models and Recursion Theory*, Proceedings of the Summer School in Mathematical Logic and Tenth Logic Colloquium Leicester, August–September, 1965, pp. 183–201.

[82] J. H. Rogers. Theory of recursive functions and effective computability. McGraw-Hill, New York, (1967). (Reprinted by MIT Press, Boston, London, 1987).

[83] E. M. A. Ronald, M. Sipper, and M. S. Capcarrère, Design, observation, surprise! a test of emergence, *Artif. Life.* **5**, 225–239, (1999).

[84] S. Rose. Guardian book review of *kluge: the haphazard construction of the human mind* by Gary Marcus, (31 May 2008).

[85] D. Ruelle, *Chance and Chaos.* Princeton University Press, Princeton, (1993).

[86] D. G. Saari and Z. Xia, Off to infinity in finite time, *Notices Amer. Math. Soc.* **5**, 538–546, (1995).

[87] R. Shaw, *The Dripping Faucet As a Model Chaotic System.* Science Frontier Express Series, Aerial Press, California, (1984).

[88] T. A. Slaman. Degree structures. In *Proc. Intl. Cong. Math., Kyoto 1990*, pp. 303–316, Springer-Verlag, Tokyo, (1991).

[89] A. Sloman and R. Chrisley, Virtual machines and consciousness, *Journal of Consciousness Studies.* **10**, 133–72, (2003).

[90] P. Smolensky, On the proper treatment of connectionism, *Behav. Brain Sci.* **11**, 1–74, (1988).

[91] L. Smolin, *The Trouble With Physics: The Rise of String Theory, the Fall of Science and What Comes Next.* Allen Lane/Houghton Mifflin, London, New York, (2006).

[92] R. I. Soare, *Recursively Enumerable Sets and Degrees.* Springer-Verlag, New

York, (1987).
[93] M. I. Soskova and S. B. Cooper, How enumeration reducibility yields extended harrington non-splitting, *J. Symbolic Logic.* **73**, 634–655, (2008).
[94] P. J. Steinhardt and N. Turok, *Endless Universe: Beyond the Big Bang.* Doubleday, New York, (2007).
[95] J. Swinton. Alan Turing and morphogenesis. Online at `www.swintons.net/jonathan/turing.htm`.
[96] M. Tegmark, The mathematical universe, *Found. Phys.* **38**, 101–50, (2008).
[97] C. Teuscher, *Turing's Connectionism – An Investigation of Neural Network Architectures.* Springer-Verlag, Berlin, Heidelberg, (2002).
[98] C. Teuscher, Ed., *Alan Turing: Life and Legacy of a Great Thinker.* Springer-Verlag, Berlin, Heidelberg, (2004).
[99] S. Torrance, R. Clowes, and R. Chrisley, *Machine Consciousness: Embodiment and Imagination.* (Special Issue of the *Journal of Consciousness Studies*, 2007).
[100] A. M. Turing, On computable numbers, with an application to the entscheidungsproblem, *Proc. London Math. Soc.* **42**, 230–265, (1936–37). (Reprinted in A. M. Turing, *Collected Works: Mathematical Logic*, pp. 18–53).
[101] A. M. Turing, Systems of logic based on ordinals, *Proc. London Math. Soc.* **45**, 161–228, (1939). (Reprinted in A. M. Turing, *Collected Works: Mathematical Logic*, pp. 81–148).
[102] A. M. Turing. Intelligent machinery. In *Machine Intelligence 5*, pp. 3–23. Edinburgh University Press, Edinburgh, (1969). (Reprinted in A. M. Turing, *Collected Works: Mechanical Intelligence*, (D. C. Ince, ed.) North-Holland, Amsterdam, New York, Oxford, Tokyo, 1992, pp. 107–127).
[103] A. M. Turing, *Collected Works of A. M. Turing: Morphogenesis (*T. Saunders Ed.*)*. North-Holland, Amsterdam, (1992).
[104] A. M. Turing, *Collected Works of A. M. Turing: Mathematical Logic (*O. Gandy and C. E. M. Yates Eds.*)*. Elsevier, Amsterdam, New York, Oxford, Tokyo, (1992).
[105] J. van Leeuwen and J. Wiedermann. The Turing machine paradigm in contemporary computing. In *Mathematics Unlimited – and Beyond.* LNCS, 2000, (2001).
[106] P. Woit, *Not Even Wrong: The Failure of String Theory and the Continuing Challenge to Unify the Laws of Physics.* Jonathan Cape, London, (2006).
[107] K. Zuse, *Rechnender Raum.* (Friedrich Vieweg & Sohn, Braunschweig). (English trans.: *Calculating Space*, MIT Technical Translation AZT-70-164-GEMIT, MIT (Proj. MAC), Boston, 1970).

# Chapter 6

# HF-Computability

Yuri L. Ershov*, Vadim G. Puzarenko†, and Alexey I. Stukachev‡

*Sobolev Institute of Mathematics,*
*Siberian Branch of the Russian Academy of Sciences,*
*Novosibirsk, 630090, Russia*

*Email:* *ershov@math.nsc.ru, †vagrig@math.nsc.ru, ‡aistu@math.nsc.ru

We survey the results on HF-computability (computability on hereditarily finite superstructures), an approach to abstract computability based on the notion of Σ-definability in admissible sets.

## Contents

## 6.1. Introduction

The notion of computability in mathematics and technics has become a subject of great interest and study. This is largely motivated by the rapid development and use of computers (in both theory and practice). An evidence of this fact is the successful realization of the European programme "Computability in Europe" (CiE), one of the aims of which is the organization of annual conferences gathering together mathematicians, specialists in computer science, biology, chemistry, and philosophy.

On the one hand, there is a generally accepted (absolute) theory of computability for (partial) functions and predicates on natural numbers – a classical computability theory. On the other hand, various proposals for generalized theories of computability have been accumulated. Such generalizations are motivated by a wish for a better understanding of the absolute theory and expansion of the possibilities of application (understanding) of computability notions to subjects (structures) far from natural numbers, in particular, to uncountable structures (such as, e.g., the field $\mathbb{R}$ of real numbers).

Development of the classical computability theory raises the following general methodological problem: How to "extend" the existing theory to a wider class of objects. One of the (successful) approaches in this direction is the theory of numberings [19, 35]. But this approach has strict cardinality limitations, since numberings are defined for countable collections of objects only. Another approach is the theory of computability on admissible sets of the form $\mathbb{HF}(\mathfrak{A})$, for reasonably "simple" structures $\mathfrak{A}$. Exactly this approach is discussed in the present paper.

The development of the theory of admissible sets began with the generalization of computability on ordinals, initially on the first nonrecursive ordinal (metarecursive theory) (Kreisel and Sacks, see [82, 83, 134]), then on arbitrary admissible (and other) ordinals (Kripke and Platek, see [84, 112]). It was completed in the papers by Barwise when he introduced admissible sets with urelements. The introduction of urelements would seem to be a technical improvement; however, now we know that just such an extension of the notion of the admissible set led to the universal theory of computability based on the notion of definability by formulas with (in a broad sense) effective semantics. Obviously, this theory generalizes nondeterministic computability unlike generalizations based on expansions of the notion of (abstract) computing devices. Therefore, we can say that this is a theory of constructively recognizable properties (predicates). Whereas the development of the classical theory of computability has shown that study of computable functions is reasonable with the partial computable functions only, the computability in arbitrary admissible sets shows that computable ($\Sigma$-) predicates are a natural environment for the study of partial computable ($\Sigma$-) functions. We can even say that the notion of the computable ($\Sigma$-) predicate is more fundamental than that of the (partial) computable function.

A general theory of admissible sets is a remarkable synthesis of the main directions in modern mathematical logic – set theory (including its classical section – descriptive set theory), model theory (infinite languages), and computability theory. The fundamental monograph of Barwise [11] is still the main source of all the indicated aspects of the theory of admissible sets. An intensive and profound study of the Turing reducibility on admissible (and not only) ordinals can be found in the monograph of Sacks [134].

In the monograph of Barwise, the class of admissible sets of the form $\mathbb{HYP}(\mathfrak{A})$ is regarded as the class of "minimal" admissible sets, probably because the author considered the admissible sets of the form $\mathbb{HF}(\mathfrak{A})$ to be too simple and trivial. The authors of the present paper think different.

We believe that, for a better understanding of the general nature of computability (constructive cognoscibility), one should develop the notion of computability in admissible sets of the form $\mathbb{HF}(\mathfrak{A})$ – the hereditarily finite superstructure over a structure $\mathfrak{A}$, where $\mathfrak{A}$ is either a model of a reasonably "simple" theory or a model of classical subjects, e.g., such as the field $\mathbb{R}$ of real numbers. It should be noted that the notion of search computability in an arbitrary structure $\mathfrak{A}$ introduced in [105], as well as the notion of abstract computability in the sense of [90], coincides (in accordance with [45]) with the notion of computability in the admissible set $\mathbb{HF}(\mathfrak{A})$. In Section 6.8, we compare HF-computability with some other closely related approaches to generalized computability, in particular, with BSS-computability. Theoretical computer science also requires the superstructures of such kind for the development of the theory of computability. In [39] an approach called semantic programming, based on the use of effective semantics as a programming language is proposed.

## 6.2. $\mathbb{HF}$-Logic

On the one hand, $\mathbb{HF}$-logic (or the weak second order logic) is a powerful tool to introduce the notion of finiteness in first order logic. On the other hand, it enables us to deal with natural numbers and, therefore, to introduce the notion of computability on arbitrary structures. By $\omega$ we denote the set of natural numbers. Also, we will often identify $\omega$ with the set of all natural ordinals. Let $M$ be an arbitrary set. We construct the collection of hereditarily finite sets over $M$ as follows:

- $HF_0(M) = \{\varnothing\}$;
- $HF_{n+1}(M) = \mathcal{P}_\omega(M \cup HF_n(M))$ (here $\mathcal{P}_\omega(X)$ is the collection of all finite subsets of $X$), $n < \omega$;
- $HF(M) = \bigcup\limits_{n<\omega} HF_n(M)$.

If $\mathfrak{M}$ is a structure of some relation signature $\sigma$ then one can define a structure $\mathbb{HF}(\mathfrak{M})$ of a signature $\sigma \cup \{U, \varnothing, \in\}$ ($\sigma \cap \{U, \varnothing, \in\} = \varnothing$) on $M \cup HF(M)$ so that

- $U^{\mathbb{HF}(\mathfrak{M})} = M$;
- $P^{\mathbb{HF}(\mathfrak{M})} = P^{\mathfrak{M}}$, $P \in \sigma$;
- $\varnothing^{\mathbb{HF}(\mathfrak{M})} = \varnothing \in HF_0(M)$;
- $\in^{\mathbb{HF}(\mathfrak{M})} = \in \subseteq ((M \cup HF(M)) \times HF(M))$.

We will consider structures of at most countable signatures only. Moreover, in most cases we shall restrict our considerations to finite signatures. As in set theory (e.g., in ZF), one can define natural ordinals and finite sequences on $\mathbb{HF}(\mathfrak{M})$. Indeed, $\mathbb{HF}(\mathfrak{M})$ is an admissible set and, therefore, we can apply methods which are used in KPU (see Section 6.9). A hereditarily finite superstructure can be considered as a structure, so we can apply usual model theoretic methods for studying it. The problem of **nonrealizability** of some type on hereditarily finite superstructures has a simple solution. We consider the following collections of formulas:

$$\theta_0(x_0) \leftrightharpoons \{\exists \text{ distinct } x_1, \ldots, x_n((x_1 \in x_0) \wedge \ldots (x_n \in x_0)) \mid n < \omega\},$$
$$\theta_1(x_0) \leftrightharpoons \{\exists \text{ distinct } x_1, \ldots, x_n((x_1 \in x_0) \wedge \ldots \wedge (x_n \in x_{n-1})) | n < \omega\}.$$

If $\theta_0(x_0)$ is satisfied on $a$ then $a$ has infinitely many elements, i.e., it has an infinite width; if $\theta_1(x_0)$ is satisfied on $a$ then $a$ has an infinite rank (in absolute sense). Thus, no hereditarily finite superstructure realizes $\theta_0(x_0)$ or $\theta_1(x_0)$. Indeed, it follows from definability of the cardinality operation on hereditarily finite superstructures that $\theta_0(x_0)$ and $\theta_1(x_0)$ are realized or not simultaneously.

**Lemma 6.1.** *Let $\mathbb{HF}(\mathfrak{M})$ be the hereditarily finite superstructure over $\mathfrak{M}$ and let $\mathrm{T}_0$ be its theory. If $\mathbb{A}$ is a structure of $\mathrm{T}_0$ on which $\theta_0(x_0)$ is not* *satisfied then* $\mathbb{A}$ *has the form* $\mathbb{HF}(\mathfrak{M}')$ *for some structure* $\mathfrak{M}' \models \mathrm{Th}(\mathfrak{M})$. *Conversely, no hereditarily finite superstructure satisfies* $\theta_0(x_0)$.

Let $T$ be a theory of signature $\sigma$. By a type of $T$ in $\sigma$ we mean a consistent (possibly incomplete) under $T$ collection of formulas of the same signature with some fixed finite number of free variables. A type $\xi(x_0, x_1, \ldots, x_{k-1})$ is called *principal* under $T$ if there exists a formula $\psi(x_0, \ldots, x_{k-1})$ such that $T \vdash \forall x_0 \ldots \forall x_{k-1}(\psi \to \varphi)$ for any $\varphi \in \xi$. Otherwise, this type is called *nonprincipal.*

Lemma 6.1 enables us to apply General Omitting Types Theorem for constructing hereditarily finite superstructures with desired properties. Namely,

**Corollary 6.1.** *Let* $\mathbb{A}$ *be an arbitrary hereditarily finite superstructure in some countable signature. Then for every countable collection* $\mathcal{S}$ *of nonprincipal types of* $\mathrm{Th}(\mathbb{A})$, *there exists a hereditarily finite superstructure* $\mathbb{HF}(\mathfrak{M}') \models \mathrm{Th}(\mathbb{A})$ *on which no type from* $\mathcal{S}$ *is satisfied.*

Since $\mathfrak{M} \preccurlyeq \mathfrak{N}$ implies $S(\mathfrak{M}) \subseteq S(\mathfrak{N})$, where $S(\mathfrak{M})$ is the collection of types of $\mathrm{Th}(\mathfrak{M})$ satisfied on $\mathfrak{M}$, the downwards Löwenheim–Skolem Theorem holds for hereditarily finite superstructures:

**Proposition 6.1.** *If* $\hbar \preccurlyeq \mathbb{HF}(\mathfrak{M})$ *then* $\hbar$ *has the form* $\mathbb{HF}(\mathfrak{M}')$ *for some* $\mathfrak{M}' \preccurlyeq \mathfrak{M}$.

In general, the upwards Löwenheim–Skolem–Mal'cev Theorem does not hold for hereditarily finite superstructures (see also Theorem 6.8). First we define the following sequence of cardinals:

- $\beth_0(\omega) = \omega$;
- $\beth_{\alpha+1}(\omega) = 2^{\beth_\alpha(\omega)}$;
- $\beth_\lambda(\omega) = \bigcup_{\gamma<\lambda} \beth_\gamma(\omega)$ if $\gamma$ is limit.

**Theorem 6.1.** [11, 118] *Let* $\mathbb{HF}(\mathfrak{M})$ *be the hereditarily finite superstructure over a structure* $\mathfrak{M}$ *in some countable signature and let* $T = \mathrm{Th}(\mathbb{HF}(\mathfrak{M}))$ *be its theory. Then the following statements are equivalent:*

*(1) for any infinite cardinal* $\beta$, *there exists* $\mathfrak{M}_\beta$ *such that* $\mathbb{HF}(\mathfrak{M}_\beta) \models T$ *and* $\mathrm{card}(\mathfrak{M}_\beta) \geqslant \beta$;
*(2) there exists* $\mathfrak{M}_0$ *such that* $\mathbb{HF}(\mathfrak{M}_0) \models T$ *and* $\mathrm{card}(\mathfrak{M}_0) = \beth_{\omega_1}(\omega)$;
*(3) there exists* $\mathfrak{M}_0$ *such that* $\mathbb{HF}(\mathfrak{M}_0) \models T$ *and there is an infinite set* $X \subseteq \mathrm{dom}(\mathfrak{M}_0)$ *of indiscernibles in some Skolem expansion of* $\mathbb{HF}(\mathfrak{M}_0)$;

(4) *there exists a countable structure* $\mathfrak{M}_0$ *such that* $\mathbb{HF}(\mathfrak{M}_0) \models T$ *and there is an infinite set* $X \subseteq \mathrm{dom}(\mathfrak{M}_0)$ *of indiscernibles in some Skolem expansion of* $\mathbb{HF}(\mathfrak{M}_0)$.

To apply this theorem for complete diagrams of hereditarily finite superstructures over countable structures, we infer the Elementary Extension Theorem. It follows from the following Theorem that $\beth_{\omega_1}(\omega)$ is the Hanf number for theories of hereditarily finite superstructures over countable structures.

**Theorem 6.2.** **[16, 118]** *For every ordinal* $\alpha < \omega_1$, *there exists a structure* $\mathfrak{M}_\alpha$ *in some finite signature such that* $\mathrm{card}(\mathfrak{M}) = \beth_\alpha(\omega)$ *and it satisfies the following conditions:*

(1) $\mathbb{HF}(\mathfrak{M}_\alpha)$ *has no proper elementary extension of kind* $\mathbb{HF}(\mathfrak{M})$;
(2) *for every* $\mathfrak{M}'$, $\mathbb{HF}(\mathfrak{M}_\alpha) \equiv \mathbb{HF}(\mathfrak{M}')$ *implies* $\mathbb{HF}(\mathfrak{M}') \leqslant \mathbb{HF}(\mathfrak{M}_\alpha)$.

Now we consider the problem of **realizability** of types on hereditarily finite superstructures. To decide this problem, we apply one more Omitting Types method. The language of hereditarily finite superstructures can be considered as the language of $\omega$-logic and a hereditarily finite superstructure can be viewed as an $\omega$-structure. In this case, ordinals of a hereditarily finite superstructure play the role of naturals.

We describe the language of *$\omega$-logic.* A signature $\sigma$ corresponds to a language $\mathcal{L}_\sigma^\omega$ which can be obtained from $\mathcal{L}_\sigma$ by adding one unary relation symbol $N$ and a collection $\{\mathbf{n} \mid n < \omega\}$ of constant symbols. Assume that $N$ and $\mathbf{n}$, $n < \omega$, don't occur in $\mathcal{L}_\sigma$. Terms and formulas of $\mathcal{L}_\sigma^\omega$ are defined just as in first order logic.

A structure $\mathfrak{A}$ is called an *$\omega$-structure* if $\{n \mid n < \omega\} \subseteq |\mathfrak{A}|$. If $\mathfrak{A}$ is an $\omega$-structure of $\sigma$, then it can be expanded to $\mathfrak{A}^\omega \leftrightharpoons \langle \mathfrak{A}, N^{\mathfrak{A}^\omega}, \langle n : n < \omega \rangle \rangle$ so that $N^{\mathfrak{A}^\omega} = \{n \mid n < \omega\}$. Let $\varphi(x_0, x_1, \ldots, x_{k-1})$ be a formula of $\mathcal{L}_\sigma^\omega$ and $a_0, a_1, \ldots, a_{k-1} \in A$. Then suppose that $\mathfrak{A} \models \varphi(\mathbf{a}_0, \mathbf{a}_1, \ldots, \mathbf{a}_{k-1})$ if $\mathfrak{A}^\omega \models \varphi(\mathbf{a}_0, \mathbf{a}_1, \ldots, \mathbf{a}_{k-1})$ in the usual sense whenever $\mathbf{n}$ and $N(x)$ are interpreted as $n$ and "$x \in N^{\mathfrak{A}^\omega}$" respectively.

Let $S$ be a collection of sentences of $\mathcal{L}_\sigma^\omega$ including $\{\neg(\mathbf{m} = \mathbf{n}) \mid m \neq n\} \cup \{N(\mathbf{n}) \mid n < \omega\}$. Then $S$ is called *$\omega$-consistent* if it is consistent in the usual sense and, for any formula $\psi(x)$ of $\mathcal{L}_\sigma^\omega$, if $S \cup \{\exists x(N(x) \wedge \psi(x))\}$ is consistent then so is $S \cup \{\psi(\mathbf{n})\}$ for some $n < \omega$. The notion of $\omega$-consistency is not finitary, i.e., there are collections $S$ of sentences that every finite subset $S_0 \subseteq S$ is $\omega$-consistent but $S$ is not.

Let $T$ be a theory of signature $\sigma \cup \langle N, \langle \mathbf{n} \mid n < \omega \rangle \rangle$. The theory $T$ is called *$\omega$-complete* if, for every formula $\psi(x)$ in $\mathcal{L}_\sigma^\omega$, $T \vdash \forall x(N(x) \to \psi(x))$ whenever $T \vdash \psi(\mathbf{n})$, for every $n < \omega$.

Given a set of sentences $T$, we write $T \models_\omega \varphi$ if $\varphi$ holds in all $\omega$-structures of $T$.

Let $T$ be a collection of sentences of $\mathcal{L}_\sigma^\omega$. A formula $\varphi$ is a *consequence* of $T$ in $\omega$-logic, written $T \vdash_\omega \varphi$, if $\varphi$ is in the smallest set of formulas containing $T$ together with the axioms of $\omega$-logic closed under the usual rules and the $\omega$-rule:

If $T \vdash_\omega \varphi(\overline{\mathbf{n}})$ for every $n < \omega$ then $T \vdash_\omega \forall v(\mathrm{N}(v) \to \varphi(v))$.

The following Existence Theorems for $\omega$-logic holds:

**Theorem 6.3.** [111] *If $S$ is a countable $\omega$-consistent collection of sentences of $\omega$-logic, then $S$ has an $\omega$-structure.*

**Theorem 6.4. ($\omega$-completeness)** *Let $\sigma$ be countable and let $T$ be a set of sentences of $\mathcal{L}_\sigma^\omega$. If $\varphi$ is a sentence of $\mathcal{L}_\sigma^\omega$, then $T \models_\omega \varphi$ iff $T \vdash_\omega \varphi$.*

Now we turn to studying hereditarily finite superstructures. Let $\mathfrak{M}$ be a structure of $\sigma$ and let $\mathbb{HF}(\mathfrak{M})$ be the hereditarily finite superstructure over $\mathfrak{M}$, $\sigma^* = \sigma \cup \{\mathrm{U}, \varnothing, \in\}$. As it is noticed above, any formula $\Psi(x_0, \ldots, x_{k-1})$, $k < \omega$, of $\mathcal{L}_{\sigma^*}^\omega$ is equivalent on $\mathbb{HF}(\mathfrak{M})$ to some formula $\Psi_0(x_0, \ldots, x_{k-1})$ of $\sigma^*$ whenever N and $\{\mathbf{n} \mid n < \omega\}$ are interpreted as the set of all ordinals and definable representations of ordinals respectively. Thus, the Orey Theorem sometimes enables us to construct a hereditarily finite superstructure on which some fixed type is satisfied.

In [115], syntactical characterizations of properties of countable categoricity and categoricity in $\mathbb{HF}$-logic are given. Recall that the theory $\mathrm{Th}(\mathbb{HF}(\mathfrak{M}))$ of the hereditarily finite superstructure $\mathbb{HF}(\mathfrak{M})$ over a countable structure $\mathfrak{M}$ is called *(countably) categorical in $\mathbb{HF}$-logic* if $\mathbb{HF}(\mathfrak{M}) \equiv \mathbb{HF}(\mathfrak{M}')$ implies $\mathbb{HF}(\mathfrak{M}) \cong \mathbb{HF}(\mathfrak{M}')$, for every (countable) structure $\mathfrak{M}'$. Hereinafter, '$\mathfrak{A} \cong \mathfrak{B}$' means that $\mathfrak{A}$ and $\mathfrak{B}$ are isomorphic where $\mathfrak{A}$ and $\mathfrak{B}$ are structures. Using these characterizations we have the following:

**Theorem 6.5.** *Let $\mathfrak{M}$ be a countable structure of some countable signature. Then $\mathrm{Th}(\mathbb{HF}(\mathfrak{M}))$ is countably categorical in $\mathbb{HF}$-logic iff every hereditarily finite superstructure of it is an atomic structure.*

**Theorem 6.6.** *Let $\mathfrak{M}$ be a countable structure in some countable signature. Then $\mathrm{Th}(\mathbb{HF}(\mathfrak{M}))$ is categorical in $\mathbb{HF}$-logic iff it is atomic and it has no pair of hereditarily finite superstructures $\mathbb{A}_0$, $\mathbb{A}_1$ such that $\mathbb{A}_0 \precneqq \mathbb{A}_1$.*

It is clear that if $\mathfrak{M}$ is a structure of (countably) categorical theory then $\mathrm{Th}(\mathbb{HF}(\mathfrak{M}))$ is also (countably) categorical in $\mathbb{HF}$–logic. We give examples which demonstrate differences between these notions.

Examples 6.2.1.

(1) Let $\mathbb{F}$ be an algebraically closed field of some finite degree of transcendency. Then $\mathrm{Th}(\mathbb{HF}(\mathbb{F}))$ is categorical in $\mathbb{HF}$–logic.
(2) Let $\mathbb{F}$ be an algebraically closed field of some infinite degree of transcendency. Then $\mathrm{Th}(\mathbb{HF}(\mathbb{F}))$ is countably categorical but not categorical in $\mathbb{HF}$–logic. In particular, $\mathrm{Th}(\mathbb{HF}(\mathbb{C}))$ is countably categorical but not categorical in $\mathbb{HF}$-logic where $\mathbb{C}$ is the field of complex numbers. Moreover, any two hereditarily finite superstructures over algebraically closed fields with infinite degrees of transcendency having the same characteristic are elementarily equivalent.
(3) It is evident that $\mathrm{Th}(\mathbb{HF}(\mathfrak{N}))$ is categorical in $\mathbb{HF}$-logic where $\mathfrak{N}$ is the standard model of arithmetic.
(4) Let $Z$ be the set of integer numbers, $0 < n < \omega$, and let $\leqslant$ be the lexicographic order on $Z^n$. Then $\mathrm{Th}(\mathbb{HF}(\langle Z^n, \leqslant\rangle))$ is categorical in $\mathbb{HF}$–logic.

In [13], an example of a finitely generated semi-group is constructed which demonstrates that the condition of a theory to be atomic in Theorem 6.6 is essential. Moreover, the following holds:

**Proposition 6.2.** *Let $\mathfrak{M}$ be a countable structure such that* $\mathrm{Th}(\mathbb{HF}(\mathfrak{M}))$ *is not atomic and has no pair of hereditarily finite superstructures $\mathbb{A}_0$, $\mathbb{A}_1$, for which $\mathbb{A}_0 \precneqq \mathbb{A}_1$. Then* $\mathrm{Th}(\mathbb{HF}(\mathfrak{M}))$ *has $2^{\aleph_0}$ pairwise non-isomorphic hereditarily finite superstructures and all of them are minimal structures.*

The notion of interpretability of one structure in another is one of the key notions in the Model Theory. For simplicity, we assume that signatures consist of relations symbols only (otherwise, we can replace all the operations by their graphs) and the equality is a signature relation.

**Definition 6.1.** Let $\mathfrak{M}$, $\mathfrak{N}$ be structures of signatures $\sigma_0$ and $\sigma_1$ respectively. We say that $\mathfrak{M}$ is *definable in* $\mathfrak{N}$ if there are

- a sequence of elements $\overline{a} = a_0, \ldots, a_{n-1}$ from $|\mathfrak{N}|$, $n < \omega$ (hereinafter, given a structure $\mathfrak{N}$, by $|\mathfrak{N}|$ we denote its universe);
- a formula $\psi(x_0, \ldots, x_{m-1}, y_0, \ldots, y_{n-1})$ of $\sigma_1$;
- a map $\nu$ from $\psi(\mathfrak{N}^m, \overline{a})$ onto $|\mathfrak{M}|$;

- a formula $\psi_P(\bar{x}_1, \ldots, \bar{x}_{\#(P)}, y_0, \ldots, y_{m-1})$ of $\sigma_1$, for every $P \in \sigma_0$ of arity $\#(P)$; $\bar{x}_k$ has length $m$ and all variables in $\bar{x}_1, \ldots, \bar{x}_{\#(P)}$ are distinct;

such that for any $P \in \sigma_0$ and $\bar{b}_0, \ldots, \bar{b}_{\#(P)}$ from $\psi(\mathfrak{N}^m, \bar{a})$, we have:

$$\mathfrak{M} \models P(\nu(\bar{b}_1), \ldots, \nu(\bar{b}_{\#(P)})) \Leftrightarrow \mathfrak{N} \models \varphi_P(\bar{b}_1, \ldots, \bar{b}_{\#(P)}, \bar{a}).$$

$\mathfrak{M}$ and $\mathfrak{N}$ are *bidefinable* if $\mathfrak{M}$ and $\mathfrak{N}$ are mutually definable.

**Definition 6.2.** Let $\mathcal{K}_0$, $\mathcal{K}_1$ be classes of structures of $\sigma_0$, $\sigma_1$ respectively. We say that $\mathcal{K}_0$ *is definable in* $\mathcal{K}_1$ if there exists a single list $S$ of formulas of $\sigma_1$ such that, for any $\mathfrak{M}_1 \in \mathcal{K}_1$, there is $\mathfrak{M}_0 \in \mathcal{K}_0$ definable in $\mathfrak{M}_1$ via $S$; for every $\mathfrak{M}_0 \in \mathcal{K}_0$, there is $\mathfrak{M}_1 \in \mathcal{K}_1$ in which $\mathfrak{M}_0$ is definable via $S$. If $\mathcal{K}_0$ and $\mathcal{K}_1$ are mutually definable then we say that $\mathcal{K}_0$ and $\mathcal{K}_1$ are *bidefinable.*

From now to the end of this section, all the signatures considered below are assumed to be finite. We consider now several examples of bidefinability of structures and hereditarily finite superstructures. Indeed, it is important that in all examples considered below, there exists a transformation of formulas of weak second order to ones of first order logic.

**Definition 6.3.** A *language* $L^w_\omega$ *of weak second order logic* consists of symbols from $L$, new variables $X_1, ..., X_n, ...$, and binary relation symbols $\in$ and $\Subset$. Formulas of $L^w_\omega$ are constructed as usual from atomic ones including atomic formulas of $L$ and $X_i \in X_j$, $v_i \Subset X_j$, where $v_i$ is an individual variable of $L$. To obtain a structure $\mathfrak{A}$ in $L^w_\omega$ we interpret symbols from $L$ as before; $X_1, ..., X_n, ...$ are interpreted on $HF(|\mathfrak{A}|) \cup |\mathfrak{A}|$ in the following way: $X_i \in X_j$ if and only if $X_i$ is an element from $X_j$ and $rnk(X_j)$=$rnk(X_i) + 1$; and $v \Subset X$ if only if there are $A_1, ..., A_n \in HF(|\mathfrak{A}|)$ such that $v \in A_1 \in ... \in A_n \in A$ and $rnk(A_1) = 1$.

All the classes considered below have the following property.

**Proposition 6.3.** *For any formula* $\Phi(v_0, ..., v_{n-1}) \in L^w_\omega(\mathcal{K})$, *there exists a formula* $\Psi(v_0, ..., v_{n-1}) \in L(\mathcal{K})$ *such that* $\mathfrak{A} \models \Phi(a_1, ..., a_n)$ *iff* $\mathfrak{A} \models \Psi(a_1, \ldots, a_n)$, *for every* $\mathfrak{A} \in \mathcal{K}$ *and* $a_1, ..., a_n \in |\mathfrak{A}|$.

### 1. *ff*–Classes.

**Definition 6.4.** [159] We say that a class of structures $\mathcal{K}$ *admits elementary definability of finite functions* ($\mathcal{K}$ is a *ff–class*) if there is $\Phi(x, y, \bar{z}) \in$

$L(\mathcal{K})$ such that, for any structure $\mathfrak{A} \in \mathcal{K}$ and function $f \subset |\mathfrak{A}| \times |\mathfrak{A}|$ whose domain is finite, there exists $\bar{a} \in |\mathfrak{A}|$ for which $f(x) = y$ if and only if $\mathfrak{A} \models \Phi(x, y, \bar{a})$.

We give now some examples of $ff$-classes [159].

**Proposition 6.4.** *The class of existentially closed groups is an $ff$-class.*

**Proposition 6.5.** *The class of unintentionally closed semi-groups is an $ff$-class.*

The proof of the following results can be found in [159].

**Proposition 6.6.** *Let $\mu : \omega \to L(\mathcal{K})$ be a Gödel numbering of terms. There exists a formula $\Phi(X, Y, z) \in L^w_\omega(\mathcal{K})$ satisfied on $\mathcal{K}$-structure $\mathfrak{A}$ if and only if $X$ is a number of a term $\tau(x_1, ..., x_k)$, $Y$ is a sequence of the length $k$, and $z$ is the value of $\tau$ from $Y$ in $\mathfrak{A}$.*

**Proposition 6.7.** *Let $\nu : \omega \to L(\mathcal{K})$ be a Gödel numbering of $\Pi_n$-formulas with free variables contained in $\{v_0, ..., v_{m-1}\}$. There exists a formula $\Phi(X, v_0, ..., v_{m-1})$ which is satisfied on $\mathcal{K}$-structure $\mathfrak{A}$ on $a_1, ..., a_m$ if and only if $X$ is a natural number and $\mathfrak{A}$ satisfies $\nu(X)$ on $a_1, ..., a_m \in |\mathfrak{A}|$.*

**Corollary 6.2.** *Let $\mathcal{K}$ be an $ff$-class and let $\{\Phi_i \mid i \in I\}$ be an arithmetical collection of $\Pi_n$-formulas of $L(\mathcal{K})$ with free variables from $\{v_0, ..., v_{m-1}\}$. Then there are formulas*

$$\Phi(v_0, ..., v_{m-1}),\ \Psi(v_0, ..., v_{m-1})$$

*such that for any $\mathcal{K}$-structure $\mathfrak{A}$ and $a_1, ..., a_m \in |\mathfrak{A}|$,*

$$\mathfrak{A} \models \bigwedge_{i \in I} \Phi_i(a_1, ..., a_m) \text{ iff } \mathfrak{A} \models \Phi(a_1, ..., a_m)$$

*and*

$$\mathfrak{A} \models \bigvee_{i \in I} \Phi_i(a_1, ..., a_m) \text{ iff } \mathfrak{A} \models \Psi(a_1, ..., a_m).$$

**2. $\varepsilon$–Fragments.** A structure $\mathbb{A} = \langle A, \varepsilon \rangle$ of signature $\{\varepsilon\}$ is called an *$\varepsilon$-fragment* if the following conditions hold:

1) regularity: for any non-empty subset $A' \subseteq A$, there exists $a' \in A'$ such that $a'' \varepsilon a'$ is not satisfied, for every $a'' \in A'$.

Let $A_u \rightleftharpoons \{a | a \in A, \forall a' \in A (\neg (a' \varepsilon a))\}$; then, by 1), $A_u \neq \varnothing$.

2) extensionality: for any $a_0, a_1 \in A \setminus A_u$, we have

$$a_0 = a_1 \iff \hat{a_0} = \hat{a_1},$$

where $\hat{a_i} \rightleftharpoons \{a | a \in A, a\varepsilon a_i\}$, $i = 0, 1$.

A structure of kind $\mathbb{A}_\varnothing \rightleftharpoons \langle A, a_0, \varepsilon\rangle$ is called a *marked $\varepsilon$-fragment* if $\langle A, \varepsilon\rangle$ is an $\varepsilon$-fragment and $a_0 \in A_u$; in this case, set $A_0 \rightleftharpoons A_u \setminus \{a_0\}$.

If $\mathbb{A}_\varnothing$ is a marked $\varepsilon$-fragment then one can define a correspondence $\varkappa(= \varkappa_{\mathbb{A}_\varnothing}) : A \to HF(A_0)$ as follows:

$\varkappa(a) \rightleftharpoons a$ if $a \in A_0$;

$\varkappa(a_0) \rightleftharpoons \varnothing$;

$\varkappa(a) \rightleftharpoons \{\varkappa(a') | a' \in A, a'\varepsilon a\}$ if $a' \in A \setminus (A_0 \cup \{a_0\})$.

A subset $A' \subseteq A$ of a marked $\varepsilon$-fragment $\mathbb{A}_\varnothing$ is called *dense* if $A' \supseteq A^0$, where $A^0 \rightleftharpoons \{a | a \in A, \forall a' \in A(a \notin a')\} \setminus \{a_0\}$.

Let $\mathbb{A}_\varnothing = \langle A, a_0, \varepsilon\rangle$ be a marked $\varepsilon$-fragment and $A' \subseteq A$ its dense subset. We say that $\langle \mathbb{A}_\varnothing, A'\rangle$ *codes* $\varkappa(\mathbb{A}_\varnothing, A') \rightleftharpoons \{\varkappa(a') | a' \in A'\} \in HF(A_0)$. In this case, $\langle \mathbb{A}_\varnothing, A'\rangle$ is called a *code*.

**Lemma 6.2.** *If $\langle \mathbb{A}_\varnothing, A'\rangle$, $\langle \mathbb{B}_\varnothing, B'\rangle$ are codes and $\varkappa(\mathbb{A}_\varnothing, A') = \varkappa(\mathbb{B}_\varnothing, B')$, then $A_0 = B_0$ and there is a unique isomorphism $\varphi : \mathbb{A}_\varnothing \to \mathbb{B}_\varnothing$ such that $\varphi \restriction A_0 = \mathrm{id}_{A_0}$, $\varphi(a_0) = b_0$, and $\varphi(A') = B'$. Conversely, if $A_0 = B_0$ and there exists an isomorphism $\varphi$, then $\varkappa(\mathbb{A}_\varnothing, A') = \varkappa(\mathbb{B}_\varnothing, B')$.*

**Lemma 6.3.** *If $B$ is infinite then, for any $S \in HF(B)$, there is a code $\langle \mathbb{A}_\varnothing, A'\rangle$ of $S$ such that $A \in B$.*

***Proof.*** Let $A_0 \rightleftharpoons \mathrm{sp}(S) \subseteq B$ where sp is the support function (see section 6.9.2). Also, let $S_* \subseteq HF(B)$ be the least end subset of $HF(B)$ containing $S \cup \{\varnothing\}$ as a subset. It is clear that $S_*$ is finite and, therefore, $S_* \in HF(B)$. The set $S_*$ can be defined as follows: $\check{s_0} \rightleftharpoons \mathrm{TC}(s_0)$ for every $s_0 \in S \setminus A_0$; $\check{s_0} \rightleftharpoons \{s_0\}$ if $s_0 \in S \cap A_0$. Then $S_* \rightleftharpoons \{\varnothing\} \cup \bigcup\limits_{s_0 \in S} \check{s_0}$ as desired. Let $\rho : S_* \to B$ be an injective map such that $\rho(a) = a$ for any $a \in A_0$. Suppose that $A \rightleftharpoons \rho(S_*)$ and $a'\varepsilon a \iff \rho^{-1}(a') \in \rho^{-1}(a)$ for any $a, a' \in A$; then $\mathbb{A}_\varnothing \rightleftharpoons \langle A, \varepsilon, a_0(= \rho(\varnothing))\rangle$ is a marked $\varepsilon$-fragment, $A' \rightleftharpoons \{\rho(s) | s \in S\}$ is a dense subset of $\mathbb{A}_\varnothing$ and, as it could be easily checked, $\varkappa(\mathbb{A}_\varnothing, A') = S$. $\square$

**Theorem 6.7.** *If there is a coding of all finite binary relations on $\mathfrak{M}$ then $\mathbb{HF}(\mathfrak{M})$ is definable in $\mathfrak{M}$.*

**3. Hereditarily listed superstructures.**

Let $M$ be an arbitrary set. We define the collection of hereditary lists over $M$ by induction on $n < \omega$:

- $HL_0(M) = M \cup \{\varnothing\}$;
- $HL_{n+1}(M) = HL_n(M) \cup \{\langle x, y\rangle \mid x, y \in HL_n(M)\}$;
- $HL(M) = \bigcup_{n<\omega} HL_n(M)$.

Natural numbers are identified with the following elements from $HL(M)$: $\varnothing$, $\langle\varnothing,\varnothing\rangle$, $\langle\langle\varnothing,\varnothing\rangle,\varnothing\rangle$ etc:

- $0 = \varnothing$;
- $n+1 = \langle n, \varnothing\rangle$;
- $\omega = \{0, 1, 2, \ldots\}$.

Every $z \in HL(M)$ corresponds to $l(z)$ and $r(z)$ inductively as follows:
$l(\varnothing) = r(\varnothing) = \varnothing$;
$l(z) = r(z) = 1$ if $z \in M$;
$l(z) = x$, $r(z) = y$ if $z = \langle x, y\rangle$.

Let $\mathfrak{M}$ be a structure of some finite relation signature $\sigma$. Then a *hereditarily listed superstructure* $\mathbb{HL}(\mathfrak{M})$ over $\mathfrak{M}$ is a structure of signature $\sigma \cup \{l, r, \langle\cdot,\cdot\rangle\}$, $HL(M)$ its domain, such that $l$ and $r$ are defined above and symbols from $\sigma$ are interpreted on $M$ only as before. Then $\mathbb{HF}(\mathfrak{M})$ and $\mathbb{HL}(\mathfrak{M})$ are bidefinable.

**4. Admissible Structures.**

Let $\mathbb{A}$ be an arbitrary admissible set (definition, examples, and basic properties of such objects are given in section 6.9). Then one can construct a directed graph $\langle|\mathbb{A}|, R\rangle$ without loops such that $\mathbb{A}$ and $\mathbb{HF}(\langle|\mathbb{A}|, R\rangle)$ are bidefinable [122]. Moreover, this transformation preserves the semilattice of $\Sigma$-degrees considered in 6.6.1, the semilattice of $m\Sigma$-degrees (see Section 6.4), and all descriptive set theoretical properties considered in Section 6.5 but quasiresolvability. However, it cannot be applied in studying $T\Sigma$- and $e\Sigma$-degrees (see Section 6.4) and semilattices of degrees of presentability (see Section 6.7).

We mention one more result which gives a natural example of bidefinability between special admissible sets and hereditarily finite superstructures. Namely, it was proved in [144] that $\mathbb{H}YP(\mathfrak{M})$ and $\mathbb{HF}(\mathfrak{M})$ are $\Sigma$-equivalent in case when $\mathfrak{M}$ is a recursively saturated model of a regular theory (in [9] it is proved that this $\Sigma$-equivalence is strong). For the definition of $\Sigma$-definability we refer the reader to Section 6.6.

In Examples 3 and 4, the property to be a $\Sigma$-subset is preserved under certain interpretations. This enables us to transfer semantic approaches to computability from one object to another.

Now we consider problems of definability of structures in hereditarily finite superstructures. First we discuss model theoretic properties.

**Definition 6.5.** A structure $\mathfrak{M}_0$ is called *saturated enough* if there exists an $\omega$-saturated structure $\mathfrak{M}_1$ such that $\mathbb{HF}(\mathfrak{M}_0) \preccurlyeq \mathbb{HF}(\mathfrak{M}_1)$.

It is well known that any structure has some elementary $\omega$-saturated extension [16]. However, there are structures which are not saturated enough. The standard model of arithmetic and the field of real numbers are examples of such structures. In [118], a series of structures of sufficiently large cardinality is given which are not saturated enough. We give a nice model theoretic property of structures saturated enough.

**Proposition 6.8.** [31, 33] *Let $\mathfrak{M}_0$ and $\mathfrak{M}_1$ be structures saturated enough. If $\mathfrak{M}_0 \equiv \mathfrak{M}_1$, then $\mathbb{HF}(\mathfrak{M}_0) \equiv \mathbb{HF}(\mathfrak{M}_1)$. If $\mathfrak{M}_0 \preccurlyeq \mathfrak{M}_1$, then $\mathbb{HF}(\mathfrak{M}_0) \preccurlyeq \mathbb{HF}(\mathfrak{M}_1)$.*

The following variant of the Löwenheim-Skolem-Mal'cev Theorem holds for structures saturated enough.

**Theorem 6.8.** [31, 33] *Let* T *be a complete $\omega$-stable or $\omega$-categorical theory, let $\mathfrak{M}$ be a structure of* T *saturated enough, and let an uncountable structure $\mathfrak{N}$ be definable in $\mathbb{HF}(\mathfrak{M})$. Then for any infinite cardinals $\alpha$ and $\beta$ such that $\alpha \leqslant \mathrm{card}(\mathfrak{N}) \leqslant \beta$, there are structures $\mathfrak{M}_\alpha$ and $\mathfrak{M}_\beta$ of $T$ saturated enough such that $\mathfrak{M}_\alpha \preccurlyeq \mathfrak{M}$ and $\mathfrak{M}_\alpha \preccurlyeq \mathfrak{M}_\beta$. The structure $\mathfrak{M}_\alpha$ contains all the parameters from $|\mathfrak{M}|$ used in the definition of $\mathfrak{N}$ in $\mathbb{HF}(\mathfrak{M})$. If $\mathfrak{N}_\alpha$ and $\mathfrak{N}_\beta$ are structures definable in $\mathbb{HF}(\mathfrak{M}_\alpha)$ and $\mathbb{HF}(\mathfrak{M}_\beta)$ respectively via the same formulas and parameters as $\mathfrak{N}$ in $\mathbb{HF}(\mathfrak{M})$ then $\mathrm{card}(\mathfrak{N}_\alpha) = \alpha$, $\mathrm{card}(\mathfrak{N}_\beta) = \beta$.*

If the theory T is categorical in some infinite power, then $\mathfrak{M}_\beta$ from Theorem 6.8 can be chosen so that $\mathfrak{M} \preccurlyeq \mathfrak{M}_\beta$. The authors do not know whether $\mathfrak{M}_\beta$ could be always chosen in such a way.

In the end of this section, we consider several examples of definability of classical structures.

**Definition 6.6.** Let $\mathbb{A}$ be an admissible structure, let $A$ be its universe, and let $S \subseteq \mathcal{P}(A)$. We say that $S$ is *definable in* $\mathbb{A}$ if there are a sequence $\overline{a}$

of elements from $A$ and a formula $\psi(x, y, \bar{z})$ such that $S \cup \{\varnothing\} = \{\psi(\mathbb{A}, b, \bar{a}) \mid b \in A\}$.

A criterion of definability of the field $\mathbb{R}$ of real numbers in admissible sets is contained in the following theorem.

**Theorem 6.9.** [117] *Let $\mathbb{A}$ be an admissible set. Then $\mathbb{R}$ is definable in $\mathbb{A}$ iff $\mathcal{P}(\omega)$ is definable in $\mathbb{A}$.*

**Corollary 6.3.** *Let* T *be a theory categorical in some infinite cardinality and $\mathfrak{M}$ a structure of* T. *Then $\mathbb{R}$ is not definable in $\mathbb{HF}(\mathfrak{M})$.*

**Corollary 6.4.** [33] *Let* T *be either the theory of dense linear orders, the theory of algebraically closed fields, or the theory of infinite sets in the empty language. If $\mathfrak{M}$ is a structure of* T, *then $\mathbb{R}$ is not definable in $\mathbb{HF}(\mathfrak{M})$.*

We give now one positive example of an application of this theorem.

**Proposition 6.9. (Puzarenko)** *For any $S \subseteq \mathcal{P}(\omega)$, there is a linearly ordered set $\mathcal{L}_S$ such that $S$ is definable in $\mathbb{HF}(\mathcal{L}_S)$.*

***Proof.*** Let $S \subseteq \mathcal{P}(\omega)$. We assume that $S \neq \varnothing$ and $\varnothing \notin S$.

We give now some method of coding of a set $A \in S$. Fix a surjective map $f : \omega + \omega^* \to A$ and define an *A-block* as follows:

- for any $n \in \omega + \omega^*$, we take a linear ordering $L_n$, containing $f(n) + 2$ elements so that if $n, m \in \omega + \omega^*$ satisfy $n < m$ and $l_0 \in L_n$, $l_1 \in L_m$, then $l_0 < l_1$;
- for any $n \in \omega + \omega^*$, we put $L_{n,n+1}$ isomorphic to the segment $[0; 1]$ of rational numbers between $L_n$ and $L_{n+1}$;
- we also put linear orderings isomorphic to $[0; 1]$ before $L_0$, $0 \in \omega$, and after $L_0$, $0 \in \omega^*$.

Now we define a structure $\mathcal{L}_S$. First we fix some surjective map $g : 2^\omega(\omega^* + \omega) \to S$. The domain of $\mathcal{L}_S$ will have the following form:

- for every $n \in 2^\omega(\omega^* + \omega)$, we take some $g(n)$-block $K_n$ so that if $n, m \in 2^\omega(\omega^* + \omega)$ satisfy $n < m$ and $l_0 \in K_n$, $l_1 \in K_m$, then $l_0 < l_1$;
- for every $n \in 2^\omega(\omega^* + \omega)$, we put a singleton $K_{n,n+1}$ between $K_n$ and $K_{n+1}$ (if $a \in K_{n-1,n}$, $b \in K_{n,n+1}$, then $\{a, b\}$ is said *to define the n-block*).

Now we show that $S$ is definable in $\mathbb{HF}(\mathcal{L}_S)$. First we give several auxiliary assumptions.

$\langle 1 \rangle$ *"$a \in K_{n,n+1}$ for some $n \in 2^\omega(\omega^* + \omega)$" is definable in $\mathcal{L}_S$ by some formula (denote this set as $R$).* Indeed, $a \in R \Leftrightarrow \exists x \exists y[(x < a < y) \wedge \forall t((t < x) \to \exists z(t < z < x)) \wedge \forall t((y < t) \to \exists z(y < z < t)) \wedge \neg\exists z(x < z < a) \wedge \neg\exists z(a < z < y)]$.

$\langle 2 \rangle$ *"$\{a, b\}$ defines an $n$-block for some $n \in 2^\omega(\omega^* + \omega)$" is definable in $\mathcal{L}_S$ by some formula (we denote this relation as $Q$, and the set $g(n)$ corresponding to it as $A_{a,b}$).* Indeed, $Q(a,b) \Leftrightarrow [R(a) \wedge R(b) \wedge (a < b) \wedge \neg\exists t((a < t < b) \wedge R(t))]$.

$\langle 3 \rangle$ *The relation "$n \in A_{a,b}$" from $\langle a, b \rangle \in Q$ and $n \in \omega$ is definable in $\mathbb{HF}(\mathcal{L}_S)$ by some formula.* Indeed, $n \in A_{a,b} \Leftrightarrow [Q(a,b) \wedge \exists x((\mathrm{card}(x) = n + 2) \wedge \forall y \in x(a < y < b) \wedge \exists u \exists v \exists f((f : \langle n+4, < \rangle \xrightarrow{\sim} \langle x \cup \{u, v\}, < \rangle) \wedge (f(0) = u) \wedge (f(n+3) = v) \wedge \forall m \in n+3((f(m) < f(m+1)) \wedge \neg\exists z(f(m) < z < f(m+1))) \wedge \forall t((t < u) \to \exists z(t < z < u)) \wedge \forall t((v < t) \to \exists z(v < z < t))))]$.

To finish the proof, it remains to note that $S = \{A_{a,b} \mid Q(a,b)\}$. □

Applying Proposition 6.9 to $\mathcal{P}(\omega)$ we have:

**Corollary 6.5.** *There is a linearly ordered set $\mathcal{L}$ such that $\mathbb{R}$ is definable in $\mathbb{HF}(\mathcal{L})$.*

In comparison with the last result, $\mathbb{R}$ is not $\Sigma$-definable in $\mathbb{HF}(\mathcal{L})$, for any linearly ordered set $\mathcal{L}$ (see Section 6.6.2).

Additional information about the $\mathbb{HF}$-logic can be found in [12, 108, 109, 158, 159].

## 6.3. Σ-Subsets on Hereditarily Finite Superstructures

Here, by computability on hereditarily finite superstructures we mean $\Sigma$-definability, and (generalized) computably enumerable sets are identified with $\Sigma$-subsets.

The class of $\Delta_0$-*formulas* is the least one containing atomic formulas which is closed under $\vee$, $\wedge$, $\to$, $\neg$ and restricted quantifiers $\forall x \in y$ and $\exists x \in y$ ($\forall x \in y\varphi$ and $\exists x \in y\varphi$ are abbreviations for $\forall x(x \in y \to \varphi)$ and $\exists x(x \in y \wedge \varphi)$ respectively.

The class of $\Sigma$-*formulas* is the least one containing $\Delta_0$-formulas closed under $\vee$, $\wedge$, restricted quantifiers $\forall x \in y$, $\exists x \in y$, and $\exists x$.

A $\Sigma_1$-*formula* is a formula of kind $\exists u\varphi_0$ where $\varphi_0$ is $\Delta_0$-formula.

It follows from $\Sigma$-Reflection Principle [11] that any $\Sigma$-formula is equivalent under KPU to some $\Sigma_1$-formula.

A *$\Sigma$-predicate* is a relation definable by some $\Sigma$-formula (possibly with parameters). A *$\Delta$-predicate* is a $\Sigma$-predicate whose complement is also $\Sigma$.

A partial operation is called a *(partial) $\Sigma$-function* if its graph is $\Sigma$.

The following fact demonstrates that computability on hereditarily finite superstructures is actually a generalization of the classical computability.

**Proposition 6.10. [11, 33]**

*(1) There exists a one-to-one correspondence $\gamma$ between $HF(\varnothing)$ and $\omega$ which is a $\Sigma$-function on $\mathbb{HF}(\varnothing)$.*

*(2) $A \subseteq \omega$ is computably enumerable iff it is a $\Sigma$-subset of $\mathbb{HF}(\varnothing)$.*

*(3) $A \subseteq \omega$ is computably enumerable iff it is a $\Sigma$-subset of $\mathbb{HF}(\mathbb{N})$ (A can be considered here as a subset of $\mathrm{Ord}(\mathbb{HF}(\mathfrak{N}))$ or as a subset of $|\mathbb{N}|$).*

Hereinafter, by $\mathbb{N}$ we denote the standard model of arithmetic.

In the study of computability on hereditarily finite superstructures, an approach of defining sets by infinite computable formulas is actively used. This method is proposed in [162].

Before stating the method we consider the problem of constructivizability of hereditarily finite superstructures. The basic notions from the constructible model theory can be found in [37]. Recall that a sequence $\{A_n\}_{n\in\omega}$ of finite subsets of natural numbers is *strongly computable* if the relation $\{\langle m,n\rangle \mid m \in A_n\}$ and the function $n \mapsto \mathrm{card}(A_n)$ are computable.

**Proposition 6.11.** *Let $(\mathfrak{M},\nu)$ be a constructivizable structure. Then there exists a constructivization $\nu_0$ of the hereditarily finite superstructure $\mathbb{HF}(\mathfrak{M})$ which satisfies the following conditions:*

*(1) $\nu \leqslant \nu_0$;*

*(2) there exists a strongly computable sequence $\{A_n\}_{n\in\omega}$ of finite sets for which $\nu_0(n) = \{\nu_0(k) \mid k \in A_n\}$;*

*(3) $\nu_0^{-1}(P)$ is computably enumerable, for every $\Sigma$-predicate $P$ on $\mathbb{HF}(\mathfrak{M})$;*

*(4) $\nu_0^{-1}(P)$ is computable, for every $\Delta$-predicate $P$ on $\mathbb{HF}(\mathfrak{M})$.*

***Proof.*** It is evident that a constructivization

$$\nu_0(n) = \begin{cases} \varnothing & \text{if } n = 0; \\ \nu(k) & \text{if } n = 2k+1; \\ \{\nu_0(k_1)\ldots,\nu_0(k_l)\} & \text{if } n = 2(2^{k_1}+\ldots+2^{k_l}),\ k_1 < \ldots < k_l; \end{cases}$$

has the desired properties. □

Indeed, 2 implies 3, 4, and, therefore, this enables us to translate elements of the hereditarily finite superstructure into finite objects on natural numbers, in particular, we can transfer restricted quantifiers on hereditarily finite superstructures into ones on the standard model of arithmetic.

Notice that there are constructivizations of hereditarily finite superstructures $\mathbb{HF}(\mathfrak{M})$ over any constructivizable structure $\mathfrak{M}$ which do not satisfy 2, 3, 4 from Proposition 6.3. To understand this, it suffices to code finite sets by computable but not strongly computable indices.

The identical map on $\omega$ is a constructivization of $\mathbb{N}$, so there exists a constructivization of $\mathbb{HF}(\mathbb{N})$ which satisfies Proposition 6.3. From now on we identify elements from $|\mathbb{HF}(\mathbb{N})|$ with their numbers under a fixed constructivization of $\mathbb{HF}(\mathbb{N})$ satisfying Proposition 6.3.

We assign every element $\varkappa \in |\mathbb{HF}(\mathbb{N})|$ a term $t_\varkappa$ in signature $\{\varnothing, \{\cdot\}, \cup\}$ for which $t_\varkappa(\overline{m}) = \varkappa$ for some sequence of pairwise distinct ur-variables $\overline{m}$ containing all the elements from $\mathrm{sp}(\varkappa)$ as follows ( $n \in \omega$ is here number of free variables in this term; if $n = 0$ then it has no free variable):

- $t_\varkappa(u_0, \ldots, u_{n-1}) = \varnothing$ if $\varkappa = \varnothing$;
- $t_\varkappa(u_0, \ldots, u_{n-1}) = u_i$ if $\varkappa = i \in N$;
- $t_\varkappa(u_0, \ldots, u_{n-1}) = \{t_{\varkappa_0}(u_0, \ldots, u_{n-1})\} \cup \{t_{\varkappa_1}(u_0, \ldots, u_{n-1})\} \cup \ldots \cup \{t_{\varkappa_k}(u_0, \ldots, u_{n-1})\}$ if $\varkappa = \{\varkappa_0, \varkappa_1, \ldots, \varkappa_k\}$ and $\varkappa_0 < \varkappa_1 < \ldots < \varkappa_k$, $k \in \omega \setminus \{0\}$.

By $\Sigma$-recursion, it is easy to check that $(\varkappa, \overline{a}) \in |\mathbb{HF}(\mathbb{N})| \times |\mathfrak{M}|^{<\omega} \mapsto t_\varkappa(\overline{a}) \in |\mathbb{HF}(\mathfrak{M})|$ will be a $\Sigma$-function on $\mathbb{HF}(\mathfrak{M})$, for any structure $\mathfrak{M}$. Notice also that the collection of permutation groups $S_\varkappa (\leftrightharpoons \{\pi \in S(\mathrm{sp}(\varkappa)) \mid t_\varkappa(\overline{u}) = t_\varkappa(\pi\overline{u})\})$ is strongly computable.

We say that $\varkappa_0, \varkappa_1 \in |\mathbb{HF}(\mathbb{N})|$ are *termally equivalent* (and denote as $\varkappa_0 \sim \varkappa_1$), if $\langle \mathrm{TC}(\{\varkappa_0\}), \in, \varnothing, \mathrm{sp}(\varkappa_0)\rangle \cong \langle \mathrm{TC}(\{\varkappa_1\}), \in, \varnothing, \mathrm{sp}(\varkappa_1)\rangle$. If $\varkappa_0 \sim \varkappa_1$ then for any hereditarily finite superstructure, there are tuples $\overline{u}$, $\overline{v}$ of urelements such that $\mathbb{HF}(\mathfrak{M}) \models \forall\overline{u}\forall\overline{v}(t_{\varkappa_0}(\overline{u}) \approx t_{\varkappa_1}(\overline{v}))$. If, in addition, we assume that elements of these tuples are distinct, then the converse assumption is also true. We will write $\varkappa_0 \widetilde{\in} \varkappa_1$ if there are $\varkappa_0' \sim \varkappa_0$ and $\varkappa_1' \sim \varkappa_1$ that $\varkappa_0' \in \varkappa_1'$.

**Remark 6.3.1.** It is convenient to use "almost" single-valued representations, i.e., elements from $|\mathbb{HF}(\mathbb{N})|$ are chosen so that every element from the hereditarily finite superstructure is in the range of some unique term with some tuple of pairwise distinct urelements. Then the values of terms are determined by some strongly computable sequence of groups of permutation of urelements. This is important in studying such principles on hereditarily finite superstructures as uniformization, reduction, the existence of a universal function etc.

For convenience, we will use different ur-variables $u_0, v_0, u_1, v_1, \ldots$ for urelements only and common variables $x_0, y_0, x_1, y_1, \ldots$ for all elements to the end of this section.

**Lemma 6.4.** *For any $\Delta_0$-formula $\Phi$ from ur-variables in signature $\sigma \cup \{U, \varnothing, \in, \cup, \{\ \}\}$, one can effectively construct $\exists-$ and $\forall-$formulas $\Phi_0$ and $\Phi_1$ respectively in signature $\sigma$ that*

$$\mathbb{HF}(\mathfrak{M}) \models \Phi \Leftrightarrow \mathfrak{M} \models \Phi_0 \Leftrightarrow \mathfrak{M} \models \Phi_1.$$

*Moreover, if $\sigma$ contains constants or $\mathrm{FV}(\Phi) \neq \varnothing$ then $\Phi_0$ can be chosen quantifier free.*

***Proof.*** Let $\Phi$ be a formula in $\sigma$. We effectively construct a quantifier-free formula $\Psi$ of $\sigma \cup \{\top, \bot\}$ equivalent to it where $\top$ and $\bot$ are logical constants "true" and "false" respectively. We prove this assumption by induction on the number of logical connectives. By an improper term in $\Phi$ we mean a term which is maximal under inclusion occurring in $\Phi$. We assume that the implications do not occur in $\Phi$, the negations appear before atomic subformulas of $\Phi$ only, and all terms of $\{\varnothing, \cup, \{\ \}\}$ improper in $\Phi$ have the form $t_\varkappa(\overline{u})$ for some $\varkappa \in |\mathbb{HF}(\mathbb{N})|$ and tuple $\overline{u}$ with pairwise distinct ur-variables. Also, all the restricted quantifiers appearing in $\Phi$ have the forms $\forall x \in \ldots$ or $\exists x \in \ldots$ where $x$ is a common variable. We consider several cases.

(1) $\Phi$ is atomic or the negation of atomic:

- if $\{U, \varnothing, \in, \cup, \{\ \}\}$ do not occur in $\Phi$ then $\Psi = \Phi$;
- if $\Phi = U(t_\varkappa(\overline{u}))$ then $\Psi = \top$ whenever $t_\varkappa(\overline{u})$ is an ur-variable; $\Psi = \bot$, otherwise;
- if $\Phi = (t_{\varkappa_0}(\overline{u}) \approx t_{\varkappa_1}(\overline{v}))$, then $\Psi = \bigvee_{\pi \in S_{\varkappa_0}} (\overline{u} \approx \pi(\overline{v}'))$ ($\overline{v}'$ satisfies $\mathbb{HF}(\mathfrak{N}) \models (t_{\varkappa_0}(\overline{v}') \approx t_{\varkappa_1}(\overline{v})))$, whenever $\varkappa_0 \sim \varkappa_1$ and $\mathrm{sp}(\varkappa_0) \neq \varnothing$; $\top$, if $\varkappa_0 \sim \varkappa_1$ and $\mathrm{sp}(\varkappa_0) = \varnothing$; otherwise, $\Psi = \bot$;
- if $\Phi = (t_{\varkappa_0}(\overline{u}) \in t_{\varkappa_1}(\overline{v}))$ then $\Psi$ is obtained by the previous rule from $\bigvee_{\varkappa_2 \in \varkappa_1} (t_{\varkappa_0}(\overline{u}) \approx t_{\varkappa_2}(\overline{v}))$ whenever $\varkappa_0 \widetilde{\in} \varkappa_1$; $\Psi = \bot$, otherwise;
- if $\Psi$ is an atomic formula of $\sigma \cup \{U, \varnothing, \in, \cup, \{\ \}\}$ which is not considered above then we let $\Psi = \bot$;
- $\neg\top$ and $\neg\bot$ are replaced with $\bot$ and $\top$ respectively;

(2) if $\Phi = (\Phi^0 \vee \Phi^1)$ or $\Phi = (\Phi^0 \wedge \Phi^1)$ then $\Psi = (\Psi^0 \vee \Psi^1)$ or $\Psi = (\Psi^0 \wedge \Psi^1)$ respectively;

(3) if $\Phi = \forall x \in t_\varkappa(\overline{u})\Phi^0$ then $\Psi$ is obtained from $\bigwedge_{\varkappa' \in \varkappa}[\Phi^0]^x_{t_{\varkappa'}(\overline{u})}$ by the previous rules (as usual, $\bigwedge \varnothing = \top$);

(4) if $\Phi = \exists x \in t_\varkappa(\overline{u})\Phi^0$ then $\Psi$ is obtained from $\bigvee_{\varkappa' \in \varkappa}[\Phi^0]^x_{t_{\varkappa'}(\overline{u})}$ by the previous rules (as usual, $\bigvee \varnothing = \bot$).

To finish the proof, it suffices to replace $\top$ and $\bot$ with some quantifier-free true and false formulas if such formulas exist; with formulas $\exists u(u \approx u)$, $\forall u(u \approx u)$, and $\exists u \urcorner (u \approx u)$, $\forall u \urcorner (u \approx u)$ respectively, otherwise. □

**Proposition 6.12.** *For every $\Sigma_1$-formula $\Phi$ of $\sigma \cup \{\mathrm{U}, \varnothing, \in, \cup, \{\ \}\}$ with ur-variables, one can effectively construct some computable disjunction $\Phi^*$ of $\exists$-formulas of the signature $\sigma$ so that*

$$\mathbb{HF}(\mathfrak{M}) \models \Phi \Leftrightarrow \mathfrak{M} \models \Phi^*.$$

***Proof.*** We assume that any unrestricted quantifier appearing in $\Phi$ acts on a common variable. If $\Phi = \exists x \Phi^0(x, \overline{v})$ then $\Psi = \bigvee_{n<\omega} \exists \overline{u} \Phi^0(t_n(\overline{u}), \overline{v})$. This transformation is effective. To finish the proof, it remains to apply Lemma 6.4. □

Thus, we have proved the following:

**Theorem 6.10.** *Let $\sigma$ be a finite relation signature. Then there exists a computable sequence $A_{m,n}$ for which the following conditions hold:*

*(1) if $\Phi(x_0, y_0, \ldots, y_{l-1})$ is a $\Sigma$-formula of $\sigma \cup \{\mathrm{U}, \varnothing, \in\}$ then $A_{\Phi,n}$ consists of $\exists$-formulas of $\sigma$;*

*(2) for any structure $\mathfrak{M}$ of $\sigma$, $A$ is definable in $\mathbb{HF}(\mathfrak{M})$ by the $\Sigma$-formula $\Phi(x_0, s_0, \ldots, s_{l-1})$ with parameters $s_0, \ldots, s_{l-1}$, $l \geqslant 0$ if and only if*

$$A = \{t_n(\overline{u}) \mid n \in \omega,\ \mathfrak{M} \models \varphi(\overline{u}, s_0, \ldots, s_{l-1}))\ \textit{for some}\ \varphi \in A_{\Phi,n}\}.$$

Moreover, an "almost" converse assumption holds:

**Theorem 6.11.** *Let $\sigma$ be a finite relation signature. For any computable sequence $\{A_n\}_{n\in\omega}$ in which every $A_n$ consists of $\exists$-formulas of signature $\sigma$,*

$$A = \{t_n(\overline{u}) \mid n \in \omega,\ \mathfrak{M} \models f(n)(\overline{u}, s_0, \ldots, s_{l-1}))\}$$

*is a $\Sigma$-subset of $\mathbb{HF}(\mathfrak{M})$ where $\mathfrak{M}$ is a structure of $\sigma$. Moreover, a $\Sigma$-formula defining $A$ is independent of a choice of $\mathfrak{M}$ and can be effectively found from $\{A_n\}_{n\in\omega}$.*

To establish this assumption, it suffices to show that the truth predicate for $\Sigma$-formulas is $\Sigma$-definable.

Let $\mathbb{HF}(\mathfrak{M})$ be the hereditarily finite superstructure over a structure $\mathfrak{M}$ of signature $\sigma$. By a *$\Sigma$-operator* we mean a map $F : \mathcal{P}(HF(M)) \to \mathcal{P}(HF(M) \cup M)$ which is defined in the following way, for every $X \subseteq HF(M)$:

$$F(X) = \{a \mid \exists b[\langle a, b\rangle \in R \wedge (b \subseteq X)]\}$$

for some $\Sigma$-predicate $R$.

The notion of $\Sigma$-operator generalizes the notion of enumeration operator. Like the enumeration operators, these operators have the following properties:

**continuity:** $x \in F(X) \Rightarrow \exists Y \subseteq X[\mathrm{card}(Y) < \omega \wedge x \in F(Y)]$;
**monotonicity:** $X_1 \subseteq X_2 \Rightarrow F(X_1) \subseteq F(X_2)$.

Now we give a series of examples of $\Sigma$-operators. Let $\Phi(x, R^+)$ be a $\Sigma$-formula of $\sigma \cup \{R\}$, $\#(R) = 1$, $R \notin \sigma$, in which $R$ occurs positively. Then for any $X \subseteq HF(M)$,

$$F_\Phi(X) = \{a \mid (\mathbb{HF}(\mathfrak{M}), X) \models \Phi(a)\}$$

is a $\Sigma$-operator.

By monotonicity, every $\Sigma$-operator has the least fixed point, namely, there is $Y_0 \subseteq HF(M)$ such that $F(Y_0) = Y_0$ and $\forall Y_1 \subseteq HF(M)[F(Y_1) \subseteq Y_1 \Rightarrow Y_0 \subseteq Y_1]$. Let $\Gamma_0 \leftrightharpoons \varnothing$; $\Gamma_{\alpha+1} \leftrightharpoons F(\Gamma_\alpha)$; $\Gamma_\eta \leftrightharpoons \bigcup_{\beta<\eta} \Gamma_\beta$ if $\eta$ is limit; then it is easy to check that $\Gamma_* \leftrightharpoons \bigcup_{\alpha<\mathrm{card}(\mathbb{HF}(\mathfrak{M}))^+} \Gamma_\alpha$ will be the least fixed point of $F$.

**Theorem 6.12. (Gandy)** *Let $\mathbb{HF}(\mathfrak{M})$ be the hereditarily finite superstructure over $\mathfrak{M}$ and $F$ be a $\Sigma$-operator on $\mathbb{HF}(\mathfrak{M})$. Then the least fixed point $\Gamma_*$ of $F$ is $\Sigma$ on $\mathbb{HF}(\mathfrak{M})$. Moreover, $\Gamma_* = \Gamma_\omega$.*

It follows from the Gandy Theorem that for any structure $\mathfrak{M}$, $\mathbb{HF}(\mathfrak{M})$ has a universal $\Sigma$-predicate. Let $\mathcal{K}$ be a class of $n$-ary relations on $\mathbb{HF}(\mathfrak{M})$. A predicate $P \subseteq |\mathbb{HF}(\mathfrak{M})|^{n+1}$ is called *universal* for $\mathcal{K}$ if $\mathcal{K} = \{\{\langle b_1, \ldots, b_n\rangle \mid \langle a, b_1, \ldots, b_n\rangle \in P\} \mid a \in |\mathbb{HF}(\mathfrak{M})|\}$. In particular, $P$ is a *universal $\Sigma$-predicate* if it is universal for the class of all $n$-ary $\Sigma$-predicates on $\mathbb{HF}(\mathfrak{M})$; a partial $\Sigma$-function $f(y, x_1, \ldots, x_n)$ is *universal* if its graph $\Gamma_f$ is universal for the class of graphs of all $n$-ary partial $\Sigma$-functions.

**Theorem 6.13.** *There exists a binary $\Sigma$-predicate* $\mathrm{Tr}_\Sigma$ *on* $\mathbb{HF}(\mathfrak{M})$ *such that for any $\Sigma$-formula $\Phi(x)$ and $a \in HF(M) \cup M$,*

$$\langle \Phi, a \rangle \in \mathrm{Tr}_\Sigma \Leftrightarrow \mathbb{HF}(\mathfrak{M}) \models \Phi(a).$$

**Theorem 6.14.** *There exists an $(n+1)$-ary universal $\Sigma$-predicate $T(e, x_1, \ldots, x_n)$ on $\mathbb{HF}(\mathfrak{M})$.*

Notice that not all hereditarily finite superstructures have universal $\Sigma$-functions [49, 101, 132, 162].

The collection of $\Sigma$-subsets of $\omega$ is one of the main computation invariants of hereditarily finite superstructures.

**Theorem 6.15. [101, 114, 133]**

*(1) For any admissible set $\mathbb{A}$, the collection of $\Delta$-subsets of $\omega$ is closed under $\oplus$ and downwards under $T$-reducibility.*
*(2) For any admissible set $\mathbb{A}$, the collection of $\Sigma$-subsets of $\omega$ is closed under $\oplus$ and downwards under e-reducibility.*
*(3) For every $T$-ideal* I, *there exists a hereditarily finite superstructure on which the class of $T$-degrees of $\Delta$-subsets coincides with* I.
*(4) For every e-ideal* I, *there exists a hereditarily finite superstructure on which the class of e-degrees of $\Sigma$-subsets coincides with* I.

At the end of this section, we consider a series of examples of hereditarily finite superstructures having universal $\Sigma$-functions. Let $\mathbb{HF}(\mathfrak{M})$ be the hereditarily finite superstructure over a structure $\mathfrak{M}$ of some finite relation signature. A sequence of its subsets

$$A_0 \subseteq A_1 \subseteq \ldots \subseteq A_n \subseteq A_{n+1} \subseteq \ldots, \; n \in \omega$$

is called a $\Sigma$-*resolution* of $\mathbb{HF}(\mathfrak{M})$ if the following hold:

(1) $A_n$ is a transitive subset of $\mathbb{HF}(\mathfrak{M})$, for any $n \in \omega$;
(2) $\bigcup_{n \in \omega} A_\alpha = A$;
(3) $\{\langle a, n \rangle \mid n \in \omega, \, a \in A_n\}$ is $\Sigma$ on $\mathbb{HF}(\mathfrak{M})$.

**Proposition 6.13.** *Let $\{A_n\}_{n \in \omega}$ be a $\Sigma$-resolution of $\mathbb{HF}(\mathfrak{M})$, let $\Phi(x)$ be a $\Sigma$-formula with parameters $m_1, \ldots, m_k \in |\mathfrak{M}|$, $k \in \omega$, and let $b \in |\mathbb{HF}(\mathfrak{M})|$. Then $\mathbb{HF}(\mathfrak{M}) \models \Phi(b)$ if and only if $\mathbb{HF}(\mathfrak{M}) \restriction A_n \models \Phi(b)$ for some $n$ satisfying $\{b, m_1, \ldots, m_k\} \subseteq A_n$.*

A sequence $D_n \rightleftharpoons \{a \mid a \in HF(M), \mathrm{rk}(a) \leqslant n\}$, $n \in \omega$, is an example of a nontrivial $\Sigma$-resolution of $\mathbb{HF}(\mathfrak{M})$. A $\Sigma$-resolution $\{A_n\}_{n\in\omega}$ of $\mathbb{HF}(\mathfrak{M})$ is called a *quasiresolution* if

$$\mathrm{Tr}^\Sigma(\{A_n\}_{n\in\omega}) \rightleftharpoons \{\langle n, \Phi, a\rangle \mid n \in \omega,\ a \in A_n,\\ \Phi(x) \text{ is a } \Sigma-\text{formula},\ \mathbb{HF}(\mathfrak{M}) \upharpoonright A_n \models \Phi(a)\}$$

is $\Delta$ on $\mathbb{HF}(\mathfrak{M})$.

**Remark 6.3.2.** If $\{A_n\}_{n\in\omega}$ is a $\Sigma$-resolution for $\mathbb{HF}(\mathfrak{M})$ then $\mathrm{Tr}^\Sigma(\{A_n\}_{n\in\omega})$ is always $\Sigma$ on $\mathbb{HF}(\mathfrak{M})$.

$\mathbb{HF}(\mathfrak{M})$ is said to be *quasiresolvable* if $\mathbb{HF}(\mathfrak{M})$ has at least one quasiresolution. If $\mathbb{HF}(\mathfrak{M})$ is quasiresolvable then it satisfies reduction and has a universal $\Sigma$-function [33, 36]. There exist hereditarily finite superstructures which are not quasiresolvable but satisfy reduction and have universal $\Sigma$-functions [49].

By a *canonical $\Sigma$-resolution* of $\mathbb{HF}(\mathfrak{M})$ we mean the following $\Sigma$-resolution:

- $B_0 \rightleftharpoons M = \mathrm{U}(\mathbb{HF}(\mathfrak{M}))$;
- $B_n \rightleftharpoons \{t_\varkappa(\overline{m}) \mid \overline{m} \in M^n,\ \varkappa \in HF_n(n)\}$, $0 < n < \omega$.

**Proposition 6.14.** *For any $\Sigma$-formula $\Phi(u_0, \ldots, u_{k-1})$ and $n \in \omega$, one can effectively find an $\exists$-formula of the signature of $\mathfrak{M}$ so that*

$$\mathbb{HF}(\mathfrak{M}) \upharpoonright B_n \models \Phi(m_0, \ldots, m_{k-1}) \Leftrightarrow \mathfrak{M} \models \Phi_n(m_0, \ldots, m_{k-1}),$$

*for each $m_0, \ldots, m_{k-1} \in |\mathfrak{M}|$.*

**Corollary 6.6.** *For every $\Sigma$-formula $\Phi(u_0, \ldots, u_{k-1})$ and $m_0, \ldots, m_{k-1} \in |\mathfrak{M}|$,*

$$\mathbb{HF}(\mathfrak{M}) \models \Phi(m_0, \ldots, m_{k-1}) \Leftrightarrow \mathfrak{M} \models \bigvee_{n\in\omega} \Phi_n(m_0, \ldots, m_{k-1}).$$

**Proposition 6.15.** *The canonical $\Sigma$-resolution of $\mathbb{HF}(\mathfrak{M})$ is a quasiresolution if and only if $\mathfrak{M}$ is 1-decidable in $\mathbb{HF}(\mathfrak{M})$, i.e.,*

$$\{\langle \Phi, \overline{a}\rangle \mid \Phi \text{ is } \exists\text{-formula},\ \overline{a} \in |\mathfrak{M}|^{<\omega}, \mathfrak{M} \models \Phi(\overline{a})\}$$

*is $\Delta$ on $\mathbb{HF}(\mathfrak{M})$.*

We give one important corollary of Proposition 6.15.

**Proposition 6.16.** *Let $\mathfrak{M}$ be such that for any $\forall$-formula $\Phi(\overline{u})$ of the signature of $\mathfrak{M}$, one can effectively find some $\exists$-formula $\Psi(\overline{u})$ satisfying*

*$\mathfrak{M} \models \forall \overline{u}(\Phi(\overline{u}) \leftrightarrow \Psi(\overline{u}))$. Then the canonical $\Sigma$-resolution of $\mathbb{HF}(\mathfrak{M})$ is a quasiresolution.*

Notice that if $\mathfrak{M}$ satisfies the conditions of Proposition 6.16 then $\mathrm{Th}(\mathfrak{M})$ is model complete. In particular, these conditions are satisfied whenever $\mathfrak{M}$ is a structure of some regular theory T. Recall that T is *regular* if it is decidable and model complete.

**Proposition 6.17.** [57] *Let $\mathfrak{M}$ be such that $\mathrm{Th}(\mathfrak{M})$ is $\omega$-categorical. Then $\mathfrak{M}$ is 1-decidable in $\mathbb{HF}(\mathfrak{M})$ iff $\mathbb{HF}(\mathfrak{M})$ is quasiresolvable.*

A structure $\mathfrak{M}$ is said to be *quasiresolvable* if $\mathbb{HF}(\mathfrak{M})$ is quasiresolvable.

**Theorem 6.16.** [58] *For any Ershov algebra $\mathfrak{A}$, the following conditions are equivalent:*

*(1) $\mathfrak{A}$ is quasiresolvable;*
*(2) $\mathfrak{A}$ is 1-decidable in $\mathbb{HF}(\mathfrak{A})$;*
*(3) $\mathfrak{A}$ is the join of a non-atomic Ershov algebra and of a finite Boolean algebra.*

**Theorem 6.17.** [58] *Let $G$ be an abelian $p$-group and let $R$, $D$ be its reducible and its divisible parts respectively. The following conditions are equivalent:*

*1) $G$ is quasiresolvable;*

*2) if $r(D) \geqslant \omega$ then $R$ is finite; if $r(D) < \omega$ then $R = R_0 \oplus R_1$, where $R_0$ is finite and there is $n \geqslant 0$, for which $R_1 \cong C^{\alpha}_{p^n}$ ($C^{\alpha}_{p^0} = 0$), $\alpha \geqslant \omega$ ($C_{p^n}$ is a cyclic group of order $p^n$ here).*

## 6.4. Reducibilities on Hereditarily Finite Superstructures

In this section, we consider generalizations of classical reducibilities on hereditarily finite superstructures. We recall some lattice-theoretic notions.

**Definition 6.7.** Let $\mathcal{L} = \langle L, \leqslant \rangle$ be a partially ordered set.

- $\mathcal{L}$ is called an *upper semilattice* if any two elements $a$ and $b$ have a least upper bound $a \sqcup b$, i.e., for any $a, b, c \in L$, we have: $a \leqslant a \sqcup b$, $b \leqslant a \sqcup b$ and $(a \leqslant c) \wedge (b \leqslant c) \Rightarrow (a \sqcup b \leqslant c)$.
- An upper semilattice $\mathcal{L}$ is called *distributive* if for any $a, b, c \in L$, there are $a_0, b_0 \in L$ such that $(c \leqslant a \sqcup b) \Rightarrow ((a_0 \leqslant a) \wedge (b_0 \leqslant b) \wedge (c = a_0 \sqcup b_0))$.
- A non-empty subset $I \subseteq L$ is called an *ideal* of an upper semilattice $\mathcal{L}$ if $I$ satisfies the following conditions:

(1) $a, b \in I \Rightarrow a \sqcup b \in I$ (it is closed under taking the least upper bound operation);
(2) $b \in I$, $a \leqslant b \Rightarrow a \in I$ (it is closed downwards under $\leqslant$).

- $a \in L$ is the *least (greatest) element* if $a \leqslant b$ $(b \leqslant a)$, for every $b \in L$.

**Definition 6.8.** Let $\mathbb{HF}(\mathfrak{M})$ be the hereditarily finite superstructure over $\mathfrak{M}$ and $B, C \subseteq |\mathbb{HF}(\mathfrak{M})|$.

- We say that $B$ is *$m\Sigma$-reducible* to $C$, written $B \leqslant_{m\Sigma} C$, if there exists a binary $\Sigma$-predicate $R$ on $\mathbb{HF}(\mathfrak{M})$ such that $\mathrm{Pr}_1(R) = |\mathbb{HF}(\mathfrak{M})|$ and $\langle a, b\rangle \in R \Rightarrow ((a \in B) \leftrightarrow (b \in C))$, for every $a, b \in |\mathbb{HF}(\mathfrak{M})|$.
- We say that $B$ is *$T\Sigma$-reducible* to $C$, written $B \leqslant_{T\Sigma} C$, if there exist two binary $\Sigma$-operators $\Phi_0$ and $\Phi_1$ on $\mathbb{HF}(\mathfrak{M})$ such that $B = \Phi_0(C, |\mathbb{HF}(\mathfrak{M})| \setminus C)$ and $|\mathbb{HF}(\mathfrak{M})| \setminus B = \Phi_1(C, |\mathbb{HF}(\mathfrak{M})| \setminus C)$.
- We say that $B$ is *$e\Sigma$-reducible* to $C$, written $B \leqslant_{e\Sigma} C$, if there exists a $\Sigma$-operator $\Phi$ on $\mathbb{HF}(\mathfrak{M})$ for which $B = \Phi(C)$.
- Let $r \in \{m\Sigma, T\Sigma, e\Sigma\}$. We say that $B$ and $C$ are *$r$-equivalent* ($B \equiv_r C$) if $B \leqslant_r C$ and $C \leqslant_r B$.
- Let $r \in \{m\Sigma, T\Sigma, e\Sigma\}$. A class w.r.t $r$-equivalence is called an *$r$-degree.*

We give some properties of these reducibilities:

(1) If $\mathbb{HF}(\mathfrak{M})$ satisfies uniformization then the notion of $m\Sigma$-reducibility can be formulated as in the classical case: $B \leqslant_{m\Sigma} C$ if and only if there exists a total $\Sigma$-function $f$ such that $a \in B \Leftrightarrow f(a) \in C$, for every $a \in |\mathbb{HF}(\mathfrak{M})|$.
(2) Let $\mathbb{HF}(S)$ be the hereditarily finite superstructure over an infinite set $S$. For any $A \subseteq \omega$, we define $S_A = \{\langle n, a\rangle \mid n \in A; a = \langle a_0, \ldots, a_{n-1}\rangle \in S^n; a_i \neq a_j, 0 \leqslant i < j < n\}$ which is $m\Sigma$-equivalent to $A$. Then there is no total $\Sigma$-function establishing $m\Sigma$-reducibility $S_A$ to $A$ whenever $A$ is not computable.

The relations of $m\Sigma$-, $e\Sigma$- and $T\Sigma$-reducibilities are preorders on $\mathcal{P}(|\mathbb{HF}(\mathfrak{M})|) \setminus \{\varnothing, |\mathbb{HF}(\mathfrak{M})|\}$ and, therefore, classes of corresponding degrees form partially ordered sets under the induced orderings. Moreover, they are upper semilattices with least elements, namely, classes containing proper $\Delta$-subsets of $\mathbb{HF}(\mathfrak{M})$. Note that $[B] \sqcup [C] = [B \oplus C]$ where $B \oplus C = (B \times \{0\}) \cup (C \times \{1\})$.

By $\mathbb{L}_{m\Sigma}(\mathbb{HF}(\mathfrak{M}))$, $\mathbb{L}_{T\Sigma}(\mathbb{HF}(\mathfrak{M}))$, $\mathbb{L}_{e\Sigma}(\mathbb{HF}(\mathfrak{M}))$ we denote upper semilattices of classes of $m\Sigma$-, $T\Sigma$- and $e\Sigma$-degrees of proper subsets of $HF(M) \cup M$ respectively. As it was shown in [114], the upper semilattice $\mathbb{L}_{m\Sigma}(\mathbb{HF}(\mathfrak{M}))$

is distributive, for every structure $\mathfrak{M}$. Furthermore, there exists a natural isomorphism between $\mathbb{L}_{m\Sigma}(\mathbb{HF}(\varnothing))$, $\mathbb{L}_{T\Sigma}(\mathbb{HF}(\varnothing))$, $\mathbb{L}_{e\Sigma}(\mathbb{HF}(\varnothing))$ and upper semilattices of $m$-,$T$-, $e$-degrees respectively.

Let $\mathfrak{M}$ be an arbitrary structure and let $r \in \{m\Sigma, T\Sigma, e\Sigma\}$.

**Definition 6.9.** A degree $\mathbf{a} \in \mathbb{L}_r(\mathbb{HF}(\mathfrak{M}))$ is called *computably enumerable* if there exists a $\Sigma$-subset $B \in \mathbf{a}$ of $\mathbb{HF}(\mathfrak{M})$.

**Definition 6.10.** A degree $\mathbf{a} \in \mathbb{L}_r(\mathbb{A})$ is called *definable* if there exists a definable subset $B \in \mathbf{a}$ of $\mathbb{HF}(\mathfrak{M})$.

Recall that a theory $T$ is called *c-simple* [33] if it is $\omega$-categorical, model complete, decidable, and has a decidable set of complete formulas.

The following theorem describes relations between semilattices on hereditarily finite superstructures over structures of $c$-simple theories and classical ones:

**Theorem 6.18.** [113, 114] *Let $\mathfrak{M}$ be a structure of a c-simple theory in some finite signature and $\imath$ the natural embedding $\mathbb{HF}(\varnothing)$ into $\mathbb{HF}(\mathfrak{M})$. Then the following conditions hold:*

*(1) $\imath$ induces an isomorphism between the upper semilattices of computably enumerable $m$-($T$-) degrees and of computably enumerable $m\Sigma$-($T\Sigma$-) degrees;*
*(2) $\imath$ induces an isomorphism between the upper semilattices of definable $m$-($T$-,$e$-) degrees and of definable $m\Sigma$-($T\Sigma$-,$e\Sigma$-) degrees;*
*(3) $\imath$ induces embedding of the upper semilattices of $m$-($T$-,$e$-) degrees into $\mathbb{L}_{m\Sigma}(\mathbb{HF}(\mathfrak{M}))$ ($\mathbb{L}_{T\Sigma}(\mathbb{HF}(\mathfrak{M}))$, $\mathbb{L}_{e\Sigma}(\mathbb{HF}(\mathfrak{M}))$) as ideals.*

If $\mathfrak{M}$ is a countable structure of some $c$-simple theory then the semilattice $\mathrm{L}_{m\Sigma}(\mathbb{HF}(\mathfrak{M}))$ is described up to isomorphism:

**Theorem 6.19.** [124, 125] *Let $\mathfrak{M}$ be a countable structure of a c-simple theory in some finite signature. Then the upper semilattices $\mathrm{L}_{m\Sigma}(\mathbb{HF}(\mathfrak{M}))$ and $L_m$ are isomorphic.*

Additional information about generalized numberings and reducibilities on admissible sets can be found in [5, 6, 33, 44, 52, 53, 116, 117, 121, 131, 160, 161, 163].

## 6.5. Descriptive Properties on Hereditarily Finite Superstructures

As in the classical computability, the existence of a universal $\Sigma$-predicate implies that the class of $\Sigma$-subsets is not closed under the complement operation. To avoid this obstacle, properties from descriptive set theory are sometimes applied. In this section, we discuss the problem of the existence of hereditarily finite superstructures with respect to relations between such properties. Recall the basic definitions.

Let $\mathbb{A}$ be a hereditarily finite superstructure and A its universe.

**Definition 6.11.**

(1) $\mathbb{A}$ is called *recursively listed* if there is a $\Sigma$-function $f : \omega \to A$ with $\rho f = A$.
(2) $\mathbb{A}$ is called *resolvable* if there exists a $\Sigma$-function $f : \omega \to A$ with $\bigcup_{n \in \omega} f(n) = A$. Such an $f$ is called a *resolution* for $\mathbb{A}$.
(3) Let $C$ be a $\Sigma$-subset of $\mathbb{A}$.
   - $\mathbb{A}$ is *projectible into* $C$ if there exists a $\Sigma$-function $f$ with $\delta f \subseteq C$ and $\rho f = A$ such that $f^{-1}(x) \in A$ for every $x \in A$.
   - $\mathbb{A}$ is *quasiprojectible into* $C$ if there exists a $\Sigma$-function $f$ with $\delta f \subseteq C$ and $\rho f = A$.

**Definition 6.12.** We say that $\mathbb{A}$ satisfies

- *reduction* if, for any $\Sigma$-subsets $B_0$ and $B_1$, there are disjoint $\Sigma$-subsets $C_0 \subseteq B_0$ and $C_1 \subseteq B_1$ such that $C_0 \cup C_1 = B_0 \cup B_1$.
- *separation* if, for any disjoint $\Sigma$-subsets $B_0$ and $B_1$, there is a $\Delta$-subset $C$ such that $B_0 \subseteq C \subseteq A \setminus B_1$.
- *extension* if, for any partial $\Sigma$-function $\varphi(x)$, there is a total $\Sigma$-function $f(x)$ such that $\Gamma_\varphi \subseteq \Gamma_f$.
- *uniformization* if, for any binary $\Sigma$-predicate $R$ on $\mathbb{A}$, there is a partial $\Sigma$-function $\varphi(x)$ with $\delta\varphi = \mathrm{Pr}_1(R)$ and $\Gamma_\varphi \subseteq R$.

The main merit of a recursively listed hereditarily finite superstructure is the existence of an effective function enumerating its range via natural numbers which enables us to transfer such principles from classical computability as reduction, uniformization, the existence of a universal $\Sigma$-function etc., [11]. Also, any infinite $\Sigma$-subset $C$ of $\mathbb{A}$ has an enumeration without repetition, namely, a one-to-one $\Sigma$-function $f$ with $\delta f = \omega$ and $\rho f = C$.

The notion of resolvable hereditarily finite superstructure is a generalization of the notion of recursively listed hereditarily finite superstructure. As for recursively listed hereditarily finite superstructures, in the study of the properties of such structures, one can apply the method of constructing $\Sigma$-subsets by using effective approximations consisting of finite subsets. As before, they satisfy reduction and have universal $\Sigma$-functions; however, in general, uniformization does not hold on them. Moreover, the following is true:

**Proposition 6.18.** *Let $\mathbb{HF}(\mathfrak{M})$ be resolvable. Then $\mathbb{HF}(\mathfrak{M})$ is recursively listed if and only if $\mathbb{HF}(\mathfrak{M})$ satisfies uniformization.*

An approach which avoids uniformization is to construct structures with a small number of types and an infinite set of indiscernibles.

**Theorem 6.20.** [57] *If $\mathfrak{M}$ is a structure of some countably categorical theory then $\mathbb{HF}(\mathfrak{M})$ does not satisfy uniformization.*

**Theorem 6.21.** [9] *If $\mathfrak{M}$ is an $\omega$-saturated structure of some uncountably categorical theory then $\mathbb{HF}(\mathfrak{M})$ does not satisfy uniformization.*

E.g., if $\mathfrak{M}$ is an algebraically closed field with characteristic zero (in other words, a structure of the theory of the field of complex numbers), then $\mathbb{HF}(\mathfrak{M})$ does not satisfy uniformization, even if $\mathfrak{M}$ is not $\omega$-saturated. However, if $\mathfrak{N}$ is a structure of the theory $\mathrm{Th}(\omega, \mathbf{0}, \mathrm{s})$ of natural numbers with zero and the successor relation, then $\mathbb{HF}(\mathfrak{N})$ does not satisfy uniformization iff $\mathfrak{N}$ is $\omega$-saturated; moreover, if $\mathfrak{N}$ is not $\omega$-saturated then $\mathbb{HF}(\mathfrak{N})$ is recursively listed.

The notion of projectibility is one more generalization of the notion of recursively listed hereditarily finite superstructure. This definition was introduced in [11]. Hereditarily finite superstructures quasiprojectible into $\omega$ seem to be interesting; such structures have properties which look like the corresponding ones on enumeration degrees. Further on, hereditarily finite superstructures quasiprojectible into $\omega$ will be called simply *quasiprojectible.*

**Example 6.1.** Let $A \subseteq \omega$. We define a structure $\mathfrak{N}_A$ of signature $\{\mathbf{0}, \mathrm{s}, \mathrm{P}\}$ as follows:

- $|\mathfrak{N}_A| \leftrightharpoons \omega \uplus \{z_a \mid a \in A\}$, whenever $z_{a_1} \neq z_{a_2}$ for $a_1 < a_2$ (hereinafter, the symbol $\uplus$ means the disjoint union);
- $\mathbf{0}^{\mathfrak{N}_A} = 0 \in \omega$; $\langle a, b\rangle \in \mathrm{s}^{\mathfrak{N}_A} \Leftrightarrow (\{a, b\} \subseteq \omega \wedge b = a + 1)$;

* $\mathrm{P}^{\mathfrak{N}_A} \leftrightharpoons \{\langle a, z_a\rangle \mid a \in A\}$.

The class of structures $\{\mathbb{HF}(\mathfrak{N}_A) \mid A \subseteq \omega\}$ consists of hereditarily finite superstructures projectible into $\omega$ only. This class has a series of nice properties [49].

Now we give sufficient conditions of satisfiability of properties from descriptive set theory on admissible sets. As is said above, we have the following:

**Theorem 6.22.** **[33, 36]** *If $\mathbb{A}$ is a quasiresolvable admissible set then $\mathbb{A}$ satisfies reduction and has a universal $\Sigma$-function.*

In section 6.6.3, a criterion for the satisfiability of uniformization on hereditarily finite superstructures structures of kind $\mathbb{HF}(\mathfrak{M})$, where $\mathrm{Th}(\mathfrak{M})$ is regular, is given (see also [142, 157]). In general, the uniformization property does not hold even on hereditarily finite superstructures of this kind.

The following theorem gives us a description of hereditarily finite superstructures with respect to relations between the properties considered here [123].

**Theorem 6.23.** *The following implications between the properties on hereditarily finite superstructures hold:*

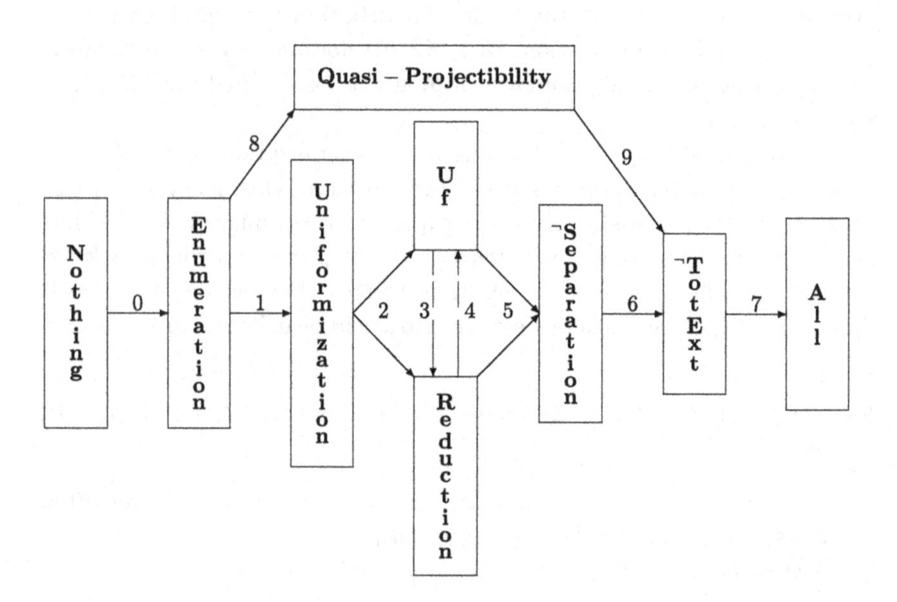

*All the implications in the diagram are proper. Hereditarily finite superstructures over computable structures can be found in (0–5, 9). There is no such structure in (6–8).*

Now we give examples of hereditarily finite superstructures over classical and (or) computable structures (if it is possible) which demonstrate differences between the properties considered above.

Examples 6.5.1. The numbering of examples coincides with the corresponding implications from Theorem 6.23.

**0** One can take here the standard model of arithmetic.

**1** Let $\mathbb{R}$ be the field of real numbers. Then $\mathbb{HF}(\mathbb{R})$ satisfies uniformization [142], however, for reasons of cardinality, this hereditarily finite superstructure is not recursively listed. Indeed, there are countable hereditarily finite superstructures which satisfy uniformization but are not recursively listed, e.g., $\mathbb{HF}(\mathfrak{N}_{\Pi_1^1})$ or a countable elementary substructure of $\mathbb{HF}(\mathbb{R})$.
However, all the structures considered above are not computable.
Now we give an example of computable real closed field $\mathbb{R}^*$ for which $\mathbb{HF}(\mathbb{R}^*)$ is not recursively listed but satisfies uniformization. Let $\mathbb{Q}(a_0, a_1, \dots, a_n, \dots)$ be a purely transcendental extension of the ordered field of rational numbers such that all the elements from $\mathbb{Q}(a_0, \dots, a_{n-1})$ are infinitesimal w.r.t. $a_n$, $n \in \omega$. Then we set $\mathbb{R}^*$ as the real closure of $\mathbb{Q}(a_0, a_1, \dots, a_n, \dots)$ [33]. Then it is computable (even decidable) and $\mathbb{HF}(\mathbb{R}^*)$ has the desired properties.

**2** If $\mathfrak{M}$ is a structure of some decidable, model complete, countably categorical theory then $\mathbb{HF}(\mathfrak{M})$ is not resolvable, uniformization does not hold on $\mathbb{HF}(\mathfrak{M})$ but this hereditarily finite superstructure satisfies reduction and has a universal $\Sigma$-function. One can take here the set of rational numbers with the natural order and a countable structure in the empty language as $\mathfrak{M}$.

**3** In [123], a series of examples of hereditarily finite superstructures including computable ones are given, however, these structures are not classical.

**4** We define a structure $\mathfrak{M}$ of signature $\{P, Q\}$, $\#(Q) = 2$, as follows:

- $P^{\mathfrak{M}}$ and $|\mathfrak{M}| \setminus P^{\mathfrak{M}}$ are infinite;

- $Q^{\mathfrak{M}}$ is the graph of a one-to-one function from $P^{\mathfrak{M}}$ into $|\mathfrak{M}| \setminus P^{\mathfrak{M}}$ such that $(|\mathfrak{M}| \setminus P^{\mathfrak{M}}) \setminus \mathrm{Pr}_2(Q^{\mathfrak{M}})$ is infinite.

Then $\mathbb{HF}(\mathfrak{M})$ does not satisfy reduction but has a universal $\Sigma$-function. Notice that $\mathrm{Th}(\mathfrak{M})$ is decidable and countably categorical, and therefore the unique countable structure of the theory is computable (even decidable).

**5** There exists a $\Sigma_2$-set $A \subseteq \omega$ such that $\mathbb{HF}(\mathfrak{N}_A)$ does not satisfy reduction and separation, and has no universal $\Sigma$-function [49].
In [132], an example of hereditarily finite superstructure over a countable structure of some decidable countably categorical theory without universal $\Sigma$-function is given. Furthermore, reduction and separation does not hold on this hereditarily finite superstructure.

**6** The hereditarily finite superstructure $\mathbb{HF}(\mathfrak{N}_{\Sigma_1^1})$ satisfies separation but does not satisfy extension. These conditions are satisfied on $\mathbb{HF}(\mathcal{G})$ for some abelian group $\mathcal{G}$ [57]. To prove this, it suffices to apply methods from [119].

**7** In [119], examples of hereditarily finite superstructures satisfying extension are constructed. At this moment, examples of hereditarily finite superstructures over classical structures have not been found.

**8** A series of quasiprojectible hereditarily finite superstructures is given in Example 6.1.

**9** The hereditarily finite superstructure $\mathbb{HF}(\mathfrak{N}_{\Sigma_1^1})$ has the desired properties.

Additional information about the properties considered above can be found in [7, 57, 58, 60, 66, 94, 114, 122, 133, 162].

### 6.6. Σ-Definability of Structures

The theory of constructive (computable) models is one of the important research areas of the classical computability theory, as well as of the model theory. Because of the evident cardinality limitations, in the classical computable model theory only countable structures are considered. The approach regarding generalized computability as $\Sigma$-definability in admissible sets allows us to consider structures with arbitrary cardinality. Hereditarily finite superstructures are the "simplest" admissible sets, from the set-theoretical point of view. Besides of this, $\Sigma$-definability in hereditary finite superstructures is one of the natural approaches generalizing classical computability theory on natural numbers to the case of computability over arbitrary structures.

Hence, for a structure $\mathfrak{M}$ the following problems naturally arise:

- to describe the structures $\Sigma$-definable in $\mathbb{HF}(\mathfrak{M})$;
- to describe the structures such that $\mathfrak{M}$ is $\Sigma$-definable in their HF-superstructures.

Let us formalize the problems stated above. Let $\mathfrak{M}$ be a structure of a finite predicate signature $\langle P_1, \dots, P_k\rangle$, where each $P_i$ is $n_i$-ary, and let $\mathbb{A}$ be an admissible set. To simplify the notations in this chapter, we write $M$ instead of $|\mathfrak{M}|$. The following notion is an effectivization of the model-theoretical notion of interpretability of one structure in another, and also a natural generalization of the notion of constructivizability of a (countable) structure on natural numbers.

**Definition 6.13.** [24, 33] $\mathfrak{M}$ is *$\Sigma$-definable in* $\mathbb{A}$ if there exist $\Sigma$-formulas

$$\Phi(x_0, y), \Psi(x_0, x_1, y), \Psi^*(x_0, x_1, y), \Phi_1(x_0, \dots, x_{n_1-1}, y),$$

$$\Phi_1^*(x_0, \dots, x_{n_1-1}, y), \dots, \Phi_k(x_0, \dots, x_{n_k-1}, y), \Phi_k^*(x_0, \dots, x_{n_k-1}, y),$$

such that for some parameter $a \in A$, and letting

$$M_0 \leftrightharpoons \Phi^{\mathbb{A}}(x_0, a), \quad \eta \leftrightharpoons \Psi^{\mathbb{A}}(x_0, x_1, a) \cap M_0^2$$

one has that $M_0 \neq \varnothing$ and $\eta$ is a congruence relation on the structure

$$\mathfrak{M}_0 \leftrightharpoons \langle M_0, P_1^{\mathfrak{M}_0}, \dots, P_k^{\mathfrak{M}_0}\rangle,$$

where $P_i^{\mathfrak{M}_0} \leftrightharpoons \Phi_i^{\mathbb{A}}(x_0, \dots, x_{n_i-1}) \cap M_0^{n_i}$ for all $1 \leqslant i < k$,

$$\Psi^{*\mathbb{A}}(x_0, x_1, a) \cap M_0^2 = M_0^2 \setminus \Psi^{\mathbb{A}}(x_0, x_1, a),$$

$$\Phi_i^{*\mathbb{A}}(x_0, \dots, x_{n_i-1}, a) \cap M_0^{n_i} = M_0^{n_i} \setminus \Phi_i^{\mathbb{A}}(x_0, \dots, x_{n_i-1})$$

for all $1 \leqslant i < k$, and the structure $\mathfrak{M}$ is isomorphic to the quotient structure ${}^{\mathfrak{M}_0}/_{\eta}$.

**Definition 6.14.** $\mathfrak{M}$ is *$\mathbb{A}$-constructivizable* if there exists a map $\nu$ from $|\mathbb{A}|$ onto $|\mathfrak{M}|$ such that $\{\langle a_0, a_1, \dots, a_{n_i-1}\rangle \mid P_i(\nu(a_0), \nu(a_1), \dots, \nu(a_{n_i-1}))\}$, $1 \leqslant i \leqslant k$, and $\{\langle a, b\rangle \mid \nu(a) = \nu(b)\}$ are $\Delta$ on $\mathbb{A}$.

**Proposition 6.19.** *Let $\mathbb{A}$ be an admissible set and $\mathfrak{M}$ a structure. Then $\mathfrak{M}$ is $\Sigma$-definable in $\mathbb{A}$ iff $\mathfrak{M}$ is $\mathbb{A}$-constructivizable.*

**Remark 6.1.** Definition 6.13 can be naturally generalized to the case of structures with infinite computable signatures. Namely, a structure $\mathfrak{M}$ with a computable predicate signature $\langle P_0, P_1, \ldots \rangle$, where each $P_i$ is $n_i$-ary, is called $\Sigma$-*definable in* $\mathbb{A}$ if there exists a computable sequence $\Phi(x_0, y)$, $\Psi(x_0, x_1, y)$, $\Psi^*(x_0, x_1, y)$, $\Phi_0(x_0, \ldots, x_{n_0-1}, y)$, $\Phi_0^*(x_0, \ldots, x_{n_0-1}, y)$, ..., $\Phi_k(x_0, \ldots, x_{n_k-1}, y)$, $\Phi_k^*(x_0, \ldots, x_{n_k-1}, y)$, ... of $\Sigma$-formulas and a parameter $a \in A$, which forms a $\Sigma$-definition of $\mathfrak{M}$ in $\mathbb{A}$, in the sense of Definition 6.13.

For structures $\mathfrak{M}$ and $\mathfrak{N}$, we denote by $\mathfrak{M} \leqslant_\Sigma \mathfrak{N}$ the fact that $\mathfrak{M}$ is $\Sigma$-definable in $\mathbb{HF}(\mathfrak{N})$. From the definition it follows that the relation $\leqslant_\Sigma$ is reflexive and transitive. We now look at the general properties of this relation, regarding it as a kind of effective reducibility on structures.

### 6.6.1. $\Sigma$-*Definability on structures: general properties*

For any infinite cardinal $\alpha$, we denote by $\mathcal{K}_\alpha$ the class of structures having a finite signature and with cardinality less than or equal to $\alpha$.

As usual, preordering $\leqslant_\Sigma$ generates on $\mathcal{K}_\alpha$ a relation of $\Sigma$-equivalence: $\mathfrak{A} \equiv_\Sigma \mathfrak{B}$ if $\mathfrak{A} \leqslant_\Sigma \mathfrak{B}$ and $\mathfrak{B} \leqslant_\Sigma \mathfrak{A}$. Classes of $\Sigma$-equivalence are called *degrees of* $\Sigma$-*definability*, or $\Sigma$-*degrees*. The poset

$$\mathcal{S}_\Sigma(\alpha) = \langle \mathcal{K}_\alpha / \equiv_\Sigma, \leqslant_\Sigma \rangle$$

is an upper semilattice with the least element, which is the degree consisting of computable structures. We denote the $\Sigma$-degree of a structure $\mathfrak{A}$ by $[\mathfrak{A}]_\Sigma$. The notion of $\Sigma$-degree of a structure is invariant from the choice of a semilattice $\mathcal{S}_\Sigma(\alpha)$, because all infinite structures of the same $\Sigma$-degree have the same cardinality. For any structures $\mathfrak{A}, \mathfrak{B} \in \mathcal{K}_\alpha$, $[\mathfrak{A}]_\Sigma \vee [\mathfrak{B}]_\Sigma = [(\mathfrak{A}, \mathfrak{B})]_\Sigma$, where $(\mathfrak{A}, \mathfrak{B})$ is a pair of $\mathfrak{A}$ and $\mathfrak{B}$ in the model-theoretical sense.

For a structure $\mathfrak{A} \in \mathcal{K}_\alpha$ and infinite cardinals $\beta \leqslant \alpha$, $\gamma \geqslant \alpha$, the sets

$$I_\beta(\mathfrak{A}) = \{[\mathfrak{B}]_\Sigma \mid \mathfrak{B} \in \mathcal{K}_\beta, \, \mathfrak{B} \leqslant_\Sigma \mathfrak{A}\}, \quad F_\gamma(\mathfrak{A}) = \{[\mathfrak{B}]_\Sigma \mid \mathfrak{B} \in \mathcal{K}_\gamma, \, \mathfrak{A} \leqslant_\Sigma \mathfrak{B}\}$$

are, correspondingly, an ideal in $\mathcal{S}_\Sigma(\beta)$ (principal for $\beta = \alpha$) and a filter in $\mathcal{S}_\Sigma(\gamma)$ (principal for any $\gamma \geqslant \alpha$). The sets $F_\gamma(\mathfrak{A})$ in semilattices $\mathcal{S}_\Sigma(\gamma)$ are natural analogues of the *spectrum* of a structure $\mathfrak{A}$. The sets $I_\beta(\mathfrak{A})$ in semilattices $\mathcal{S}_\Sigma(\beta)$ consist of $\Sigma$-degrees of structures $\Sigma$-presentable over $\mathfrak{A}$.

A *presentation* of a structure $\mathfrak{M}$ in an admissible set $\mathbb{A}$ is any structure $\mathcal{C}$ which is isomorphic to $\mathfrak{M}$ and whose domain $C$ is a subset of $A$ (the relation $=$ is treated as a congruence relation on $\mathcal{C}$, and it may differ from the standard equality relation $C$). In what follows, we will identify the presentation $\mathcal{C}$ (more precisely, its atomic diagram) with some subset of $A$, fixing a Gödel numbering of atomic formulas of the signature $\sigma_{\mathfrak{M}}$.

**Definition 6.15.** A *problem of presentability* of a structure $\mathfrak{M}$ in $\mathbb{A}$ is the set $\mathrm{Pr}(\mathfrak{M}, \mathbb{A})$ consisting of all possible presentations of $\mathfrak{M}$ in $\mathbb{A}$.

Denote by $\underline{\mathfrak{M}}$ the set $\mathrm{Pr}(\mathfrak{M}, \mathbb{HF}(\varnothing))$ of presentations of $\mathfrak{M}$ in the least admissible set.

Since $\Sigma$-definability in $\mathbb{HF}(\varnothing)$ is equivalent to classical computability on natural numbers, we get the following:

**Proposition 6.20.** *Let $\mathfrak{M}$ be a countable structure. The following are equivalent:*

*1) $\mathfrak{M}$ is constructivizable;*
*2) $\underline{\mathfrak{M}}$ is $\Sigma$-definable in $\mathbb{HF}(\varnothing)$.*

Moreover, there exist natural embeddings of the semilattices $\mathcal{D}$ of Turing degrees and $\mathcal{D}_e$ of degrees of enumerability of sets of natural numbers into the semilattice $\mathcal{S}_\Sigma(\omega)$ (and hence into any semilattice $\mathcal{S}_\Sigma(\alpha)$) via the mappings $i : \mathcal{D} \to \mathcal{S}_\Sigma(\omega)$ and $j : \mathcal{D}_e \to \mathcal{S}_\Sigma(\omega)$ defined below. These definitions show that the notion of $\Sigma$-degree of a structure, which is total, i.e., defined for any structure, no matter countable or not, is a natural generalization of the (partial) notion of a degree of a countable structure, introduced in [127]. Also, we get that the semilattices $\mathcal{S}_\Sigma(\alpha)$ extend in a natural way the semilattices $\mathcal{D}$ and $\mathcal{D}_e$.

**Definition 6.16.** Let $\mathfrak{M}$ be a countable structure. We say that $\mathfrak{M}$ has a *degree (e-degree)* if there exists the least degree in the set of $T$-degrees (e-degrees) of all possible presentations of $\mathfrak{M}$ on natural numbers.

Using the equivalence of "$\forall$-recursiveness" and "$\exists$-definability", in the sense of [85] and [104] (see also [4] and [3]), we get:

**Theorem 6.24.** [150] *For a countable structure $\mathfrak{M}$, the following are equivalent:*

*1) $\mathfrak{M}$ has a degree (e-degree);*
*2) there exists a presentation $C \in \underline{\mathfrak{M}}$ which is a $\Delta$-subset ($\Sigma$-subset) of $\mathbb{HF}(\mathfrak{M})$.*

We define mappings $i : \mathcal{D} \to \mathcal{S}_\Sigma(\omega)$ and $j : \mathcal{D}_e \to \mathcal{S}_\Sigma(\omega)$ in the following way: for every degree $\mathbf{a} \in \mathcal{D}$, put

$$i(\mathbf{a}) = [\mathfrak{M}_{\mathbf{a}}]_\Sigma, \text{ where } \mathfrak{M}_{\mathbf{a}} \text{ is any structure having degree } \mathbf{a}.$$

Similarly, for every $e$-degree $\mathbf{b} \in \mathcal{D}_e$, put

$$j(\mathbf{b}) = [\mathfrak{M}_{\mathbf{b}}]_\Sigma, \text{ where } \mathfrak{M}_{\mathbf{b}} \text{ is any structure having } e\text{-degree } \mathbf{b}.$$

**Lemma 6.5.** *The mappings $i$ and $j$ are well defined: For any (e-)degree* $\mathbf{a}$ *there are structures having (e-)degree* $\mathbf{a}$. *Moreover, for any countable structures $\mathfrak{M}$ and $\mathfrak{N}$, if $\mathfrak{M}$ has (e-)degree* $\mathbf{a}$ *and $\mathfrak{M} \equiv_\Sigma \mathfrak{N}$, then $\mathfrak{N}$ also has (e-)degree* $\mathbf{a}$.

Note, however, that the property of having a ($e$-)degree is not closed downwards w.r.t. $\leqslant_\Sigma$.

**Definition 6.17. [152, 153]** For a structure $\mathfrak{A}$, a *jump* of the $\Sigma$-degree $[\mathfrak{A}]_\Sigma$ (in the semilattice $\mathcal{S}_\Sigma(\mathrm{card}(\mathfrak{A}))$) is the $\Sigma$-degree of the structure

$$\mathfrak{A}' = (\mathbb{HF}(\mathfrak{A}), \Sigma\text{-Sat}_{\mathbb{HF}(\mathfrak{A})}),$$

where $\Sigma\text{-Sat}_{\mathbb{HF}(\mathfrak{A})}$ denotes the satisfiability relation for the set of $\Sigma$-formulas in $\mathbb{HF}(\mathfrak{A})$.

The definition of $\Sigma$-jump is correct: For any structures $\mathfrak{A}$ and $\mathfrak{B}$, from $\mathfrak{A} \equiv_\Sigma \mathfrak{B}$ it follows that $\mathfrak{A}' \equiv_\Sigma \mathfrak{B}'$. It seems to be an open problem whether the inequality $\mathfrak{A} <_\Sigma \mathfrak{A}'$ holds for every structure $\mathfrak{A}$.

**Remark 6.2.** In a similar way the jump operation was introduced in [10] for the semilattice of $s$-degrees of countable structures. Also, in the same way a notion of the jump of an admissible set with respect to various effective reducibilities was introduced in [96, 122]. One more definition of the jump of a structure, closely related to the notion of $\Sigma$-jump, was given in [91].

The jump operation for $\Sigma$-degrees agrees with the jump operations for Turing and enumeration degrees w.r.t. the natural embeddings: If a structure $\mathfrak{A}$ has a ($e$-)degree $\mathbf{a}$, then the structure $\mathfrak{A}'$ has ($e$-)degree $\mathbf{a}'$. Henceforth, we have the following:

**Proposition 6.21.** *The mappings $i : \mathcal{D} \to \mathcal{S}_\Sigma$ and $j : \mathcal{D}_e \to \mathcal{S}_\Sigma$ are embeddings preserving* $0$, $\vee$ *and the jump operation.*

The existence of an embedding of $\mathcal{D}$ in $\mathcal{S}_\Sigma$ was first noted in [59].

The jump inversion theorem from the classical computability theory can also be generalized to the case of the semilattices of $\Sigma$-degrees of structures. There is:

**Theorem 6.25. [152, 153]** *Let $\mathfrak{A}$ be a structure such that $i(\mathbf{0}') \leqslant_\Sigma \mathfrak{A}$. Then there exists a structure $\mathfrak{B}$ such that*

$$\mathfrak{B}' \equiv_\Sigma \mathfrak{A}.$$

**Remark 6.3.** Relation of $\Sigma$-reducibility, being defined on structures of arbitrary cardinality, in the case of countable structures can be viewed as the strongest reducibility in the hierarchy of effective reducibilities on structures [150, 151] (see Section 6.7). One of the weak reducibilities in this hierarchy is the Muchnik reducibility. In [139, 140], the jump inversion theorem for the semilattices of degrees of presentability of countable structures with respect to the Muchnik reducibility is proved. As a corollary of Theorem 6.25, we get the jump inversion theorem for all known effective reducibilities on countable structures (see Section 6.7).

### 6.6.2. $\Sigma$-*Definability on special structures*

As has already been mentioned, cardinality boundaries are unavoidable in the classical theory of computability (CTC). Numberings allow us to use CTC for countable objects. Admissible sets of the form $\mathbb{HF}(\mathfrak{M})$ can have an arbitrary cardinality. Hence, the following question naturally arise: Does there exists a "reasonably good" theory $T$ such that the class of admissible sets of the form $\mathbb{HF}(\mathfrak{M})$, with $\mathfrak{M} \models T$, allows to extend, in some natural way, the classical theory CTC to the case of objects with an arbitrary cardinality?

Recall that a theory $T$ of a finite signature is called *regular* [33] if it is decidable and model complete. Recall also, that a theory $T$ is called *c-simple* (constructively simple) [33] if it is regular, $\omega$-categorical, and has a decidable set of the complete formulas.

**Remark 6.4.** In [33] such theories were called simple, but this terminology was simultaneously used in the model theory for a different notion.

In the definition of a $c$-simple theory, $\omega$-categoricity gives the uniqueness, up to an isomorphism, of a countable model of such theory. Model completeness, decidability of a theory, and decidability of the set of its complete formulas, guarantee the autostability of every constructivization of this countable theory, i.e., the uniqueness of the "computability" on its countable models.

Furthermore, if $T$ is a $c$-simple theory, $\mathfrak{M}_0$ and $\mathfrak{M}_1$ are any models of $T$ ($\mathfrak{M}_i \models T$, $i = 0, 1$), then $\mathbb{HF}(\mathfrak{M}_0) \equiv \mathbb{HF}(\mathfrak{M}_1)$, since the models of $\omega$-categorical theories are saturated enough ([33]).

Henceforth, for a $c$-simple theory $T$, the class of admissible sets of the form $\mathbb{HF}(\mathfrak{M})$, $\mathfrak{M} \models T$, extends "uniformly" the classical theory of computability for arbitrary infinite cardinalities.

An example of a $c$-simple theory is the theory $T_E$ of infinite structures with the empty signature. But this theory is too "weak", if we regard a theory $T$ being "strong" in case there are many uncountable structures $\Sigma$-definable in $\mathbb{HF}(\mathfrak{M})$, $\mathfrak{M} \models T$. The reason of the "weakness" of $T_E$ is the following property: For an arbitrary set $X$ and arbitrary permutation $f$ on $X$, $f$ can be extended (in a unique way) to an automorphism $f^*$ of $\mathbb{HF}(X)$.

Another example of a $c$-simple theory is the theory $T_{DLO}$ of dense linear orders (without endpoints). This theory seems to be quite reasonable candidate for a "correct extension of CTC for arbitrary cardinalities". Below we present two different characterizations of the theories having uncountable models which are $\Sigma$-definable in $\mathbb{HF}(\mathfrak{L})$, $\mathfrak{L} \models T_{DLO}$.

We now formalize a desired property of $T_{DLO}$ to be the "strongest" in the class of $c$-simple theories.

**Conjecture 6.1.** [34] *Suppose a theory $T$ has an uncountable model which is $\Sigma$-definable in $\mathbb{HF}(\mathfrak{M})$, for some structure $\mathfrak{M}$ with a c-simple theory. Then $T$ has an uncountable model which is $\Sigma$-definable in $\mathbb{HF}(\mathfrak{L})$ for some $\mathfrak{L} \models T_{DLO}$.*

It is an open question whether this conjecture is equivalent to the following one (which is its formal consequence).

**Conjecture 6.2.** *Any c-simple theory has an uncountable model which is $\Sigma$-definable in $\mathbb{HF}(\mathfrak{L})$ for some $\mathfrak{L} \models T_{DLO}$.*

It is known that Conjecture 6.2 is true for rather a "rich" class of $c$-simple theories (see Theorem 6.29 below).

Following [23, 32, 33], we present a characterization of the theories having uncountable models which are $\Sigma$-definable in $\mathbb{HF}(\mathfrak{L})$ for $\mathfrak{L} \models T_{DLO}$.

The category $^*\omega$ is defined as follows: Its objects are the sets of the form $[\mathbf{n}] \rightleftharpoons \{0, 1, \ldots, n-1\}$, $n \in \omega$ ($[\mathbf{0}] \rightleftharpoons \varnothing$), and its morphisms are order-preserving embeddings. It should be noted that there is a unique morphism from $[\mathbf{0}]$ into $[\mathbf{n}]$ for any $n \in \omega$.

**Definition 6.18.** By a *$^*\omega$-spectrum* we mean any functor $S$ from the category $^*\omega$ into the category $\mathrm{Mod}_\sigma^*$ of structures (of some fixed signature $\sigma$), whose morphisms are all possible embeddings.

To define a $^*\omega$-spectrum $S$, it is necessary to give an infinite sequence $\mathfrak{M}_0, \mathfrak{M}_1, \ldots, \mathfrak{M}_n, \ldots$, $n \in \omega$, of structures of signature $\sigma$, and associate with each order-preserving embeddings $\mu : [\mathbf{n}] \to [\mathbf{m}]$ an embedding $\mu_* : \mathfrak{M}_n \to \mathfrak{M}_m$ so that, if $\mu_0 : [\mathbf{n}] \to [\mathbf{m}]$ and $\mu_1 : [\mathbf{m}] \to [\mathbf{k}]$, $n \leqslant m \leqslant k \in \omega$, are morphisms of the category $^*\omega$, then $(\mu_1\mu_0)_* = \mu_{1*}\mu_{0*}$, and if $\mu : [\mathbf{n}] \to [\mathbf{n}]$ is the unique morphism from $[\mathbf{n}]$ into $[\mathbf{n}]$ $(= \mathrm{id}_{[\mathbf{n}]})$, then $\mu_* = \mathrm{id}_{\mathfrak{M}_n} : \mathfrak{M}_n \to \mathfrak{M}_n$, $n \in \omega$.

If the $^*\omega$-spectrum $S=\{\mathfrak{M}_n, \mu_* | n \in \omega, \mu \in \mathrm{Mor}^*\omega\}$ has been defined, then for any linearly ordered set $\mathfrak{L}$, it is possible to define the structure $\mathfrak{M}_\mathfrak{L}(\mathfrak{M}_\mathfrak{L}^S)$ as a direct limit $\varinjlim_{\mathfrak{L}_0} \mathfrak{M}'_{\mathfrak{L}_0}$ of the spectrum

$$\{\mathfrak{M}'_{\mathfrak{L}_0}, \varphi_{\mathfrak{L}_0,\mathfrak{L}_1} \mid \mathfrak{L}_0 \subseteq \mathfrak{L}_1 \subseteq \mathfrak{L},\ \mathfrak{L}_1 \text{ is finite}\},$$

where $\mathfrak{M}'_{\mathfrak{L}_0} \rightleftharpoons \mathfrak{M}_n$, if $\mathfrak{L}_0 \subseteq \mathfrak{L}$ is finite and $|\mathfrak{L}_0| = n$, and the embedding $\varphi_{\mathfrak{L}_0,\mathfrak{L}_1} : \mathfrak{M}'_{\mathfrak{L}_0} \to \mathfrak{M}'_{\mathfrak{L}_1}$ is defined for finite $\mathfrak{L}_0 \subseteq \mathfrak{L}_1 (\subseteq \mathfrak{L})$ as follows: If $\mathfrak{L}_1 = \{l_0 < l_1 < \ldots < l_{m-1}\}$ and $\mathfrak{L}_0 = \{l_{i_0} < l_{i_1} < \ldots < l_{i_{n-1}}\}$ (in which case $0 \leqslant i_0 < i_1 < \ldots < i_{n-1} \leqslant m$) and $\mu : [\mathbf{n}] \to [\mathbf{m}]$ is defined as $\mu(j) \rightleftharpoons i_j$, $j < n$, then

$$\varphi_{\mathfrak{L}_0,\mathfrak{L}_1} \rightleftharpoons \mu_* : \mathfrak{M}'_{\mathfrak{L}_0} = \mathfrak{M}_n \to \mathfrak{M}_m = \mathfrak{M}'_{\mathfrak{L}_1}.$$

If $\mathfrak{L} \subseteq \mathfrak{L}'$ are linearly ordered sets, then the structure $\mathfrak{M}_\mathfrak{L}$ can be identified with a substructure of $\mathfrak{M}_{\mathfrak{L}'}$ in a natural way.

Any isomorphism between linearly ordered sets $\mathfrak{L}$ and $\mathfrak{L}'$ induces an isomorphism between $\mathfrak{M}_\mathfrak{L}$ and $\mathfrak{M}_{\mathfrak{L}'}$. Also if $\mathfrak{L} \subseteq \mathfrak{L}'$ are dense linear orders without endpoints, then $\mathfrak{M}_\mathfrak{L} \preccurlyeq \mathfrak{M}_{\mathfrak{L}'}$. As a corollary, if $\mathfrak{L}$ and $\mathfrak{L}'$ are dense linear orders without endpoints, then $\mathfrak{M}_\mathfrak{L} \equiv \mathfrak{M}_{\mathfrak{L}'}$.

Let $\mu_0$ and $\mu_1$ be morphisms from $[\mathbf{1}]$ into $[\mathbf{2}]$ such that $\mu_0(0) = 0$ and $\mu_1(0) = 1$. The condition

$$\mu_{0*} \neq \mu_{1*} \quad (*)$$

is sufficient for $|\mathfrak{M}_\mathfrak{L}^S| \geqslant |\mathfrak{L}|$ to hold for any linearly ordered set $\mathfrak{L}$.

**Definition 6.19.** A system of numberings $\nu_n : \omega \to M_n$, $n \in \omega$, is called a *computable sequence of constructivization*

$$(\mathfrak{M}_0, \nu_0), (\mathfrak{M}_1, \nu_1), \ldots, (\mathfrak{M}_n, \nu_n), \ldots, \quad n \in \omega,$$

if the following conditions hold (we assume that the signature $\sigma$ of the structures $\mathfrak{M}_0, \mathfrak{M}_1, \ldots$ is finite and without function symbols):

1) $E \rightleftharpoons \{\langle n, m_0, m_1\rangle | n, m_0, m_1 \in \omega, \nu_n(m_0) = \nu_n(m_1)\}$ is a $\Delta$-predicate on $\omega$;
2) $N_P \rightleftharpoons \{\bar{n} = \langle n_0, n_1, \ldots, n_k\rangle | \bar{n} \in \omega^{k+1}, \langle \nu_{n_0}(n_1), \ldots, \nu_{n_0}(n_k)\rangle \in P^{\mathfrak{M}_{n_0}}\}$ is a $\Delta$-predicate on $\omega$ for any ($k$-ary) predicate symbol $P \in \sigma$;
3) for any constant symbol $c \in \sigma$ there exists a $\Sigma$-function $f_c : \omega \to \omega$ such that $c^{\mathfrak{M}_n} = \nu_n f_c(n)$.

Every morphism $\mu : [\mathbf{n}] \to [\mathbf{m}]$ of the category $^*\omega$ is uniquely defined by the number $m$ and the subset $\mu([\mathbf{n}]) \subseteq [\mathbf{m}]$. This remark allows us to define a one-to-one correspondence $\mu^* : \Delta \to \mathrm{Mor}^*\omega$ between the subset $\Delta \rightleftharpoons \{n | n \in \omega, r(n) < 2^{l(n)}\} \subseteq \omega$ and the set $\mathrm{Mor}^*\omega$, provided that $n \in \Delta$ is assumed to code the morphism $\mu : [\mathbf{k}] \to [\mathbf{l}]$ such that $l = l(n)$ and $r(n)$ is the number of the subset $\mu([\mathbf{k}]) \subseteq [\mathbf{l}] = [\mathbf{l(n)}]$ in some standard listing of the finite subsets of $\omega$. It is evident that $\Delta$ is a $\Delta$-subset of $\omega$.

**Definition 6.20.** Let $S = \{\mathfrak{M}_n, \mu_* | n \in \omega, \mu \in \mathrm{Mor}^*\omega\}$ be a $^*\omega$-spectrum. By a *constructivization* of $S$ we mean any computable sequence of constructivizations

$$(\mathfrak{M}_0, \nu_0), (\mathfrak{M}_1, \nu_1), \ldots, (\mathfrak{M}_n, \nu_n), \ldots, \quad n \in \omega,$$

together with a $\Sigma$-function $f : \Delta \times \omega \to \omega$ such that, for any $n, m, k \in \omega$ and $\mu : [\mathbf{n}] \to [\mathbf{m}] \in \mathrm{Mor}^*\omega$, if $n^* \in \Delta$ is such that $\mu^*(n^*) = \mu$, then $\mu_*\nu_n(k) = \nu_m f(n^*, k)$.

A $^*\omega$-spectrum $S$ is called *constructivizable* if there exists a constructivization for it.

**Theorem 6.26.** [33] *Let $\mathfrak{L}$ be a dense linear order without endpoints. A theory $T$ has an uncountable model $\Sigma$-definable in $\mathbb{HF}(\mathfrak{L})$ if and only if there exists a constructivizable $^*\omega$-spectrum $S$, satisfying condition $(*)$, and such that $\mathfrak{M}^S_{\mathfrak{L}} \models T$.*

One of the important corollaries of this theorem is the first part of the following result, showing that the field $\mathbb{C}$ of complex numbers is rather "simple". The second part shows that $\mathbb{C}$ is not "too simple".

**Theorem 6.27.** [33]

1) $\mathbb{C}$ *is* $\Sigma$*-definable in* $\mathbb{HF}(\mathfrak{L})$ *for any dense linear order* $\mathfrak{L}$ *of size continuum;*
2) $\mathbb{C}$ *is not* $\Sigma$*-definable in* $\mathbb{HF}(\mathcal{S})$ *for any structure* $\mathcal{S}$ *with empty signature.*

A structure $\mathfrak{A}$ is called *locally constructivizable* [33] if $\mathrm{Th}_\exists(\mathfrak{A}, \bar{\mathrm{a}})$ is c.e. for every $\bar{a} \in A^{<\omega}$. It is easy to verify that a structure $\mathfrak{A}$ is locally constructivizable if and only if, for any $\bar{a} \in A^{<\omega}$, there exist a constructivizable structure $\mathfrak{B}$ and a tuple $\bar{b} \in B^{<\omega}$ such that $(\mathfrak{A}, \bar{a}) \equiv_1 (\mathfrak{B}, \bar{b})$ (or, which is the same, $\mathbb{HF}(\mathfrak{A}, \bar{a}) \equiv_1 \mathbb{HF}(\mathfrak{B}, \bar{b})$). Symbol $\equiv_\alpha$, here and further on, denotes elementary equivalence w.r.t. the class of formulas with less than $\alpha$ groups of alternating groups of quantifiers in the prenex normal form $(0 \leqslant \alpha \leqslant \omega)$. Henceforth, the next definition is a generalization of the notion of local constructivizability.

**Definition 6.21.** [151] A structure $\mathfrak{A}$ is called *locally constructivizable of level* $\alpha$ $(0 < \alpha \leqslant \omega)$ if for any $\bar{a} \in A^{<\omega}$ there exists a constructivizable structure $\mathfrak{B}$ and a tuple $\bar{b} \in B^{<\omega}$ such that

$$\mathbb{HF}(\mathfrak{A}, \bar{a}) \equiv_\alpha \mathbb{HF}(\mathfrak{B}, \bar{b}).$$

Local constructivizability of any level is preserved by $\Sigma$-definability. There is:

**Proposition 6.22.** [151] *Let* $\mathfrak{A}$ *and* $\mathfrak{B}$ *be such that* $\mathfrak{A} \leqslant_\Sigma \mathfrak{B}$ *and* $\mathfrak{B}$ *is locally constructivizable of level* $\alpha$, $0 < \alpha \leqslant \omega$*. Then* $\mathfrak{A}$ *is also locally constructivizable of level* $\alpha$*.*

Any structure with a *c*-simple theory is saturated enough [33] and locally constructivizable of level $\omega$. Moreover, its countable "computable simulation", in the terminology from [89], is unique up to the computable isomorphism. The situation is different in the case of regular theories: There are structures with a regular theory, which are not locally constructivizable even of level 1. For example, consider the fields $\mathbb{R}$ and $\mathbb{Q}_p$ of real and *p*-adic numbers.

**Corollary 6.7.** [33] *For any linear order* $\mathfrak{L}$*, fields* $\mathbb{R}$ *and* $\mathbb{Q}_p$ *are not* $\Sigma$*-definable in* $\mathbb{HF}(\mathfrak{L})$*.*

In some cases there are more simple criteria, for a given theory, of the existence of uncountable models $\Sigma$-definable in $\mathbb{HF}(\mathfrak{L})$, $\mathfrak{L} \models T_{DLO}$. We now present such a criterion for $c$-simple theories.

The next definition is a generalization of the model-theoretical notions of order and total indiscernibility.

**Definition 6.22.** For structures $\mathfrak{A}$, $\mathfrak{B}$ and some $k > 0$, a set $I \subseteq A^k \cap B$ is called a *set of $\mathfrak{A}$-indiscernibles in $\mathfrak{B}$* (with dimension $k$) if for any pair of tuples $\bar{i}, \bar{i}' \in I^{<\omega}$ with the same length,

$$\langle \mathfrak{A}, \bar{i} \rangle \equiv \langle \mathfrak{A}, \bar{i}' \rangle \text{ implies } \langle \mathfrak{B}, \bar{i} \rangle \equiv \langle \mathfrak{B}, \bar{i}' \rangle.$$

**Proposition 6.23.** *Suppose $\mathfrak{A}$ is uncountable structure, structure $\mathfrak{B}$ is saturated enough and locally constructivizable of level $\omega$, and let $\mathfrak{A} \leqslant_\Sigma \mathfrak{B}$. There exist computable structures $\mathfrak{A}_0$ and $\mathfrak{B}_0$ such that $\mathfrak{A}_0 \equiv \mathfrak{A}$, $\mathfrak{B}_0 \equiv \mathfrak{B}$, and there is an infinite computable set of $(\mathfrak{B}_0, \bar{b}_0)$-indiscernibles in $\mathfrak{A}_0$ with a dimension $k$, for some $k > 0$ and $\bar{b}_0 \in (B_0)^{<\omega}$.*

For certain $c$-simple theories this necessary condition of $\Sigma$-definability of uncountable models can be simplified (by assuming the dimension to equal 1), and turns out also to be sufficient. Namely, for theory $T_{DLO}$ of dense linear orders without endpoints, and theory $T_E$ of infinite structures with empty signature, there is

**Theorem 6.28.** [145] *Let $T$ be a c-simple theory, and let $\mathfrak{A}$ be any computable model of $T$. Then*

1) *there exists an uncountable $\mathfrak{M} \models T$ such that $\mathfrak{M} \leqslant_\Sigma \mathfrak{L}$, $\mathfrak{L} \models T_{DLO}$, if and only if there exists an infinite computable set of order indiscernibles in $\mathfrak{A}$ (with dimension 1);*
2) *there exists an uncountable $\mathfrak{M} \models T$ such that $\mathfrak{M} \leqslant_\Sigma \mathcal{S}$, $\mathcal{S} \models T_E$, if and only if there exists an infinite computable set of total indiscernibles in $\mathfrak{A}$ (with dimension 1).*

**Remark 6.5.** This result is not true in the case theory $T$ is not $c$-simple. For example, there exists a computable algebraically closed field (with characteristic 0) with an infinite computable set of total indiscernibles (see [62]), but there are no uncountable algebraically closed fields (with characteristic 0) $\Sigma$-definable in $\mathbb{HF}(\mathcal{S})$, $\mathcal{S} \models T_E$.

We now present some applications of Theorem 6.28.

**Definition 6.23.** Let $n \in \omega$. A (first-order) theory $T$ is called *$n$-discrete* if every finite type of $T$ is uniquely determined by its $n$-subtypes.

A theory $T$ is called *discrete* if it is $n$-discrete for some $n \in \omega$. If $T$ is $n$-discrete and has a finite number of $n$-types then $T$ is $\omega$-categorical and submodel complete in some expansion by a finite number of definable predicates. Any regular $n$-discrete theory with a finite number of $n$-types is $c$-simple. Also, any submodel complete theory of a finite relational signature is $n$-discrete with a finite number of $n$-types, for some $n \in \omega$, and any $\omega$-categorical submodel complete theory of a finite signature is $n$-discrete with a finite number of $n$-types, for some $n \in \omega$.

A theory $T$ is called *sc-simple* [154] if it is $\omega$-categorical, submodel complete, decidable, and has a decidable set of complete formulas. Henceforth, a theory (of a finite signature) is *sc*-simple if it is $c$-simple and submodel complete.

From the Ehrenfeucht–Mostowski Theorem we get

**Proposition 6.24.** [154] *If $T$ is a sc-simple theory of a finite signature then, in any computable model of $T$, there exists an infinite computable set of order indiscernibles.*

As a corollary of the above fact, we get

**Theorem 6.29.** [154] *Let $T$ be sc-simple theory of a finite signature. There exists an uncountable model $\mathfrak{A}$ of $T$ such that $\mathfrak{A} \leqslant_\Sigma \mathfrak{L}$, $\mathfrak{L} \models T_{DLO}$.*

In the case of an infinite signature there is a counterexample. Using a construction from [63] together with Theorem 6.28, the following result was proved in [145]:

**Theorem 6.30.** *There is an sc-simple theory of an infinite computable signature, such that, for any uncountable $\mathfrak{A} \models T$ and any $\mathfrak{L} \models T_{DLO}$, we have $\mathfrak{A} \not\leqslant_\Sigma \mathfrak{L}$.*

We now present some examples of *sc*-simple theories. For a $\omega$-categorical theory $T$, by a *Ryll-Nardzewski function of $T$* we mean the function $r_T : \omega \to \omega$ defined as follows: for any $n \in \omega$, $r_T(n)$ is the number of (complete) $n$-types of theory $T$.

It is easy to check that, for any $\omega$-categorical decidable theory $T$, the following are equivalent:

1) $T$ has a decidable set of complete formulas;
2) $T$ has a decidable Ryll-Nardzewski function.

One of the methods for constructing $\omega$-categorical theories is the Fraïssé construction [46]. Let $K$ be a class of finitely generated structures of some fixed signature. $K$ is said to satisfy, respectively,

1) the *hereditary property* ($K \models \mathrm{HP}$) if, for any $\mathfrak{A} \in K$ and $\mathfrak{B}$, $\mathfrak{B} \subseteq \mathfrak{A}$ implies that $\mathfrak{B} \in K$;
2) the *joint embedding property* ($K \models \mathrm{JEP}$) if, for any $\mathfrak{A}, \mathfrak{B} \in K$, there is $\mathfrak{C} \in K$ such that there exist embeddings $\mathfrak{A} \hookrightarrow \mathfrak{C}$ and $\mathfrak{B} \hookrightarrow \mathfrak{C}$;
3) *amalgamation property* ($K \models \mathrm{AP}$) if, for any $\mathfrak{A}, \mathfrak{B}, \mathfrak{C} \in K$ and embeddings $f_1 : \mathfrak{C} \hookrightarrow \mathfrak{A}$, $f_2 : \mathfrak{C} \hookrightarrow \mathfrak{B}$, there are $\mathfrak{D} \in K$ and embeddings $g_1 : \mathfrak{A} \hookrightarrow \mathfrak{D}$, $g_2 : \mathfrak{B} \hookrightarrow \mathfrak{D}$ such that $f_1 g_1 = f_2 g_2$;
4) the property of *uniform local finiteness* ($K \models \mathrm{ULF}$) if there is a function $f : \omega \to \omega$ such that, for any $\mathfrak{A} \in K$ with no more than $n$ generators, the cardinality of $\mathfrak{A}$ is no more than $f(n)$.

If a class $K$ of finitely generated structures satisfy the properties HP, JEP and AP, then there is a unique, up to the isomorphism, submodel complete countable structure $\mathfrak{A}$, the class of finitely generated substructures of which is equal to $K$, up to the isomorphism (see, for example, [46]). We call such structure $\mathfrak{A}$ a *Fraïssé limit* of $K$ (denoted as $\mathfrak{A} = \lim_{\mathrm{F}} K$).

**Theorem 6.31 (see [46]).** *Let $K$ be a countable class of finitely generated structures of some fixed finite signature, satisfying the properties* HP, JEP, AP, *and* ULF. *Then $\lim_{\mathrm{F}} K$ is $\omega$-categorical.*

We present some examples of *sc*-simple theories constructed via Fraïssé limits (see [62, 63] for the details related to decidability).

Let *FinGraph* be the class of all finite symmetric graphs. It is easy to check that this class satisfies the properties HP, JEP, AP, and ULF.

**Definition 6.24.** A symmetric graph $\mathfrak{A}$ is called *random* if, for any finite $X, Y \subseteq A$ such that $X \cap Y = \emptyset$, there is a vertex $v \in A \setminus (X \cup Y)$ such that $v$ is adjacent with all vertexes from $X$ and not with vertexes from $Y$.

**Proposition 6.25 (see [46]).** *If $\mathfrak{A}$ is the Fraïssé limit of the class FinGraph then $\mathfrak{A}$ is a random graph. Moreover,* $\mathrm{Th}(\mathfrak{A})$ *is sc-simple.*

**Corollary 6.8.** [154] *There is an uncountable random graph $\mathfrak{A}$ such that $\mathfrak{A} \leqslant_{\Sigma} \mathfrak{L}$, $\mathfrak{L} \models T_{DLO}$.*

Let $\sigma$ be a finite predicate signature. The class $Fin(\sigma)$ of all finite structures of signature $\sigma$ satisfies the properties HP, JEP, AP, and ULF.

**Definition 6.25.** Let $\sigma$ be a finite predicate signature. A *random structure* $Ran(\sigma)$ of signature $\sigma$ is the Fraïssé limit of the class $Fin(\sigma)$.

**Corollary 6.9.** [154] *There is an uncountable structure* $\mathfrak{A} \equiv Ran(\sigma)$ *such that* $\mathfrak{A} \leqslant_\Sigma \mathfrak{L}$, $\mathfrak{L} \models T_{DLO}$.

For other computability properties of Fraïssé limits we refer the reader to [17].

### 6.6.3. *Special cases of Σ-definability*

In some cases, for structures $\mathfrak{A}$ and $\mathfrak{B}$ one can say more than just state the fact that $\mathfrak{A} \leqslant_\Sigma \mathfrak{B}$. For example, it is obvious that $\mathbb{HF}(\mathfrak{A}) \leqslant_\Sigma \mathfrak{A}$ for any $\mathfrak{A}$, but, in case of the standard model of arithmetic $\mathbb{N}$, much stronger result is true: $\mathbb{HF}(\mathbb{N})$ is $\Sigma$-definable within $\mathbb{N}$, not using the elements of the superstructure.

In particular, a natural additional restriction on $\Sigma$-definability of structures in admissible sets is the restriction on the rank of elements used in this process. To describe the situation formally, we now give some definitions.

Fix some signature $\sigma$, and let $P$ be an unary predicate symbol not in $\sigma$. For any formula $\Phi$ of the signature $\sigma \cup \{\in\}$, with the bounded quantifiers of the form $\forall x \in t$ and $\exists x \in t$, we define by induction the *relativization* $\Phi^P$ of $\Phi$ by $P$:

– if $\Phi$ is an atomic formula, put $\Phi^P = \Phi$;
– if $\Phi = (\Phi_1 * \Phi_2)$, $* \in \{\wedge, \vee, \to\}$, put $\Phi^P = (\Phi_1^P * \Phi_2^P)$;
– if $\Phi = \neg\Psi$, put $\Phi^P = \neg\Psi^P$;
– if $\Phi = (Qx \in y)\Psi$, $Q \in \{\forall, \exists\}$, put $\Psi^P = (Qx \in y)\Psi^P$;
– if $\Phi = \exists x\Psi$, put $\Phi^P = \exists x(P(x) \wedge \Psi^P)$;
– if $\Phi = \forall x\Psi$, put $\Phi^P = \forall x(P(x) \to \Psi^P)$.

Let now $\mathbb{A}$ be an admissible set, $B \subseteq A$ be some transitive subset of $\mathbb{A}$, and $\Phi(x_0, \ldots, x_{n-1})$ be a formula of the signature $\sigma_\mathbb{A}$. Define the set

$$(\Phi(x_0, \ldots, x_{n-1}))^B = \{\langle a_0, \ldots, a_{n-1}\rangle \in A^n \mid \langle \mathbb{A}, B\rangle \models \Phi^P(a_0, \ldots, a_{n-1})\}.$$

**Definition 6.26.** [146] Let $\mathbb{A}$ be an admissible set, $B \subseteq A$ be some transitive subset of $\mathbb{A}$. A structure of a computable predicate signature $\langle P_0, P_1, \ldots\rangle$, where each $P_i$ is $n_i$-ary, is called $\Sigma$-*definable in* $\mathbb{A}$ *inside* $B$ if there exist a computable sequence

$$\Phi(x_0, y), \Psi(x_0, x_1, y), \Psi^*(x_0, x_1, y), \Phi_0(x_0, \ldots, x_{n_0-1}, y),$$

$$\Phi_0^*(x_0,\ldots,x_{n_0-1},y),\ldots,\Phi_k(x_0,\ldots,x_{n_k-1},y),\Phi_k^*(x_0,\ldots,x_{n_k-1},y),\ldots$$

of $\Sigma$-formulas of $\sigma_{\mathbb{A}}$, and a parameter $b \in B$, such that, for the sets

$$M_0 \leftrightharpoons \Phi^B(x_0,b),\ M_0 \subseteq B,\ \ \eta \leftrightharpoons \Psi^B(x_0,x_1,b) \cap M_0^2,$$

the following holds: $M_0 \neq \varnothing$, $\eta$ is a congruence relation on the structure

$$\mathfrak{M}_0 \leftrightharpoons \langle M_0, P_0^{\mathfrak{M}_0},\ldots,P_k^{\mathfrak{M}_0},\ldots\rangle,$$

where $P_k^{\mathfrak{M}_0} \leftrightharpoons (\Phi_k(x_0,\ldots,x_{n_k-1}))^B \cap M_0^{n_k},\ k \in \omega,$

$$(\Psi^*(x_0,x_1,a))^B \cap M_0^2 = M_0^2 \setminus (\Psi(x_0,x_1,a))^B,$$

$$(\Phi_k^*(x_0,\ldots,x_{n_k-1},a))^B \cap M_0^{n_k} = M_0^{n_k} \setminus (\Phi_k(x_0,\ldots,x_{n_k-1}))^B$$

for any $k \in \omega$, and the quotient structure $\mathfrak{M}$ is isomorphic to $\mathfrak{M}_0/\eta$.

For an admissible set $\mathbb{A}$ and a subset $B \subseteq A$, define the ordinal $\mathrm{rnk}(B)$ as follows:

$$\mathrm{rnk}(B) = \sup\{\mathrm{rnk}(b) | b \in B\}.$$

**Definition 6.27.** [146] The *rank of inner constructivizability* of an admissible set $\mathbb{A}$ is the ordinal

$$\mathrm{cr}(\mathbb{A}) = \inf\{\mathrm{rnk}(B) \mid \mathbb{A} \text{ is } \Sigma\text{-definable in } \mathbb{A} \text{ inside } B\}.$$

The next theorem gives the precise estimate for the rank of inner constructivizability of hereditarily finite superstructures. It can be viewed as an effective analogue of some results from [90] on definability in higher order languages.

**Theorem 6.32.** [146] *Let $\mathfrak{M}$ be a structure of a computable signature.*
*1) If $\mathfrak{M}$ is finite then* $\mathrm{cr}(\mathbb{HF}(\mathfrak{M})) = \omega$.
*2) If $\mathfrak{M}$ is infinite then* $\mathrm{cr}(\mathbb{HF}(\mathfrak{M})) \leqslant 2$.

As a corollary of Theorem 6.32 we get the following. For structures $\mathfrak{M}$, $\mathfrak{N}$, and a natural number $n \in \omega$, we denote by $\mathfrak{M} \leqslant_\Sigma^n \mathfrak{N}$ the fact that $\mathfrak{M}$ is $\Sigma$-definable in $\mathbb{HF}(\mathfrak{N})$ inside the subset consisting of all elements with the rank less or equal $n$. If $\mathfrak{N}$ is an infinite structure then

$$\mathfrak{M} \leqslant_\Sigma^n \mathfrak{N} \text{ if and only if } \mathfrak{M} \leqslant_\Sigma \mathfrak{N}$$

for any $\mathfrak{M}$ and any $n \geqslant 2$.

Typical examples of structures $\mathfrak{M}$ with $\mathrm{cr}(\mathbb{HF}(\mathfrak{M})) = 2$ are infinite structures with the empty signature, dense linear orders, and, more interesting one, the structure $\langle \omega, s \rangle$ of natural numbers with the successor function. This fact follows from the next proposition, taking into account the decidability of $\mathrm{Th}_{\mathrm{WM}}(\langle \omega, s \rangle)$, where $\mathrm{Th}_{\mathrm{WM}}(\mathfrak{M})$ is the weak monadic second-order theory of $\mathfrak{M}$.

**Proposition 6.26.** [146] *If* $\mathrm{Th}_{\mathrm{WM}}(\mathfrak{M})$ *is decidable then* $\mathrm{cr}(\mathbb{HF}(\mathfrak{M})) = 2$.

An example of a structure $\mathfrak{M}$ with $\mathrm{cr}(\mathbb{HF}(\mathfrak{M})) = 0$ is, obviously, the standard model of arithmetic. An example of a structure which hereditary finite superstructure has rank of inner constructivizability 1 is the field $\mathbb{R}$ of real numbers. There is the following:

**Proposition 6.27.** [146] $\mathrm{cr}(\mathbb{HF}(\mathbb{R})) = 1$.

Another natural special type of a $\Sigma$-presentation of a structure $\mathfrak{M}$ in an admissible set $\mathbb{A}$, s.t. $M \subseteq U(\mathbb{A})$, is a $\Sigma$-presentation preserving the domain of a structure. For a signature $\sigma$ and an ordinal $n \leqslant \omega$, we denote by $Form_n(\sigma)$ the set of (finite first-order) formulas of the signature $\sigma$, which have a prenex normal form with no more than $n$ alternating groups of quantifiers.

We assume that, for any signature considered, some Gödel numbering $\ulcorner \cdot \urcorner$ of its terms and formulas is fixed.

**Definition 6.28.** Let $\mathfrak{M}$ be a structure of a finite signature $\sigma$, $\mathbb{A}$ an admissible set, and let $M \subseteq U(A)$. The structure $\mathfrak{M}$ is *n-decidable in* $\mathbb{A}$ $(n \leqslant \omega)$ if

$$\{\langle \ulcorner \varphi \urcorner, \overline{m} \rangle \mid \varphi \in Form_n(\sigma), \overline{m} \in M^{<\omega}, \mathfrak{M} \models \varphi(\overline{m})\}$$

is $\Delta$-definable in $\mathbb{A}$.

A structure $\mathfrak{M}$ is *computable in* $\mathbb{A}$ if $\mathfrak{M}$ is 0-decidable in $\mathbb{A}$, and *decidable in* $\mathbb{A}$ if $\mathfrak{M}$ is $\omega$-decidable in $\mathbb{A}$.

**Proposition 6.28.** *If* $\mathrm{Th}(\mathfrak{M})$ *is regular then* $\mathfrak{M}$ *is decidable in* $\mathbb{HF}(\mathfrak{M})$.

The decidability is rather a strong condition. For example, there is:

**Proposition 6.29.** *A liner order* $\mathfrak{L}$ *is 1-decidable in* $\mathbb{HF}(\mathfrak{L})$ *if and only if* $\mathfrak{L}$ *is a sum of a finite number of dense linear orders and points.*

A structure $\mathfrak{M}$ of signature $\sigma$ is *n-complete* [37] $(n \leqslant \omega)$ if for any formula $\varphi(\overline{x}) \in Form_n(\sigma)$ and for any $\overline{m} \in M^{<\omega}$ such that $\mathfrak{M} \models \varphi(\overline{m})$

there exists a $\exists$-formula $\psi(\overline{x})$ such that $\mathfrak{M} \models \psi(\overline{m})$ and $\mathfrak{M} \models \forall\overline{x}(\psi(\overline{x}) \to \varphi(\overline{x}))$. The following proposition follows immediately from the definitions.

**Proposition 6.30.**

1) *Suppose $\mathfrak{M}$ is n-decidable in $\mathbb{HF}(\mathfrak{M})$ $(n \leqslant \omega)$. Then $\mathfrak{M}$ is n-complete in some expansion of $\mathfrak{M}$ by a finite number of constants.*
2) *Suppose $\mathfrak{M}$ is n-complete and* Th($\mathfrak{M}$) *is decidable. Then $\mathfrak{M}$ is n-decidable in $\mathbb{HF}(\mathfrak{M})$.*

Suppose $\mathfrak{M}$ is 1-decidable in $\mathbb{HF}(\mathfrak{M})$. Then $\mathbb{HF}(\mathfrak{M})$ is quasiresolvable, and hence has a universal $\Sigma$-function and satisfies reduction, but not necessarily uniformization.

Let $\mathfrak{M}$ be a structure of signature $\sigma$ and let signature $\sigma_{\exists\text{-}Skolem}$ consist of all symbols of $\sigma$ and new functional symbols $f_\varphi(x_1, \ldots, x_n)$ for all $\exists$-formulas $\varphi(x_0, x_1, \ldots, x_n)$ of signature $\sigma$. The structure $\mathfrak{M}'$ of signature $\sigma_{\exists\text{-}Skolem}$ is called an *$\exists$-Skolem expansion of $\mathfrak{M}$* if $M' = M$, $\mathfrak{M} \upharpoonright_\sigma = \mathfrak{M}' \upharpoonright_\sigma$, and for any $\exists$-formula $\varphi(x_0, x_1, \ldots, x_n)$ of signature $\sigma$

$$\mathfrak{M}' \models \forall x_1 \ldots \forall x_n (\exists x \varphi(x, x_1, \ldots, x_n) \to \varphi(f_\varphi(x_1, \ldots, x_n), x_1, \ldots, x_n)).$$

**Theorem 6.33.** [142] *If $\mathbb{HF}(\mathfrak{M})$ satisfies uniformization then some $\exists$-Skolem expansion of $\mathfrak{M}$ is computable in $\mathbb{HF}(\mathfrak{M})$.*

In some cases, this necessary condition is also sufficient.

Skolem expansion $\mathfrak{M}^S$ of a structure $\mathfrak{M}$ is *well defined* if for every $\varphi(x_0, x_1, \ldots, x_n) \in Form(\sigma)$, every $\overline{m} \in M^n$, and every permutation $\rho$ of the set $\{1, \ldots, n\}$,

$$\mathfrak{M} \models (\varphi(x_0, \overline{m}) \leftrightarrow \varphi(x_0, \rho(\overline{m}))) \text{ implies } \mathfrak{M}^S \models (f_\varphi(\overline{m}) = f_\varphi(\rho(\overline{m}))),$$

where $\rho(\overline{m}) = \langle m_{\rho(1)}, \ldots, m_{\rho(n)} \rangle$.

The next theorem is a reformulation (and correction) of the main result from [142] (unfortunately, the property of well-definedness for Skolem expansions was not explicitly stated there, yet it was implicitly used in the text).

**Theorem 6.34.** [142, 157] *Suppose* Th($\mathfrak{M}$) *is regular. Then $\mathbb{HF}(\mathfrak{M})$ satisfies uniformization if and only if some well-defined $\exists$-Skolem expansion $\mathfrak{M}^S$ of $\mathfrak{M}$ is computable in $\mathbb{HF}(\mathfrak{M})$.*

**Remark 6.6.** As it was recently noted (see [157]), this theorem admits a natural reformulation in terms of the $s$-reducibility on structures [10] and

Proposition 6.31 can be viewed as a natural (and non-trivial) example of $s$-equivalence.

One of the important corollaries of this criterion follows from the next result.

**Proposition 6.31.** [142] *There exist well-defined Skolem expansions $\mathbb{R}^S$ and $(\mathbb{Q}_p)^S$, of the fields $\mathbb{R}$ and $\mathbb{Q}_p$, respectively, such that $\mathbb{R}^S$ and $(\mathbb{Q}_p)^S$ are computable in $\mathbb{HF}(\mathbb{R})$ and $\mathbb{HF}(\mathbb{Q}_p)$, respectively.*

**Corollary 6.10.** [142] *Structures $\mathbb{HF}(\mathbb{R})$ and $\mathbb{HF}(\mathbb{Q}_p)$ satisfy uniformization and have a universal $\Sigma$-function.*

For $\mathbb{HF}(\mathbb{R})$, the uniformization property and existence of a universal $\Sigma$-function was independently proved in [141] and [66].

The role of parameters in the $\Sigma$-definition of a structure is rather important. For example, as it is easy to see, any countable structure is $\Sigma$-definable in $\mathbb{HF}(\mathbb{R})$, where $\mathbb{R}$ is the field of real numbers. The case of $\Sigma$-definability without parameters turned out to be more interesting, as it was shown recently in [100].

**Theorem 6.35.** [100] *Suppose a countable structure $\mathfrak{M}$ is $\Sigma$-definable in $\mathbb{HF}(\mathbb{R})$ without parameters. Then $\mathfrak{M}$ has a hyperarithmetic presentation.*

This estimate is precise, as follows from the next theorem:

**Theorem 6.36.** [100] *For any $\delta < \omega_1^{CK}$ there is a countable structure $\mathfrak{M}$ such that*

1) *$\mathfrak{M}$ is $\Sigma$-definable in $\mathbb{HF}(\mathbb{R})$ without parameters;*
2) *for any $H \subseteq \omega$ such that $\mathfrak{M}$ has an $H$-computable presentation, holds $0^{(\delta)} \leqslant_T H$.*

In case we fix some restrictions on the cardinality of the congruence classes, the estimate of complexity becomes much lower.

**Theorem 6.37.** [100] *Let $\mathfrak{M}$ be a countable structure with a finite signature. The following are equivalent:*

1) *$\mathfrak{M}$ is $\Sigma$-definable without parameters in $\mathbb{HF}(\mathbb{R})$, and all equivalence classes are at least countable;*
2) *$\mathfrak{M}$ is computable.*

For other results on computability ($\Sigma$-definability) on the reals and on some topological spaces, we refer the reader to [65–70, 99], and [71–80, 117].

## 6.7. Semilattices of Degrees of Presentability of Structures

Relation $\leqslant_\Sigma$ of $\Sigma$-reducibility, being defined on structures of arbitrary cardinality, in the case of countable structures can be viewed as the strongest reducibility in the hierarchy of effective reducibilities on structures, as it was shown in [150, 151]. We overview briefly some of the results in this field.

Let $\mathbb{A}$ be an admissible set. We define uniform reducibilities on families of subsets of $A$, which are the direct generalizations of the Medvedev, Muchnik, and Dyment reducibilities on mass problems. Let $\mathcal{X}, \mathcal{Y} \subseteq P(A)$. Then,

(1) $\mathcal{X}$ is *Medvedev reducible* to $\mathcal{Y}$ ($\mathcal{X} \leqslant \mathcal{Y}$) if there are binary $\Sigma$-operators $F_0$ and $F_1$ such that, for all $Y \in \mathcal{Y}$, $\langle Y, A \setminus Y\rangle \in \delta_c(F_0) \cap \delta_c(F_1)$, and for some $X \in \mathcal{X}$, $X = F_0(Y, A \setminus Y)$ and $A \setminus X = F_1(Y, A \setminus Y)$;

(2) $\mathcal{X}$ is *Dyment reducible* to $\mathcal{Y}$ ($\mathcal{X} \leqslant_e \mathcal{Y}$) if there is a unary $\Sigma$-operator $F$ such that $Y \in \delta_c(F)$ for all $Y \in \mathcal{Y}$, and $F(\mathcal{Y}) \subseteq \mathcal{X}$;

(3) $\mathcal{X}$ is *Muchnik reducible* to $\mathcal{Y}$ ($\mathcal{X} \leqslant_w \mathcal{Y}$) if for every $Y \in \mathcal{Y}$ there are binary $\Sigma$-operators $F_0$ and $F_1$ such that $\langle Y, A \setminus Y\rangle \in \delta_c(F_0) \cap \delta_c(F_1)$, and for some $X \in \mathcal{X}$, $X = F_0(Y, A \setminus Y)$ and $A \setminus X = F_1(Y, A \setminus Y)$;

(4) $\mathcal{X}$ is *weakly Dyment reducible* to $\mathcal{Y}$ ($\mathcal{X} \leqslant_e \mathcal{Y}$) if there is a unary $\Sigma$-operator $F$ such that $Y \in \delta_c(F)$ for every $Y \in \mathcal{Y}$, and $F(Y) \in \mathcal{X}$.

For any admissible set $\mathbb{A}$ and for any $r \in \{e, \ , w, ew\}$ (here $r =$ ' ' is used to denote the Medvedev reducibility), we denote by $\mathcal{M}_r(\mathbb{A})$ the degree structure $\langle P(P(A))/\equiv_r, \leqslant_r\rangle$. We will write $\mathcal{M}_r$ instead of $\mathcal{M}_r(\mathbb{HF}(\varnothing))$ for brevity. All structures of the form $\mathcal{M}_r(\mathbb{A})$ are lattices with 0 and 1, and $\mathcal{M}$, $\mathcal{M}_e$, and $\mathcal{M}_w$ are isomorphic to the Medvedev, Dyment, and Muchnik lattices, respectively.

For a countable structure $\mathfrak{M}$, we consider the following classes consisting of structures that are effectively reducible to $\mathfrak{M}$:

$\mathcal{K}_\Sigma(\mathfrak{M}) = \{\mathfrak{N} \mid \mathfrak{N} \leqslant_\Sigma \mathfrak{M}\}$,

$\mathcal{K}_e(\mathfrak{M}) = \{\mathfrak{N} \mid \underline{\mathfrak{N}} \leqslant_e \underline{(\mathfrak{M}, \bar{m})}$ for some $\bar{m} \in M^{<\omega}\}$,

$\mathcal{K}(\mathfrak{M}) = \{\mathfrak{N} \mid \underline{\mathfrak{N}} \leqslant \underline{(\mathfrak{M}, \bar{m})}$ for some $\bar{m} \in M^{<\omega}\}$,

$\mathcal{K}_{ew}(\mathfrak{M}) = \{\mathfrak{N} \mid \underline{\mathfrak{N}} \leqslant_{ew} \underline{\mathfrak{M}}\}$,

$\mathcal{K}_w(\mathfrak{M}) = \{\mathfrak{N} \mid \underline{\mathfrak{N}} \leqslant_w \underline{\mathfrak{M}}\}$.

It is known [151] that for any structure $\mathfrak{M}$, the following inclusions hold:

$$\mathcal{K}_\Sigma(\mathfrak{M}) \subseteq \mathcal{K}_e(\mathfrak{M}) \subseteq \mathcal{K}(\mathfrak{M}) \subseteq \mathcal{K}_w(\mathfrak{M}),$$

and

$$\mathcal{K}_e(\mathfrak{M}) \subseteq \mathcal{K}_{ew}(\mathfrak{M}) \subseteq \mathcal{K}_w(\mathfrak{M}).$$

In general, all these inclusions are proper [48].

For any $r \in \{e, \ , w, ew\}$, we define a relation $\leqslant_r$ on $\mathcal{K}_\omega$ by setting $\mathfrak{M} \leqslant_r \mathfrak{N}$ iff $\mathcal{K}_r(\mathfrak{M}) \subseteq \mathcal{K}_r(\mathfrak{N})$ and letting $\mathcal{S}_r = \langle \mathcal{K}_\omega / \equiv_r, \leqslant_r \rangle$ be the structure of degrees of presentability corresponding to this relation.

**Theorem 6.38.** *For any $r \in \{e, \ , w, ew\}$, the structure $\mathcal{S}_r$ is an upper semilattice with 0, and the following embeddings ($\hookrightarrow$) and homomorphisms ($\rightarrow$) hold:*

$$\mathcal{D} \hookrightarrow \mathcal{D}_e \hookrightarrow \mathcal{S}_\Sigma \rightarrow \mathcal{S}_e \rightarrow \mathcal{S} \hookrightarrow \mathcal{M}.$$

As a corollary from this result and the Jump Inversion Theorem for the semilattices of $\Sigma$-degrees we get:

**Theorem 6.39.** [152, 153] *Let $r$ be an effective reducibility, i.e., $r \in \{e, \ , w, ew\}$. If $\mathfrak{A}$ is a structure with $\mathbf{0}' \leqslant_r \mathfrak{A}$ then there exists a structure $\mathfrak{B}$ such that*

$$\mathfrak{B}' \equiv_r \mathfrak{A}.$$

This result can be generalized to the case of degrees of presentability of structures in arbitrary admissible sets, see [156].

For arbitrary structures $\mathfrak{M}$ and $\mathfrak{M}'$ with the same signature and any $n \in \omega$, we denote by $\mathfrak{M} \equiv_n^{\mathrm{HF}} \mathfrak{M}'$ the fact that $\mathbb{HF}(\mathfrak{M}) \equiv_n \mathbb{HF}(\mathfrak{M}')$. It is clear that for $n < 2$, $\mathfrak{M} \equiv_n^{\mathrm{HF}} \mathfrak{M}'$ if and only if $\mathfrak{M} \equiv_n \mathfrak{M}'$. In case $n = 2$, $\mathfrak{M} \equiv_2^{\mathrm{HF}} \mathfrak{M}'$ if and only if, for any computable sequence $\{\varphi_{mn}(\overline{x}_m, \overline{y}_n) | m, n \in \omega\}$ of quantifier-free formulas of signature $\sigma_{\mathfrak{M}}$,

$$\mathfrak{M}' \models \bigvee_{m\in\omega} \exists \overline{x}_m \bigwedge_{n\in\omega} \forall \overline{y}_n \varphi_{mn}(\overline{x}_m, \overline{y}_n)$$

if and only if the same sentence is true in $\mathfrak{M}$.

For arbitrary structures $\mathfrak{M}$ and $\mathfrak{N}$, we denote by $\mathfrak{M} \leqslant_\exists \mathfrak{N}$ the fact that, for any tuple $\overline{m} \in M^{<\omega}$, there exists a tuple $\overline{n} \in N^{<\omega}$ such that $\mathrm{Th}_\exists(\mathfrak{M}, \overline{m}) \leqslant_e \mathrm{Th}_\exists(\mathfrak{N}, \overline{n})$. In particular, if $\mathfrak{M}$ is locally constructivizable then $\mathfrak{M} \leqslant_\exists \mathfrak{N}$ for any structure $\mathfrak{N}$. As was noted in [33], if $\mathfrak{M} \leqslant_\Sigma \mathfrak{N}$ and

$\mathfrak{N}$ is locally constructivizable then $\mathfrak{M}$ is also locally constructivizable. A straightforward generalization of this fact is as follows: $\mathfrak{M} \leqslant_\Sigma \mathfrak{N}$ implies $\mathfrak{M} \leqslant_\exists \mathfrak{N}$.

**Definition 6.29.** A structure $\mathfrak{M}$ is *uniformly locally constructivizable of level* $n$ $(1 < n \leqslant \omega)$ if there exists a constructivizable structure $\mathfrak{N}$ for which $\mathfrak{M} \preccurlyeq_n^{\mathrm{HF}} \mathfrak{N}$.

For instance, the structure $\langle \omega_1^{CK}, \leqslant \rangle$ is uniformly locally constructivizable of level $\omega$ since $\langle \omega_1^{CK}, \leqslant \rangle \preccurlyeq^{\mathrm{HF}} \langle \omega_1^{CK}(1+\eta), \leqslant \rangle$, where the last ordering (known as the *Harrison ordering*) is constructivizable.

**Proposition 6.32.** *If* $\mathfrak{M} \leqslant_\Sigma \mathfrak{N}$ *and a structure* $\mathfrak{N}$ *is (uniformly) locally constructivizable of level* $n$ $(1 < n \leqslant \omega)$, *then* $\mathfrak{M}$ *is also (uniformly) locally constructivizable of level* $n$.

The next proposition states that a class of locally constructivizable (of level 1) countable structures is downward closed w.r.t. $\leqslant_w$, which is weakest among the reducibilities under consideration.

**Proposition 6.33.** *Let* $\mathfrak{M}$ *and* $\mathfrak{N}$ *be structures. Then* $\mathfrak{N} \leqslant_\exists \mathfrak{M}$ *if* $\mathfrak{N} \in \mathcal{K}_w(\mathfrak{M})$. *In particular, if* $\mathfrak{M}$ *is locally constructivizable, then every structure* $\mathfrak{N} \in \mathcal{K}_w(\mathfrak{M})$ *is also locally constructivizable.*

A pair $(\mathfrak{M}, \mathfrak{N})$ is locally constructivizable iff so are $\mathfrak{M}$ and $\mathfrak{N}$; therefore, a set of degrees generated by locally constructivizable structures is an ideal in semilattices $\mathcal{S}_r$, $r \in \{\Sigma, e, \ , w, ew\}$. Classes of locally constructivizable structures of level $n$, $n > 1$, however, are downward closed w.r.t. $\leqslant_\Sigma$ only (so they form initial segments in $\mathcal{S}_\Sigma$). For weaker reducibilities, this is not the case. For example, we have:

**Theorem 6.40.** *There exists a countable structure* $\mathfrak{M}_0$ *which is locally constructivizable of level 1 (strictly) and is such that* $\mathfrak{M}_0 \leqslant \mathfrak{M}$ *for every nonconstructivizable countable structure* $\mathfrak{M}$. *Specifically, if* $\mathfrak{M}$ *is locally constructivizable of level* $n > 1$ *but is not constructivizable, then* $\mathcal{K}_\Sigma(\mathfrak{M}) \subsetneq \mathcal{K}(\mathfrak{M})$.

The proof makes use of the result (obtained by T. Slaman [137], and, independently, S. Wehner [164]) which states that there exists a structure whose problem of presentability belongs to the least nonzero degree of the Medvedev lattice (which, in particular, means that a semilattice $\mathcal{S}$ of degrees of presentability has a least nonzero element). Every such structure is locally constructivizable. Namely, in [150] was proved the following:

**Theorem 6.41.** *There exist a countable structure* $\mathfrak{M}$ *and a unary relation* $P \subseteq M$ *for which* $\underline{(\mathfrak{M}, P)} \equiv \underline{\mathfrak{M}}$ *but* $(\mathfrak{M}, P) \not\leqslant_\Sigma \mathfrak{M}$.

Theorem 6.41 is of interest in connection with the following result in [4]: For any countable structure $\mathfrak{M}$, a relation $P \subseteq M^n$, $n \in \omega$, is $\Sigma$-definable in $\mathbb{HF}(\mathfrak{M})$ iff $P^{\mathcal{C}}$ is $\mathcal{C} \upharpoonright \sigma_{\mathfrak{M}}$-c.e. for every $\mathcal{C} \in \underline{(\mathfrak{M}, P)}$.

The next result from [150] gives some sufficient conditions for the equality of the principal ideals generated by a structure $\mathfrak{M}$ with respect to different effective reducibilities.

**Theorem 6.42.** *If* $\mathfrak{M}$ *has a degree then* $\mathcal{K}_\Sigma(\mathfrak{M}) = \mathcal{K}_e(\mathfrak{M}) = \mathcal{K}(\mathfrak{M}) = \mathcal{K}_w(\mathfrak{M})$. *If* $\mathfrak{M}$ *has an e-degree then* $\mathcal{K}_\Sigma(\mathfrak{M}) = \mathcal{K}_e(\mathfrak{M}) = \mathcal{K}_{ew}(\mathfrak{M})$.

A natural (open) question is, Are these sufficient conditions also necessary?

For structures $\mathfrak{M}$ and $\mathfrak{N}$ with card$(M) \leqslant$ card$(N)$, consider the class

$$\mathcal{K}(\mathfrak{M}, \mathfrak{N}) = \{\mathfrak{M}' \mid \mathrm{Pr}(\mathfrak{M}', \mathbb{HF}(\mathfrak{N})) \leqslant \mathrm{Pr}((\mathfrak{M}, \bar{m}), \mathbb{HF}(\mathfrak{N})),\ \bar{m} \in M^{<\omega}\}.$$

Classes $\mathcal{K}_e(\mathfrak{M}, \mathfrak{N})$, $\mathcal{K}_w(\mathfrak{M}, \mathfrak{N})$, and $\mathcal{K}_{ew}(\mathfrak{M}, \mathfrak{N})$ are defined similarly.

**Proposition 6.34.** *Let* $\mathfrak{M}$ *and* $\mathfrak{N}$ *be countable structures and let* $\mathfrak{N}$ *be a structure of the empty signature, or dense linear order. Then* $\mathcal{K}_\Sigma(\mathfrak{M}) = \mathcal{K}_e(\mathfrak{M}, \mathfrak{N}) = \mathcal{K}(\mathfrak{M}, \mathfrak{N})$.

As a consequence, there exist natural isomorphisms between a semilattice $\mathcal{S}_\Sigma$ of degrees of $\Sigma$-definability and semilattices $\mathcal{S}(\mathbb{HF}(\mathfrak{N}))$ of degrees of presentability, where $\mathfrak{N}$ is a countable structure of the empty signature, or dense linear order.

One more result on the equivalence of "$\forall$-recursiveness" and "$\exists$-definability", in the sense of [85] and [104] (see also [4] and [3]), is the following:

**Theorem 6.43.** *For any countable structures* $\mathfrak{M}$ *and* $\mathfrak{N}$ *and any relation* $R \subseteq \mathbb{HF}(\mathfrak{N})$, *the following conditions are equivalent:*

*1)* $R \leqslant_{e\Sigma} \mathcal{C}$ *for every presentation* $\mathcal{C}$ *of* $\mathfrak{M}$ *in the admissible set* $\mathbb{HF}(\mathfrak{N})$;
*2)* $R$ *is* $\Sigma$*-definable in* $\mathbb{HF}(\mathfrak{M}, \mathfrak{N})$.

**Definition 6.30.** Let $\mathfrak{M}$ and $\mathfrak{N}$ be countable structures. Structure $\mathfrak{M}$ *has a degree* (an *e-degree*) *over structure* $\mathfrak{N}$ if there exists a least degree among all $T\Sigma$-degrees ($e\Sigma$-degrees) of all possible presentations of $\mathfrak{M}$ in $\mathbb{HF}(\mathfrak{N})$.

An immediate consequence of 6.43 is a generalization of 6.24:

**Theorem 6.44.** *Let $\mathfrak{M}$ and $\mathfrak{N}$ be countable structures. Then the conditions below are equivalent:*

*1) $\mathfrak{M}$ has a degree (an e-degree) over $\mathfrak{N}$;*
*2) some presentation $C \subseteq HF(N)$ of $\mathfrak{M}$ is a $\Delta$-subset ($\Sigma$-subset) in $\mathbb{HF}(\mathfrak{M}, \mathfrak{N})$.*

Obviously, for $\mathfrak{M} \leqslant_{\exists} \mathfrak{N}$, the structure $\mathfrak{M}$ has a degree, and also an e-degree, over $\mathfrak{N}$ iff $\mathfrak{M} \leqslant_{\Sigma} \mathfrak{N}$. It is also clear that if $\mathfrak{M}$ has a degree, and also an e-degree, over $\mathfrak{N}$, and $\mathfrak{N} \leqslant_{\Sigma} \mathfrak{N}'$, then $\mathfrak{M}$ has a degree, and also an e-degree, over $\mathfrak{N}'$. Furthermore, we have for any countable structure $\mathfrak{A}$, there exists a structure $\mathfrak{M}$ which has a degree but is not $\Sigma$-definable in $\mathbb{HF}(\mathfrak{A})$.

As in the nonrelativized case, we have:

**Theorem 6.45.** *Let $\mathfrak{M}$ and $\mathfrak{N}$ be countable structures. If $\mathfrak{M}$ has a degree over $\mathfrak{N}$, then $\mathcal{K}_{\Sigma}(\mathfrak{M}, \mathfrak{N}) = \mathcal{K}_e(\mathfrak{M}, \mathfrak{N}) = \mathcal{K}(\mathfrak{M}, \mathfrak{N})$. If $\mathfrak{M}$ has an e-degree over $\mathfrak{N}$, then $\mathcal{K}_{\Sigma}(\mathfrak{M}, \mathfrak{N}) = \mathcal{K}_e(\mathfrak{M}, \mathfrak{N})$.*

## 6.8. Closely Related Approaches to Generalized Computability

Now, we overview some of the approaches to the computability over abstract structures, looking for the differences and similarities of a given approach and the approach based on HF-computability.

### 6.8.1. BSS-*computability*

All results of this section are from [8], and we use the original terminology from this paper, saying "recursive" instead of "computable". The following definition is a generalization of the main definition from [14]. Let $\mathfrak{M}$ be a structure of a finite signature $\sigma$.

**Definition 6.31.** A BSS-machine contains following:

1) a triple of positive integers $\langle m, n, k\rangle$, which are called *input, working*, and *output dimensions*, respectively, and are denoted by $m{=}dim_I M$, $n{=}dim_W M$, and $k{=}dim_O M$;

2) a flow chart of a program.

A *flow chart of a program* is a connected directed graph having 4 types of nodes, with each of which, either a tuple of terms or an atomic formula of signature $\sigma$ is associated.

(1) There exists a unique node without incoming edges. It has just one outgoing edge and the associated tuple of terms

$$\langle t_1(x_1,\ldots,x_m),\ldots,t_n(x_1,\ldots,x_m)\rangle, \qquad (1)$$

which is called *an input node.* Here $m$ and $n$ are the input and working dimensions, respectively. We call this node *an input node.*

(2) There exists at least one node without outgoing edges.With each such node we associate a tuple of terms

$$\langle t_1(x_1,\ldots,x_n),\ldots,t_k(x_1,\ldots,x_n)\rangle, \qquad (2)$$

and we call it *an output tuple.* Here $n$ and $k$ are the working and output dimensions, respectively. We call such nodes *output nodes.*

(3) A *computation node* has several incoming and one outgoing edge. Associated with this node is a tuple of terms

$$\langle t_1(x_1,\ldots,x_n),\ldots,t_n(x_1,\ldots,x_n)\rangle, \qquad (3)$$

where $n$ is the working dimension.

(4) A *branch node* has several incoming and two outgoing edges. One of the outgoing edges is labeled by "0", the other by "1". Associated with this node is an atomic formula $\varphi(x_1,\ldots,x_n)$, where $n$ is the working dimension, in the signature $\sigma$.

Note that a flow chart may have no computation and branch nodes.

Each term $t(x_1,\ldots,x_r)$ of signature $\sigma$ defines a *term function* $f : M^r \to M$ as follows: $f(m_1,\ldots,m_r) = t(m_1,\ldots,m_r)$ for $m_1,\ldots,m_r \in M$. Each tuple of terms $\langle t_1(x_1,\ldots,x_r),\ldots,t_s(x_1,\ldots,x_r)\rangle$ defines a *term function* $f : M^r \to M^s$ similarly.

We define an arbitrary BSS-machine $S$ over a structure $\mathfrak{M}$. The sets $\bar{I} = A^m$, $\bar{S} = A^n$, $\bar{O} = A^k$ are called, respectively, *input*, *working*, and *output* spaces.

Given any $x \in \bar{I}$, a BSS-machine does computations which either never halt or halt and produce $y \in \bar{O}$. First, the machine sends $x$ into an input node, which computes the term function $I(x)$ defined by the associated tuple of terms (1). The resulting value $z = I(x)$ goes along the outgoing edge to the next node. At a computation node, the term function $g$ defined by the tuple (3) is applied to $z$, and $g(z)$ is sent along the outgoing edge. When $z \in \bar{S}$ reaches (if ever) a branch node, the truth value of associated formula $\varphi(z)$ is computed. If $\varphi(z)$ is true, the element $z$ goes to the next node along the edge labeled "1"; if not, it is sent along the edge labeled "0". At an output node, the element $z \in \bar{S}$ is converted to $y = O(z)$ of the output space, where $O$ is the term function defined by the associated tuple of terms (2), and the machine $S$ halts and produces $y \in \bar{O}$. If the machine never reaches some output node, we say that the result is undefined.

If the machine $S$ with input $x \in \bar{I}$ outputs $y$, we write $y = S(x)$.

The set

$$\Omega(S, \mathfrak{M}) = \{x \in \bar{I} \mid S \textit{ halts on input } x\}$$

is called the *halting set* of a machine $S$ in the structure $\mathfrak{M}$.

**Definition 6.32.** A function $f : \Omega \to M^k$, $\Omega \subseteq M^m$, is said to be *BSS-computable* if there exists a BSS-machine $S$ such that $\Omega = \Omega(S, \mathfrak{M})$ and $f(x) = S(x)$ for all $x \in \Omega$.

**Definition 6.33.** A set $X \subseteq M^n$ is called *recursively enumerable over* $\mathfrak{M}$ if and only if it is the domain of some BSS-computable function over $\mathfrak{M}$. A set $X \subseteq HL(M)$ is called *recursively enumerable (r.e.)* if it is r.e. over $\mathbb{HL}(\mathfrak{M})$.

**Definition 6.34.** A set $X \subseteq M^n$ is called *recursive over* $\mathfrak{M}$ if $X$ itself and its complement $M^n \setminus X$ are r.e. over $\mathfrak{M}$. Recursive sets $X \subseteq HL(M)$ over $\mathbb{HL}(\mathfrak{M})$ are called *recursive*.

**Definition 6.35.** A set $X \subseteq M^n$ is called an *output set over* $\mathfrak{M}$ if $X$ is the range of some BSS-computable function over $\mathfrak{M}$. Output sets $X \subseteq HL(M)$ over $\mathbb{HL}(\mathfrak{M})$ are called (simply) *output sets*.

**Lemma 6.6.** *Each recursive set over $\mathfrak{M}$ is r.e. over $\mathfrak{M}$. Each r.e. set over $\mathfrak{M}$ is an output set over $\mathfrak{M}$.*

**Proposition 6.35.** *The following statements are valid:*

*1) each r.e. set $X \subseteq HL(M)^n$ is the projection of some recursive set over $\mathbb{HL}(\mathfrak{M})$;*

*2) $X \subseteq HL(M)^n$ is the output set over $\mathbb{HL}(\mathfrak{M})$ if and only if $X$ is the projection of a recursive set over $\mathfrak{M}$.*

**Theorem 6.46.** *Each recursively enumerable set $X \subseteq M^n$ over $\mathbb{HL}(\mathfrak{M})$ is defined in $\mathfrak{M}$ by a formula of the form*

$$\bigvee_{i\in\omega} \varphi_i(x_1,\ldots,x_n),$$

*where $\{\varphi_i \mid i \in \omega\}$ is a recursive set of quantifier-free formulas in the signature $\sigma$. Conversely, each set $X \subseteq M^n$ defined by a formula*

$$\bigvee_{i\in\omega} \varphi_i(x_1,\ldots,x_n),$$

*where $\{\varphi_i \mid i \in \omega\}$ is a recursively enumerable set of quantifier-free formulas in the signature $\sigma$, is recursively enumerable over $\mathbb{HL}(\mathfrak{M})$.*

**Theorem 6.47.** *Each output set $X \subseteq M^n$ over $\mathbb{HL}(\mathfrak{M})$ is defined in $\mathfrak{M}$ by a formula of the form*

$$\bigvee_{i\in\omega} (\exists \overline{x}_i)\varphi_i(\overline{x}_i, y_1,\ldots,y_n),$$

*where $\{\varphi_i \mid i \in \omega\}$ is a recursive set of quantifier-free formulas in the signature $\sigma$. Conversely, each set $X \subseteq M^n$ defined by a formula*

$$\bigvee_{i\in\omega} (\exists \overline{x}_i)\varphi_i(\overline{x}_i, y_1,\ldots,y_n),$$

*where $\{\varphi_i \mid i \in \omega\}$ is a recursively enumerable set of quantifier-free formulas in the signature $\sigma$, is an output set over $\mathbb{HL}(\mathfrak{M})$.*

For other results on BSS-computability (and similar machine-style approaches), see [5–7] and [15].

### 6.8.2. *Search computability*

We recall some of the central notions of the theory introduced in [105, 106], together with the relationships with $\Sigma$-definability established in [45].

Let $\mathfrak{M}$ be a structure of a finite signature, and let $\mathbb{HL}(\mathfrak{M})$ denote the hereditarily listed superstructure over $\mathfrak{M}$. The central notion is that of a *partial multi-valued function (p.m.f)* from $HL(M)^k$ to the set of subsets $HL(M)$, where $k < \omega$. We use the following notations (here $\mathbf{u} \in HL(M)^k$):

- $f(\mathbf{u}) \to z$, if $z \in f(\mathbf{u})$ (we say that $f(\mathbf{u})$ *produces* $z$);
- $f(\mathbf{u})\downarrow$, if $f(\mathbf{u}) \neq \varnothing$ (we say that $f(\mathbf{u})$ *is defined*);
- $f \subseteq g$, if $\forall \mathbf{u}(f(\mathbf{u}) \subseteq g(\mathbf{u}))$;
- $f = g$, if $(f \subseteq g) \wedge (g \subseteq f)$;
- $f(\mathbf{u}) = z$, if $f(\mathbf{u}) = \{z\}$.

Substitution (superposition) of p.m.f. is defined in the natural way:

$$f(\mathbf{x}, g(\mathbf{x}, \mathbf{y}), \mathbf{y}) \to z \Leftrightarrow \exists u[(g(\mathbf{x}, \mathbf{y}) \to u) \wedge (f(\mathbf{x}, u, \mathbf{y}) \to z)].$$

Simultaneous substitution is interpreted as a successful substitution. In particular,

$$[f(g(\mathbf{x}), g(\mathbf{x})) \to z] \Leftrightarrow \exists u \exists v[(g(\mathbf{x}) \to u) \wedge (g(\mathbf{x}) \to v) \wedge (f(u, v) \to z)],$$

so that (in effect) a multi-valued term that occurs more than once in a formula may have different denotations for each of its occurrences.

A $\nu$-operator is a nondetermined analogue of the minimization operator, and is defined as follows:

$$\nu y[g(y, \mathbf{x}) \to 0] \to z \Leftrightarrow (g(z, \mathbf{x}) \to 0).$$

Now, we consider the construction schemes for multi-valued functions. Let $\overline{\varphi} = \varphi_1, \ldots, \varphi_l$ be a finite (possibly empty) list of p.m.f. on $HL(M)$, $\varphi_i$ is $n_i$-ary, $1 \leqslant i \leqslant l$. In the schemes C0–C10, $\mathbf{x} \in HL(M)^n$, $\mathbf{y} \in HL(M)^m$ $n, m \in \omega$ (possibly $n = 0$ or $m = 0$). We explain shortly the expressions in the right parts.

| | |
|---|---|
| C0. $f(t_1, \ldots, t_{n_i}, \mathbf{x}) = \varphi_i(t_1, \ldots, t_{n_i})$, | $\langle 0, n_i + n, i \rangle$ |
| C1. $f(\mathbf{x}) = y$ | $\langle 1, n, y \rangle$ |
| C2. $f(y, \mathbf{x}) = y$ | $\langle 2, n+1 \rangle$ |
| C3. $f(s, t, \mathbf{x}) = \langle s, t \rangle$ | $\langle 3, n+2 \rangle$ |
| C4$_0$. $f(y, \mathbf{x}) = l(y)$ | $\langle 4, n+1, 0 \rangle$ |
| C4$_1$. $f(y, \mathbf{x}) = r(y)$ | $\langle 4, n+1, 1 \rangle$ |
| C5. $f(\mathbf{x}) = g(h(\mathbf{x}), \mathbf{x})$ | $\langle 5, n, g, h \rangle$ |
| C6. $f(y, \mathbf{x}) = g(y, \mathbf{x})$, if $y \in M$; $f(\langle s, t \rangle, \mathbf{x}) = h(f(s, \mathbf{x}), f(t, \mathbf{x}), s, t, \mathbf{x})$ | $\langle 6, n+1, g, h \rangle$ |
| C7. $f(\mathbf{x}) = g(x_{j+1}, x_1, \ldots, x_j, x_{j+2}, \ldots, x_n)$ | $\langle 7, n, j, h \rangle$ |
| C8. $f(e, \mathbf{x}, \mathbf{y}) = \{e\}(\mathbf{x})$ | $\langle 8, n+m+1, n \rangle$ |
| C9. $f(\mathbf{x}) = \nu y[g(y, \mathbf{x}) \to 0]$ | $\langle 9, n, g \rangle$. |

All schemes, besides C8, were defined previously. Schemes C0–C4 define basic operations; C5, C7 corresponds to the superposition; C6 corresponds to the primitive recursion; and C9 to the minimization. Scheme C8 corresponds to the universal machine, with the expressions in the left playing the role of function indices. More exactly,

C0$'$. If $\varphi_i(t_1, \ldots, t_{n_i}) \to z$, then $\{\langle 0, n_i + n, i \rangle\}(t_1, \ldots, t_{n_i}, \mathbf{x}) \to z$.

C1$'$. $\{\langle 1, n, y \rangle\}(\mathbf{x}) \to y$.

C2$'$. $\{\langle 2, n+1 \rangle\}(y, \mathbf{x}) \to y$.

C3$'$. $\{\langle 3, n+2 \rangle\}(s, t, \mathbf{x}) \to \langle s, t \rangle$.

C4$_0$'. $\{\langle 4, n+1, 0\rangle\}(y, \mathbf{x}) \to l(y)$.
C4$_1$'. $\{\langle 4, n+1, 1\rangle\}(y, \mathbf{x}) \to r(y)$.
C5'. If there exists a $u$ such that $\{g\}(\mathbf{x}) \to u$ and $f(u, \mathbf{x}) \to z$, then $\{\langle 5, n, g, h\rangle\}(\mathbf{x}) \to z$.
C6'. If $y \in M$ and $g(y, \mathbf{x}) \to z$, then $\{\langle 6, n+1, g, h\rangle\}(y, \mathbf{x}) \to z$; if there exist $u$, $v$ such that $\{\langle 6, n+1, g, h\rangle\}(s, \mathbf{x}) \to u$, $\{\langle 6, n+1, g, h\rangle\}(t, \mathbf{x}) \to v$ and $\{h\}(u, v, s, t, \mathbf{x}) \to z$, then $\{\langle 6, n+1, g, h\rangle\}(\langle s, t\rangle, \mathbf{x}) \to z$.
C7'. If $\{g\}(x_{j+1}, x_1, \ldots, x_j, x_{j+2}, \ldots, x_n) \to z$, then $\{\langle 7, n, j, h\rangle\}(\mathbf{x}) \to z$.
C8'. If $\{e\}(\mathbf{x}) \to z$, then $\{\langle 8, n+m+1, n\rangle\}(e, \mathbf{x}, \mathbf{y}) \to z$.
C9'. If $\{g\}(y, \mathbf{x}) \to 0$, then $\{\langle 9, n, g\rangle\}(\mathbf{x}) \to y$.

A p.m.f. $f$ is called *search computable* relative to $\overline{\varphi}$, if it is constructed with C0–C9, where C0 may contain functions from $\overline{\varphi}$. A predicate $R(\mathbf{u})$ on $HL(M)$ is called *search computable* relative to $\overline{\varphi}$, if its characteristic function is search computable relative to $\overline{\varphi}$. A predicate $R(\mathbf{u})$ on $HL(M)$ is called *semi-search computable* relative to $\overline{\varphi}$, if there exists a search computable (relative to $\overline{\varphi}$) predicate $R_0(y, \mathbf{u})$ such that $R(\mathbf{u}) \Leftrightarrow \exists y R_0(y, \mathbf{u})$.

If a structure $\mathfrak{M}$ is defined on the set $M$ then (if not stated overwise), the list $\overline{\varphi}$ consists exactly of characteristic functions of the signature predicates of $\mathfrak{M}$.

**Theorem 6.48.** [45] *Let $\mathfrak{M}$ be a structure of a finite predicate signature, and $R$ be a relation on $\mathbb{HL}(\mathfrak{M})$.*

*(1) $R$ is semi-search computable on $\mathbb{HL}(\mathfrak{M})$ if and only if $R$ is a $\Sigma$-predicate on $\mathbb{HL}(\mathfrak{M})$;*
*(2) $R$ is search computable on $\mathbb{HL}(\mathfrak{M})$ if and only if $R$ is a $\Delta$-predicate on $\mathbb{HL}(\mathfrak{M})$.*

In conclusion, we present an approach to relative computability of abstract countable structures, introduced by I.N.Soskov in the framework of search computability. Let us consider algebraic structures of the form

$$\mathfrak{A} = \langle U, \mathbb{N}, =_U, \neq_U, R_1, \ldots, R_n\rangle,$$

where $U$ is an infinite countable set, $\mathbb{N}$ is the set of the natural numbers, and $R_i \subseteq U^{a_i} \times \mathbb{N}^{b_i}$, $a_i, b_i \in \mathbb{N}$, $1 \leqslant i \leqslant n$, $a_i + b_i \geqslant 1$, are partial predicates, which take only value true, whenever defined.

We use the so called *Moschovakis enrichment.* Let $U_0 = U \cup \{o\}$, where $o \notin U$ and let $\langle \cdot, \cdot \rangle$ be an injective binary function defined on $U_0$ with values outside of $U_0$. Let $U^*$ be the closure of $U_0$ with respect to $\langle \cdot, \cdot \rangle$.

Let $R_i^*(\overline{s}, \overline{z})$ be true if and only if $R_i(\overline{s}, \overline{z})$ is true, for any $1 \leqslant i \leqslant n$ and $(\overline{s}, \overline{z}) \in U^{a_i} \times \mathbb{N}^{b_i}$. Also define partial predicates $\mathcal{U}$, $\mathcal{O}$ and $\Pi$ on the set $U^*$ in the following way: $\mathcal{U}(s)$ is true if and only if $s \in U$ for each $s \in U^*$; $\mathcal{O}(s)$ is true if and only if $s{=}o$ for each $s \in U^*$; and $\Pi(s,t,r)$ is true if and only if $s = \langle t, r\rangle$ for every $s$, $t$, $r \in U^*$.

By $\overline{\mathcal{U}}$, $\overline{\mathcal{O}}$ and $\overline{\Pi}$ denote the complement predicates of $\mathcal{U}$, $\mathcal{O}$ and $\Pi$, for example $\overline{\mathcal{U}}(s)$ is true if and only if $\mathcal{U}(s)$ is false for each $s \in U^*$. *Moschovakis enrichment* of $\mathfrak{A}$ ($*$-structure of $\mathfrak{A}$) is

$$\mathfrak{A}^* = \langle U^*, \mathbb{N}, =_{U^*}, \neq_{U^*}, \mathcal{U}, \overline{\mathcal{U}}, \mathcal{O}, \overline{\mathcal{O}}, \Pi, \overline{\Pi}, R_1^*, \ldots, R_n^*\rangle.$$

We write $(\mathfrak{A}^*, R)$ to denote the structure that is obtained by adding $R$ to $\mathfrak{A}$.

The predicate $R \subseteq U^k \times \mathbb{N}^m$ is called *SC-definable* in $\mathfrak{A}$ (write $R \leqslant_{SC} \mathfrak{A}$) if and only if there exists a primitive recursive $(m+1)$-ary function $\gamma$ and $t_1, \ldots, t_q \in U$ such that for all $(\overline{s}, \overline{x}) \in U^k \times \mathbb{N}^m$ the following equivalence holds:

$$R(\overline{s}, \overline{x}) \text{ is true} \iff \exists n \in \mathbb{N}(\nu(\gamma(n, \overline{x}))(\overline{t}, \overline{s}) \text{ is true}),$$

where $\nu$ is some Gödel numeration of positive $\exists$-formulas.

**Definition 6.36.** For structures $\mathfrak{A} = \langle U, \mathbb{N}, =_U, \neq_U, R_1^{\mathfrak{A}}, \ldots, R_n^{\mathfrak{A}}\rangle$ and $\mathfrak{B} = \langle U, \mathbb{N}, =_U, \neq_U, R_1^{\mathfrak{B}}, \ldots, R_n^{\mathfrak{B}}\rangle$, $\mathfrak{A}$ is said to be *SC-reducible* to $\mathfrak{B}$ ($\mathfrak{A} \leqslant_{SC} \mathfrak{B}$), if $R_i^{\mathfrak{A}} \leqslant_{SC} \mathfrak{B}$ for each $1 \leqslant i \leqslant n$.

The relation $\leqslant_{SC}$ is reflexive and transitive, and induces an equivalence relation $\equiv_{SC}$ in the class of all algebraic structure with the abstract sort $U$. The respective equivalence classes are called *s-degrees*, and they form an upper semilattice. For the results in this field we refer the reader to [10, 140].

### 6.8.3. *Montague computability*

The results from this section describe one of the very first generalizations of computability theory over the natural numbers to the case of computability over arbitrary structures. It was presented by R. Montague in [90] as an attempt to look at the computability theory as a part of the model theory, considering computability as definability in higher order logics. The connections with the search computability introduced by Y.N. Moschovakis [104, 105], another one of the first generalizations of computability theory, are due to C. Gordon [45].

Let $\mathfrak{A}$ be a structure of a finite predicate signature $\langle R_1, \ldots, R_k \rangle$, where each $R_i$ is $n_i$-ary, and let $\kappa$ be a cardinal. Define

$$S^{0,\kappa} = A,$$
$$S^{n+1,\kappa} = \{x \subset S^{n,\kappa} \mid card(x) < \kappa\}.$$

Consider a language with relation symbols for the relations of $\mathfrak{A}$ and the membership symbol $\in$ and *variables of type n* to range over $S^{n,k}$.

**Definition 6.37.** $S = \bigcup_{n \in \omega} S^n$, where $S^n$ is defined inductively:
$S^0 = A$,
$S^{n+1} = \{x \mid x \text{ is a finite subset of } S^n\}$.
The elements of $S^n$ are called *objects of type n*.

**Definition 6.38.** A system $\mathfrak{A}^t = \langle S, \in, R_1, \ldots, R_k, R_1^*, \ldots, R_k^* \rangle$ is called a *t-extension* of the system $\mathfrak{A}$, where $R_i^*$ is the complement of $R_i$ relative to $A^{n_i}$.

**Definition 6.39.** The language $\Sigma^t$ (for the structure $\mathfrak{A}^t$) has the following symbols:

**(a)** For each $n \in \omega$, a countable sequence $v_{0,n}, v_{1,n}, \ldots,$ of variables of type $n$;
**(b)** Relation symbols $R_1, \ldots, R_k, R_1^*, \ldots, R_k^*$;
**(c)** The symbols $\wedge$, $\vee$, $\forall$, $\exists$, $\in$, (, ), and , .

The formulas of $\Sigma^t$ are defined inductively by:

**(d)** For $i$=$1, \ldots, k$, if $x_1, \ldots x_{n_i}$ are type 0 variables then $R_i(x_1, \ldots, x_{n_i})$ and $R_i^*(x_1, \ldots, x_{n_i})$ are formulas;
**(e)** If $\varphi$ and $\psi$ are formulas then $(\varphi \wedge \psi)$ and $(\varphi \vee \psi)$ are formulas;
**(f)** If $\varphi$ is a formula, $x$ is a variable of type $n$ and $y$ is a variable of type $n+1$ then $(\exists x \in y)\varphi$, $(\forall x \in y)\varphi$ and $\exists x \varphi$ are formulas.

(Notice that $x \in y$ is not a formula of $\Sigma^t$).

The interpretation of $\Sigma^t$ in $\mathfrak{A}^t$ is the obvious one with variables of type $n$ ranging over objects of type $n$.

The relations on $\mathfrak{A}$ which are $\Sigma^t$ definable in $\mathfrak{A}^t$ are those which are considered in [90] as analogs of the recursively enumerable relations.

**Theorem 6.49.** [45] *Any $\Sigma^t$-relation on $\mathfrak{A}$ is semi-search computable.*

**Theorem 6.50.** [45] *If the equality relation on $\mathfrak{A}$ and its complement are $\Sigma^t$-relations in $\mathfrak{A}$ then any semi-search computable relation on $\mathfrak{A}$ is a $\Sigma^t$-relation.*

## 6.9. KPU. Examples of Admissible Structures

Now we give some general information about admissible sets. As it is said above, it can be used in the $\mathbb{HF}$-computability because any hereditarily finite superstructure is an admissible set.

### 6.9.1. *Elements of* KPU

Recall the axioms of Kripke-Platek Theory with Urelements (KPU). Let $\sigma$ be a signature which contains a binary symbol $\in$ and a unary symbol U. They are interpreted as the membership relation and as the set of urelements respectively.

**Extensionality** $\forall x \forall y((\neg \mathrm{U}(x) \wedge \neg \mathrm{U}(y)) \rightarrow (\forall z((z \in x) \leftrightarrow (z \in y)) \rightarrow (x \approx y)))$;
**Pair** $\forall x \forall y \exists z((x \in z) \wedge (y \in z))$;
**Union** $\forall x \exists y(\neg \mathrm{U}(y) \wedge \forall z \forall w(((z \in x) \wedge (w \in z)) \rightarrow (w \in y)))$;
**Urelements** $\forall x(\mathrm{U}(x) \rightarrow \forall y \neg (y \in x))$;
**Empty Set Existence** $\exists x(\neg \mathrm{U}(x) \wedge \forall y \neg (y \in x))$;
**Foundation Scheme** $\forall \overline{z}(\exists x \varphi(x, \overline{z}) \rightarrow \exists x(\varphi(x, \overline{z}) \wedge \forall y((y \in x) \rightarrow \neg \varphi(y, \overline{z}))))$, for any formula $\varphi$ of $\sigma$ in which $y$ does not occur free.

It follows from Extensionality that a set without elements (i.e., an empty set) is unique.

To formulate the remaining axioms, we need a definition of $\Delta_0$-formula:

**Definition 6.40.** The class of $\Delta_0$-*formulas* of signature $\sigma$ is the least one which contains atomic formulas and is closed under the following logical connectives: $\rightarrow$, $\vee$, $\wedge$, $\neg$, $\forall y \in t$, $\exists y \in t$, where $t$ is a term of $\sigma$ and $y$ is a variable (as before, $\forall y \in t \ldots$ and $\exists y \in t \ldots$ are abbreviations for $\forall y((y \in t) \rightarrow \ldots)$ and $\exists y((y \in t) \wedge \ldots)$ respectively).

$\Delta_0$ **Separation Scheme** $\forall \overline{z} \forall x(\neg \mathrm{U}(x) \rightarrow \exists y(\neg \mathrm{U}(y) \wedge \forall w((w \in y) \leftrightarrow ((w \in x) \wedge \varphi(w, \overline{z})))))$, for every $\Delta_0$-formula $\varphi$ of the signature $\sigma$ in which $y$ does not occur free;
$\Delta_0$ **Collection Scheme** $\forall \overline{z} \forall x(\neg \mathrm{U}(x) \rightarrow (\forall w \in x \exists y \varphi(w, y, \overline{z}) \rightarrow \exists u \forall w \in x \exists y \in u \varphi(w, y, \overline{z})))$, for every $\Delta_0$-formula $\varphi$ of the signature $\sigma$ in which $u$ does not occur free.

It follows from these axioms that, for any elements $x, y$, there exist the pair $\{x, y\}$, the ordered pair $\langle x, y\rangle \rightleftharpoons \{\{x\}, \{x, y\}\}$, the union $\bigcup x$, and the results of the usual set theoretic operations $x \cup y$, $x \cap y$, $x \setminus y$.

Structures of the theory KPU are denoted as $\mathbb{A}$, $\mathbb{B}$, $\mathbb{C}$, ... (possibly with indices); their domains are denoted as $A$, $B$, $C$, ... respectively (with corresponding indices). Given a structure $\mathbb{A}$ of KPU, elements from $\mathrm{U}(\mathbb{A})$ are called *urelements* and elements from $A \setminus \mathrm{U}(\mathbb{A})$ are called *sets*. The axioms of KPU enables us to prove the existence of the Cartesian product $a \times b$ for any sets $a$ and $b$. A structure with operations and relations can be given on the set of urelements.

The theory KPU can be considered as a fragment of the theory ZF with urelements and, therefore, we can define the notions of a transitive set as a set containing all its elements as subsets and that of an ordinal as a transitive set consisting of transitive sets only. Notice that for any set $x$ there exists the transitive closure $\mathrm{TC}(x)$, i.e., the least transitive set under inclusion containing $x$ as a subset. Moreover, $\mathrm{TC}(x)$ is a $\Sigma$-function. By using foundation [11, 33] one can prove that ordinals on structures of KPU are linearly ordered by the membership relation and every non-empty definable subset of ordinals has the least element. A structure $\mathbb{A}$ of KPU is called an *admissible set* [33] if the set $\mathrm{Ord}(\mathbb{A})$ of ordinals of the structure is well ordered under the membership relation. Such a definition is more abstract than the definition from [11] because it is closed under all isomorphic images. However, any admissible set is isomorphic to some admissible set in the sense [11]. An ordinal $\alpha$ is called *admissible* if $\mathrm{Ord}(\mathbb{A}) = \alpha$, for some admissible set $\mathbb{A}$.

As it is said above, hereditarily finite superstructures are admissible sets. We give now a series of other examples of admissible structures:

(1) Any standard model of ZF with urelements is an admissible set.
(2) Let $\varkappa$ be an infinite cardinal and let $\mathfrak{M}$ be a structure (possibly, empty) of some signature $\tau$. Then a structure $\mathbb{H}_{\varkappa}(\mathfrak{M})$ of $\tau \cup \{\mathrm{U}, \in\}$ with its domain $\{a \in \mathbb{V}_M \mid \mathrm{card}(\mathrm{TC}(a)) < \varkappa\}$, where $\mathbb{V}_M$ is the universe over $M$(II.1 [11])), is an admissible set with $\mathrm{Ord}(\mathbb{H}_{\varkappa}(\mathfrak{M})) = \varkappa$. Thus, any infinite cardinal is admissible.
(3) Let $\mathfrak{M}$ be a structure. Then there exists the least admissible set $\mathbb{HYP}(\mathfrak{M})$ under inclusion containing $\mathfrak{M}$ as an element. Moreover, its domain can be found constructively in any such admissible set, namely, there is a $\Sigma$-function $\mathrm{L}(a, \alpha)$ that it coincides with

$\bigcup_{\alpha\in\mathrm{Ord}(\mathbb{HYP}(\mathfrak{M}))} \mathrm{L}(M,\alpha)$ [11, 33]. Consider $\mathbb{HYP}(\mathfrak{N})$ where $\mathfrak{N}$ is the standard model of arithmetic. Then $\mathrm{Ord}(\mathbb{HYP}(\mathfrak{N})) = \omega_1^{CK}$ is the first non-constructible ordinal and the collections of $\Delta$- and $\Sigma$-subsets are exactly $\Delta_1^1$ and $\Pi_1^1$ respectively. The properties of this admissible set are studied in detail in [11].

(4) Given an admissible set $\mathbb{A}$, $\{a \in A \mid \mathrm{TC}(\{a\}) \cap \mathrm{U}(\mathbb{A}) = \varnothing\}$ is the domain of an admissible set which is called the *pure part* of $\mathbb{A}$. Generally, admissible sets without urelements are said to be *pure*. As a corollary, an ordinal $\alpha_0$ is admissible if and only if $\mathbb{L}_{\alpha_0} = \langle \bigcup_{\beta<\alpha_0} \mathrm{L}(\varnothing,\beta), \in\rangle$ is admissible. Admissible sets of such kind are called *constructible.*

Indeed, the pure part of any admissible set whose ordinal is $\omega$ coincides with $\mathbb{HF}(\varnothing)$, i.e., the least admissible set under inclusion(II.2.12 [11]).

Notice that if $\mathbb{A}$ is an admissible set over $\mathfrak{M}$ then $HF(M)$ is exactly the closure of the set $M$ of urelements together with $\{\varnothing\}$ under values of set-theoretic terms $\{\cdot\}$ and $\cup$.

### 6.9.2. $\Sigma$-*subsets*

In comparison with classical computability, an effectively presented relation is the main object of study here, not a function. The main interest in these relations lies in the method of defining them, as well as in the general absence of a universal effective function.

The notions of $\Sigma$-formulas, $\Sigma$- and $\Delta$-subsets, and $\Sigma$-functions on structures of KPU are defined like these for hereditarily finite superstructures.

We give examples of basic $\Delta$-predicates and $\Sigma$-functions used here:

- $\mathrm{Ord}(x)$ ($x$ is an ordinal);
- $\mathrm{Nat}(x)$ ($x$ is a natural number; we often denote the set of finite ordinals in admissible sets as $\omega$);
- $\mathrm{TC}(x)$ is the least transitive set containing $x$ as a subset;
- $\mathrm{sp}(x) \leftrightharpoons \{y \in \mathrm{TC}(x) \mid \mathrm{U}(y)\}$ is the support of $x$;
- $\mathrm{rk}(x) = \sup\{\mathrm{rk}(y) + 1 \mid y \in x\}$ is the rank of $x$.

As usual, $\langle x,y\rangle \leftrightharpoons \{\{x\},\{x,y\}\}$, $\langle x\rangle \leftrightharpoons x$, $\langle x_1,x_2,\ldots,x_{n-1},x_n\rangle \leftrightharpoons \langle\langle x_1,x_2,\ldots,x_{n-1}\rangle,x_n\rangle$. As in the classical case, it suffices to consider subsets of admissible sets only because the ordered pair operation is definable by some $\Delta_0$-formula. Moreover, this formula is independent of choice of a structure of KPU. We give now several equivalent definitions of $\Sigma$-subsets in any structure of KPU.

**Proposition 6.36.** *Let $\mathbb{A}$ be a structure of* KPU *and let $B \subseteq A$. Then the following conditions are equivalent:*

**(i)** *$B$ is a $\Sigma$-subset of $\mathbb{A}$;*
**(ii)** *$B$ is a $\Sigma_1$-subset of $\mathbb{A}$;*
**(iii)** *$B = \delta F$ for some partial $\Sigma$-function $F$;*
**(iv)** *$B = \rho F$ for some partial $\Sigma$-function $F$;*
**(v)** *$B = \varnothing$ or $B = \rho F$ for some total $\Sigma$-function $F$.*

***Proof.*** (i) $\to$ (ii) follows from the Reflection Principle [11]. (ii) $\to$ (v) Let $B$ be a nonempty $\Sigma_1$-subset and let $\exists y \varphi_0(x,y)$ define $B$, where $\varphi_0$ is a $\Delta_0$-formula. Take $b_0 \in B$ and define a $\Sigma$-formula $\psi(x,y)$ as follows:

$$(\exists u \exists v((x = \langle u,v \rangle) \wedge ((\varphi_0(u,v) \wedge (y = u)) \vee (\neg \varphi_0(u,v) \wedge (y = b_0))))\vee$$
$$(\neg(x \text{ is an ordered pair}) \wedge (y = b_0))).$$

It is easy to check that a $\Sigma$-formula $\psi(x,y)$ defines the graph of some total function $f$ with $B = \rho f$. (v) $\to$ (iv) If $B = \varnothing$ then a $\Sigma$-formula $\neg(x = x)$ defines the graph of nowhere converged function, in particular, the range of it is empty. If $B \neq \varnothing$, then it is evident that (iv) is true. (iv) $\to$ (i), (iii) $\to$ (i) Let a $\Sigma$-formula $\phi(x,y)$ define the graph of $F$. Then $\exists x \phi(x,y)$ and $\exists y \phi(x,y)$ define $B$ in (iv) and (iii) respectively. (i) $\to$ (iii) Suppose that $B$ is definable by $\Sigma$-formula $\theta(x)$. Then $(\theta(x) \wedge (x = y))$ defines the graph of some function $f$ whose domain coincides with $B$. □

An infinite $\Sigma$-subset $B$ of $\mathbb{A}$ needs not have total $\Sigma$-functions "enumerating" it without repetitions, i.e., one-to-one correspondences from $A$ onto $B$. Several examples are given.

Examples 6.9.1.

(1) Any admissible set has always a countable $\Delta$-subset $\omega \subseteq \mathrm{Ord}(\mathbb{A})$.
(2) If an admissible set $\mathbb{A}$ satisfies $\omega < \mathrm{Ord}(\mathbb{A})$ then $\omega$ cannot be enumerated without repetitions via a total $\Sigma$-function, otherwise $A \in A$, by $\Sigma$-Replacement (I.4.6 [11]). Moreover, if $a \in A$ then $a$ cannot be enumerated without repetitions via a total $\Sigma$-function.
(3) There exists a hereditarily finite superstructure over a countable structure of some finite signature which has an infinite $\Sigma$-subset $B \subseteq \omega$ such that any coinfinite $\Sigma$-subset of $B$ is finite (theorem 2.1 [101]). In particular, $B$ cannot be enumerated without repetitions via a total $\Sigma$-function or even a partial $\Sigma$-function with domain $\omega$.

### 6.9.3. *Gandy's Theorem*

An approximation by some strongly computable sequence of finite sets is one of the universal methods of defining computably enumerable sets in classical computability. In general, this method cannot be applied in admissible sets because $\Sigma$-subsets cannot be constructed by ordinal steps in some admissible sets. However, nondeterministic analogues can be used here if elements of a certain kind play the role of steps.

**Definition 6.41.** Let $\mathfrak{M}$, $\mathfrak{N}$ be structures of some signature $\sigma \supseteq \{\in\}$. A structure $\mathfrak{N}$ is called an *end extension* of $\mathfrak{M}$ (we write $\mathfrak{M} \leqslant_{\mathrm{end}} \mathfrak{N}$) if $\{b \mid b \in^{\mathfrak{M}} a\} = \{b \mid b \in^{\mathfrak{N}} a\}$ for each $a \in M$.

If $\mathbb{A}$ is a structure of KPU in some relation signature and $a \in A$ is transitive then $\mathbb{A} \upharpoonright a \leqslant_{\mathrm{end}} \mathbb{A}$. Any embedding of one structure into another is extendible to some end extension, that is, given two structures $\mathfrak{M}$ and $\mathfrak{N}$ such that $\mathfrak{M} \leqslant \mathfrak{N}$, there is an embedding $\imath : \mathbb{HF}(\mathfrak{M}) \hookrightarrow \mathbb{A}_{\mathfrak{N}}$ such that $\imath(\mathbb{HF}(\mathfrak{M})) \leqslant_{\mathrm{end}} \mathbb{A}_{\mathfrak{N}}$, for every admissible set $\mathbb{A}_{\mathfrak{N}}$ over $\mathfrak{N}$.

Since $\Sigma$-formulas are preserved under end extensions (I.8.4 [11]) we have:

**Proposition 6.37.**

(1) *Let $\mathbb{A}$ be a structure of* KPU *in some relation signature, $\Phi(x)$ be a $\Sigma$-formula in the signature with a parameter $a_0 \in A$, and $b \in A$. Then $\mathbb{A} \models \Phi(b)$ if and only if $\mathbb{A} \upharpoonright c \models \Phi(b)$ for some transitive set $c \in A$, $\{a_0, b\} \subseteq c$.*

(2) *Let $\mathbb{HF}(\mathfrak{M})$ be a hereditarily finite superstructure in some relation signature, $\Phi(x)$ be a $\Sigma$-formula in the signature with parameters $m_0, \ldots, m_{k-1}$ from $\mathfrak{M}$, and $b \in HF(M)$. Then $\mathbb{HF}(\mathfrak{M}) \models \Phi(b)$ if and only if $\mathbb{HF}(\mathfrak{M}_0) \models \Phi(b)$ for some finite substructure $\mathfrak{M}_0 \leqslant \mathfrak{M}$, $\{m_0, \ldots, m_{k-1}\} \cup \mathrm{sp}(b) \subseteq M_0$.*

An important circumstance is that both the approximations are defined by some $\Sigma$-formulas which can be effectively found from $\Phi$. It is convenient to use variations of proposed approaches in practice.

Now we describe Gandy's method of construction of a $\Sigma$-predicate as the least fixed point of some $\Sigma$-operator. In section 6.3, this method was defined on hereditarily finite superstructures.

Let $\mathbb{A}$ be a structure of KPU. We define two topologies on $\mathcal{P}(A)$.

- The *strong topology* $\tau_s$ is defined by an open basis consisting of sets of kind $V_a \rightleftharpoons \{M \mid M \subseteq A, a \subseteq M\}$, $a \in A \setminus U(\mathbb{A})$.

- The *weak topology* $\tau_w$ is defined by an open pre-basis consisting of sets of kind $V_{\{a\}}$, $a \in A$. In other words, sets of kind $V_a$, where $a \in A \backslash U(\mathbb{A})$ is a finite set, form an open basis of this topology.

Note that these topologies coincide on hereditarily finite superstructures. A continuous map $F : \langle \mathcal{P}(A), \tau_s \rangle \to \langle \mathcal{P}(A), \tau_w \rangle$ is called a *weakly continuous operator*. Every weakly continuous operator $F$ is monotonic, i.e., $M \subseteq N \subseteq A \Rightarrow F(M) \subseteq F(N)$ and, therefore, it has the least fixed point which can be found in the following way: $\Gamma_0 \leftrightharpoons \varnothing$; $\Gamma_{\alpha+1} \leftrightharpoons F(\Gamma_\alpha)$; $\Gamma_\eta \leftrightharpoons \bigcup_{\beta<\eta} \Gamma_\beta$, if $\eta$ is limit; then, as it is easily checked, $\Gamma_* \leftrightharpoons \bigcup_{\alpha<\mathrm{card}(\mathbb{A})^+} \Gamma_\alpha$ is the least fixed point of $F$.

A weakly continuous operator $F$ is called a $\Sigma$*-operator* if $\Gamma_F^* \leftrightharpoons \{\langle a, b\rangle \mid a \in A \setminus U(\mathbb{A}),\ b \in F(a)\}$ is $\Sigma$ on $\mathbb{A}$.

**Theorem 6.51.** [Gandy] *Let $\mathbb{A}$ be an admissible set and $F$ be a $\Sigma$-operator on $\mathbb{A}$. Then the least fixed point $\Gamma_*$ of the operator $F$ is a $\Sigma$-subset of $\mathbb{A}$. Moreover, $\Gamma_* = \Gamma_{\mathrm{Ord}(\mathbb{A})}$.*

We illustrate some applications of this theorem.

Let $\Psi(x, P^+)$ be a $\Sigma$-formula and $F_\Psi(M) = \{b \mid \langle \mathbb{A}, M\rangle \models \Psi(b)\}$, for every subset $M$ of an admissible set $\mathbb{A}$. Then $F_\Psi$ is a $\Sigma$-operator on $\mathbb{A}$. Thus, the Gandy Theorem can be viewed as a generalization of $\Sigma$-Recursion Principles.

**Proposition 6.38.** *Let $\mathbb{A}$ be an admissible structure over $\mathfrak{M}$. Then $HF(M)$ is a $\Sigma$-subset of $\mathbb{A}$.*

***Proof.*** Let $\Psi(x, P^+)$ be
$U(x) \vee \exists y \exists z (P(y) \wedge P(z) \wedge ((x = \{y\}) \vee (x = y \cup z)))$. □

Indeed, $\Sigma$- cannot be replaced by $\Delta$- in 6.38(V.2.6 [11]). However, the following holds:

**Proposition 6.39.** *Let $\mathbb{A}$ be an admissible set. Then $HF(\varnothing)$ will be a $\Delta$-subset of $\mathbb{A}$, $\mathbb{HF}(\varnothing) \leqslant_{\mathrm{end}} \mathbb{A}$ and hence every $\Sigma$-($\Delta$-)predicate on $\mathbb{HF}(\varnothing)$ is $\Sigma(\Delta)$ on $\mathbb{A}$.*

**Proposition 6.40.** *Let $\mathbb{A}$ be an admissible set and let $M$ be a $\Delta$-($\Sigma$-)subset of $\mathbb{A}$. Then $\{\langle n, a\rangle \mid a \in M^n,\ n < \omega\}$ will be $\Delta(\Sigma)$ on $\mathbb{A}$.*

***Proof.*** Let $\Psi_0(x, y, P^+)$ be
$((y = 1) \wedge (x \in M)) \vee \exists u \exists v[(x = \langle u, v\rangle) \wedge (v \in M) \wedge \exists z(\mathrm{Nat}(z) \wedge (y = z + 1) \wedge (z > 0) \wedge P(u, z))]$

and let $\Psi_1(x, y, Q^+)$ be
$\neg\mathrm{Nat}(y) \vee (y = 0) \vee ((y = 1) \wedge (x \notin M)) \vee \exists z(\mathrm{Nat}(z) \wedge (y = z + 1) \wedge (z > 0) \wedge \forall u \in \mathrm{TC}(x) \forall v \in \mathrm{TC}(x)(\neg(x = \langle u, v\rangle) \vee Q(u, z) \vee (v \notin M)))$. □

**Corollary 6.11.** *Let $\mathbb{A}$ be an admissible set and let $M$ be a $\Delta$-($\Sigma$-)subset of $\mathbb{A}$. Then $M^{<\omega} \rightleftharpoons \bigcup_{n<\omega} M^n$ is also a $\Delta$-($\Sigma$-)subset of $\mathbb{A}$.*

Gandy's Theorem implies the existence of a universal $\Sigma$-predicate on any admissible set. Let $\mathbb{A}$ be an admissible set and $\mathcal{K}$ a class of $n$-ary relations on $\mathbb{A}$. A predicate $P \subseteq A^{n+1}$ is *universal* for $\mathcal{K}$ if $\mathcal{K} = \{\{\langle b_1, \ldots, b_n\rangle \mid \langle a, b_1, \ldots, b_n\rangle \in P\} \mid a \in A\}$. In particular, $P$ is a *universal $\Sigma$-predicate* if it is universal for the class of all $n$-ary $\Sigma$-predicates on $\mathbb{A}$; a partial $\Sigma$-function $f(y, x_1, \ldots, x_n)$ is a *universal $\Sigma$-function* if its graph $\Gamma_f$ is universal for the class of graphs of all $n$-ary partial $\Sigma$-functions.

We identify formulas with their Gödel numbers.

**Theorem 6.52.** *There is a binary $\Sigma$-predicate $\mathrm{Tr}_\Sigma$ on $\mathbb{A}$ such that, for every $\Sigma$-formula $\Phi(x)$ and $a \in A$,*

$$\langle \Phi, a\rangle \in \mathrm{Tr}_\Sigma \Leftrightarrow \mathbb{A} \models \Phi(a).$$

**Theorem 6.53.** *There exists a universal $(n+1)$-ary $\Sigma$-predicate $T(e, x_1, \ldots, x_n)$ on $\mathbb{A}$.*

As is mentioned above (see Sections 6.3, 6.5), there are admissible sets without universal $\Sigma$-functions [49, 101, 132, 162].

## Acknowledgements

The research was partially supported by the Russian Foundation for Basic Research (grants 06-01-04002-NNIOa (joint with DFG), 08-01-00442a, and 09-01-12140ofim) and by the State Maintenance Program for the Leading Scientific Schools of the Russian Federation (grants NSh-3606.2010.1, NSh-3669.2010.1).

## References

[1] Adamson, A. (1978). Admissible sets and the saturation of structures, *Ann. Math. Logic* **14**, 2, pp. 111–157.

[2] Adamson, A. (1980). Saturated srtuctures, unions of chains and preservation theorems, *Ann. Math. Logic* **18**, 1/2, pp. 67–96.

[3] Ash, C. and Knight, J. F. (2000). *Computable Structures and the Hyperarithmetical Hierarhy*, North-Holland, Amsterdam–London.

[4] Ash, C., Knight, J. F., Manasse, M., and Slaman, T. (1989) Generic copies of countable structures, *Ann. Pure Appl. Logic* **42**, pp. 195-205.

[5] Ashaev, I. V. (1995). Computability in fields (Russian), *Current problems od modern mathematics. Collection of scientific works*, vol. 1, Novosibirck: NII MIOO NGU, pp. 10–18.

[6] Ashaev, I. V. (1999). Priority method in generalized computability, *De Gruyter Series in Logic and its application 2*, Walter de Gruyter, pp. 1–13.

[7] Ashaev, I. V. (2003). On the reduction and uniformization principles in generalized computability (Russian), *Vestn. Omsk. Univ.* **3**, pp. 12–14.

[8] Ashaev, I. V., Belyaev, V. Ya., and Myasnikov, A. G. (1993). Toward a Generalized Computability Theory, *Algebra and Logic* **32**, 4, pp. 183–205.

[9] Avdeev, R.(to appear). On admissible sets of kind $\mathbb{HYP}(\mathfrak{M})$ over recursively saturated structures.

[10] Baleva, V. (2006). The jump operation for structure degrees, *Arch. Math. Logic* **45**, pp. 249–265.

[11] Barwise, J. (1975). *Admissible Sets and Structures*, Springer, Berlin–Heidelberg–New York.

[12] Belyaev, V. Ya., and Tajtslin, M. A. (1979). On elementary properties of existentially closed systems, *Russ. Math. Surveys* **34**, 2, pp. 43–107.

[13] Belyaev, V. Ya., Lyutikova, E. E., and Remeslennikov, V. N. (1995). Categoricity of Finitely Generated Algebraic Systems in HF-Logic, *Algebra and Logic* **34**, 1, pp. 6–17.

[14] Blum, L., Shub, M., and Smale, S. (1989). On a theory of computation and complexity over the real numbers: Np-completeness, recursive functions and universal machines, *Bull. Amer. Math. Soc.* **21**, 1, pp. 1–46.

[15] Calvert, W. (to appear). On three notions of effective computation over R, *Log. J. IGPL.*

[16] Chang, C. C. and Keisler, H. J. (1973). *Model Theory*, North-Holland, Amsterdam–London.

[17] Csima, B. F., Harizanov, V. S., Miller, R., and Montalbán, A. (to appear). Computability of Fraïssé limits.

[18] Cooper, S. B. (2003). *Computability Theory*, Chapman Hall, London–New York.

[19] Ershov, Yu. L. (1977). *Numbering Theory*, Nauka, Moscow.

[20] Ershov, Yu. L. (1980). *Decidability Problems and Constructivizable Models*, Nauka, Moscow.

[21] Ershov, Yu. L. (1983). The principle of Σ-enumeration, *Sov. Math. Dokl.* **27**, pp. 670–672.

[22] Ershov, Yu. L. (1983). Dynamic Logic over admissible sets, *Sov. Math. Dokl.* **28**, pp. 739–742.

[23] Ershov, Yu. L. (1985). Σ-predicates of finite types over an admissible set, *Algebra and Logic* **24**, pp. 327–351.

[24] Ershov, Yu. L. (1985). Σ-definability in admissible sets, *Sov. Math. Dokl.* **32**, pp. 767–770.

[25] Ershov, Yu. L. (1986). $f_A$-spaces, *Algebra and Logic* **25**, pp. 336–343.

[26] Ershov, Yu. L. (1986). The language of Σ-expressons (Russian), *Vychisl.*

*Sist.* **114**, pp. 3–10.

[27] Ershov, Yu. L. (1986). Σ-admissible sets (Russian), *Vychisl. Sist.* **114**, pp. 35–39.

[28] Ershov, Yu. L. (1987). Generatability of admissible sets, *Algebra and Logic* **26**, 5, pp. 346–361.

[29] Ershov, Yu. L. (1989). Each family of subsets of the urelements generates an admissible set, *Sib. Math. J.* **30**, 6, pp. 883–885.

[30] Ershov, Yu. L. (1990). Forcing in admissible sets, *Algebra and Logic* **29**, 6, pp. 424–430.

[31] Ershov, Yu. L. (1993). The Löwenheim–Skolem–Mal'tsev theorem for definable models (Russian), *Vychisl. Sist.* **148**, pp. 9–17.

[32] Ershov, Yu. L. (1995). Definability in hereditarily finite manifolds, *Dokl. Math.* **51**, 1, pp. 8–10.

[33] Ershov, Yu. L. (1996). *Definability and Computability*, Consultants Bureau, New York–London–Moscow.

[34] Ershov, Yu. L. (1998). Σ-definability of algebraic structures, in *Handbook of Recursive Mathematics*, Elsevier, vol. 1, pp. 235–260.

[35] Ershov, Yu. L. (1999). Theory of numberings, in *Handbook of Computability*, Stud. in Logic and Foundations of Mathematics, vol. 140, pp.473–511.

[36] Ershov, Yu. L. (2000). *Definability and Computability*, Ekonomika, Nauch. Kniga, Moscow–Novosibirsk.

[37] Ershov,Yu. L. and Goncharov, S. S. (2000). *Constructive Models*, Consultants Bureau, New York–London–Moscow.

[38] Ershov, Yu. L. and Palutin, E. A. (1987). *Mathematical Logic*, Mir Publishers, Moscow.

[39] Ershov, Yu. L., Goncharov, S. S., and Sviridenko, D. I. (1986). Semantic programming, *Inform. Processing* **86**, pp. 1093–1100.

[40] Friedman, H. (1969). Algorithmic procedures, generalized Turing algorithms, and elementary recursion theory, in R.O. Gandy and C.E.M. Yates, eds., *Logic Colloquium 1969.*

[41] Gandy, R. O. (1974). Inductive Definitions, *Generalized Recursion Theory*, pp. 265–300.

[42] Goncharov, S. S. and Sviridenko, D. I. (1985). Σ-programming (Russian), *Vychisl. Sist.* **107**, pp. 3–29.

[43] Goncharov, S. S. and Sviridenko, D. I. (1987). Σ-programs and their semantics (Russian), *Vychisl. Sist.* **120**, pp. 24–52.

[44] Goncharov, S. S., Harizanov, V. S., Knight, J. F., Morozov, A. S., and Romina, A. V. (2005). On automorphic tuples of elements in computable models, *Sib. Math. J.* **46**, 3, pp. 405–412.

[45] Gordon, C. (1970). Comparisons between some generalizations of recursion theory, *Compos. Math.* **22**, pp. 333–346.

[46] Hodges, W. (1993). *Model Theory*, Cambridge University Press, Cambridge.

[47] Kalimullin, I. S. (2006). The Dyment reducibility on the algebraic structures and on the families of subsets of $\omega$, *Logical Approaches to Computational Bariers, CiE2006, Report Series*, Swansea, pp. 150–159.

[48] Kalimullin, I. S. (2009). Uniform reducibility of representability problems

for algebraic structures, *Sib. Math. J.* **50**, 2, pp. 265–271.

[49] Kalimullin, I. S. and Puzarenko, V. G. (2005). Computability principles on admissible sets, *Sib. Adv. Math.* **15**, 4, pp. 1–33.

[50] Kalimullin, I. S. and Puzarenko, V. G. (2009). Reducibility on families, *Algebra and Logic* **48**, 1, pp. 20–32.

[51] Kfoury, A. J., Stolboushkin, A. P., and Urzyczyn, P. (1989). Some open questions in the theory of program schemes and dynamic logic, *Russ. Math. Surveys* **44**, 1, pp. 43–68.

[52] Khisamiev, A. N. (1996). $\Sigma$-enumeration and $\Sigma$-definability in $\mathrm{HF}_{\mathcal{M}}$ (Russian), *Vychisl. Sist.* **156**, pp. 44–58.

[53] Khisamiev, A. N. (1997). Numberings and definability in the hereditarily finite superstructure of a model, *Sib. Adv. Math.* **7**, 3, pp. 63–74.

[Khisamiev1998] Khisamiev, A. N. (1998). On definability of a model in a hereditarily finite admissible set (Russian), *Vychisl. Sist.* **161**, pp. 15–20.

[54] Khisamiev, A. N. (1998). Strong $\Delta_1$-definability of a model in an admissible set, *Sib. Math. J.* **39**, 1, pp. 168–175.

[55] Khisamiev, A. N. (1999). On resolvable and internally enumerable models (Russian), *Vychisl. Sist.* **165**, pp. 31–35.

[56] Khisamiev, A. N. (2000). The Intrinsic Enumerability of Linear Orders, *Algebra and Logic* **39**, 6, pp. 423–428.

[57] Khisamiev, A. N. (2001). Quasiresolvable Models and $B$-Models, *Algebra and Logic* **40**, 4, pp. 272–280.

[58] Khisamiev, A. N. (2004). Quasiresolvable Models, *Algebra and Logic* **43**, 5, pp. 346–354.

[59] Khisamiev, A. N. (2004). On the Ershov upper semilattice $\mathfrak{L}_E$, *Sib. Math. J.* **45**, 1, pp. 173–187.

[60] Khisamiev, A. N. (2006). On $\Sigma$-subsets of naturals over abelian groups, *Sib. Math. J.* **47**, 3, pp. 574–583.

[61] Khisamiev, A. N. (to appear). $\Sigma$-bounded structures and universal $\Sigma$-functions.

[62] Kierstead, H. A. and Remmel, J. B. (1983). Indiscernibles and decidable models, *J. Symbolic Logic* **48**, 1, pp. 21–32.

[63] Kierstead, H. A. and Remmel, J. B. (1985). Degrees of indiscernibles in decidable models, *Trans. Amer. Math. Soc.* **289**, 1, pp. 41–57.

[64] Kirpotina, N. A. (1993). Elementary equivalence in the language of list superstructures, *Sib. Adv. Math.* **3**, 4, pp. 46–52.

[65] Korovina, M. V. (1992). Generalised computability of real functions, *Sib. Adv. Math.* **2**, 4, pp. 1–18.

[66] Korovina, M. V. (1996). On the universal recursive function and on abstract machines on real numbers with the list superstructure (Russian), *Vychisl. Sist.* **156**, pp. 24–43.

[67] Korovina, M. V. (2002). Fixed points on the real numbers without the equality test, *Electron. Notes Theor. Comput. Sci.* **66**, 1.

[68] Korovina, M. V. (2003). Computational aspects of Sigma-definability over the real numbers without the equality test, in CSL, *Lect. Notes Comput. Sci.* **2803**, pp. 330–344.

[69] Korovina, M. V. (2003). Gandy's theorem for abstract structures without the equality test, in LPAR, *Lect. Notes Comput. Sci.* **2850**, pp. 290–301.

[70] Korovina, M. V. (2003). Recent advances in S-definability over continuous data types, in Ershov Memorial Conference, *Lect. Notes Comput. Sci.* **2890**, pp. 238–247.

[71] Korovina, M. V. and Kudinov, O. V. (1996). A new approach to computability of real-valued function (Russian), *Vychisl. Sist.* **156**, pp. 3–23.

[72] Korovina, M. V. and Kudinov, O. V. (1998). Characteristic properties of majorant-computability over the reals, in CSL, *Lect. Notes Comput. Sci.* **1584**, pp. 188–203.

[73] Korovina, M. V. and Kudinov, O. V. (1998). New approach to computability, *Sib. Adv. Math.* **8**, 3, pp. 59–73.

[74] Korovina, M. V. and Kudinov, O. V. (1999). A logical approach to specification of hybrid systems, in Ershov Memorial Conference, *Lect. Notes Comput. Sci.* **1755**, pp. 10–16.

[75] Korovina, M. V. and Kudinov, O. V. (2000). Formalisation of computability of operators and real-valued functionals via domain theory, in CCA, *Lect. Notes Comput. Sci.* **2064**, pp. 146–168.

[76] Korovina, M. V. and Kudinov, O. V. (2001). Semantic characterisations of second-order computability over the real numbers, in CSL, *Lect. Notes Comput. Sci.* **2142**, pp. 160–172.

[77] Korovina, M. V. and Kudinov, O. V. (2001). Generalised computability and applications to hybrid systems, in Ershov Memorial Conference, *Lect. Notes Comput. Sci.* **2244**, pp. 494–499.

[78] Korovina, M. V. and Kudinov, O. V. (2005). Towards computability of higher type continuous data, in CiE2005, *Lect. Notes Comput. Sci.* **3526**, pp. 235–241.

[79] Korovina, M. V. and Kudinov, O. V. (2008). Effectively enumerable topological spaces (Russian), *Vestn. Novosib. Gos. Univ., Ser. Mat. Mech. Inform.* **8**, 2, pp. 74–83.

[80] Korovina, M. V. and Kudinov, O. V. (2008). *Basic principles of Σ-definability and abstract computability*, Bericht Nr. 08-01, Fachbereich Mathematik, D-57068, Siegen.

[81] Korovina, M. V. and Vorobjov, N. (2004). Pfaffian hybrid systems, in CSL, *Lect. Notes Comput. Sci.* **3210**, pp. 430–441.

[82] Kreisel, G. (1971). Some reasons for generalizing recursion theory, in R.O. Gandy and C.E.M. Yates eds., *Logic Colloquium 69*, pp. 139–198, North-Holland, Amsterdam–London.

[83] Kreisel, G. and Sacks, G. E. (1965). Metarecursive sets, *J. Symbolic Logic* **30**, pp. 318–338.

[84] Kripke, S. (1964). Transfinite recursion on admissible ordinals, I,II (abstracts), *J. Symbolic Logic* **29**, pp. 161–162.

[85] Lacombe,D. (1964). Deux généralizations de la notion de récursivité relative, *C. R. Acad. Sci., Paris* **258**, pp. 3410–3413.

[86] Levy, A. (1965). A hierarchy of formulas in set theory, *Mem. Amer. Math. Soc.* **57**.

[87] Makkai, M. (1982). Admissible sets and infinitary logic, in *Handbook of Mathematical Logic*, ed. J. Barwise, North-Holland, Amsterdam–London, pp. 233–281.

[88] Miller, R. (2007). Locally computable structures, in CiE2007, *Lect. Notes Comput. Sci.* **4497**, pp. 575–584.

[89] Miller, R. and Mulcahey, D. (2008). Perfect local computability and computable simulations, in CiE2008, *Lect. Notes Comput. Sci.* **5028**, pp. 388–397.

[90] Montague, R. (1967). Recursion theory as a branch of model theory, in *Proceedings of the Third International Congress for Logic, Methodology and Philosophy of Science*, North-Holland, Amsterdam–London, pp. 63–86.

[91] Montalbán, A. (2009). Notes on the jump of a structure, *Mathematical Theory and Computational Practice*, pp. 372–378.

[92] Morozov, A. S. (1993). Functional trees and automorphisms of models, *Algebra and Logic* **32**, 1, pp. 28–38.

[93] Morozov, A. S. (1998). *Groups of Σ-permutations of admissible ordinals*, Mathematisches Institut Universitaet Heidelberg, preprint 36, Heidelberg.

[94] Morozov, A. S. (2000). A Σ subset of natural numbers which is not enumerable by natural numbers, *Sib. Math. J.* **41**, 6, pp. 1162–1165.

[95] Morozov, A. S. (2002). Presentability of groups of Σ-presentable permutaions over admissible sets, *Algebra and Logic* **41**, 4, pp. 254–266.

[96] Morozov, A. S. (2004). On the relation of Σ-reducibility between admissible sets, *Sib. Math. J.* **45**, 3, pp. 522–535.

[97] Morozov, A. S. (2005). About the admissible predicates on admissible sets, *Sib. Math. J.* **46**, 4, pp. 668–674.

[98] Morozov, A. S. (2006). Elementary submodels of parametrizable models, *Sib. Math. J.* **47**, 3, pp. 491–504.

[99] Morozov, A. S. (2008). On the index sets of Σ-subsets of the real numbers, *Sib. Math. J.* **49**, 6, pp. 1078–1084.

[100] Morozov, A. S. and Korovina, M. V. (2008). Σ-definability of countable structures over real numbers, complex numbers and quaternions, *Algebra and Logic* **47**, 3, pp. 193–209.

[101] Morozov, A. S. and Puzarenko, V. G. (2004). Σ-subsets of natural numbers, *Algebra and Logic* **43**, 3, pp. 162–178.

[102] Morozov, A. S. and Samokhvalov, K. F. (2002). On possible types of time in generalized computability theory (Russian), *Vychisl. Sist.* **170**, pp. 45–51.

[103] Moschovakis, Y. N. (1969). Axioms for computation theories – first draft, in R.O. Gandy and C.E.M. Yates, eds., *Logic Colloquium 1969.*

[104] Moschovakis, Y. N. (1969). Abstract computability and invariant definability, *J. Symbolic Log.* **34**, pp. 605–633.

[105] Moschovakis, Y. N. (1969). Abstract first order computability I, *Trans. Amer. Math. Soc.* **138**, pp. 427–464.

[106] Moschovakis, Y. N. (1969). Abstract first order computability II. *Trans. Amer. Math. Soc.* **138**, pp. 465–504.

[107] Moschovakis, Y. N. (1974). *Elementary Induction on Abstract Structures*, North-Holland, Amsterdam–London.

[108] Myasnikov, A. G. and Remeslennikov, V. N. (1992). Weak second order logic in group theory, *Contemp. Math.* **131**, part 1, pp. 273–278.

[109] Myasnikov, A. G. and Remeslennikov, V. N. (1992). Admissible Sets in Group Theory, *Algebra and Logic* **31**, 4, pp. 248–261.

[110] Odifreddi, P. (1989). *Classical Recursion Theory*, North–Holland, Amsterdam–London.

[111] Orey, S. (1956). On $\omega$-consistency and related properties, *J. Symbolic Log.* **21**, pp. 246–252.

[112] Platek, R. (1966). Foundations of recursion theory, Doctoral Dissertation and Supplement, Stanford University, Stanford, CA.

[113] Puzarenko, V. G. (1999). On computability over models of decidable theories, *Algebra and models theory 2*, Novosibirsk, NSTU, pp. 94–103.

[114] Puzarenko, V. G. (2000). On computability over models of decidable theories, *Algebra and Logic* **39**, 2, pp. 98–113.

[115] Puzarenko, V. G. (2002). On model theory in hereditarily finite superstructures, *Algebra and Logic* **41**, 2, pp. 111–122.

[116] Puzarenko, V. G. (2002). Decidable computable $\mathbb{A}$-numberings, *Algebra and Logic* **41**, 5, pp. 314–322.

[117] Puzarenko, V. G. (2003). Generalized numberings and definability of $\mathbb{R}$ in admissible sets (Russian), *Vestn. Novosib. Gos. Univ., Ser. Mat. Mech. Inform.* **2**, 3, pp. 107–117.

[118] Puzarenko, V. G. (2004). The Löwenheim–Skolem–Mal'tsev Theorem for $\mathbb{HF}$-structures, *Algebra and Logic* **43**, 6, pp. 418–423.

[119] Puzarenko, V. G. (2005). Computability in special models, *Sib. Math. J.* **46**, 1, pp. 148–165.

[120] Puzarenko, V. G. (2006). Definability of the field of reals in admissible sets, *Logical Approaches to Computational Bariers, CiE2006, Report Series*, Swansea, pp. 236–240.

[121] Puzarenko, V. G. (2008). On collection of all computable subsets on admissible sets (Russian, English abstract), *Sib. Elektron. Mat. Izv.* **5**, pp. 1–7.

[122] Puzarenko, V. G. (2009). About a certain reducibility on admissible sets, *Sib. Math. J.* **50**, 2, pp. 330–340.

[123] Puzarenko, V. G. (2010). Descriptive properties on admissible sets, *Algebra and Logic* **49**, 2, pp. 160–176.

[124] Puzarenko, V. G. (2010). On a semilattice of numberings, *Sib. Adv. in Math.* **20**, 2, pp. 128–154.

[125] Puzarenko, V. G. (2010). On a semilattice of numberings, II, *Algebra and Logic* **49**, 4.

[126] Ressayre, J. P. (1977). Models with compactness properties relative to an admissible language, *Ann. Math. Logic* **11**, 1, pp. 31–56.

[127] Richter, L. (1981). Degrees of structures, *J. Symbolic Log.* **46**, pp. 723–731.

[128] Rogers, H. (1972). *Theory of recursive functions and effective computability*, McGraw-Hill Book Company, New York.

[129] Romina, A. V. (1998). Hyperarithmetical stability of Boolean algebras (Russian), *Vychisl. Sist.* **161**, pp. 21–27.

[130] Romina, A. V. (2000). Autostability of hyperarithmetical models, *Algebra and Logic* **39**, 2, pp. 114–118.

[131] Romina, A. V. (2000). Definability of Boolean algebras in $\mathbb{HF}$-superstructures, *Algebra and Logic* **39**, 6, pp. 407–411.

[132] Rudnev, V. A. (1986). A universal recursive function on admissible sets, *Algebra and Logic* **25**, 4, pp. 267–273.

[133] Rudnev, V. A. (1988). Existence of an inseparable pair in the recursive theory of admissible sets, *Algebra and Logic* **27**, 1, pp. 33–39.

[134] Sacks, G. E. (1990). *Higher Recursion Theory*, Springer, Berlin–Heidelberg–New York.

[135] Sazonov, V. Yu. and Sviridenko, D. I. (1986). Denotational semantics of the language of $\Sigma$-expressions (Russian), *Vychisl. Sist.* **114**, pp. 16–34.

[136] Schmerl, J. H. (1980). Decidability and $\aleph_0$-categoricity of theories of partially ordered sets, *J. Symbolic Log.* **45**, pp. 585–611.

[137] Slaman, T. A. (1998). Relative to any non-recursive set, *Proc. Amer. Math. Soc.* **126**, pp. 2117–2122.

[138] Soskov, I. N. (2004). Degree spectra and co-spectra of structures, *Ann. Univ. Sofia* **96**, pp. 45–68.

[139] Soskova, A. A. (2007). A jump inversion theorem for the degree spectra, in CiE2007, *Lect. Notes Comput. Sci.* **4497**, pp. 716–726.

[140] Soskova, A. A. and Soskov, I. N. (2009). A jump inversion theorem for the degree spectra, *J. Logic Comput.* **19**, 1, pp. 199–215.

[141] Stukachev, A. I. (1996). Uniformization Theorem for HF(R) (Russian), *Proceedings of the XXXIV International Scientific Student Conference 'Student i Nauchno-Tehnicheskij Progress: Matematica'*, Novosibirsk, p. 83.

[142] Stukachev, A. I. (1997). Uniformization property in hereditary finite superstructures, *Sib. Adv. Math.* **7**, 1, pp. 123–132.

[143] Stukachev, A. I. (1998). Uniformization property in hereditary finite superstructures (Russian), *Vychisl. Sist.* **161**, pp. 3–14.

[144] Stukachev, A. I. (2002). $\Sigma$-admissible families over linear orders, *Algebra and Logic* **41**, 2, pp. 127–139.

[145] Stukachev, A. I. (2004). $\Sigma$-definability in hereditary finite superstructures and pairs of models, *Algebra and Logic* **43**, 4, pp. 258–270.

[146] Stukachev, A. I. (2005). On inner constructivizability of admissible sets (Russian), *Vestn. Novosib. Gos. Univ., Ser. Mat. Mech. Inform.* **5**, 1, pp. 69–76.

[147] Stukachev, A. I. (2005). Presentations of structures in admissible sets, in CiE2005, *Lect. Notes Comput. Sci.* **3526**, pp. 470–478.

[148] Stukachev, A. I. (2006). On mass problems of presentability, in TAMC2006, *Lect. Notes Comput. Sci.* **3959**, pp. 774–784.

[149] Stukachev, A. I. (2006). On inner constructivizability of admissible sets, *Logical Approaches to Computational Bariers, CiE2006, Report Series*, Swansea, pp. 261–267.

[150] Stukachev, A. I. (2007). Degrees of presentability of structures, I, *Algebra and Logic* **46**, 6, pp. 419–432.

[151] Stukachev, A. I. (2008). Degrees of presentability of structures, II, *Algebra*

*and Logic* **47**, 1, pp. 65–74.

[152] Stukachev, A. I. (2009). A Jump Inversion Theorem for the semilattices of $\Sigma$-degrees (Russian), *Sib. Elektron. Mat. Izv.* **6**, pp. 182–190.

[153] Stukachev, A. I. (2010). A Jump Inversion Theorem for the semilattices of $\Sigma$-degrees, *Sib. Adv. Math.* **20**, 1, pp. 68–74.

[154] Stukachev, A. I. (2010). $\Sigma$-definability of uncountable structures of $c$-simple theories, *Sib. Math. J.* **51**, 3, pp. 515–524.

[155] Stukachev, A. I. (to appear). $\Sigma$-definability of uncountable structures of $c$-simple theories, II.

[156] Stukachev, A. I. (to appear). Semilattices of $\Sigma$-degrees of structures.

[157] Stukachev, A. I. (to appear). Effective model theory via the $\Sigma$-definability approach.

[158] Tajtslin, M. A. (1979). *Descriptions of algebraic systems in weak $\omega$-logic and program logic, Theory of nonregular curves in various geometric spaces*, Alma-Ata, pp. 91–98.

[159] Trofimov, M. Yu. (1975). Definability in algebraically closed systems, *Algebra and Logic* **14**, 3, pp. 198–202.

[160] Vajtsenavichyus, R. Yu. (1987). Recursive enumerations of recursive functionals on recursively listed (RL) admissible sets (Russian, English abstract), *Mat. Logika Primen.* **5**, pp. 123–132.

[161] Vajtsenavichyus, R. Yu. (1989). On admissible sets with inner resolutions (Russian, English abstract), *Mat. Logika Primen.* **6**, pp. 9–20.

[162] Vajtsenavichyus, R. Yu. (1989). On necessary conditions for the existence of a universal function on an admissible set (Russian, English abstract), *Mat. Logika Primen.* **6**, pp. 21–37.

[163] Vajtsenavichyus, R. Yu. (1990). Principal numerations of functionals on admissible sets, *Algebra and Logic* **29**, 4, pp. 262–279.

[164] Wehner, S. (1998). Enumerations, countable structures and Turing degrees, *Proc. Amer. Math. Soc.* **126**, pp. 2131–2139.

# Chapter 7

# The Mathematics of Computing between Logic and Physics

Giuseppe Longo and Thierry Paul *

*Département d'Informatique UMR 8548 et CNRS*
*École Normale Supérieure*
*F 75730 Paris, France*
*Email: longo@di.ens.fr*

*Département de Mathématiques et Applications UMR 8553 et CNRS*
*École Normale Supérieure*
*Paris Cedex 05, France*
*Email: paul@dma.ens.fr*

Do physical processes compute? And what is a computation? These questions have gained a revival of interest in recent years due to new technologies in physics, new ideas in computer sciences (for example quantum computing, networks, non-deterministic algorithms), and new concepts in logic. In this chapter we examine a few directions, as well as the problems they bring to the surface.

## Contents



*This is a largely expanded version of a previous paper in French appeared in the proceedings of the LIGC colloquium, Cerisy, September 2006 (Joinet et al., eds), Hermann, 2009 (see http://www-philo.univ-paris1.fr/Joinet/ligc.html).

## 7.1. Introduction

Digital machines, by their extraordinary logical and computational capability, are changing the world. They are changing it with their power and their originality, but also with the image of the world they reflect: they help perform thousands of tasks and enable radically new ones, they are an indispensable tool for scientific research, but they also project their own mathematical structure upon the processes they are involved in.

The aim of this paper is to present several situations (in a non-exhaustive and rather kaleidoscopic way) where a precise confrontation of digital capacities with real settings in natural sciences is possible, and, in particular, to show how, in these situations, the computer science's concept of computability has to be carefully handled and sometimes not pertinent.

Digital machines are not neutral, as they have a complex history, based on several turning points in terms of the thinking which enabled their invention. They synthesize a vision and a science which is very profound. They are "alphabetic" in the specific sense of the encoding of human language, produced by a bagpipe over strings, by means of discrete and meaningless letter-units, an incredible invention which dates back 5,000 years. They are Cartesian in their software/hardware duality and in their reduction of thought to the elementary and simple steps of arithmetic calculus. They are logical by stemming from a logico-arithmetical framework, in the tradition of Frege and Hilbert, during the 1930s ("proofs are programs"). And this by the final remarkable invention, by Gödel: the number-theoretic encoding of any alphabetic writing. For all of these reasons, they contribute to a reading of nature based on the computable discrete, from the alphabet to arithmetic, on a space–time framed within discrete topology, of which the access and the measurement are *exact*, just like in digital databases.

We will see why confounding physics, despite its great "mathematicity", with computations and calculus, in any form whatsoever, seems a mistake to us. First, the idea that physics "reduces to solving" equations is an erroneous idea. To be assured of this, one needs only to consider that a great part of physics concerns variational problems in which the search of a geodesic differs greatly from the search for the solution to an equation. And this, without mentioning the singular quantum situation, to be discussed below, nor the life sciences, which are not very mathematized and for which the notions of invariant and of the transformation which preserves it, central to mathematics, are far from being "stabilized".

The new importance of digital machines, in particular in the natural sciences, requires a thorough analysis of the relationship between computations and natural processes. We will focus here on the relationships between computations and, among the physical processes, those which we consider as "natural", that is, those that occur somehow "independently" of human intervention (because a machine also produces, or even is, a physical process, but it is a result of a human construction which is extremely original and theoretically rich). We will then ask the question: Do physical processes compute?

The paper is organized as follows: Section 2 is devoted to a topological discussion of the link between computability and continuity. It leads to Section 3 where mathematics, especially computational mathematics, is confronted to physics endowed with its peculiar "reality" property. We show in particular how physics deals with a lot of concepts which escape from any sense of "calculus". Section 4 gives an epistemological example of a mathematical object which, with the evolution of physics, lost its computational flavour after entering the game of modern physics. Sections 5, 6, 7, and 8 are somewhat the core of the paper. We first discuss the concept of predictability in the mirror of chaoticity in dynamical systems. Then we come back to topological remarks and consider the problem of determinism, a fashionable subject in computer sciences nowadays. We then look at the case of quantum mechanics, also a subject which entered strongly into computer sciences lately. Section 9 discusses the position of randomness inside dynamical systems, and we end up with some final remarks.

*Let us mention once again that the scope of this paper is by no means to present a general theory of non-adequacy of computer sciences in natural philosophy, but rather to present warnings concerning a general temptation of overusing computational ideas in physics and mathematics, given the major role of computing in today's science.*

## 7.2. Computability and Continuity

The naive, and unfortunately highly widespread response to the question above is that yes, everything can be seen in terms of alphanumeric information and its computational elaboration. This thesis, under different forms, is often called the "Physical Church Thesis". So let's return briefly to Church's thesis in its original form, which is purely logico-mathematical and in no way physical.

Church's thesis, introduced in the 1930s after the functional equivalence proofs of various formal systems for computability (and concerning only computability over integers), is an extremely robust thesis: it ensures that any *finitistic formal system* over integers (a Hilbertian-type logico-formal system) computes at best the recursive functions, as defined by Gödel, Kleene, Church, Turing.... This thesis, therefore, emerged within the context of mathematical logic, as grounded on formal systems for arithmetic/discrete computations: the lambda-calculus (Church, 1932), a system for the functional encoding of logical deductions, and Turing's Logical Computing Machine[a], were the motors of various equivalence proofs[b].

The very first question to ask is the following: If we broaden the formal framework, what happens? For example, if we consider as basic support for computation a set "greater" than the natural integers, is this invariance of formalisms preserved? Of course, if we want to refer to continuous (differentiable) physics-mathematics, an extension to consider may be the following: What about the computational processing of these computable "limit" numbers which are the computable real numbers? Are the various formalisms for computability over real numbers equivalent, when they are maximal? An affirmative response could suggest a sort of Church thesis "extended" to this sort of computational "continuity". Of course, the computable reals are countably many, but they are dense in the "natural" topology over Cantor's reals, a crucial difference as we shall see.

With this question, we then begin to near physics, all the while remaining in a purely mathematical framework, because mathematics on the continuum of real numbers constitutes a very broad field of application to physics, since Newton and Leibniz. In particular, it is within spatial and often also temporal continuity that we represent dynamical systems, that

[a]1936: "A man provided with paper, pencil and rubber, and subject to a strict discipline, is in effect a Universal (Turing) Machine", [31]. In fact, the reader/writer needs only to know how to read/write 0 and 1 on an endless length of tape, then to move one notch to the right or to the left, according to given instructions (write, erase, move right, move left) to compute any formally computable function (see the next note).

[b]The other definitions of computability are more "mathematical": they propose, in different ways, arithmetic function classes which contain the constant function 0, the identity and the successor functions +1, and which are closed by composition, by primitive recursion (in short: $f(x+1) = h(f(x), x)$) and by minimalization (that is, $f(x) = min_y[g(x,y) = 0]$). It is a mathematically non obvious remark that by reading/writing/moving 0s and 1s left and right on a tape it is possible to calculate all of these functions: there lies the genius of Turing and the origin of the 0 and 1 machine which will change the world.

is, most mathematical models (in logical terms: mathematical formalisms) for classical physics. This does not imply that the world is continuous, but only that we have said many things thanks to continuous tools as very well specified by Cantor (but his continuum is not the only possible one: Lawvere and Bell, [6], say, proposed another without points, but one which is unfortunately not richer for the moment in mathematical terms – although some may hope to use it to better address the geometry of quantum physics; so, let's rest on Cantor for the time being).

Now, from this equivalence of formalisms, at the heart of Church's thesis, there remains nothing regarding computability over real numbers: the models proposed, in their original structure, are demonstrably different, in terms of computational expressiveness (the classes of defined functions).

Today, it is possible to roughly group different formal systems into four main groups (however not exhaustive ones), in order to perform computations over real numbers:

- recursive analysis, which develops the approach to Turing's computable real numbers, or even the Turing Machine itself, by an infinite extension recently formalized by Weihrauch (two tapes, one which can encode a computable real hence infinite number, and the other which encodes the program, see [35]; from the mathematical standpoint, the idea was first developed by Lacombe and Grezgorzcyk, in 1955–57);

- the Blum, Shub, and Smale BSS model (an infinite tape and a little control system, see [7]);

- the Moore-type recursive real functions (defined in a more mathematical manner: a few basic functions, and closure by composition, projection, integration, and search for the zero, see [24]);

- different forms of "analog" systems, among which threshold neurons, the GPAC (General Purpose Analog Computer, attributable to Shannon, [30], of which a first idea preceded classical recursivity: V. Bush, M.I.T., 1931, [10]).

Each of these systems has its own interest. Besides, they confirm the solidity of Church's original thesis, since the restriction to integers of all known models of computability over continua again produces classical recursivity (or no more than that). What else could we say, concerning inclusions, links, demonstrable passages, as for these formalisms for computability on continua?

Of course, it is a matter of "relative" continuity: computable real numbers do not form a Cantor-type continuum, as we said; they are a denumerable set of measure 0. However, their "natural" (interval) topology is *not* the discrete topology (and mathematicians know what "natural" means: the discrete topology over Cantor reals is not natural; one does nothing with it). This is the crucial mathematical difference of computations on reals from computability over the isolated points of the countable discrete: the natural topology is not the discrete topology, but the induced one, by intervals.

The difference is crucial with regard to physical modelling for the following reasons. In physics, the (Cartesian) dimension of space is fundamental. By dimension we mean both the number of independent variables in functions and their "physical meaning" (the dimension of energy, say, is different from that of force). Relativity and string theory, to use some examples, make it into a constitutive issue, as for the dimension of space-time; but also, the propagation of heat, or the mean field theory, to remain in classical physics, depends in an essential way upon the dimension under consideration, see [3]. Now, computability over integers is "indifferent" to the Cartesian dimension: the expressivity of the machine does not change by changing the dimensions of its databases, but only the polynomial efficiency. This is due to the computable isomorphism $< .,. >$ between $N^2$ and $N$. One may therefore define, without difficulty and for any discrete formalism, the universal function $U$ within the very class of computable functions (that is, once the computable functions have been enumerated, $(f_i)_{i \in N}$, function $U(i,n) = f_i(n)$ belongs to such class by the coding $< .,. >$).

These properties, quite interestingly, are a consequence of the rather general fact that discrete topology does not force a dimension. In short, in the discrete universe (the category of sets), any infinite set (integers, in particular) is isomorphic to all of its finite (Cartesian) products. But when discrete topology is no longer "natural", within a continuum, say, with Euclidean (or real) topology, for example, the spaces having different dimensions are no longer isomorphic. We then say that the dimension is a topological invariant, for topologies which derive from the interval of physical measurement (Euclidean, typically). A remarkable relationship between geometry and physics: the metrics (and the topology induced) of the sphere (or interval) indeed corresponds to the "natural" physical measurement, that of the intervals, and it "forces" the dimension, a crucial notion in physics. So here is a fundamental difference for continuous mathematics (and for computability over continua, would they be just dense): any bijective encoding of spaces with different dimensions is necessarily non-continuous and, in order to define, typically, the universal function, it is necessary to change dimension, hence to leave the given class.

So let's return to our question, which is, in our view, a rigorous way to address the extensions of the Church thesis to the *mathematics* of physics: Can we correlate different formalisms for computability over a continuum, these being adequate for physical systems and which, therefore, make the Cartesian dimension into a fundamental issue, even if they are non-equivalent? There are no extensions today of the Church thesis to computable continua and just partial answers are provided by many authors: [9, 21] present an overview and recent results which, by the addition of functions and operators which are highly relevant from the physical standpoint, enable us to establish inclusions under certain conditions, these being rather informative links. On the basis of these works, we should arrive at a notion of a "standard system" for computability over the set of computable real numbers which represent a reasonable extension of Church's thesis to computable continuity (all "standard" systems would be equivalent, modulo the fundamental issue of dimensions), and therefore also find an interesting link with the mathematics of physics.

However, for a large enough class, this standardization is not obvious and we are far from having a Church-like equivalence between systems. Moreover, it is clear that we remain, as in the case of the logico-formal Church thesis, within *mathematical formalisms*[c]. And what about physical processes?

## 7.3. Mathematical Computability and the Reality of Physics

Let's ask a preliminary question to asking if nature computes: What could nature actually compute? If we look at the object before looking at the method, things may not be so simple. Vladimir Arnol'd recalls in his book [2] the formula attributed to Newton: "It is useful to solve differential equations". From another perspective, physics could very well be expressed according to another formula, provided this time by Galileo[d].

[c]In what concerns the extension of the Church thesis to computer networks and to concurrent systems in general, systems which are perfectly discrete but distributed over space–time, this being better understood by means of continuous tools, we refer to [1] and to its introduction: in this text, it is noted that this thesis, in such a context, is not only false, but also completely misleading (the processes are not input-output relationships and their "computational path" – modulo homotopy, for instance – is the true issue of interest).

[d]"La filosofia scritta in questo grandissimo libro che continuamente ci sta aperto innanzi agli occhi (io dico l'Universo) non si puó intendere se prima non s'impara a intender la lingua, e conoscer i caratteri, nei quali è scritto. Egli è scritto in lingua matematica, e i caratteri son triangoli, cerchi, ed altre figure geometriche, senza i quali mezzi è

And from Galileo's standpoint which is, however, far from being formalistic or number-theoretic but rather "geometric", and which continues to perceive "filosofia" as an intermediary between ourselves and the world, the question asked above could very well be natural.

Newton–Arnold's view point seems more modern. However, it is now necessary to observe that the importance of an equation, or more generally, of a mathematical conceptual structure used in physics is often more important *in abstracto* than its numerical solutions. But let's nevertheless look at what happens upwards to this.

*Is there something to solve, to compute?*

The description of a physical phenomenon takes place within a framework of "modelling", that is, within a fundamentally "perturbational" framework. The isolation of a phenomenon, its intrinsic comprehension, supposes that we neglect its interaction with the rest of the world. But to neglect does not mean to annihilate: the rest of the world exists and creates perturbations at this isolation. From this point of view, a model must be immersed in an "open set" of models.

The isolation of a concept upon which one is working, for instance, results from the choice of a given scale. Neighboring scales are then supposed to be either inaccessible (smaller scales), or processable by averaging (larger scales). In both cases, they can influence the model and the equation which yields it. Asking the question whether something which we compute, physically, fits into a framework of computability, in the classical sense, commands having precautions at least.

In particular, are there equations and only equations? A great part of classical physics rests upon variational principles. The trajectory appears not only as the solution to an equation, but as a solution that is chosen because it optimizes, extremizes a quantity (action). Of course, this is equivalent to resolving equations (Euler–Lagrange), but this is only an equivalence. Let's recall that Feynman [14] preferred solutions to equations for quantum mechanics. In this case, no more equations: all possible trajectories (minimizing or not the action functional) are involved. This is possible, but is so thanks to the functional integral, in *an infinite dimensional space.* And what about computability in this case?

impossibile a intenderne umanamente parola; senza questi è un aggirarsi vanamente per un oscuro laberinto." (**Il Saggiatore**, 1623.)

Let's look at another example: quantum field theory, a physical theory which is not mathematically well founded yet, but which has been phenomenally successful in terms of precision, is based entirely on perturbative calculations [26].

Thus it is obvious that, even without considering the lack of precision of classical measurement, which we will address later on, the true situation is somewhat fuzzy, largely perturbative, and hence that the problem of computability in physics is multiple and complex.

Nevertheless let's suppose that there actually *are* equations. And let's suppose that the true issue is really the solution, which is predictive. We will then be compelled to remark that the situations where the solution's values are important are rare. A simple example: physicists like to draw curves, even when a formula providing the solution is available. But what is left of computability when the "result" is smoothed by the graphical process, where only the general "trends" are important, not the exact values?

Let's take a look at the dynamical systems provided by maps, the case of the "baker's map", for instance. In principle, there is no mapping in physics; there are flows. A map appears when we compute a flow at time 1 (which we will later iterate), but this flow at time 1 is actually computed from equations. The Poincaré first recurrence map, and the dynamical systems which followed, were invented as simpler tools, qualitatively and quantitatively more manageable, but it would be wise to not identify them too much with the initial systems.

In conclusion let us see whether it is possible to consider an isolated equation in physics. As we observed, if equations come in families within which (possibly continuous) parameters change, how must one apprehend the problem of computability, so carefully defined within a discrete and countable space? Maybe nature does compute, but knowledge, our theory of nature, fundamentally rests in huge, infinite spaces (spaces of parametrized equations, typically), which could very well escape any computationalistic approach.

Let us examine carefully the example of the epistemological evolution of the classical concept of "action".

## 7.4. From the Principle of Least Action to the Quantum Theory of Fields

The concept of differential equation is not the only one which provides a way for computing dynamics in physics. As we mentioned, an alternative approach consists in minimizing a certain functional (the action) among different candidates for the trajectory. More precisely to any path $\gamma$ going from an initial point to a final one is associated a number, $S(\gamma)$, and the "true" trajectory, the one that the particle is going effectively to follow, is the one which provides the lowest (in fact any extremal) value of $S(\gamma)$. This principle of "least action" does not ask to solve an equation, it just asks to evaluate the functional $S$ at any possible path $\gamma$, and select the extremal one. If it asks to compute something it doesn't ask to compute a finite number or set of numbers, it asks to evaluate a huge set of numbers, and to find the smallest.

As a matter of fact it is true that the principle of least action is, in many situations at least, equivalent to the so-called Euler–Lagrange equations, therefore shown to be embedded in the operational setting. But the Feynman "path integral" formulation of quantum mechanics creates a revival of this idea of evaluating instead of computing. The quantum amplitude of probability is obtained by summing expressions of the form $e^{i\frac{S(\gamma)}{\hbar}}$ over all paths $\gamma$: the process of minimizing the action of the path disappeared completely. Here again this formalism is shown to be equivalent to the Schrödinger equation, getting back once more to the operational level.

The situation drastically changes with quantum theory of fields, a mixing of quantum mechanics and partial differential equations. This theory, conceptual basis of our deep understanding of elementary particles, is a generalization of quantum mechanics to infinite dimension. The formalism of quantum theory of fields is an extension of the "path integral" method to the case where the "paths" $\gamma$ sit in infinite dimensional spaces. This is the theory which provides nowadays the most accurate numerical agreement with experimental data. It "lives" in an extremely huge space (the space of infinite dimensional paths), and has, up to now, no equivalent operational setting. Once more we are very far from any form of (extended) Turing computability. Of course computers were of definitive usefulness in quantum theory of fields, as heavy computations were involved. But this was inside a perturbative approach (see [26]), and not at a conceptual level, as the conceptual frames radically differ. The close analysis of this difference is one of the enriching challenges (and the interest) of computing, today, in physics.

## 7.5. Chaotic Determinism and Predictability

In what concerns the relationships between dynamical systems and their capacity to predict physical evolutions, there is often a great confusion between mathematics and physical processes. The notion of deterministic chaotic system is purely mathematical and is given, in a standard way, by three formal properties (sensitivity to initial conditions, topological transitivity, and density of periodic trajectories, see [12]). However, it is legitimate to speak of a physical process and to say that it is deterministic (and chaotic, if such is the case): what is meant by that is that it is possible, or believed to be so, to write a system of equations, or even an evolution function, which determines its evolution (in time or regarding the relevant control parameter). Chaos pops out when the formal properties above are realized in deterministic systems (yet, weaker forms of chaos are possible, "mixing systems" for example, [12])

Unpredictability is then a property which arises at the interface between physical and mathematical processes. One gives oneself a physical process and a mathematical system, which is supposed to "model" it (a system of equations, typically, or even an evolution function – an iterated system thus a discrete-time system, within a continuous space). Then the process *with regard to the system* (or even with regard to any reasonable system which we consider to modelize the given process) is said to be unpredictable. A physical process "as such" is not unpredictable: one must attempt to *state* or even *pre*dict, usually by mathematical writing, for there to be unpredictability. Likewise, a mathematical system is not unpredictable, as such: it is written and, if fed values, it computes.

And this is where computability comes into play. It happens that any "reasonable" mathematical system would be characterized by effective writing: save a pathology (feeding a polynomial with non-computable coefficients, Chaitin's Omega for instance!), we normally write evolution functions which are computable (we will however see some counter-examples). More specifically, any Cauchy problem (a very broad class of differential equations) admits computable solutions (if solutions there are), in one of the known systems for continuous computability. Interesting pathologies, or counter-examples, do exist; for the moment, it suffices to mention some solutions of the Poisson equation in [27], the boundary of a Julia set, in [5].

But the problem is not only there (not really there, as a matter of fact): the choice of scale, of perturbative method, of phase space, (or of hidden variables, or those which were explicitly or unconsciously excluded) shows the constituted autonomy of mathematical language, because mathematics is constructed within a friction contingent to the world and then detaches from it by its symbolic autonomy. And this construction is a highly non-computable historical decision, often an infinitary transition towards a limit concept. By this, mathematics is not arbitrary, but the result of a constructed objectivity.

In summary, when we write a formalism, we give ourselves something "computable" (*grosso modo*, because the different continuous systems for computability are not yet unified, as we recalled above) but this is obtained by an historical choice or limit process, which singles out *finitistic* symbolic construction from the world. So the fact of the computability of an evolution function, which we *suppose* to be adequate regarding the description (modelling) of a physical system, is the evidence which we deduct from its writing. The logistic function, for instance, see [12, 22], is a simple and important chaotic system; a computable bilinear function, with a coefficient $k$ (well, only if one takes a non-computable $k$, a crazy choice, it is not computable). A very famous variant of the logistic function is also given by the "tent" function, a continuous but non-differentiable deformation which preserves many of its interesting properties. This function modelizes, *grosso modo*, the movements of stretching and mixing of a piece of dough by a baker who is a little stiff and repetitive in his movements. These systems, as in the case of any formal writing, are effective and are in no way unpredictable, as such. We give them values (computable ones) and they compute: within the limits of the available (finite) machine, they produce outputs. However, any *physical system* which is considered to be modelized (formalized) by one of these functions is unpredictable, even if by one of their non-differentiable variants (an ago-antagonistic system – chemical action–reaction oscillations, for example, or the baker's transformation, in the differentiable or non-differentiable case of which we were speaking). As soon as we give the result of a *physical measurement*, that is, an interval, to the function in question, this interval is mixed and exponentially widened, quickly preventing any prediction of the evolution. Of course, the machine which computes these non-linear functions can also help appreciate chaos:

1 - it provides *images* of "dense" trajectories (sequences of points) in the definition domain;

2 - a difference (at the 16th decimal, for instance) in the numeric input gives very different values after few iterations (about 50 in our logistic cases, see below).

However, if it is relaunched with the *same initial values*, in a *discrete* context (and this is fundamental) it will always return the same trajectory (sequence of numbers). The point is that, in *discrete state machines*, access to data is *exact*: this is the crucial difference w.r. to access to the world by (classical) physical measure, which is given by the interval of (approximated) measurement, by *principle* (there is at least the thermal fluctuation).

And there lies also the advantage of the discrete state machine, of which the access to the database is exact: it *iterates identically*, because it is, firstly, an iterating machine. Iteration founds Gödel-type primitive recursion, which is iteration and +1 in a register (see the note above). It enables the portability of the software and hence its identical transferal and iteration at will (and it works – without portability and iterability of software, there would be no computing, nor market for software). You may launch a program hundreds of times, thousands of times and it iterates.

Computer scientists are so good that they have been able to produce reliable and portable software (that is iterating identically) even for networks of concurrent computers, embedded in continuous space–time, with no absolute clock. Yet, the discrete data types allow this remarkable performance. Note that identical iteration of a process is very rare in nature (fortunately, otherwise we would still be with the universe of the origin or with the early protozoans). We, humans, along our history, invented the discrete state machines, which iterate. A remarkable human construction, in our space of humanity, using the alphabet, Descartes dualism (software/hardware), Hilbert's systems, Gödel's numberings, Turing's ideas..., and a lot of discrete state physics. Computing, programs and alike are not "already there" in nature. Unfortunately, some miss the point and do not appreciate the originality nor the founding principles of computing and claim, for example, that "sometimes they do not iterate", like nature. Of course, there may be hardware problems, but these are *problems*, usually (and easily) fixed. Instead, non-iteration, identically, is part of the *principles* of non-linear dynamics, it is not "a problem". Let alone life sciences where the main invariant is... variability, even within "structural stability", which is not phenotypic *identity*.

But let's go back to the interface mathematics/physics. The passage from the physical process to the formal system is done by means of measurement. If the only formalization/determination we have, or which we consider to be relevant for a given process, is of the deterministic but chaotic type, the (classical) physical measurement, which is always an interval (and which we describe, in general, within a context of continuity) enables us to only give an interval as input for the computation. And this has a further fundamental connection with physics, that we already mentioned: the interval topology yields the topological invariance of dimension, a fundamental property of the continua of mathematical physics. Now, given that non-linear dynamics are *mixing* (the extremes and the maximum and minimum points of any interval are "mixed" at each step) and have an "exponential drift" as Turing puts it, this is a nice way, Turing's way, to say what we observed: the interval of measurement soon occupies in a chaotic – mixing – way an increasing part of space and it is impossible to further predict the evolution of the physical process. If we were to use as input not an interval, but a rounded value, this would obviously not help prediction: the result of the computation may have nothing to do with the physical evolution – for the logistic function, with $k = 4$, a rounded value at the 16th decimal makes any physical process unpredictable approximately from the 50th iteration, – this is calculated using the value of the Lyapounov exponent, [12].

To return to the baker's dough, a very simple and common example, it is a physical process determined by a demonstrably chaotic evolution function, thus unpredictable. It is a mistake to say, as we sometimes hear, that it is non-deterministic; it is unpredictably deterministic, which is quite different (the error, in this case, is exactly Laplace's error, for whom determination should imply predictability). In physical terms, the forces at play are all known; the "tent" function determines its evolution well, just as the logistic determines that of the ago-antagonistic processes or as the equations of Newton–Laplace determine the evolution of Poincaré's three-body. In classical physics everything is deterministic, even a toss of dice! But sometimes, it is impossible to predict or calculate evolutions because of the approximation of physical measurement *in conjunction with* the sensitivity to contour conditions, proper to the intended, modelling, mathematical systems (or with the excess of relevant but hidden variables in the process: Einstein hoped to transfer this very paradigm to quantum physics).

So, in general, the mathematical systems which we write are computable and predictable, at the formal level; some of these systems, being chaotic, refer to unpredictable *physical* processes. In principle, the latter, as such, do not "compute", in the sense of the Church thesis or of its continuous versions. Let's specify this point once more.

Computation is an issue of numbers, in fact of the (re-)writing of integer numbers: lambda-calculus, Turing Machines, are actually a paradigm of it. Now, to associate a number to a physical system, it is necessary to have recourse to measurement, a challenge and major issue regarding principles in physics, as has been realized since Poincaré and Planck, extraneous to the logic of arithmetic and, thus, largely forgotten by computationalists (the world is a large "digital computer"). Classically, if we were to decide that a certain state of a physical process constitutes the input, and another the output, and that we associate these states to measurement *intervals* and if all we know of this process is mathematically unpredictable, then it will be impossible, in general and after a sufficient amount of time (if time acts as a control parameter), to compute or predict an output interval from the input interval *of the order of magnitude of the given physical measurement.* In short, if we launch a good old physical double pendulum, if we manipulate a baker's dough, it will be impossible to compute, within the limits of measurement, its position after five or six oscillations or foldings, although they may themselves be determined by two equations or by an evolution function in which all is computable. So the double pendulum, the stretched dough, as a physical machine/process, does not compute a computable function. As for quantum mechanics, we will return to this below.

But do they define a function, in the usual sense of a single-valued relation? Because in the same initial (physical) conditions, they do not generally iterate, and therefore do not even define a mathematical function of an argument (which one?) within the initial interval of measurement, that is a function which would always return the same value. In short, in mathematics, $f(x) = y$, when $x$ is not time, is $f(x) = y$ also tomorrow; while in chaos, even the intervals are not preserved. It would therefore be necessary to parameter them across time according to a physical reference system: at best they would define a function with multiple variables of which one is the time of the chosen reference system. This makes them rather useless as machines for defining non-computable functions: they cannot even be re-used, in time, to compute the same function, because at each different moment we would have different values which are *a priori* non-repeatable. And no one would buy them as "non-Turing" machines.

And here we are confronted once more with another common error: expecting that if the physical Church thesis were to be false, then the counter-example should return a process which computes more than Turing. But such is not the case. This is an error because a "wild" physical process (as biologists would put it), in general, does not even define a function, that is, a single-valued argument/value relation. The very idea that a process could be reiterated suggests that it could be redone in the same (identical, as within a discrete framework) initial conditions. And this, which is so trivial (in both the English and French senses of the term) for a *discrete state* machine, is unachievable in nature, except in very rare or artificial cases, save the extension of the parameters to an additional temporal dimension which considers the counting of the experience performed. In what concerns life phenomena, do not by any means try to make the halting of a Turing Machine computed by a paramecium and the movements of its two thousand cilia: quite upstream to computation, paramecia do not define functions by their activities (between the paramecium and computation there is the "wall" of measurement: how to measure, what to measure, using which level of approximation?).

Quite thankfully, we have invented an alpha-numeric machine that is not wild at all, but well domesticated and *exact*. It comes with its own reference system and clock (hence the problem in concurrent networks, where a spatiotemporal absolute is lacking). Thanks to its structure as a discrete state machine, as Turing emphasized from the moment he produced his invention[e], this machine enables an access to the data and computations and... it iterates, identically, when made to: there are the two reasons for its strength. And even within computer networks, thanks to the discrete aspect of databases, we manage to iterate processes, as we said, despite the challenges entailed by concurrency within physical space–time.

## 7.6. Return to Computability in Mathematics

Let's return to the issue of computability beyond the measurement which we just addressed.

Mathematically, chaos is a long-time phenomenon: as for the sensitivity to the initial conditions, it is the long-time asymptotic behavior which differs between chaos and integrability. What is the evolution of the baker's dough in the case of an infinite number of iterations? Let's be more specific and look at the case of ergodicity, a property of chaos which is actually weak (and non-characteristic). A system is ergodic when, for almost all points (the "ergodic" ones), the temporal and spatial averages of any observable coincide at the infinite limit. This is a property "in measure" (measure meant here in the mathematical sense) and it requires, in its "time" component, an integration over an infinite time.

Clearly, the question of computability of average up to time $t$ for any value of $t$ makes sense, and has a clear answer in terms of properties of computability of ordinary differential equations, but the passing to the limit $t \to \infty$ shifts us towards these limits of which it was questioned earlier

[e]Or shortly after: in 1936, it was nothing more than a logical machine, "a man in the act of computing"; it is only after 1948 that Turing viewed it also as a physical process – a discrete state one, as he called it in [32] and [33].

and which we will return to now. In particular, the rate of (mathematical) convergence will intervene in the answer to the first question, and obtaining information on the rate of this convergence is a very delicate problem *especially in what concerns real, practical flows, those which nature provides us with.*

One must nevertheless not forget the huge contribution of computer science: the computer, however fundamentally non-chaotic, "shows" chaos amazingly well, suggests it, presents it to our eyes in a very spectacular and now completely indispensable way. And this by the (approximated) images of the density of trajectories, by the very different results in the change of the 16th decimal or so etc. By developing turbulences of any sort in an otherwise unfeasible way and showing them on a screen, a fantastic help to scientific insight is achieved.

*The passing to continuity*

The passing from rational numbers to real numbers poses more problems than it may seem: a quantum system in a finite volume is indeed represented by a vector space of finite dimension. Yet, some caution is required; not only must this space be bounded, but so must the momentum dimension, that is, the phase space, of which the *standard of measurement* is Plank's constant. But the superposition principle immediately makes the *number* of states infinite (to the power of the continuum): this is precisely the "vectorial aspect" of the theory. Quantum mechanics resides in vector spaces and the "finiteness" of space entails the finiteness of the *dimensions* of these spaces, not their cardinals. It is impossible, for a set value of the Planck constant, to put anything but a finite number, $d = \frac{V}{\hbar}$, of *independent* vectors (states) within a finite volume $V$, but thanks to (because of) the superposition principle, it is in fact possible to put an infinite number of vectors, as many as there are points in $\mathbb{C}^d$. This doesn't mean of course that, for certain definitions of information, the "quantity" of information could not remain bounded as the system remains confined in a finite volume, but this shows the difference of the concepts of space in classical and quantum situations (for a discussion of this discrete/continuous dichotomy see e.g. [25]).

One must then evoke the Rolls-Royce of mathematical physics: the theory of partial differential equations (PDEs). A PDE can be seen as an ordinary differential equation in infinite dimension, it is like a system of ordinary differential equations, each of them labelled by a continuous parameter (by the way, it is precisely this aspect which the computer retains before discretizing this continuous variable): each point in space "carries" a dynamic variable of which the evolution depends on immediately (even infinitesimally) neighboring points. Contrary to ordinary differential equations which, in general, have a solution for all values of time, we can say that a PDE has (still generally speaking, in the "hyperbolic" case) a limited life span, in the sense that its solution can explode in a finite amount of time. We therefore witness the emergence of two pitfalls: one passing to infinity for space, and one passage to "finiteness" for time. This is another example where the very notion of computability does not apply well to the physico-mathematical phenomenon.

Let's now ask ourselves why chaos was invented. The sensitivity to the initial conditions has appeared as a negative result, preventing integrability. The negation of integrability aims to be perceptible in a finite amount of time (since integrability places us in front of eternity). But it is very difficult to demonstrate that a system is not integrable. An alternative result consists in looking for a totally inverse paradigm: instead of stability, one looks at instability. The theory of chaos, an extreme and antipodic point of integrability, offers powerful and realistic results and shows, by this inversion of paradigm and its qualitative (and negative, yet very informative) fall-out, its limits with regard to computability.

## 7.7. Non-determinism?

In computer science, we often define non-functional relations as being "non-deterministic"; in short, when we associate a number to a set. Let's first examine the case of so-called "non-deterministic" Turing Machines, of which the transition functions have precisely this nature (from a value to a set of values). Calling them non-deterministic may be reasonable, as an *a priori* as long as we remain within logico-computational formalisms, but makes no physico-mathematical sense. Is there an underlying physical process which will associate to an input number a set or an element of the set in question? Not necessarily. So, if deterministic (classical) means (potentially) determined by equations or evolution functions, a "non-deterministic" Turing Machine is indeed *determined* by a function which associates an output set to an input value (an issue of asymmetrical typing, nothing more).

If there is indeed a choice of value to be made among a set, quantum physics could certainly propose one: it is legitimate to say that quantum measurement, by giving probabilities within possible values, performs such an operation. Can we use a classical process for the same association? Why not: we can take a physical double pendulum, determined by two equations or the baker's dough, of which the evolution is described by the "tent" function – so there is nothing more deterministic than these two objects and their evolutions. We give them an input number; the evolution starts off on an interval of measurement which is roughly centered around this number, but the result, which is unpredictable after a few iterations, can take a value among all those within the space. This is what deterministic unpredictability is. Yet, with a playful use of language (and a little bit of confusion), computer scientists also say that this association (one value/one set) produces a non-functional relation and so consider it as non-deterministic. But contextual clarity, necessary to the good relationship between mathematics and physics, then disappears: all is grey and that which is not functional (nor calculable) is the same, as there is no more difference between classical unpredictable determination and quantum indetermination, typically.

In what concerns concurrent systems, the situation is more interesting. Over the course of a process, which occurs within physical space–time, choices are made among possible values, following the interaction with other processes. In concrete machines, these choices can depend on classical, relativistic, quantum, or even human phenomena which intervene within a network. In the first two cases, everything is deterministic, although described by non-singled-valued relations and although there may be classical unpredictability (which value within the determined set? A lesser temporal discrepancy can produce different choices). In the other two cases (quantum and "human"), the choice of value will be intrinsically non-deterministic, but, in principle, for different reasons (not being able to give an appropriate physical name to the will of humans acting upon a network). In some cases, authors in concurrency, by non-determinism, refer to a "do-not-care" of the physical "determination": whatever is your hardware and your (compatible) operating system or compiler, my program for the network must work identically. A new concept of "non-determination" a very interesting one, probably with no analogy in natural sciences (my soul doesn't work independently of my body, this was Descartes' mistake, nor it is portable – this would be a form of metempsychosis).

It would be preferable to introduce a notion of "indeterminacy" specific to computer science corresponding to the absence of univocity of the input–output relation *with choices*, in particular that which can be found in "multitasking", in the concurrence of network processes, etc.

*The discrete and the "myth" of continuity.*

This loss of meaning of continuous physics can be found in Gandy's reflections on Church's Thesis, for instance (he was one of the pioneers of the *physical* Church thesis, [19]). He posits among other things a physical world within which information is finite, because it is part of a finite universe. So it is made to be discrete, all the while remaining within a classical framework, and then deterministic chaos disappears, as happens with the Turing Machine (Turing says this very clearly in [32], see also [23], and the discussion on finiteness in quantum mechanics in the preceding paragraph).

Firstly, the mathematical definitions of chaos use continuity (to represent the interval of measurement); they will lose their meaning when the natural topology of space considered is discrete topology (we keep returning to this, because it is important: the access to the measurement of the process will then become exact, because isolated points are accessed *exactly*, mathematically – another way to summarize all which we have just said).

Now Gandy does not appear to have followed his master Turing, the inventor of the "Discrete State Machine" (which is theoretically predictable, says Turing, [32], though it may be practically hard to predict – very long programs), in the adventure of the continuity of non-linear dynamics (theoretically unpredictable, Turing remarks, this being their most interesting property, [33], see [23] for a discussion).

As a matter of fact, Turing had a deep understanding of this issue in the later years of his life, making a remarkable contribution to the development of what he called "continuous systems" (the name which he gives to the linear and non-linear models of morphogenesis, [33], and which he already uses in [32] in contraposition to his machine). In fact, continuity is currently the best tool we have for addressing classical determination. It is the "myth" of an underlying or abstract space, a mathematical continuum, which leads us to think that any classical trajectory is deterministic: it is "filiform" (widthless) and stems from a Euclid–Cantor point (dimensionless, said Euclid). It is a "myth" in the sense of Greek mythology, because it constructs knowledge, but is removed from the world. This limit, the point, and Euclid's widthless line are not given by measurement, our only access to the physical world. The myth is at the asymptotic limit, like the thermodynamic integral which gives us the irreversibility of diffusion at the infinite limit (that is, which demonstrates the second principle, by supposing an infinity of trajectories for the molecules of a gas within a large, but finite volume). The mythical (conceptual, if the reader prefers) limit makes us understand: how audacious this beginning of a science, this imagination of the widthless line, of the point of no dimension. Without those limit (infinitary!) concepts, which are not in the world, there would have been no *theory* of the measurement of surfaces: it is necessary to have "widthless" edges and dimensionless points at the intersections of lines, in order to propose a general theory of areas, the Greek extraordinary invention (how thick should otherwise be the border of a triangle?). Myths, as the invention of something "which is not there", are necessary to enhance theories, beginning with Euclid's continuum, lines, points... finiteness, as the discrete of a naive and pre-scientific, pre-Greek perception, entails machine-like stupidity.

In this context and since Einstein, we have gone further and have even come to say that finite, for the universe, does not mean limited. Think of the relativistic model of the Riemann sphere: it is finite but unlimited, contrary to the notion of finiteness as limitation to be found with Euclid (infinite = a-peiron = without limits). Why would the information on the Riemann sphere be "finite" in such a model? Of which type of finiteness would we be speaking of? Euclidean finiteness or modern unlimited finiteness? Be it relativistic or quantum, "finiteness" contains infinity, as unboundedness, by measure.

Except for great thinkers such as Turing, logicians and computer scientists tend to have a culture of the finite/discrete/Laplacian, as Turing said of his machine, which is difficult to escape. Its origin is the arithmetizing perspective of Frege with regard to the "delirium" of Riemannian geometry, says he in 1884. But it is also in the philosophical incomprehension of Hilbert, one of the great figures of mathematical physics, concerning unpredictability, of even Poincaré's type of undecidability (it is impossible to calculate – decide – the position of three planets after a sufficiently long period of time), when he speaks of mathematics: 20 years later, he will launch one decidability conjecture after the other, all of them being false (Arithmetics, Choice, Continuum Hypothesis), despite the highly justified objections from Poincaré (Mr Hilbert thinks of mathematics as a sausage-making machine!). Poincaré had already experimented with undecidability, as unpredictability, though in the friction between mathematics and physics (not of purely mathematical statements, Hilbert's question). However, this culture of predictable (and integrable!), of the determination within a universe (a discrete, finite, and limited database), has given us marvelous Laplacian machines. Let's just make an effort to better correlate them to the world, today. A good practice, and theory, of modelling and of networks, that of concurrency, impose them. They evolve within a space–time which we understand better, for the moment, thanks to continuity. Thankfully, there are also hybrid systems and continuous computability which propose quite different perspectives. And likewise for the work of Girard which tries to enrich logic with concepts that are central to the field of the physico-mathematical: symmetries, operator algebra, quantum non-commutativity.

But let's return to quantum mechanics.

## 7.8. The Case of Quantum Mechanics

The quantum issue could at first glance present a perfect symbiosis between the two preceding sections: we are dealing with a fundamentally fundamental equation, Schrödinger's equation, which derives from nothing, which must be at the center of any fundamental process, and of which the mathematization is perfect, depending on only a single parameter (actually, is the value of Planck's constant a computable number?). Moreover, "measurement" takes on a whole new dimension. The interval, as such, no longer exists and intrinsic randomness is introduced.

Let's now mention the importance, particularly in the field of the physics of elementary particles, of the role played by computers. The computation of precise numeric values, for instance the calculation of the electron's magnetic moment, and their literally "phenomenal" concordance with experience has doubtlessly had a crucial importance for the development of the theory. And this precisely in the very field where computers have become irreplaceable: numeric computation. Associating a number to hundreds, to thousands of Feynman diagrams is an operation beyond human capability and which computer science bravely accomplishes.

The results provided by quantum physics are precise, and have a level of precision which any other physical theory has yet to attain. They are also discrete, meaning that the richness of continuity has been lost, and that we are facing a (discrete) play of possibilities. Of course, what we are actually measuring is a classical object, a classical trace (bubble chamber, photographic plate, etc.) with a quantum value. We are indeed at the heart of the problem: a quantum measurement provides values belonging to a discrete set (set of values specific to the Hamiltonian), hence a certain rigidity that is a source of stability and therefore of precision (those of discrete topology). Seen from this angle, quantum "precision" seems tautological in a way; we allow ourselves no leeway around discrete values which would enable us to extend into the voluptuousness of imprecision. We could even say: let's provide ourselves, once and for all, with all the values specific to all the Hamiltonians of the world and we will have a field of "outputs" which is discrete in its very essence.

But this is precisely forgetting that the result of such measurement is obtained upon a classical object from which the result of the measure is accessible to us. The atomic spectral lines appear on a photographic plate. Therefore the classical continuum is, *a posteriori*, the locus of the quantum result, together with its virtues, harmful because prone to introduce imprecision. And so, what the fact that quantum mechanics is incredibly precise really means is that, during experiments, it leaves *classical* traces of an extreme level of precision, practically exhibiting a discrete sub-structure of the continuum.

And this is not tautological at all.

In addition to this discreteness, and precision, quantum mechanics has caused some difficulties by conferring a random aspect to the result of measurement. Let's say right now that something had to happen, because the principle of quantum superposition prohibits a direct access, beyond measurement, to the quantum space of states (we do not "see" superposed states, or entangled states); more accurately, we "look" at them, and they must be looked at to be seen, by getting measured, they "de-superpose" themselves, they de-entangle. This random aspect immediately escapes any computational system of... computability. No more determinism, no more equations. Of course, it is possible to talk about statistics, and to wonder whether these statistics are computable. We then return to the non-deterministic algorithms of the preceding section, but with a different problem.

Quantum algorithms are a perfect illustration of this. Let's recall that a quantum algorithm consists in a quantum system evolving from an initial piece of data having, in a way, a classical "input". By principle of superposition, entanglement, at the end of an evolution, has done its job and the final state is typically quantum, superposed in several states, of which a single one contains the "output" sought. To get it, we then perform a measurement that is supposed, by construction, to produce the good result with a maximal probability.

What is Turing computable in all of this?

We can wonder regarding the first part of the quantum evolution related to quantum "equational" evolution modulo the remarks made at the end of Section 7.6 concerning PDEs (Schrödinger's equation is a PDE after all, but a linear and not an hyperbolic one), and could possibly answer: yes, this part of the quantum evolution is computable. But the last phase, that of measurement, again escapes computational reduction: *the random aspect of measurement, let us rest assured, will never enable a quantum computer to decode a credit card at the desired moment with certainty.*

*Quantum algorithms versus non-deterministic algorithms*

It could be advisable to specify the important difference between quantum and non-deterministic algorithms, a source, it appears, of many confusions. Indeed, one could confuse two very different "parallel" aspects.

A quantum algorithm, in a way, works well in parallel; computation is fundamentally vectorial because of the very nature of quantum dynamics. But the final result, that which needs to be extracted from the final quantum state, is a single one of the components present within the latter. The other components, the whole "final state" vector, has no interest as such: firstly because it is inaccessible, then because the other components (other than the component containing the results) do not carry any information related to the initial problem. So it is not an issue of dispersing the information in order to parcel it out and hence increase the power of the computation and then "patching the pieces back together", in a way, but rather of placing oneself within a space (a quantum space, and again, one that has not yet been satisfactorily achieved experimentally) from which one needs to suddenly return in order to finish the computation.

Because the essential is indeed there: the "computation", the "process" is finished *only once the ultimate measurement is taken.* It is this total process which must be placed in the view of computability, and not the purely quantum part which conveys no information. It is exactly the same idea which is responsible for there being no "EPR paradox" because, although we are acting from a distance upon the entangled vector, no information is transmitted.

Let us mention also that logic based on the quantum mechanics paradigm has been recently introduced by J.Y. Girard, without explicit motivation in the direction of quantum calculus [20]. We conclude by saying that the randomness of quantum mechanics is intrinsic, it escapes computation. What about classical randomness?

## 7.9. Randomness, Between Unpredictability and Chaos

In [4], classical randomness and deterministic unpredictability are identified, from the point of view of mathematical physics. Randomness would present itself, we observed, at the interface between mathematics (or, more generally, between language) and physical processes. It must however not be ignored that, in certain probabilistic, purely mathematical frameworks (measure theory), we can also speak of randomness, away from physical processes. By computation theoretic tools, Per Martin-Löf advanced, 40 years ago, a purely mathematical notion of randomness. More specifically, one can, by means of computability, tell when an *infinite sequence of integers* (of 0s and 1s for example) is random, without reference to an eventual physical generative process. In short, a random sequence is Martin-Löf (ML) computable if it is "strongly" non-computable, a definition which requires a little bit of work (see [29] for a recent overview). In a sense, formal computability/predictability can tell us when we leave its domain: this is like Gödel who, in his proof of incompleteness, never left the formal, and who was yet able to give a formula which escapes the formal (which is formally unprovable, jointly to its negation).

Moreover, what interests us here, this purely mathematical randomness, is "at infinity", exactly like the randomness *within* chaotic classical dynamics is asymptotic: a random Martin-Löf sequence is infinite (the initial segments are at best incompressible).

What can then be said of the relationship between this notion, purely mathematical, and physics? From the statistical viewpoint, which was the preoccupation of Martin-Löf at the time, every thing is fine: the distribution of the probabilities of a ML-random sequence, for a good probability measurement, is that of the toss of a coin, to infinity. But what about the relationship to the physico-mathematical of dynamical systems? How can one pass directly, by mathematical means, without reference to the physical processes that the two approaches modelize, from ML-randomness to unpredictable determinism (systems of equations or evolution functions)? We can see possible correlations in the recent PhD theses by M. Hoyrup and C. Rojas (in Longo's team): the points and the trajectories within chaotic systems are analyzed in terms of ML-randomness, all the while using suitable notions of measure, of mathematical entropy and Birkhoff ergodicity. In the two cases, those of sequences of integers and of continuous dynamics, we work to infinity.

Let's be more precise. A dynamical system, as a purely mathematical formalism for physics, is said to be "mixing" if the correlation of a given pair of observables decreases at least polynomially with time. Like ergodicity, this is an asymptotic property of "disorder", a weak form of chaos. What was recently proved is that, in mixing dynamics, ergodic points coincide with ML-random ones (in fact for a slightly different definition of ML-randomness, due to Schnorr). Thus deterministic unpredictability, as ergodicity in mixing dynamics, overlap with a strong form of undecidability, that is algorithmic randomness. In other words, if we want to relate physical processes to effective computations, which is an issue of elaboration of numbers, we can, but, at the limit: all processes that are modeled by a somewhat chaotic system, produce non-computable, actually random, sequences, within the mathematical system. Or, also, (strong) non-computability (as algorithmic randomness) may be found in formal writings of the physical world (dynamical systems are perfectly formalisable, of course). That is, at the limit, we may say "no" to Laplace's conjecture of predictability of deterministic systems and, this, in terms of (a strong form of) undecidability, à la Gödel. Predicting, in physics, is a matter of "saying" (pre-dicere, to say in advance) by a formal language or system about a physical process in finite time, as we said several times: by these results, instead, Poincaré's finite unpredictability joins undecidability, asymptotically. In conclusion, deterministic ergodic and mixing dynamics, which model "weakly chaotic" physical processes, generate (highly) non-computable features[f].

## 7.10. General Conclusions

The reader might have felt that the authors have a point of view "against" a vision of nature that was too organized around computations. Once again computers have brought so much to science that it is not necessary to recall the benefit provided. It seems to us that this situation, where a given viewpoint invades a whole field of science, happened several times in the past. An example is the case of mathematical analysis at the turn of the last century, a period where many new objects in mathematics were born, such as nowhere differentiable functions, Cantor sets, summation methods of diverging series. To focus on the latter let us quote Emile Borel, in the introduction of his famous book on diverging series [8]. Borel discusses the fact that analysis "à la Cauchy", based on convergent Taylor expansions of analytic functions, although it brought a considerable amount

[f]See [16–18], [15] and http://www.di.ens.fr/~longo/ for ongoing work. Connections between algorithmic and quantum randomness are analyzed in [11].

of progress in mathematics, fixed also into rigidity a lot of non-rigorous methods used by the geometers (in the sense of physicists) of the older time: "This revolution[g] was necessary: nevertheless one might ask if dropping the less rigorous methods of the geometers (...) was good or not: (...) but this period [of rigor] being passed, the study of former methods might be wealthy...".

Let's see what we have done so far. We have reviewed certain aspects of computation in physics and in mathematics. We have seen that many situations in physics, even classical physics, cause processes which are "beyond computation" (in the sense of "calculus resolving equations") to intervene. We have also mentioned the calculatory contribution of computer science and its essential role. Now, let's not forget the importance as such of the plurality of "visions" for understanding the natural sciences, a plurality which has always existed in the sciences. The new perspective proposed by the discrete, in great part due to the contribution of computer science, is a conceptual and technical resource, which adds itself to the differentiable physico-mathematical continuum, from Newton to Schrödinger (or even consider, for example, the importance of computer modelling in biology, to mention another discipline, [34]). On the other hand, the reduction to a conceptual and mathematical dimension that is too "computational" (in the excessively naive sense of the term) would, in our view, lead us to sterile boredom, in which even the "nuances" of the post-Laplacian continuum would be absent. Finally and in particular, within an "equational" framework for the play between the continuum and the discrete, we have discussed notions that appear to be fundamental to modern science, such as those of determinism and of predictability, from where emerges the notion of uncertainty. But let's take a further look.

As compounded in [4], classical physical randomness is of an "epistemic" nature, whereas that of quantum measurement is intrinsic or "objective": a distinction which should be solely an instrument of clarity, of conceptual clarity if possible, and nothing more. By this we refer to several aspects among which the one of interest to us is the following: classical randomness can be analyzed by means of different methods. In short, it is possible to address dice, the double pendulum, the baker's dough, etc. in terms of statistically random sequences and of probability distributions (central limit theorem, etc.), but also by means of the mathematics of chaotic de-

[g]The Cauchy and Abel rigorous vision of Analysis based on convergence of expansions of Taylor series.

terminism (if we have the courage to write the several equations needed for the movement of dice; it is easy for the double pendulum and the baker's dough). Some people, mainly in the field of computer science as we have seen, say that the toss of dice or that the baker's dough (or even the three bodies?) are non-deterministic because, by using the approximation of measurement, it is possible to associate several numeric outputs to an exact input number and the same wording is used for computational non-determinism. It is an abuse of language which ignores the specifications brought by the broadening by Poincaré of the field of determination, which includes classical randomness in the field of chaotic determination (the non-linearity of "continuous systems", and the related "exponential drift", says Turing in 1952), and by the indetermination of quantum physics. This is specific to the culture of the discrete, which is wonderful for our discrete state machines, but which misses the 120 years of geometrization of physics (geometry of dynamic and relativistic systems) and which fails to appreciate the role of measurement (classical/quantum).

We thus see the apparition of three idealizations thanks to which we could think it possible to discover and understand the world (classical).

1. The digital, discrete ideal which (possibly) shows nature as *computing* and only as computing. Computing, iterating, and reiterating to infinity with a wonderful and misleading precision.

2. The ideal of continuous mathematics, where nature (mathematics) solves equations. In itself, this vision is perfectly deterministic, the equations have solutions.

3. The ideal of the equation, for which nature *divides* itself into different scales, impenetrable to each other – for example the quantum world, the classical one, hydrodynamics, celestial mechanics, cosmology etc.

These ideals (1,2,3) are placed in anti-chronological order: historically, equations were the first to appear, followed by their mathematical models, and finally by their digital simulation.

To conclude, let's look at the connections and anti-connections between these three worldviews, these three tiers that we could compare to Girard's three basements. This would be the result of the present work.

At first glance, we could easily go up from the third to the second and then to the first level. Continuous mathematics seem perfect for equations, and digital approximation has become so commonplace that one must almost hide to criticize it. But the elevator does not work properly: between the third and second levels, Poincaré shakes things up (non-integrability and sensitivity to the initial conditions, as it is, make difficult the practical idea of a trajectory within continua), and between the second and first levels we have lost, by climbing to the level of the discrete, a few aspects that were important to continuity (the fluctuations below the threshold of discretization as well as the discrete blackness of milk). If we take the stairs to go down, we get dizzy: lack of computational equivalence for the passing to computational continua (Section 2), and loss of reliability with the introduction of the interval of imprecision when passing from the second to the third levels...

And there is quantum mechanics with its intrinsic randomness. Ideal 3 is then shattered during measurement: no more equations. Of course, physics can make do without *individual* measurement processes: we have not (yet) experimentally observed the reduction of wave packets during unique events, all we can observe are averages, statistics. But recent physics pushes towards the study and observation of simple quantum physical systems which are always better at conducting the "gedenken Experiment"[h] of the founding fathers [28], and in any case the reduction of the wave packet during measurement is, we believe, a necessary component of quantum formalism, an axiom which makes it coherent.

This situation is not new in physics: we do not observe Newtonian mechanics in a mole of gas. And yet it is thanks to such mechanics that we can reconstruct the dynamics of gases and thermodynamics. Mind though, this reconstruction is the result of the passing to *infinity* (the thermodynamic integral) from a finite non-observable model.

## Acknowledgements

We would like to thank Eugène Asarin, Olivier Bournez, Mathieu Hoyrup, and Cristobal Rojas for their critical reading of the manuscript. The authors' papers may be downloaded from http://www.di.ens.fr/~longo/ and http://www.dma.ens.fr/~paul/.

## References

[1] L. Aceto, G. Longo, and B. V., eds., The difference between sequential and concurrent computations, *Math. Structures Comput. Sci. (Special issue).* **13** (4–5), (2003).

[2] V. I. Arnol'd, *Geometrical Methods in Ordinary Differential Equations.* Springer, Berlin, (1987).

[h]EPR paradox, Schrödinger's cat for example.

[3] F. Bailly and G. Longo, *Mathématiques et sciences de la nature. La singularité physique du vivant.* Hermann, Paris, (2006). English translation: in Imperial College Press/World Scientific, 2011.
[4] F. Bailly and G. Longo, Randomness and determination in the interplay between the continuum and the discrete, *Math. Structures Comput. Sci.* **17** (2), 289–307, (2007).
[5] M. Baverman and M. Yampolski, (2005). Non-computable Julia sets. Preprint. Available at `http://www.math.toronto.edu/yampol/`, [Accessed October 2010].
[6] J. Bell, *A Primer in Infinitesimal Analysis.* Cambridge University Press, (1998).
[7] L. Blum, L. Cucker, M. Shub, and S. Smale, *Complexity and Real Computation.* Springer, Berlin, (1998).
[8] E. Borel, *Leçons sur les séries divergentes.* Gauthier-Villars, Paris, (1928).
[9] O. Bournez. Modèles continus. calculs. Habilitation DR, LORIA, Nancy, (2006).
[10] V. Bush, The differential analyzer, *J. Franklin Inst.* **212**(4), 447–488, (1931).
[11] C. Calude and M. Stay, From Heisenberg to Gödel via Chaitin, *Internat. J. Theoret. Phys.* **44**(7), 1053–1065, (2005).
[12] R. L. Devaney, *An Introduction to Chaotic Dynamical Systems.* Addison-Wesley, New York, (1989).
[13] G. Dowek. La forme physique de la thèse de Church et la sensibilité aux conditions initiales. In eds. J.-B. Joinet and Tronçon, *Ouvrir la logique au monde. Philosophie et mathématique de l'interaction, actes de l'école thématique interdisciplinaire "Logique, Sciences, Philosophie" (Cerisy-la-Salle, 19-26 septembre 2006)*, Hermann, Paris, (2009).
[14] R. Feynman, *Quantum mechanics and Path Integrals.* McGrawth-Hill, New York, (1965).
[15] P. Gacs, M. Hoyrup, and C. Rojas, Randomness on computable probability spaces – A dynamical point of view, *Theory Comput. Syst. (Special issue STACS 09).* (2010). doi: 10.1007/s00224-010-9263-x.
[16] S. Galatolo, M. Hoyrup, and C. Rojas, A constructive Borel–Cantelli Lemma. Constructing orbits with required statistical properties, *Theoret. Comput. Sci.* **410**, 2207–2222, (2009).
[17] S. Galatolo, M. Hoyrup, and C. Rojas, Effective symbolic dynamics, random points, statistical behavior, complexity and entropy, *Inform. and Comput.* **208**(1), 23–41, (2010).
[18] S. Galatolo, M. Hoyrup, and C. Rojas. Dynamics and abstract computability: computing invariant measures. To appear in *Disc. Cont. Dyn. Sys.*, (arXiv:0903.2385).
[19] R. Gandy. Church's Thesis and the principles for mechanisms. In eds. J. Barwise, J. Keisler, and K. Kunen, *The Kleene Symposium.* North–Holland, (1980).
[20] J. Y. Girard, *Le point aveugle.* Hermann, Paris, (2007).
[21] E. Hainry. Modèles de calculs sur les réels. Thèse. LORIA, Nancy, (2006).
[22] M. Hoyrup, A. Kolcak, and G. Longo, Computability and the morphological

complexity of some dynamics on continuous domains, *Theoret. Comput. Sci.* **398**(1–3), 170–182, (2008).
[23] G. Longo. Laplace, Turing and the "imitation game" impossible geometry: randomness, determinism and programs in Turing's test. In eds. R. Epstein, G. Roberts, and G. Beber, *Parsing the Turing Test*, pp. 377–411. Springer, Berlin, (2009).
[24] C. Moore, Recursion theory on the reals and continuous-time computation, *Theoret. Comput. Sci.* **162**, 23–44, (1996).
[25] T. Paul, Discrete-continuous and classical-quantum, *Math. Structures Comput. Sci.* **17**, 177–183, (2007).
[26] T. Paul, On the status of perturbation theory, *Math. Structures Comput. Sci.* **17**, 277–288, (2007).
[27] M. B. Pour-El and J. I. Richards, *Computability in Analysis and Physics.* Perspectives in mathematical logic, Springer, Berlin, (1989).
[28] J.-M. Raimond. Complémentarité, intrication, décohérence: les expériences de pensées réalisées (conference). In *Qu'est-ce qui est réel*, École Normale Supérieure, (2005). Organizers: C. Debru, and G. Longo, T. Paul and G.Vivance. Available at: `http://www.diffusion.ens.fr/index.php?res=conf\&idconf=807`, [Accessed October 2010].
[29] C. Rojas, Computability and information in models of randomness and chaos, *Math. Structures Comput. Sci.* **18**, 291–307, (2008).
[30] C. Shannon, Mathematical theory of the differential analyzer, *J. Math. Phys.* **20**, 337–354, (1941).
[31] A. M. Turing. Intelligent machinery. In eds. B. Meltzer and D. Michie, *Machine Intelligence 5*, pp. 3–23. Edinburgh University Press, Edinburgh, (1969). National Physical Laboratory Report, 1948.
[32] A. M. Turing, Computing machines and intelligence, *Mind.* **LIX**(236), 433–460, (1950).
[33] A. M. Turing, The chemical basis of morphogenesis, *Phil. Trans. R. Soc. London.* **B237**, 37–72, (1952).
[34] F. Varenne, *Du modèle à la simulation informatique.* Vrin, Paris, (2007).
[35] K. Weihrauch, *Computable Analysis.* Texts in Theoretical Computer Science, Springer, Berlin, (2000).

# Chapter 8

# Liquid State Machines: Motivation, Theory, and Applications

Wolfgang Maass

*Institute for Theoretical Computer Science*
*Graz University of Technology*
*A-8010 Graz, Austria*
*E-mail: maass@igi.tugraz.at*

The Liquid State Machine (LSM) has emerged as a computational model that is more adequate than the Turing machine for describing computations in biological networks of neurons. Characteristic features of this new model are (i) that it is a model for adaptive computational systems, (ii) that it provides a method for employing randomly connected circuits, or even "found" physical objects for meaningful computations, (iii) that it provides a theoretical context where heterogeneous, rather than stereotypical, local gates, or processors increase the computational power of a circuit, (iv) that it provides a method for multiplexing different computations (on a common input) within the same circuit. This chapter reviews the motivation for this model, its theoretical background, and current work on implementations of this model in innovative artificial computing devices.

## Contents



## 8.1. Introduction

The Liquid State Machine (LSM) had been proposed in [26] as a computational model that is more adequate for modelling computations in cortical microcircuits than traditional models, such as Turing machines or attractor-based models in dynamical systems. In contrast to these other models, the LSM is a model for real-time computations on continuous streams of data (such as spike trains, i.e., sequences of action potentials of neurons that provide external inputs to a cortical microcircuit). In other words: both inputs and outputs of an LSM are streams of data in continuous time. These inputs and outputs are modelled mathematically as functions $u(t)$ and $y(t)$ of continuous time. These functions are usually multi-dimensional (see Fig. 8.1, Fig. 8.2, and Fig. 8.3), because they typically model spike trains from many external neurons that provide inputs to the circuit, and many different "readouts" that extract output spike trains. Since an LSM maps input streams $u(\cdot)$ onto output streams $y(\cdot)$ (rather than numbers or bits onto numbers or bits), one usually says that it implements a functional or operator (like a filter), although for a mathematician it simply implements a function from and onto objects of a higher type than numbers or bits. A characteristic feature of such higher-type computational processing is that the target value $y(t)$ of the output stream at time $t$ may depend on the values $u(s)$ of the input streams at many (potentially even infinitely many) preceding time points $s$.

Another fundamental difference between the LSM and other computational models is that the LSM is a model for an *adaptive* computing system. Therefore its characteristic features only become apparent if one considers it in the context of a learning framework. The LSM model is motivated by the hypothesis that the learning capability of an information processing device is its most delicate aspect, and that the availability of sufficiently many training examples is a primary bottleneck for goal-directed (i.e., supervised or reward-based) learning. Therefore its architecture is designed to make the learning as fast and robust as possible. It delegates the primary load of goal-directed learning to a single and seemingly trivial stage: the output, or readout stage (see Fig. 8.4), which typically is a very simple computational component. In models for biological information processing each readout usually consists of just a single neuron, a projection neuron in the terminology of neuroscience, which extracts information from a local microcircuit and projects it to other microcircuits within the same or other brain areas. It can be modelled by a linear gate, a perceptron (i.e., a linear gate with a threshold), by a sigmoidal gate, or by a spiking neuron. The bulk of the LSM (the "Liquid") serves as pre-processor for such readout neuron, which amplifies the range of possible functions of the input streams $u(t)$ that it can learn. Such division of computational processing into Liquid and readout is actually quite efficient, because the same Liquid can serve a large

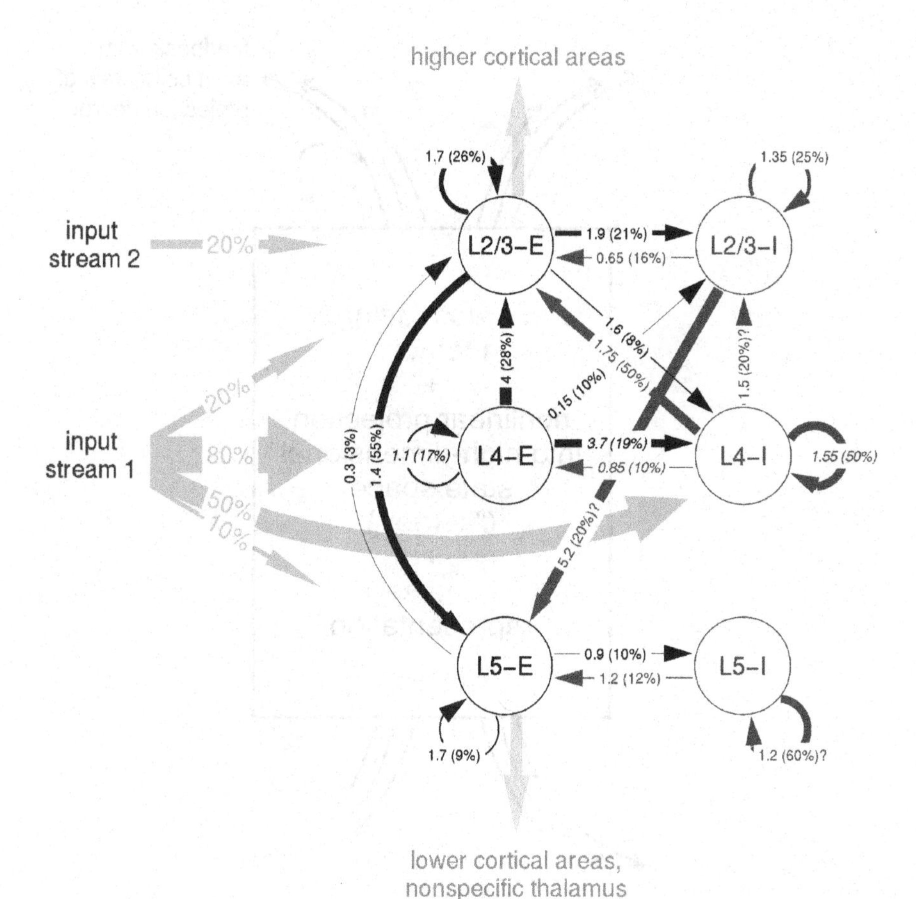


Figure 8.1. Modelling a generic cortical microcircuit by an LSM. Template for a generic cortical microcircuit based on data from [33], see [9, 10] for details. The width of arrows indicates the product of connection probabilities and average strength (i.e., synaptic weight) between excitatory (left hand side) and inhibitory (right hand side) neurons on three cortical layers. Input stream 1 represents sensory inputs, input stream 2 represents inputs from other cortical areas. Arrows toward the top and toward the bottom indicate connections of projection neurons ("readouts") on layer 2/3 and layer 5 to other cortical microcircuits. In general these projection neurons also send axonal branches (collaterals) back into the circuit.

number of different readout neurons, that each learn to extract a different "summary" of information from the same Liquid. The need for extracting different summaries of information from a cortical microcircuit arises from different computational goals (such as the movement direction of objects versus the identity of objects in the case where $u(t)$ represents visual in-

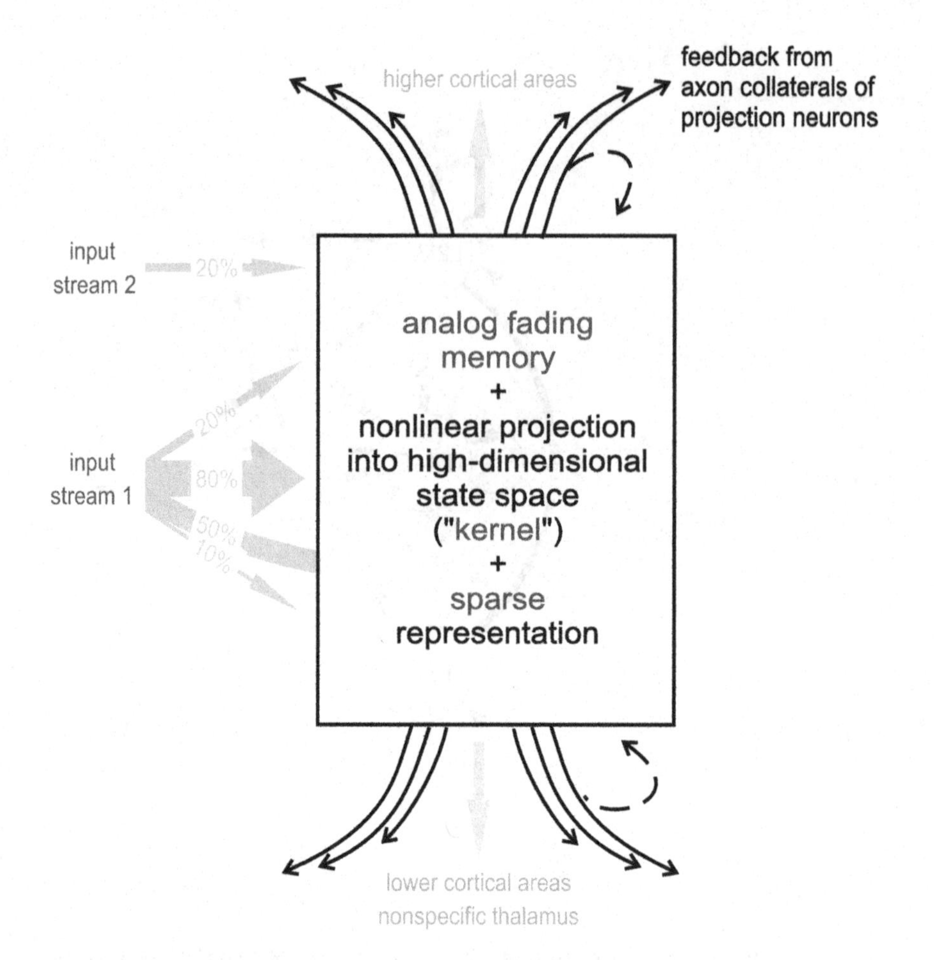


Figure 8.2. Hypothetical computational function of a generic cortical microcircuit in the context of the LSM model. In general the projection neurons also provide feedback back into the microcircuit (see Theorem 8.2 in Section 3).

puts) of different projection targets of the projection neurons. Data from neurophysiology show in fact that for natural stimuli the spike trains of different projection neurons from the same column tend to be only weakly correlated. Thus the LSM is a model for multiplexing diverse computations on a common input stream $u(t)$ (see Fig. 8.1, Fig. 8.2, and Fig. 8.3).

One assumes that the Liquid is not adapted for a single computational task (i.e., for a single readout neuron), but provides computational preprocessing for a large range of possible tasks of different readouts. It could also

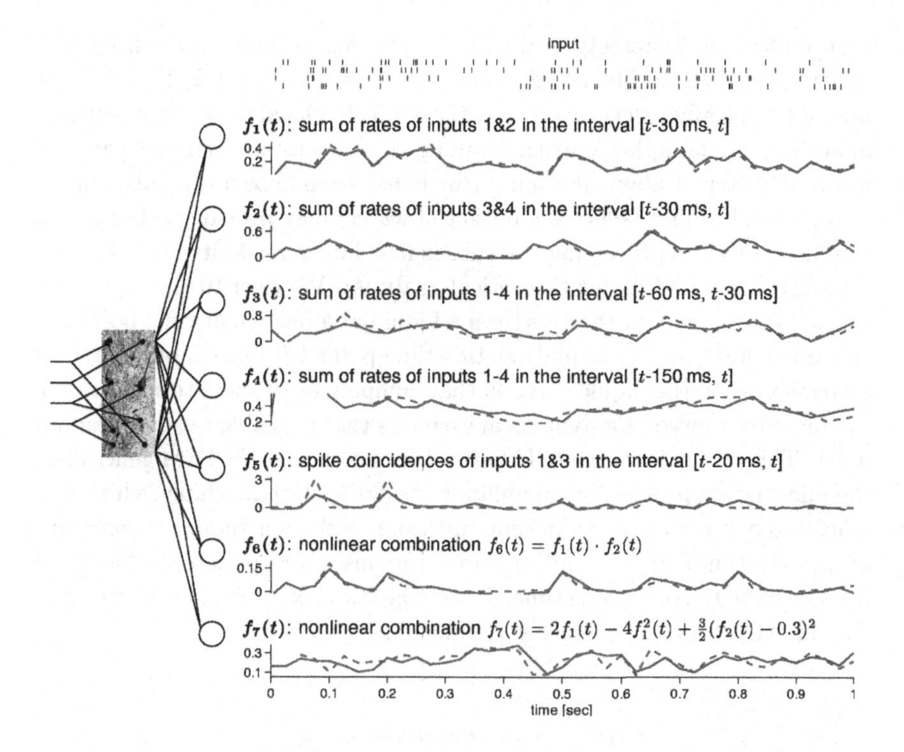


Figure 8.3. Multi-tasking in real-time. Below the 4 input spike trains (shown at the top) the target outputs (dashed curves) and actual outputs (solid curves) of 7 linear readout neurons are shown in real-time (on the same time axis). Targets were to output every 30 ms the sum of the current firing rates of input spike trains 1 and 2 during the preceding 30 ms ($f_1$), the sum of the current firing rates of input spike trains 3 and 4 during the preceding 30 ms ($f_2$), the sum of $f_1$ and $f_2$ in an earlier time interval [$t$-60 ms, $t$-30 ms] ($f_3$) and during the interval [$t$-150 ms, $t$] ($f_4$), spike coincidences between inputs 1&3 ($f_5(t)$ is defined as the number of spikes which are accompanied by a spike in the other spike train within 5 ms during the interval [$t$-20 ms, $t$]), a simple nonlinear combination $f_6$ (product) and a randomly chosen complex nonlinear combination $f_7$ of earlier described values. Since all readouts were linear units, these nonlinear combinations are computed implicitly within the generic microcircuit model (consisting of 270 spiking neurons with randomly chosen synaptic connections). The performance of the model is shown for test spike inputs that had not been used for training (see [27] for details).

be adaptive, but by other learning algorithms than the readouts, for example by unsupervised learning algorithms that are directed by the statistics of the inputs $u(t)$ to the Liquid. The Liquid is in more abstract models a generic dynamical system – preferentially consisting of diverse rather than uniform and stereotypical components (for reasons that will become apparent below). In biological models (see Fig. 8.1, Fig. 8.2, Fig. 8.3) the Liquid is typically a generic recurrently connected local network of neurons, modelling for example a cortical column which spans all cortical layers and has a diameter of about 0.5 mm. But it has been shown that also an actual physical Liquid (such as a bucket of water) may provide an important computational preprocessing for subsequent linear readouts (see [7] for a demonstration, and [8] for theoretical analysis). We refer to the input vector $\mathbf{x}(t)$ that a readout receives from a Liquid at a particular time point $t$ as the liquid state (of the Liquid) at this time point $t$ (in terms of dynamical systems theory, this liquid state is that component of the internal state of the Liquid – viewed as a dynamical system – that is visible to some readout unit). This notion is motivated by the observation that the LSM generalizes the information processing capabilities of a finite state machine (which also maps input functions onto output functions, although these are functions of discrete time) from a finite to a continuous set of possible values, and from discrete to continuous time. Hence the states $\mathbf{x}(t)$ of an LSM are more "liquid" than those of a finite state machine.

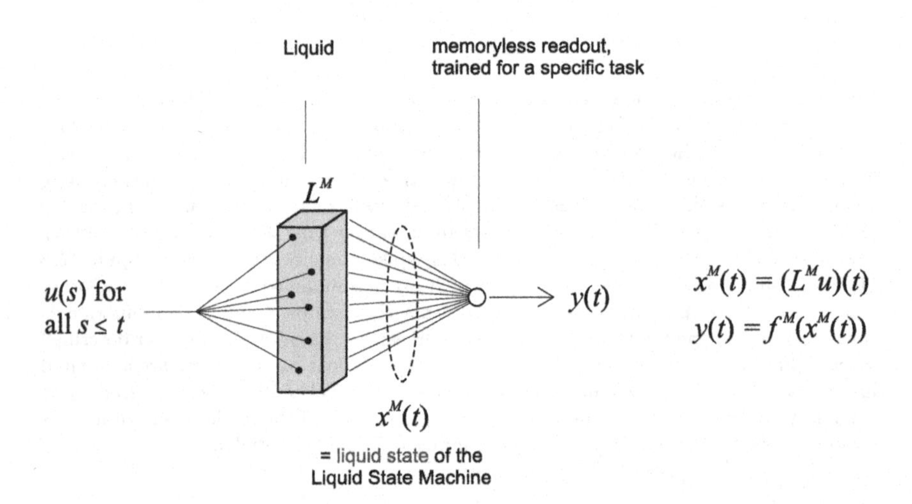


Figure 8.4. Structure of a Liquid State Machine (LSM) $M$, which transforms input streams $u(\cdot)$ into output streams $y(\cdot)$. $L^M$ denotes a Liquid (e.g., some dynamical system), and the "liquid state" $\mathbf{x}^M(t) \in \mathbb{R}^k$ is the input to the readout at time $t$. More generally, $\mathbf{x}^M(t)$ is that part of the current internal state of the Liquid that is "visible" for the readout. Only one input and output channel are shown for simplicity.

This architecture of a LSM, consisting of Liquid and readouts, makes sense, because it turns out that in many contexts there exist common computational preprocessing needs for many different readouts with different computational goals. This can already be seen from the trivial fact that computing all pairwise products of a set of input numbers (say: of all components of a multi-dimensional input $u(t')$ for a fixed time point $t'$) gives any subsequent linear readout the virtual expressive power of any quadratic computation on the original input $u(t')$. A pre-processor for a linear readout is even more useful if it maps more generally any frequently occurring (or salient) different input streams $u(\cdot)$ onto linearly independent liquid states $\mathbf{x}(t)$ [21], similarly as an RBF-kernel for Support Vector Machines. A remarkable aspect of this more general characterization of the pre-processing task for a Liquid is that it does not require that it computes precise products, or any other concrete nonlinear mathematical operation. Any "found" analog computing device (it could even be very imprecise, with mismatched transistors or other more easily found nonlinear operations in physical objects) consisting of sufficiently diverse local processes, tends to approximate this requirement quite well. A closer look shows that the actual requirement on a Liquid is a bit more subtle, since one typically only wants that the Liquid maps "saliently" different input streams $u(\cdot)$ onto linearly independent liquid states $\mathbf{x}(t)$, whereas noisy variations of the "same" input stream should rather be mapped onto a lower dimensional manifold of liquid states, see [20, 21] for details.

An at least equally important computational pre-processing task of a Liquid is to provide all temporal integration of information that is needed by the readouts. If the target value $y(t)$ of a readout at time $t$ depends not only on the values of the input streams at the same time point $t$, but on a range of input values $u(s)$ for many different time points $s$ (say, if $y(t)$ is the integral over one component of $u(s)$ for a certain interval $[t-1,\ t]$), then the Liquid has to collect all required information from inputs at preceding time points $u(s)$, and present all this information simultaneously in the liquid state $\mathbf{x}(t)$ at time point $t$ (see Fig. 8.3 and Fig. 8.4). This is necessary, because the readout stage has, by assumption, no temporal integration capability of its own, i.e., it can only learn to carry out "static" computations that map $\mathbf{x}(t)$ onto $y(t)$. A readout does not even know what the current time $t$ is. It just learns a map $f$ from input numbers to output numbers. Hence it just learns a fixed recoding (or projection) $f$ from liquid states into output values. This severe computational limitation of the readout of an LSM is motivated by the fact, that learning a static map $f$ is so much simpler than learning a map from input streams to output streams. And a primary goal of the LSM is to make the learning as fast and robust as possible. Altogether, an essential prediction of LSM-theory for information processing in cortical microcircuits is that they accumulate information over time. This prediction has recently been verified for cortical microcircuits in the primary visual cortex [28] and in the primary auditory cortex [18].

The advantage of choosing for a LSM the simplest possible learning device is twofold: Firstly, learning for a single readout neuron is fast, and cannot get stuck in local minima (like backprop or EM). Secondly, the simplicity of this learning device entails a superior – in fact, arguably optimal – generalization capability of learned computational operations to new inputs streams. This is due to the fact that its VC-dimension (see [2] for a review) is equal to the dimensionality of its input plus 1. This is the smallest possible value of any nontrivial learning device with the same input dimension.

It is *a priori* not clear that a Liquid can carry the highly nontrivial computational burden of not only providing all desired nonlinear preprocessing for linear readouts, but simultaneously also all temporal integration that they might need in order to implement a particular mapping from input streams $u(\cdot)$ onto output streams $y(\cdot)$. But there exist two basic mathematical results (see Theorems 8.1 and 8.2 in Section 8.3) which show that this goal can in principle be achieved, or rather approximated, by a concrete physical implementation of a Liquid which satisfies some rather general property. A remarkable discovery, which had been achieved independently and virtually simultaneously around 2001 by Herbert Jaeger [14], is that there are surprisingly simple Liquids, i.e., generic preprocessors for a subsequent linear learning device, that work well independently of the concrete computational tasks that are subsequentially learned by the learning device. In fact, naturally found materials and randomly connected circuits tend to perform well as Liquids, which partially motivates the interest of the LSM model both in the context of computations in the brain, and in novel computing technologies.

Herbert Jaeger [14] had introduced the name Echo State Networks (ESNs) for the largely equivalent version of the LSM that he had independently discovered. He explored applications of randomly connected recurrent networks of sigmoidal neurons without noise as Liquids (in contrast to the biologically oriented LSM studies, that assume significant internal noise in the Liquid) to complex time series prediction tasks, and showed that they provide superior performance on common benchmark tasks. The group of Benjamin Schrauwen (see [31, 32, 35, 36]) introduced the term Reservoir Computing as a more general term for the investigation of LSMs, ESNs, and variations of these models. A variety of applications of these models can be found in a special issue of Neural Networks 2007 (see [15]). All these groups are currently collaborating in the integrated EU-project ORGANIC (= Self-organized recurrent neural learning for language processing) that investigates applications of these models to speech understanding and reading of handwritten text (see `http://reservoir-computing.org`). An industrial partner in this project, the company PLANET (`http://english.planet.de`) had already good success in applications of Reservoir Computing to automated high-speed reading of hand-written postal addresses.

We will contrast these models and their computational use with that of Turing machines in the next section. In Section 8.3 we will give a formal definition of the LSM, and also some theoretical results on its computational power. We will discuss applications of the LSM and ESN model to biology and new computing devices in Section 8.4 (although the discussion of its biological aspects will be very short in view of the recent review paper [5] on this topic).

## 8.2. Why Turing Machines are Not Useful for Many Important Computational Tasks

The computation of a Turing machine always begins in a designated initial state $q_0$, with the input $x$ (some finite string of symbols from some finite alphabet) written on some designated tape. The computation runs until a halt-state is entered (the inscription $y$ of some designated tape segment is then interpreted as the result of the computation). This is a typical example for an *offline computation* (Fig. 8.5A), where the complete input $x$ is available at the beginning of the computation, and no trace of this computation, or of its result $y$, is left when the same Turing machine subsequently carries out another computation for another input $\tilde{x}$ (starting again in state $q_0$). In contrast, the result of a typical computation in the neuronal system of a biological organism, say the decision about the location $y$ on the ground where the left foot is going to be placed at the next step (while walking or running), depends on several pieces of information: on information from the visual system, from the vestibular system which supports balance control, from the muscles (proprioceptive feedback about their current state), from short term memory (how well did the previous foot placement work?), from long-term memory (how slippery is this path in the current weather conditions?), from brain systems that have previously decided where to go and at what speed, and on information from various other parts of the neural system. In general these diverse pieces of information arrive at different points in time, and the computation of $y$ has to start before the last one has come in (see Fig. 8.5B). Furthermore, new information (e.g., visual information and proprioceptive feedback) arrives continuously, and it is left up to the computational system how much of it can be integrated into the computation of the position $y$ of the next placement of the left foot (obviously those organisms have a better chance to survive which also can integrate later arriving information into the computation). Once the computation of $y$ is completed, the computation of the location $y'$ where the right foot is subsequently placed is not a separate computation, that starts again in some neutral initial state $q_0$. Rather, it is likely to build on pieces of inputs and results of subcomputations that had already been used for the preceding computation of $y$.

The previously sketched computational task is a typical example for an *online computation* (where input pieces arrive all the time, not in one batch, see Fig. 8.5B). Furthermore it is an example for a *real-time computation*, where one has a strict deadline by which the computation of the output $y$ has to be completed (otherwise a two-legged animal would fall). In fact, in some critical situations (e.g., when a two-legged animal stumbles, or hits an unexpected obstacle) a biological organism is forced to apply an *anytime algorithm*, which tries to make optimal use of intermediate results of computational processing that has occurred up to some externally given time point $t_0$ (such forced halt of the computation could occur at "any time"). Difficulties in the control of walking for two-legged robots have taught us how difficult it is to design algorithms which can carry out this seemingly simple computational task. In fact, this computational problem is largely unsolved, and humanoid robots can only operate within environments for which they have been provided with an accurate model. This is perhaps surprising, since on the other hand current computers can beat human champions in seemingly more demanding computational tasks, such as winning a game of chess. One might argue that one reason, why walking in a new terrain is currently a computationally less solved task, is that computation theory and algorithm design have focused for several decades on offline computations, and have neglected seemingly mundane computational tasks such as walking. This bias is understandable, because evolution had much more time to develop a computational machinery for

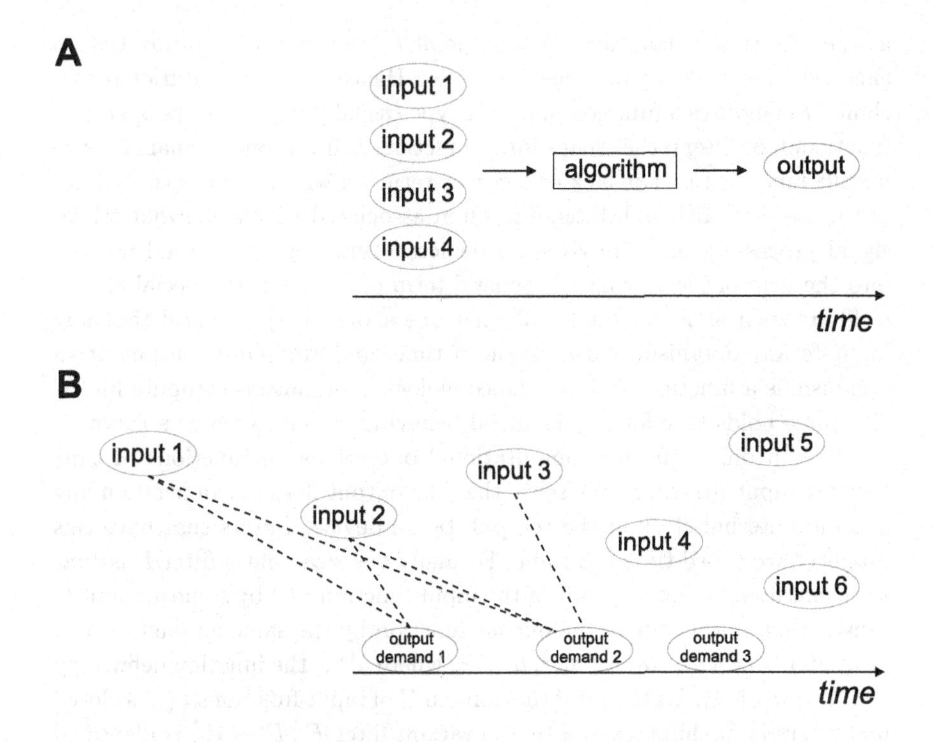


Figure 8.5. Symbolic representation of offline and online computations. **(A)** In an offline computation all relevant input computations are available at the start of the computation, and the algorithm may require substantial computation time until the result becomes available. **(B)** In online computations additional pieces of information arrive all the time. The most efficient computational processing scheme integrates as many preceding input pieces as possible into its output whenever an output demand arises. In that sense computations by a LSM are optimally efficient.

the control of human walking, and this computational machinery works so well that we don't even notice anymore how difficult this computational task is.

## 8.3. Formal Definition and Theory of Liquid State Machines

A computation machine $M$ that carries out online computations typically computes a function $F$ that does not map input numbers or (finite) bit strings onto output numbers or bit strings, but input streams onto output streams. These input and output streams are usually encoded as functions $u : \mathbb{Z} \to \mathbb{R}^n$ or $u : \mathbb{R} \to \mathbb{R}^n$, where the argument $t$ of $u(t)$ is interpreted as the (discrete or continuous) time point $t$ when the information that is encoded by $u(t) \in \mathbb{R}^n$ becomes available. Hence such computational machine $M$ computes a function of higher type (usually referred to as operator, functional, or filter), that maps input functions $u$ from some domain $U$ onto output functions $y$. For lack of a better term we will use the term "filter" in this section, although filters are often associated with somewhat trivial signal processing or preprossessing devices. However, one should not fall into the trap of identifying the general term of a filter with special classes of filters such as linear filters. Rather one should keep in mind that any input to any organism is a function of time, and any motor output of an organism is a function of time. Hence biological organisms compute filters. The same holds true for any artificial behaving system, such as a robot.

We will only consider computational operations on functions of time that are input-driven, in the sense that the output does not depend on any absolute internal clock of the computational device. Filters that have this property are called time invariant. Formally one says that a filter $F$ is *time invariant* if any temporal shift of the input function $u(\cdot)$ by some amount $t_0$ causes a temporal shift of the output function by the same amount $t_0$, i.e., $(Fu^{t_0})(t) = (Fu)(t+t_0)$ for all $t, t_0 \in \mathbb{R}$, where $u^{t_0}$ is the function defined by $u^{t_0}(t) := u(t+t_0)$. Note that if the domain $U$ of input functions $u(\cdot)$ is closed under temporal shifts, then a time invariant filter $F : U \to \mathbb{R}^{\mathbb{R}}$ is identified uniquely by the values $y(0) = (Fu)(0)$ of its output functions $y(\cdot)$ at time 0. In other words: in order to identify or characterize a time invariant filter $F$ we just have to observe its output values at time 0, while its input varies over all functions $u(\cdot) \in U$. Hence one can replace in the mathematical analysis such filter $F$ by a functional, i.e., a simpler mathematical object that maps input functions onto real values (rather than onto functions of time).

Various theoretical models for analog computing are of little practical use because they rely on hair-trigger decisions, for example they allow that the output is 1 if the value of some real-valued input variable $u$ is $\geq 0$, and 0 otherwise. Another unrealistic aspect of some models for computation on functions of time is that they automatically allow that the output of the computation depends on the full infinitely long history of the input function $u(\cdot)$. Most practically relevant models for analog computation on continuous input streams degrade gracefully under the influence of noise, i.e., they have a fading memory. *Fading memory* is a continuity property of filters $F$, which requires that for any input function $u(\cdot) \in U$ the output $(Fu)(0)$ can be approximated by the outputs $(Fv)(0)$ for any other input functions $v(\cdot) \in U$ that approximate $u(\cdot)$ on a sufficiently long time interval $[-T, 0]$ in the past. Formally one defines that $F : U \to \mathbb{R}^{\mathbb{R}}$ has fading memory if for every $u \in U^n$ and every $\varepsilon > 0$ there exist $\delta > 0$ and $T > 0$ so that $|(Fv)(0) - (Fu)(0)| < \varepsilon$ for all $v \in U$ with $\|u(t) - v(t)\| < \delta$ for all $t \in [-T, 0]$. Informally, a filter $F$ has fading memory if the most significant bits of its current output value $(Fu)(0)$ depend just on the most significant bits of the values of its input function $u(\cdot)$ in some finite time interval $[-T, 0]$. Thus, in order to compute the most significant bits of $(Fu)(0)$ it is not necessary to know the *precise* value of the input function $u(s)$ for any time $s$, and it is also not necessary to have knowledge about values of $u(\cdot)$ for more than a finite time interval back into the past.

The universe of time-invariant fading memory filters is quite large. It contains all filters $F$ that can be characterized by Volterra series, i.e., all filters $F$ whose output $(Fu)(t)$ is given by a finite or infinite sum (with $d = 0, 1, \ldots$) of terms of the form $\int_0^\infty \ldots \int_0^\infty h_d(\tau_1, \ldots, \tau_d) \cdot u(t - \tau_1) \cdot \ldots \cdot u(t - \tau_d) d\tau_1 \ldots d\tau_d$, where some integral kernel $h_d$ is applied to products of degree $d$ of the input stream $u(\cdot)$ at various time points $t - \tau_i$ back in the past. In fact, under some mild conditions on the domain $U$ of input streams the class of time invariant fading memory filters coincides with the class of filters that can be characterized by Volterra series.

In spite of their complexity, all these filters can be uniformly approximated by the simple computational models $M$ of the type shown in Fig. 8.4, which had been introduced in [26]:

**Theorem 8.1.** (based on [3]; see Theorem 3.1 in [24] for a detailed proof). *Any filter $F$ defined by a Volterra series can be approximated with any desired degree of precision by the simple computational model $M$ shown in Fig. 8.1 and Fig. 8.2.*

- *if there is a rich enough pool* **B** *of basis filters (time invariant, with fading memory) from which the basis filters $B_1, \ldots, B_k$ in the filterbank $L^M$ can be chosen (***B** *needs to have the pointwise separation property) and*
- *if there is a rich enough pool* **R** *from which the readout functions $f$ can be chosen (***R** *needs to have the universal approximation property, i.e., any continuous function on a compact domain can be uniformly approximated by functions from* **R***).*

**Definition 8.1.** *A* class **B** of basis filters has the pointwise separation property if there exists for any two input functions $u(\cdot), v(\cdot)$ with $u(s) \neq v(s)$ for some $s \leq t$ a basis filter $B \in \mathbf{B}$ with $(Bu)(t) \neq (Bv)(t)$.

It turns out that many real-world dynamical systems (even a pool of water) satisfy (for some domain $U$ of input streams) at least some weak version of the pointwise separation property, where the outputs $\mathbf{x}^M(t)$ of the basis filters are replaced by some "visible" components of the state vector of the dynamical system. In fact, many real-world dynamical systems also satisfy approximately an interesting kernel property[a], which makes it practically sufficient to use just a *linear* readout function $f^M$. This is particularly important if $L^M$ is kept fixed, and only the readout $f^M$ is selected (or trained) in order to approximate some particular Volterra series $F$. Reducing the adaptive part of $M$ to the *linear* readout function $f^M$ has the unique advantage that a learning algorithm that uses gradient descent to minimize the approximation error of $M$ cannot get stuck in local minima of the mean-squared error. The resulting computational model can be viewed as a generalization of a finite state machine to continuous time and continuous ("liquid") internal states $\mathbf{x}^M(t)$. Hence it is called a Liquid State Machine.

If the dynamical systems $L^M$ have fading memory, then only filters with fading memory can be represented by the resulting LSMs. Hence they cannot approximate arbitrary finite state machines (not even for the case of discrete time and a finite range of values $u(t)$). It turns out that a large jump in computational power occurs if one augments the computational model from Fig. 8.4 by a feedback from a readout back into the circuit (assume it enters the circuit like an input variable).

**Theorem 8.2.** [23]. *There exists a large class $S_n$ of dynamical systems $C$ with fading memory (described by systems of $n$ first order differential equations) that acquire through feedback universal computational capabilities for analog computing. More precisely: through a proper choice of a (memoryless) feedback function $K$ and readout $h$ they can simulate any given dynamical system of the form $z^{(n)} = G(z, z', \ldots, z^{(n-1)}) + u$ with a sufficiently smooth function $G$ (see Fig. 8.6). This holds in particular*

[a]A kernel (in the sense of machine learning) is a nonlinear projection $Q$ of $n$ input variables $u_1, \ldots, u_n$ into some high-dimensional space. For example all products $u_i \cdot u_j$ could be added as further components to the $n$-dimensional input vector $< u_1, \ldots, u_n >$. Such nonlinear projection $Q$ boosts the power of any *linear* readout $f$ applied to $Q(\mathbf{u})$. For example in the case where $Q(\mathbf{u})$ contains all products $u_i \cdot u_j$, a subsequent linear readout has the same expressive capability as quadratic readouts $f$ applied to the original input variables $u_1, \ldots, u_n$. More abstractly, $Q$ should map all inputs $\mathbf{u}$ that need to be separated by a readout onto a set of linearly independent vectors $Q(\mathbf{u})$.

*for neural circuits C defined by differential equations of the form* $x_i'(t) = -\lambda_i x_i(t) + \sigma(\sum_{j=1}^n a_{ij}x_j(t)) + b_i \cdot \sigma(v(t))$ *(under some conditions on the* $\lambda_i, a_{ij}, b_i$*).*

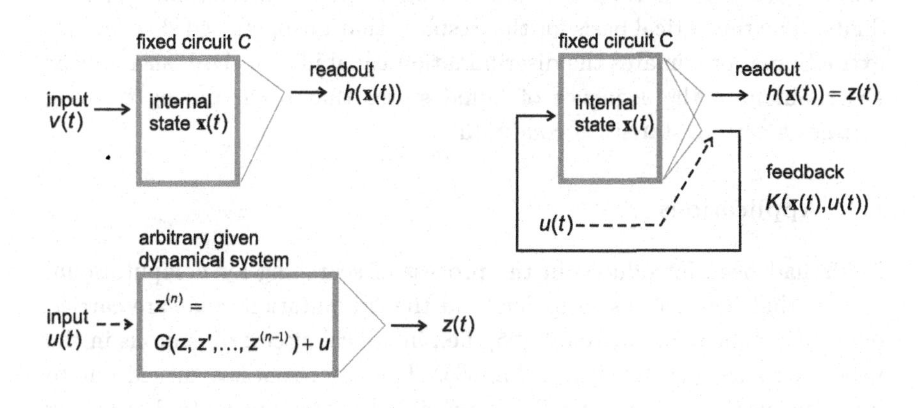


Figure 8.6. Illustration of the notation and result of Theorem 8.2.

If one allows several feedbacks $K$, such dynamical systems $C$ become universal for $n^{\text{th}}$ order dynamical systems defined by a system consisting of a corresponding number of differential equations. Since such systems of differential equations can simulate arbitrary Turing machines [4], these dynamical systems $C$ with a finite number of feedbacks become (according to the Church–Turing thesis) also universal for *digital computation.*

Theorem 8.2 suggests that even quite simple neural circuits with feedback have, in principle, unlimited computational power[b]. This suggests that the main problem of a biological organism becomes the *selection* (or learning) of suitable feedback functions $K$ and readout functions $h$. For dynamical systems $C$ that have a good kernel-property, already *linear* feedbacks and readouts endow such dynamical systems with the capability to emulate a fairly large range of other dynamical systems (or "analog computers").

Recent theoretical work has addressed methods for replacing supervised training of readouts by reinforcement learning [22] (where readout neurons explore autonomously different settings of their weights, until they find some which yield outputs that are rewarded) and by completely unsupervised learning (where not even rewards for successful outputs are available).

[b]Of course, in the presence of noise this computational power is reduced to that of a finite state machine, see [23] for details.

It is shown in [19] that already the repeated occurrence of certain trajectories of liquid status enables a readout to classify such trajectories according to the type of input which caused them. In this way a readout can for example learn without supervision to classify (i.e., "understand") spoken digits. The theoretical basis for this result is that unsupervised slow feature extraction approximates the discrimination capability of the Fisher Linear Discriminant if the sequence of liquid states that occur during training satisfies a certain statistical condition.

## 8.4. Applications

LSMs had been introduced in the process of searching for computational models that can help us to understand the computations that are carried out in a "cortical microcircuit" [25], i.e., in a local circuit of neurons in the neocortex (say in a "cortical column"). This approach has turned out to be quite successful, since it made it possible to carry out quite demanding computations with circuits consisting of reasonably realistic models for biological neurons ("spiking neurons") and biological synapses ("dynamical synapses"). Note that in this model a large number of different readout neurons can learn to extract different information from the same circuit. One concrete benchmark task that has been considered was the classification ("recognition") of spoken digits [12]. It turned out that already an LSM where the "Liquid" consisted of a randomly connected circuit of just 135 spiking neurons performed quite well. In fact, it provided a nice example for "anytime computations", since the linear readout could be trained effectively to guess at "any time", while a digit was spoken, the proper classification of the digit [26, 27]. More recently it has been shown that with a suitable transformation of spoken digits into spike trains one can achieve with this simple method the performance level of state-of-the-art algorithms for speech recognition [36].

A number of recent neurobiological experiments *in vivo* has lead many biologists to the conclusion that also for neural computation in larger neural systems than cortical microcircuits a new computational model is needed (see the recent review [30]). In this new model certain frequently occurring trajectories of network states – rather than attractors to which they might or might not converge – should become the main carriers of information about external sensory stimuli. The review [5] examines to what extent the LSM and related models satisfy the need for such new models for neural computation.

It has also been suggested [16] that LSMs might present a useful framework for modeling computations in gene regulation networks. These networks also compute on time varying inputs (e.g., external signals) and produce a multitude of time varying output signals (transcription rates of genes). Furthermore these networks are composed of a very large number of diverse subprocesses (transcription of transcription factors) that tend to have each a somewhat different temporal dynamics (see [1]). Hence they exhibit characteristic features of a Liquid in the LSM model. Furthermore there exist perceptron-like gene regulation processes that could serve as readouts from such Liquids (see chapter 6 in [1]).

In the remainder of this section we will review a few applications of the LSM model to the design of new artificial computing system. In [7] it had been demonstrated that one can use a bucket of water as Liquid for a physical implementation of the LSM model. Input streams were injected via 8 motors into this Liquid and video-images of the surface of the water were used as "liquid states" $\mathbf{x}(t)$. It was demonstrated in [7] that the previously mentioned classification task of spoken digits could in principle also be carried out with this – certainly very innovative – computing device. But other potential technological applications of the LSM model have also been considered. The article [32] describes an implementation of a LSM in FPGAs (Field Programmable Gate Arrays). In the US a patent was recently granted for a potential implementation of a LSM via nanoscale molecular connections [29]. Furthermore work is in progress on implementations of LSMs in photonic computing, where networks of semiconductor optical amplifiers serve as Liquid (see [35] for a review).

The exploration of potential engineering applications of the computational paradigm discussed in this article is simultaneously also carried out for the closely related echo state networks (ESNs) [14], where one uses simpler non-spiking models for neurons in the "Liquid", and works with high numerical precision in the simulation of the "Liquid" and the training of linear readouts. Research in recent years has produced quite encouraging results regarding applications of ESNs and LSMs to problems in telecommunication [14], robotics [11], reinforcement learning [6], natural language understanding [34], as well as music-production and -perception [13].

## 8.5. Discussion

We have argued in this article that Turing machines are not well suited for modeling computations in biological neural circuits, and proposed Liquid state machines (LSMs) as a more adequate modeling framework. They are designed to model real-time computations (as well as anytime computations) on continuous input streams. In fact, it is quite realistic that an LSM can be trained to carry out the online computation task that we had discussed in Section 8.2 (see [17] for a first application to motor control). A characteristic feature of practical implementations of the LSM model is that its "program" consists of the weights $\mathbf{w}$ of a linear readout function. These weights provide suitable targets for learning (while all other parameters of the LSM can be fixed in advance, based on the expected complexity and precision requirement of the computational tasks that are to be learnt). It makes a lot of sense (from the perspective of statistical learning theory) to restrict learning to such weights $\mathbf{w}$, since they have the unique advantage that gradient descent with regard to some mean-square error function $E(\mathbf{w})$ cannot get stuck in local minima of this error function (since $\nabla_{\mathbf{w}} E(\mathbf{w}) = \mathbf{0}$ defines an affine – hence connected – subspace of the weight space for a linear learning device).

One can view these weights $\mathbf{w}$ of the linear readout of a LSM as an analog to the code $< M >$ of a Turing machine $M$ that is simulated by a universal Turing machine. This analogy makes the learning advantage of LSMs clear, since there is no efficient learning algorithm known which allows us to learn the program $< M >$ for a Turing machine $M$ from examples for correct input/output pairs of $M$. However the examples discussed in this chapter show that an LSM can be trained quite efficiently to approximate a particular map from input to output streams.

We have also shown in Theorem 8.2 that LSMs can overcome the limitation of a fading memory if one allows feedback from readouts back into the "Liquid". Then not only all digital, but (in a well-defined sense) also all analog computers can be simulated by a fixed LSM, provided that one is allowed to vary the readout functions (including those that provide feedback). Hence these readout functions can be viewed as program for the simulated analog computers (note that all "readout functions" are just "static" functions, i.e., maps from $\mathbb{R}^n$ into $\mathbb{R}$, whereas the LSM itself maps input streams onto output streams). In those practically relevant cases that have been considered so far, these readout functions could often be chosen to be linear. A satisfactory characterization of the computational power that can be reached with linear readouts is still missing. But obviously the kernel-property of the underlying "Liquid" can boost the richness of the class of analog computers that can be simulated by a fixed LSM with linear readouts.

The theoretical analysis of computational properties of randomly connected circuits and other potential "Liquids" is still in its infancy. We refer to [8, 20, 21, 31, 37] for useful first steps. The qualities that we expect from the "Liquid" of an LSM are completely different from those that one expects from standard computing devices. One expects diversity (rather than uniformity) of the responses of individual gates within a Liquid (see Theorem 8.1), as well as diverse local dynamics instead of synchronized local gate operations. Achieving such diversity is apparently easy to attain by biological neural circuits and by new artificial circuits on the molecular or atomic scale. It is obviously much easier to attain than an emulation of precisely engineered and synchronized circuits of the type that we find in our current generation of digital computers. These only function properly if all local units are identical copies of a small number of template units that respond in a stereotypical fashion. For a theoretician it is also interesting to learn that sparse random connections within a recurrent circuit turn out to provide better computational capabilities to an LSM than those connectivity graphs that have primarily been considered in earlier theoretical studies, such as all-to-all connections (Hopfield networks) or a two-dimensional grid (which is commonly used for cellular automata). Altogether one sees that the LSM and related models provide a wide range of interesting new problems in computational theory, the chance to understand biological computations, and new ideas for the invention of radically different artificial computing devices that exploit, rather than suppress, inherent properties of diverse physical substances.

## Acknowledgements

Partially supported by the Austrian Science Fund FWF, project P17229 and project S9102-N13, and by the European Union, project # FP6-015879 (FACETS), project FP7-216593 (SECO) and FP7-231267 (ORGANIC).

## References

[1] U. Alon, *An Introduction to Systems Biology: Design Principles of Biological Circuits*. Chapman & Hall, Netherlands, (2007).

[2] P. L. Bartlett and W. Maass. Vapnik-Chervonenkis dimension of neural nets. In ed., M. A. Arbib, *The Handbook of Brain Theory and Neural Networks*, pp. 1188–1192. MIT Press, Cambridge, 2nd edition, (2003).

[3] S. Boyd and L. O. Chua, Fading memory and the problem of approximat-

ing nonlinear oparators with Volterra series, *IEEE Trans. on Circuits and Systems.* **32**, 1150–1161, (1985).

[4] M. S. Branicky, Universal computation and other capabilities of hybrid and continuous dynamical systems, *Theoret. Comput. Sci.* **138**, 67–100, (1995).

[5] D. Buonomano and W. Maass, State-dependent computations: Spatiotemporal processing in cortical networks. *Nature Reviews Neuroscience.* **10**(2), 113–125, (2009).

[6] K. Bush and C. Anderson. Modeling reward functions for incomplete state representations via echo state networks. In *Proceedings of the International Joint Conference on Neural Networks, Montreal, Quebec,* (2005).

[7] C. Fernando and S. Sojakka. Pattern recognition in a bucket: a real liquid brain. In *Proceedings of ECAL.* Springer, Berlin–Heidelberg, (2003).

[8] S. Ganguli, D. Huh, and H. Sompolinsky, Memory traces in dynamical systems, *Proc. Natl. Acad. Sci. USA.* **105**, 18970–18975, (2008).

[9] S. Haeusler and W. Maass, A statistical analysis of information processing properties of lamina-specific cortical microcircuit models, *Cereb. Cortex.* **17** (1), 149–162, (2007).

[10] S. Haeusler, K. Schuch, and W. Maass, Motif distribution and computational performance of two data-based cortical microcircuit templates, *J. Physiology - Paris.* (2009). in press.

[11] J. Hertzberg, H. Jäger, and F. Schönherr. Learning to ground fact symbols in behavior-based robot. In ed. F. van Harmelen ed., *Proc. of the 15th European Conference on Artificial Intelligence,* pp. 708–712, IOS Press, Amsterdam, (2002).

[12] J. J. Hopfield and C. D. Brody, What is a moment? Transient synchrony as a collective mechanism for spatio-temporal integration, *Proc. Natl. Acad. Sci. USA.* **98**(3), 1282–1287, (2001).

[13] H. Jaeger and D. Eck. Can't get you out of my head: A connectionist model of cyclic rehearsal. In eds., I. Wachsmuth and G. Knoblich, *Modeling Communication with Robots and Virtual Humans,* (2008).

[14] H. Jaeger and H. Haas, Harnessing nonlinearity: predicting chaotic systems and saving energy in wireless communication, *Science.* **304**, 78–80, (2004).

[15] H. Jaeger, W. Maass, and J. Principe, Introduction to the special issue on echo state networks and liquid state machines, *Neural Networks.* **20**(3), 287–289, (2007).

[16] B. Jones, D. Stekel, J. Rowe, and C. Fernando, Is there a liquid state machine in the bacterium escherichia coli?, *Artif. Life.* ALIFE'07, IEEE Symposium, 187–191, (2007).

[17] P. Joshi and W. Maass, Movement generation with circuits of spiking neurons, *Neural Comput.* **17**(8), 1715–1738, (2005).

[18] S. Klampfl, S. V. David, P. Yin, S.A. Shamma, and W. Maass, Integration of stimulus history in information conveyed by neurons in primary auditory cortex in response to tone sequences, *39th Annual Conference of the Society for Neuroscience, Program 163.8, Poster T6.* (2009).

[19] S. Klampfl and W. Maass, A neuron can learn anytime classification of trajectories of network states without supervision, *submitted for publication.*

(Feb. 2009).

[20] R. Legenstein and W. Maass. What makes a dynamical system computationally powerful? In eds., S. Haykin, J. C. Principe, T. Sejnowski, and J. McWhirter, *New Directions in Statistical Signal Processing: From Systems to Brains*, pp. 127–154. MIT Press, Cambridge, MA, (2007).

[21] R. Legenstein and W. Maass, Edge of chaos and prediction of computational performance for neural microcircuit models, *Neural Networks.* **20**(3), 323–334, (2007).

[22] R. Legenstein, D. Pecevski, and W. Maass, A learning theory for reward-modulated spike-timing-dependent plasticity with application to biofeedback, *PLoS Computational Biology.* **4**(10), 1–27, (2008).

[23] W. Maass, P. Joshi, and E. D. Sontag, Computational aspects of feedback in neural circuits, *PLoS Computational Biology.* **3**(1), e165, 1–20, (2007).

[24] W. Maass and H. Markram, On the computational power of recurrent circuits of spiking neurons, *J. Comput. System Sci.*. **69**(4), 593–616, (2004).

[25] W. Maass and H. Markram. Theory of the computational function of microcircuit dynamics. In eds. S. Grillner and A. M. Graybiel, *The Interface between Neurons and Global Brain Function*, Dahlem Workshop Report 93, pp. 371–390. MIT Press, (2006).

[26] W. Maass, T. Natschlaeger, and H. Markram, Real-time computing without stable states: A new framework for neural computation based on perturbations, *Neural Comput.* **14**(11), 2531–2560, (2002).

[27] W. Maass, T. Natschlaeger, and H. Markram, Fading memory and kernel properties of generic cortical microcircuit models, *J. Physiology – Paris.* **98** (4–6), 315–330, (2004).

[28] D. Nikolic, S. Haeusler, W. Singer, and W. Maass, Distributed fading memory for stimulus properties in the primary visual cortex, *PLoS Biology.* **7**(12), 1–19, (2009).

[29] A. Nugent, *Physical neural network liquid state machine utilizing nanotechnology.* (US-Patent 7 392 230 32, June 2008).

[30] M. Rabinovich, R. Huerta, and G. Laurent, Transient dynamics for neural processing, *Science.* **321**, 45–50, (2008).

[31] B. Schrauwen, L. Buesing, and R. Legenstein. On computational power and the order-chaos phase transition in reservoir computing. In *Proceeding of NIPS 2008, Advances in Neural Information Processing Systems.* MIT Press, Cambridge, MA, (2009). In press.

[32] B. Schrauwen, M. D'Haene, D. Verstraeten, and D.Stroobandt, Compact hardware liquid state machines on FPGA for real-time speech recognition, *Neural Networks.* **21**, 511–523, (2008).

[33] A. M. Thomson, D. C. West, Y. Wang, and A. P. Bannister, Synaptic connections and small circuits involving excitatory and inhibitory neurons in layers 2–5 of adult rat and cat neocortex: triple intracellular recordings and biocytin labelling in vitro, *Cereb. Cortex.* **12**(9), 936–953, (2002).

[34] M. Tong, A. Bickett, E. Christiansen, and G. Cotrell, Learning grammatical structure with echo state networks, *Neural Networks.* **20**(3), 424–432, (2007).

[35] K. Vandoorne, W. Dierckx, B. Schrauwen, D. Verstraeten, R. Baets, P. Bi-

enstman, and J. V. Campenhout, Toward optical signal processing using photonic reservoir computing, *Opt. Express.* **16**(15), 11182–11192, (2008).

[36] D. Verstraeten, B. Schrauwen, D. Stroobandt, and J. V. Campenhout, Isolated word recognition with the liquid state machine: a case study., *Inform. Process. Lett.* **95**(6), 521–528, (2005).

[37] O. L. White, D. D. Lee, and H. Sompolinsky, Short-term memory in orthogonal neural networks, *Phys. Rev. Letters.* **92**(14), 148102, (2004).

# Chapter 9

# Experiments on an Internal Approach to Typed Algorithms in Analysis

Dag Normann
*Department of Mathematics,*
*The University of Oslo,*
*Blindern, NO-0316 Oslo, Norway*
*E-mail: dnormann@math.uio.no*

The chapter consists of four sections. First we discuss aspects of generalized computability theory with a focus of how various approaches to abstract computability theory relate to computational analysis. Emphasis is put on the distinction between internal and external algorithms. Then we prove some old and some new results related to the typed hierarchy of hereditarily total objects over complete and separable normed vectorspaces, with the aim of carrying out the arguments within the framework of Kuratowski limit spaces.

In the final section we prove a topological consequence of an assumption that the total continuous functions from one complete, separable metric space to another is dense in the sense of domain theory. It turns out that this assumption will have consequences for how the connectedness properties of the two metric spaces relate.

## Contents

## 9.1. Introduction

This chapter will consist of four sections. In Sections 1 and 2 we will survey some of the history of generalized computability theory with the partial aim of discussing the elements from generalized computability theory that may be relevant for computational analysis.

One of the aims of this introductory part is to clarify the distinction between *internal* and *external* approaches to computability over a mathematical structure in general and over a structure appearing in analysis in particular. The objective is, however, wider. On a general basis we will discuss the motivations for generalizing computability theory.

In Section 3, we will investigate spaces of functionals of higher finite types in the category of Kuratowski limit spaces, where the base types are interpreted as complete and separable normed vector spaces. We prove a new theorem about the topological embeddability of some of these hierarchies into others, and give a proof of a density theorem not stated in its present form elsewhere, but nevertheless provable using known methods from domain theory. One important aspect of Section 3 is that we only use concepts related to the limit space structure of the spaces at hand, and no domain representation or other kinds of superstructure. What we aim to learn from this is which tools may be available and needed in order to study aspects of computability on such spaces without bringing in the computational structure of representing spaces.

One observation we have made while this work was in progress, is that when we try to restrict the means we can use in proofs, the results we obtain are often slightly better. The reason is that we need to formulate sharper theorems in order to carry out, for example, proofs by induction. In contrast, the proofs often turn out to be simpler. As an example of this observation, if we prove the density theorem for the Kleene–Kreisel functionals in the traditional way, we simply get that there is a recursive enumeration of a dense set of total functionals at every type, and to extract further properties requires further work. If we prove the density theorem in the setting of limit spaces, the proof is actually simpler and we get for free how to approximate any functional by a sequence from this countable dense subset, (see Normann [18]).

In Section 4 we consider the standard domain representation of the set of continuous functions from one metric space to another. We show that if the total objects in the domain representation of $X \to Y$ is dense in the underlying domain, then $Y$ is what we call *compactly saturated over* $X$. This result has no consequence for the rest of the paper, and is included partly to prevent other researchers to look for strong density theorems based on domain theory in a naive way, and partly because the concept of compactly saturated may be of independent interest.

### 9.1.1. *Classical computability theory*

By *classical computability theory* we mean the study of the concept of computability induced by *Turing Machines* on sets of words over a prefixed alphabet, or of any of the equivalent reformulations. Classical computability theory is simple in the sense that the basic definitions are well understood, but complex in the sense that it offers deep results with occasionally very hard proofs. Classical computability theory appears in many guises, the authors's favorites are via Kleene schemes giving an elegant proof of the recursion theorem, and set recursion over the set $HF$ of hereditarily finite sets, as we consider $HF$ to be the ultimate data-type (of finite data) inside which all other genuine data-types live.

### 9.1.2. *Generalizing computability theory*

There are several reasons for generalizing computability theory. This was discussed in depth by Kreisel [11], and anyone working with generalized computability in any sense should consult [11]. One of the reasons suggested by Kreisel is that we may find applications to the rest of the mathematical world and the world of science in general. Applications of metarecursion theory to descriptive set theory will be an example of this. Another reason is that we may learn something about the concepts used in computability theory and which properties of these concepts we actually use by generalizing them.

We will not give a complete historical survey of generalized computability theory, but mention a few directions it has taken and how this may have some impact on today's research. As a simple example, let us consider the set $\mathbb{Q}$ of rational numbers. We may view $\mathbb{Q}$ as constructed from $\mathbb{N}$ via $\mathbb{Z}$ or we may view $\mathbb{Q}$ as a spontaneously given field or even as an ordered field. In the latter case, it is well known that $\mathbb{N}$ is not definable over $\mathbb{Q}$ in the sense of first order logic. However, $\mathbb{N}$ is an inductively defined substructure of $\mathbb{Q}$, and then $\mathbb{Q}$ itself is an inductively definable substructure of $\mathbb{Q}$.

Building $\mathbb{Q}$ from 0 and 1 and the algebraic operations as an inductively defined structure gives us the tools needed in order to perform induction, selection, etc. In the case of $\mathbb{Q}$ we may even let the identity relation be inductively defined. If we accept positive induction and corresponding recursion as basic elements of computability, it does not really matter if we consider $\mathbb{Q}$ as constructed or given. What is to be considered as computable or semicomputable over $\mathbb{Q}$ does not depend on how the elements of $\mathbb{Q}$ are represented as data.

If we replace $\mathbb{Q}$ with its algebraic completion $C(\mathbb{Q})$, i.e. we hereditarily add solutions to all polynomial equations, the situation is different. We of course have an effective enumeration of $C(\mathbb{Q})$, i.e. a surjective map

$$\nu : \mathbb{N} \to C(\mathbb{Q})$$

such that all algebraic operations on $C(\mathbb{Q})$ have their computable counterparts over $\mathbb{N}$. Given $\nu$, we may even find a computable function

$$sqrt : \mathbb{N} \to \mathbb{N}$$

that represents a kind of *square root* on $C(\mathbb{Q})$ in the following sense:

$$\nu(sqrt(n)) \cdot \nu(sqrt(n)) = \nu(n)$$

for all $n \in \mathbb{N}$. There is, however, no square root function definable over $C(\mathbb{Q})$, even if we accept higher order definitions. In order to have one we need some kind of external representations of the objects of $C(\mathbb{Q})$ and we need to be allowed to compute on these external representations. It will of course suffice to identify $i = \sqrt{-1}$, but the structure offers no distinction between $i$ and $-i$. If we increase the ambition and aim at finding a function solving polynomial equations in general, we know that we cannot hope to do so from within.

This example illustrates in a nutshell our distinction between *internal* and *external* concepts of computability; the internal concepts must grow out of the structure at hand, while external concepts may be inherited from computability over superstructures via, for example, enumerations, domain representations, or in other ways. This distinction was first made explicit in Normann [15]. A similar distinction between *abstract* and *concrete* notions of computations was discussed in Tucker and Zucker [26].

### 9.1.3. *Generalizing finiteness*

The step from standard computability theory to computability in analysis has to take the step from the discrete to the continuous, and the step from locally finiteness to locally continuum, into account. One of the ways computability theory was generalized was by assuming or axiomatizing that certain infinite sets share some of the properties of finite sets. Following Kreisel [11] this helps us understand which of the properties finite sets have that we actually make use of in, for example, degree theory. The question is if we have to make similar steps in order to make sense of internal computational analysis.

The original example of this kind of generalization is metrarecursion theory, or hyperarithmetical theory.

At first, *computable* was replaced by $\Delta^1_1$ and *semicomputable* was replaced by $\Pi^1_1$. This made a poor analogue of the classical theory. Then one replaced *finite* with *hyperarithmetic*, *computable* with $\Delta^1_1$ on the set of hyperarithmetical sets and *semicomputable* with $\Pi^1_1$ on the set of hyperarithmetical sets.

This led in turn to $\alpha$-recursion theory, $\beta$-recursion theory, computability relative to higher type functionals, preferably *normal* ones, computability over admissible structures, and to set recursion. See Sacks [21] for an introduction to this area, known as *higher recursion theory*.

The recent investigations of the so called *hypercomputations*, i.e. where Turing Machines and Register Machines are allowed to work in transfinite time and occasionally with transfinite memory stores, fits well into this tradition. We will not discuss possible motivations for this renewal of higher recursion theory, but advise anyone wanting to enter the field to use [11] for the calibration of motives.

With several examples of generalized computability, one naturally wanted to axiomatize the theory. It is worthwhile to consult the contributions from Moschovakis [14] and Fenstad [6] in order to get two different perspectives on what a computation might be in a general setting.

The kind of generalizations we find in higher computability theory are too far from the classical Turing model to be of any relevance to questions of internal computability in analysis, where we after all must have as a requirement that there is at least one suitable digitalization of the data at hand, and then that what is computable in an internal sense must also be computable in the external sense. This is a soundness criterion for any concept of computability.

We will use the expression *Extended Computability Theory* for the situation where our concepts try to capture genuine algorithms in a setting going beyond the Turing world, but where we at least, through some kind of digitalization, may reduce our concepts to the classical ones. This does not mean that we believe that in all situations, using digitalization to define the concept of computability is the best approach. We will return to this discussion later.

### 9.1.4. *Computability at higher types*

One direction of generalized computability of interest to both logic and computer science is the study of computations relative to continuous functionals of higher types. The starting point was the equivalent constructions of the typed hierarchy of hereditarily total continuous functionals by Kleene [9] and Kreisel [10]. An extensive survey of what Kleene and Kreisel achieved, and of the significance of the related work of Scott [23, 24], Plotkin [20], and Milner [13] is given in Normann [17]. Kleene and Kreisel worked with hereditarily total objects, i.e. with natural numbers, functions sending natural numbers to natural numbers, continuous functionals sending such functions to natural numbers, and so on. Kleene showed how his internal definition of *computations* using the $S1$-$S9$ – schemes from Kleene [8] makes sense for the continuous functionals. Kleene's computations are essentially well-founded trees, in most cases of the cardinality of the continuum. Kreisel [10] objected to Kleene's concept for this reason, and preferred an external notion essentially based on digitalization and Turing machines working on oracles. The facts that all Kleene-computable functionals are Kreisel-computable, and that all Kreisel-computable functionals of interest in [10] are Kleene-computable, indicate that all applications of the continuous functionals one had in mind in 1959 could be made using internal concepts. Of course, the very definitions of the continuous functionals by the two authors involved an element of digitalizability of the objects, and thus the external definition of the computability of a continuous functional may seem the most natural one. Later characterizations of this typed hierarchy makes internal approaches to computations in higher types more natural, (see e.g. Normann [18]).

The *Scott Model* is a typed hierarchy of partial, continuous functionals, given in the form of *Scott domains*. Since each domain in this hierarchy is an effective Scott domain, each domain accepts an external concept of computability. The hierarchy is also the original domain for the denotational semantics of $LCF$ (Scott [23, 24]) and the equivalent $PCF$ (Plotkin [20]). We consider $LCF$ and $PCF$ as defining internal concepts of computability for the Scott hierarchy, but the inherited concepts for the classical Kleene-Kreisel functionals must be considered to be external since it is based on a computability concept for a superstructure. When it comes to domain representations in general, we will modify our views here to some extent later.

Grilliot [7] showed that we sometimes, in a computable way, may decide problems expressed with number quantifiers. Under certain circumstances we may decide if a given functional $F : \mathbb{N}^{\mathbb{N}} \to \mathbb{N}$ is continuous with respect to a given convergent sequence $f = \lim_{n\to\infty} f_n$ from $\mathbb{N}^{\mathbb{N}}$. We may actually find a term in Gödel's $T$ accepting $F$, the sequence, its limit and a modulus function for the sequence as inputs, and the output (when we use the full typed hierarchy of all total functionals as the base for the denotational semantics) will answer if $F(f) = \lim_{n\to\infty} F(f_n)$ or not. In case of local continuity, we may use $\mu$-recursion to find the modulus of the limit, and in case of local discontinuity there is (uniformly) another $T$-term defining the functional ${}^2E$ from the data at hand.
Here ${}^2E$ is quantification over $\mathbb{N}$ as the total functional of type 2:

$$ {}^2E(f) = \begin{cases} 0 \text{ if } \forall x \in \mathbb{N}(f(x) = 0) \\ 1 \text{ if } \exists x \in \mathbb{N}(f(x) > 0. \end{cases} $$

This must not be confused with the continuous existential quantifier $\exists_\omega$ defined by

$$ \exists_\omega(f) = \begin{cases} 0 \text{ if } \quad f(\bot) = 0 \\ 1 \text{ if } \exists x \in \mathbb{N}(f(x) > 0. \end{cases} $$

We call the computational machinery originating from Grilliot [7] the *Grilliot Theory*, and Grilliot theory shows that even internal concepts of computability not aimed at generalizing what is considered to be finite to some extent is strong enough to make quantification over certain infinite sets computable in some sense. In [5] Escardó explores another example of such phenomena.

## 9.2. Computational Analysis

The term *computational analysis* covers the study of problems in mathematics and theoretical computer science where one is interested in the computational content of phenomena in analysis. It is mainly an area for foundational research, and not so practically oriented as numerical analysis. One of the main motivations is to make computational analysis relate to numerical analysis in the same way as classical computability theory relates to the practical use of digital computers. With our terminology we may say that computational analysis is classical computability theory extended to structures appearing in analysis. This does not mean that we have to take an external view on computational analysis and that there is only one valid concept of computability in a given relevant context. There are at least three levels on the scale from external to internal that will be of interest.

### 9.2.1. *Type two enumerability*

It is well known how we may extend classical computability theory to computations relative to function oracles. Thus the structure $(\mathbb{N}, \mathbb{N}^{\mathbb{N}})$ accepts a natural, internal concept of computability. In computational algebra it is standard to tie a concept of computability to effective enumerations of the algebraic structure at hand. In analysis, the structures are mostly uncountable, but often of the power of the continuum. Weihrauch suggested that for many such structures, an effective "enumeration" over a subset of $\mathbb{N}^{\mathbb{N}}$ could be used instead. See Weihrauch [27] for the carrying out of this approach.

Representing objects as elements of $\mathbb{N}^{\mathbb{N}}$ is essentially the one natural way to "digitalize" the same objects, so borrowing concepts of computability from a type two enumeration is a very external approach to computational analysis. If the purpose is just to capture "in principle computable by a digital computer in one way or another", the $TTE$-approach is both sound and natural. If the aim is to find algorithms that in a natural way fall out of the structure at hand, the $TTE$-approach may be of less help.

### 9.2.2. *Domain representability*

As an alternative to the $TTE$-approach by Weihrauch and others, representations over effective domains have been attempted as a foundation for computational analysis. Among those initiating this approach we mention Stoltenberg-Hansen and Tucker [25] and Edalat [2]. This approach is so general that unless the domain in question is carefully chosen, the approach is more or less as external as the $TTE$-approach. However, a carefully chosen domain may capture a reasonable notion of "partial object" for the structure at hand, like the objects in the Scott hierarchy does. If there is a natural internal approach to computability for the extended set of partial objects, then the derived notions for the original structure will be more internal than if the $TTE$-approach is used.

If we go back to one of the origins of domain theory, Scott [23], the motivation was to construct a structure that could provide denotational interpretations of programs or algorithmic terms of some sort. This indicates that if we start with a mathematical structure that we for some reason would like to consider as a data-type and then form a programming language suitable for dealing with data of this kind (and not primarily with digitalized representations of these data), then a cleverly chosen domain representation, where the finitary objects of the domain actually represent partial objects of the data-type in question, together with some internal notion of computability on the representing domain reflecting the algebra of the given data-type, may lead to a more fine tuned and less external approach to computability on the original data-type.

How can we judge if a domain representation is cleverly chosen, in the sense of leading to more internal concepts than the $TTE$-approach? One criterion is that if two finitary objects have extensions representing the same original data-object, then they should have a joint extension representing the same object. In an abstract way, this would mean that the domain object directly approximates the data-objects, not just digitalized versions of them. Another criterion will be that the algebra of the data-structure has a natural extension to an algebra on the representing domain. This is, for instance, the case for Escardó's *Real PCF* [3]. Actually, we will consider real $PCF$ as a purely internal way of defining computable functions from reals to reals, but slightly external if it is used to define, for example, computable operators. The reason is that $\mathbb{R}$ is a substructure of its domain representation, so when a real number is considered as an input of an algorithm in Real $PCF$, it is really that number, and not some representative for it, that is the input.

Now, why should we be interested in whether an approach is internal or not, why not use the strongest concept of computability that makes sense in a given context?

Our main reason is analogue to the reason why logicians should try to prove theorems in weak systems, the weaker tools we use to obtain a result, the more extra knowledge can be obtained from the process of obtaining the result. We would like to claim that an internal approach to computability in analysis will result in easy-to-use, high level, programming languages for computing in analysis, but the development cannot support this claim yet. The possibility of finding support for such a claim, together with basic curiosity, is nevertheless the motivation behind trying to find out what internally based algorithms might look like.

### 9.2.3. *Quotients of countably based spaces*

There has been a renewed interest in the spaces that may have $TTE$ representations or domain representations, and in exploring the possible categories of such spaces without actually always doing it within domain theory or over Baire space $\mathbb{N}^{\mathbb{N}}$. *qcb*-spaces, quotients of countably based spaces, forms an interesting category $QCB$ of topological spaces. As a category, $QCB$ is at present too general to accept a uniform approach to internal computability, but all spaces considered in this paper will be *qcb*-spaces or various sorts of representations for them. This category originates from Menni and Simpson [12], but was characterized as *qcb*-spaces by Schröder [22], who independently characterized them as the spaces with admissible $TTE$-representations..

### 9.2.4. *A purely internal approach?*

As an example, let us consider the Banach-space $l^2$ of functions $f : \mathbb{N} \to \mathbb{R}$ such that

$$\sum_{i=0}^{\infty}(f(n))^2 < \infty.$$

This is a normed vector space, and we may ask for the set of total and partial functions

$$F : l^2 \to \mathbb{R}$$

that may be considered to be computable.

Taking a strictly external point of view, we may construct admissible representations of $l^2$ and $\mathbb{R}$ over $\mathbb{N}^{\mathbb{N}}$, and then consider those functions $F : l^2 \to \mathbb{R}$ that can be lifted to partial computable functions $\hat{F} : \mathbb{N}^{\mathbb{N}} \to \mathbb{N}^{\mathbb{N}}$. The problem with this approach is that there is no natural structure on the set of computable functions defined this way that we can use for further investigations.

Our approach will be quite the opposite. We will see what can be achieved accepting the internal structure (algebra, norm etc.) as computable, and then use general principles for creating new computable operators from old ones. $l^2$ is a separable space. If we let $f_n$ be the object in $l^2$ that takes the value 1 on $n$ and 0 elsewhere, it is reasonable to consider the sequence

$$\{f_n\}_{n\in\mathbb{N}}$$

as a part of the basis for defining computability over $l^2$. Clearly the set of finite, rational linear combinations of the $f_n$'s is dense in $l^2$, and since $\mathbb{Q}$ is inductively definable over $\mathbb{R}$, there is a dense, internally computably enumerable subset of $l^2$. Moreover, all coordinate functions

$$n, f \mapsto f(n)$$

will be computable, since they can be calculated from $f_n$ as defined above. This again shows that the identity function, seen as a function from $l^2$ to $\mathbb{R}^{\mathbb{N}}$ is computable. The inverse is not continuous, and should not be computable.

In order to be able to define functions via some sort of recursive constructions, and at the same time make use of the fact that $l^2$ is a topological space, it is natural to include some natural limit process as basically computable. Since we are just providing an example of how it can be done, and not of how it has to be done, we suggest the principle developed below:

**Definition 9.1.**

a) Let $\Delta : \mathbb{R} \to [0, 1]$ be the projection of $\mathbb{R}$ to $[0, 1]$ and for each $n \in \mathbb{N}$, let

$$\Delta_n(x) = 2^{-n}\Delta(2^n x).$$

b) If $u$ and $v$ are distinct vectors in a normed vector space $V$ like $l^2$ and $n \in \mathbb{N}$, we let

$$mod_n(u, v) = u + \frac{\Delta_n(||v - u||)}{||v - u||}(v - u),$$

and we let $mod_n(u, u) = u$ for all $u$.

c) We define $mod_{seq} : V^{\mathbb{N}} \to V^{\mathbb{N}}$ by recursion

- $mod_{seq}(\{v_n\}_{n\in\mathbb{N}})_0 = v_0$
- $mod_{seq}(\{v_n\}_{n\in\mathbb{N}})_{k+1} = mod_{k+1}(mod_{seq}(\{v_n\}_{n\in\mathbb{N}})_k, v_{k+1})$.

We consider $mod_n$ as computable uniformly in $n$ since it is the single-valued interpretation of the following nondeterministic algorithm:

**If** $||v - u|| < 2^{-n}$ **then** $mod_n(u, v) = u$
**AND**
**If** $||v - u|| > 0$ **then** $mod_n(u, v) = u + \frac{\Delta_n(||v-u||)}{||v-u||}(v - u)$.

The function $mod_n(u, v)$ demonstrates that the $2^{-n}$-ball around $u$ is a retract of the full space, and the recursively defined $mod_{seq}$ then retracts all sequences to a subset of uniformly converging Cauchy sequences.

The point is not that we propose to include general nondeterministic or parallel algorithms like the one used in "computing" $mod_n$ in a construction of internal computability principles for analysis, but that some basic functions justifiable by the use of such or other principles may be included. The danger is, as we learned from Escardó, Hofmann, and Streicher [4], that we may introduce an unwanted amount of nondeterministic processes just by adding one basic function. They showed that including a continuous extension of $+$ to the domain representation of $\mathbb{R}$ into an otherwise deterministic calculus, gives us the full power of the *weak parallel or*.

**Remark 9.1.** In this example, we have shown (the well known fact) that for a normed vector space $V$, the space of Cauchy sequences converging at least as fast as $\{2^{-n}\}_{n\in\mathbb{N}}$ is a retract of the space of all sequences, the point being that we have used well known computable functions over $\mathbb{R}$ and primitive recursion in combination with the internal algebra. We do not consider the use of $\Delta_n$ as being an example of external algorithms, since we do not replace the data-objects in $V$ with something else representing them.

The important choice to make is which infinite steps we may take in describing internal algorithms. Our lesson from the Grilliot theory of functionals in higher types will be that if we have a sequence with a known modulus of convergence, then passing to the limit is in essence computable. Thus, as an example, we propose to add the following principle to our definition of internal computability over $V$:

$$LIM : V^{\mathbb{N}} \to V$$

defined by

$$LIM(\{v_n\}_{n\in\mathbb{N}}) = \lim_{k\to\infty} \{mod_{seq}(\{v_n\}_{n\in\mathbb{N}})_k\}_{k\in\mathbb{N}}.$$

In the sequel, we are not going to propose a rigid definition of internally computable analysis, but prove some nontrivial results using internal, in some sense effective, means only.

## 9.3. Some Typed Hierarchies of Limit Spaces

### 9.3.1. *Total versus partial functionals*

In this section we will consider typed hierarchies of total continuous functionals where the base spaces are certain complete, separable metric spaces. From the point of view of computational analysis it may be more natural to consider hereditarily partial functionals that in some way are continuous. However, given effective metric spaces as base spaces, it is a challenging task to explore what we might mean by a functional of finite type that is partial and hereditarily computable relative to an object of the base space. This would involve finding the proper analogue of the sequential functionals over $\mathbb{N}_\perp$ for other base spaces than $\mathbb{N}$. It is by no means obvious that if a project like this is successfully carried out, then each hereditarily total object we study in this paper will be represented by hereditarily relative computable functionals.

Not ignoring the importance of partiality in computational analysis, our starting point is that the structures we study are of independent interest, and that it thus is of interest to find internal approaches to computability over these structures.

An effective metric space is normally given as the completion of a computable metric on $\mathbb{N}$, and though we may consider spaces with additional computational structure, the enumeration of the dense set and how it relates to its completion will form one basis for how to compute internally. For other separable topological spaces, an enumeration of a countable dense subset and some additional structure relating the other elements to this dense set will be a natural tool in an internal approach to computability. In this section we will see how this infrastructure may be established for some hierarchies of functionals.

### 9.3.2. *The problem with density*

All spaces we will consider in this section will be by default separable Hausdorff spaces, and they will be sequential spaces. The latter means that the topology is the finest one with exactly the present set of convergent sequences with limits.

If $X_1, \ldots, X_n$ are Polish spaces and

$$\sigma = \sigma(x_1, \ldots, x_n)$$

is a type expression in the grammar

$$\sigma ::= x_1 \mid \cdots \mid x_n \mid (\sigma \to \sigma)$$

(where we will drop parentheses according to standard conventions) we may interpret $\sigma$ as a space $\sigma(X_1, \ldots, X_n)$ using the category of Kuratowski limit spaces.

Then we know that $\sigma(X_1, \ldots, X_n)$ is a sequential, separable space. However, there is no general way of *constructing* a dense countable subset, even when $X_1, \ldots, X_n$ are effective metric spaces. As a general problem, this is related to understanding how the connectedness properties of the spaces in question relate, and even how local connectedness of the domain space relates to global connectedness of the image space. For a few cases, one has used domain theory to obtain effective density theorems. Indeed, in Section 4 we will show that this method has major limitations unless combined with some insight on the connectedness properties of the spaces involved.

### 9.3.3. *Probabilistic projections*

The rational numbers is a dense subset of the real numbers, so every real can be approximated by a sequence of rationals. This fact is used in most of the standard ways we represent reals as digitalized data objects. There is, however no way we continuously in a real $r$ may select a sequence of rationals converging to $r$, since any such continuous selection must be constant. It turns out that if we consider sequences of probability distributions on $\mathbb{Q}$ with finite support instead of sequences of rationals, we may use the sequences of probability distributions to many of the computational tasks for which we initially would like to be able to select a convergent sequence of rationals in a continuous way. In this section we will introduce the concepts of probabilistic projections and probabilistic selection that has turned out to be useful in this respect.

**Definition 9.2.** Let $Y$ be a sequential space, let $A = \cup_{n \in \mathbb{N}} A_n$ be the union of a family of finite subsets $A_n$ in $Y$ and let $X \subseteq Y$ be a subspace with $A \subseteq X$.
A *probabilistic projection* from $Y$ to $X$ will be a sequence of continuous maps

$$y \mapsto \mu_{y,n},$$

where $\mu_{y,n}$ is a probability distribution on $A_n$, such that whenever $x = \lim_{n \to \infty} x_n$, $x \in X$ and $a_{k_n} \in A_n$ with $\mu_{x_n,n}(a_{k_n}) > 0$ for each $n \in \mathbb{N}$, then

$$x = \lim_{n \to \infty} a_{k_n}.$$

**Remark 9.2.** It follows that $A$ is dense in $X$.
Though our definition depends on $A$ and its enumeration, we will only require that there exists an $A$ as above when we say that there is a probabilistic projection from $Y$ to $X$.

**Lemma 9.1.** *Let $Y$ be a complete, separable metric space, $A = \bigcup_{n\in\mathbb{N}} A_n$ an increasing union of finite subsets of $Y$ and let $X$ be the closure of $A$ in $Y$.*
*Then we may construct a probabilistic projection from $Y$ to $X$.*

***Proof.*** Let $d$ be the metric on $Y$. For any $y \in Y$, let $d(y, A_n)$ be the minimal distance from $y$ to an element of $A_n$. This function is continuous in $y$.
For $u, v \in \mathbb{R}_{\geq 0}$ we let $u \dot{-} v = max\{u - v, 0\}$.
This function is also continuous.
For each $y \in Y$ and $a \in A_n$ we let

$$\mu_{y,n}(a) = \frac{d(y, A_n) + 2^{-n} \dot{-} d(y, a)}{\sum_{b\in A_n} [d(y, A_n) + 2^{-n} \dot{-} d(y, b)]}.$$

If $d(y, A_n) = d(y, b)$, then

$$d(y, A_n) + 2^{-n} \dot{-} d(y, b) = 2^{-n} > 0$$

so the denominator is positive. Thus $\mu_{y,n}(a)$ is well defined.
Clearly $0 \leq \mu_{y,n}(a) \leq 1$ and

$$\sum_{a\in A_n} \mu_{y,n}(a) = 1$$

by trivial calculation.
Thus $\mu_{y,n}$ is a probability distribution on $A_n$, and $y \mapsto \mu_{y,n}$ will be continuous by construction.

It remains to prove that $\{A_n\}_{n\in\mathbb{N}}$ and $\{y \mapsto \mu_{y,n}\}_{n\in\mathbb{N}}$ satisfy the requirement of the definition.
Let $x = \lim_{n\to\infty} x_n$ where $x \in X$, each $x_n \in Y$ and let $b_n \in A_n$ such that $\mu_{x_n,n}(b_n) > 0$ for each $n \in \mathbb{N}$.
Let $\epsilon > 0$. We will find $n_0$ such that if $n \geq n_0$, then $d(x, b_n) < \epsilon$.

Let $n_0$ satisfy the following requirements:

i) If $n \geq n_0$ then $d(x, x_n) < \frac{\epsilon}{4}$.
ii) For some $a \in A_{n_0}$ we have that $d(x, a) < \frac{\epsilon}{4}$.
iii) $2^{-n_0} < \frac{\epsilon}{4}$.

Let $n \geq n_0$.
Then $d(x, A_n) < \frac{\epsilon}{4}$ by ii), so $d(x_n, A_n) < \frac{\epsilon}{2}$ by i).
Since $\mu_{x_n,n}(b_n) > 0$, we have that

$$d(x_n, b_n) < \frac{\epsilon}{2} + 2^{-n} < \frac{3\epsilon}{4}.$$

Consequently, $d(x, b_n) < \frac{3\epsilon}{4} + \frac{\epsilon}{4} = \epsilon$.
This ends the proof of the lemma. □

We will prove a combined embedding and density theorem for hierarchies of limit spaces using normed vector spaces at base level. We need:

**Lemma 9.2.** *For each $n$, let $V$ be a complete, normed vector space, and let $v \in V$. For each $n \in \mathbb{N}$, let $X_n \subset V$ be finite, and assume that $v = \lim_{n\to\infty} v_n$ whenever $v_n \in X_n$ for all $n \in \mathbb{N}$.*
*For each $n$, let $\mu_n$ be a probability distribution on $X_n$. Then*

$$v = \lim_{n\to\infty} \sum_{u \in X_n} \mu_n(u) \cdot u.$$

The proof is trivial.

**Definition 9.3.** Let $X$ be a sequential separable Hausdorff space. We say that $X$ *admits uniform probabilistic selection* if there is a probabilistic projection from $X$ to $X$.

**Definition 9.4.** Let $X$ be a metric space, and let $x = \lim_{n\to\infty} x_n$ from $X$. A *probabilistic modulus of convergency* for the sequence is a sequence $\{\nu_k\}_{k\in\mathbb{N}}$ of probability distributions on $\mathbb{N}$ such that for all $n$, $m$ and $k$ in $\mathbb{N}$, if

$$\nu_k(n) > 0$$

and $n \leq m$, then $d(x_n, x) < 2^{-k}$.

The proof of Lemma 9.1 is constructive in the following sense:

**Observation 9.1.** Let $Y$ be a complete metric space, and let $A = \bigcup_{n\in\mathbb{N}} A_n$ and $X$ be as in Lemma 9.1. Then, uniformly in $k$, we may compute a function $\nu_k$ mapping $x \in X$ into a probability distribution $\nu_k(x)$ on $\mathbb{N}$ such that for all $n$, $m$, $k$ in $\mathbb{N}$ and all $a \in A_m$, if $\nu_k(x)(n) > 0$, $n \leq m$ and $\mu_{x,m}(a) > 0$ then $d(a, x) < 2^{-k}$.

***Proof.*** That $\mu_{x,m}(a) > 0$ simply means that

$$d(x, A_m) + 2^{-m} > d(x, a)$$

and this means that $d(x, a) < 2^{-k}$ whenever $d(x, A_m) < 2^{-(k+1)}$ and $m > k$.
Let $r \in [1, \infty\rangle$ and let $r = m + \lambda$ where $m \in \mathbb{N}$ and $0 \leq \lambda < 1$.
Let

$$f_x(r) = (1 - \lambda)d(x, A_m) + \lambda d(x, A_{m+1}) + \frac{1}{r}.$$

Then $f_x$ is strictly decreasing with 0 as its limit value when $r \to \infty$, and $x, r \mapsto f_x(r)$ is continuous. Let $z = f_x^{-1}(2^{-k})$.
Then $z = n + \xi$ for some $n \in \mathbb{N}$ and $0 \leq \xi < n$.
Let $\nu_k(x)(n+1) = 1 - \xi$ and $\nu_k(x)(n+2) = \xi$.
This does the trick. □

In order to prove the combined density and embedding theorem, we need to extend our pool of concepts.

**Definition 9.5.** Let $X$ and $Y$ be sequential separable Hausdorff spaces, $\pi : X \to Y$ a topological embedding.
Let $\{A_n\}_{n\in\mathbb{N}}$ be a family of finite subsets of $X$, and for each $n \in \mathbb{N}$, let $y \mapsto \mu_{y,n}$ be a continuous map from $Y$ to the set of probability distributions on $A_n$.

a) We define the auxiliary equivalence relation $\sim$ on $Y$ by $y \sim z$ when

$$\forall n \in \mathbb{N}(\mu_{y,n} = \mu_{z,n}).$$

We will let $\overline{X}$ be a closed subset of $Y$ extending the image $\pi[X]$ such that

$$\forall y \in \overline{X} \exists x \in X(y \sim \pi(x)).$$

We call the sequence of maps $y \mapsto \mu_{y,n}$ a *probabilistic projection* with respect to $(X, \pi, \overline{X}, Y)$ if whenever $y \in \overline{X}$, $x \in X$, $y \sim \pi(x)$, $y = \lim_{n\to\infty} y_n$ and $a_n \in A_n$ such that $\mu_{y,n}(a_n) > 0$ for each $n \in \mathbb{N}$, then $x = \lim_{n\to\infty} a_n$.

b) A *control* will be a continuous map $y \mapsto h_y : \mathbb{R}_{\geq 0} \to \mathbb{R}_{\geq 0}$ such that
   i) Each $h_y$ is strictly increasing.
   ii) $h_y(0) = 0$ for all $y$.
   iii) If $y \in \overline{X}$ then $h_y$ is bounded by $\frac{1}{2}$.
   iv) If $y \notin \overline{X}$ then $h_y$ is unbounded.

We may use the terminology and notation introduced in this definition without explicit reference, when it is clear from the context that they apply.

Since $X$ is Hausdorff, if $y \in \overline{X}$ there is a unique $x_y \in X$ such that $y \sim \pi(x_y)$. From now on, we will always assume that $y \mapsto x_y$ is continuous, and we view this as a partial projection commuting with the embedding $\pi$.

We will apply these concepts to hierarchies of typed functionals, where the base types will be separable, complete, normed vector spaces over $\mathbb{R}$.

As an induction start, we will show that in the case of metric spaces, our construction can be modified in order to satisfy all the extra properties.
So let $X$ and $Y$ be complete metric spaces, and let $\pi : X \to Y$ be isometric. Then $\overline{X} = \pi[X]$ will be closed. Let $A = \bigcup_{n\in\mathbb{N}} A_n$ be as in the proof of Lemma 9.1.
We modify the construction of the probability distributions to this new situation, and we let

$$\mu_{y,n}(a) = \frac{d_Y(\pi[A_n], y) + 2^{-n} \dot{-} d(y, \pi(a))}{\sum_{b\in A_n} d_Y(\pi[A_n], y) + 2^{-n} \dot{-} d_Y(y, \pi(b))}.$$

We define the control

$$h_y(r) = \frac{1}{2}(1 - 2^{-r}) + r \cdot d_Y(y, \overline{X}).$$

It is easy, but tedious, to see that all properties are satisfied, partly based on the proof of Lemma 9.1.

Now, in order to create an induction step, we will assume that $X$, $Y$, $\pi : X \to Y$, $\{y \mapsto \mu_{y,n}\}_{n\in\mathbb{N}}$, and $y \mapsto h_y$ satisfy that $X$ and $Y$ are sequential Hausdorff spaces, $\pi$ is a topological embedding, $\{y \mapsto \mu_{y,n}\}_{n\in\mathbb{N}}$ is a corresponding probabilistic projection with control $y \mapsto h_y$ over $\overline{X} \subseteq Y$. We will let $\{A_n\}_{n\in\mathbb{N}}$ be the sequence of finite subsets of $X$ supporting each $\mu_{y,n}$

Let $U$ and $V$ be normed, complete separable vector spaces such that $U$ is a closed subspace of $V$.

We aim to construct an embedding $\pi^+$ of $X \to U$ into $Y \to V$ together with a corresponding probability projection and control. To this end, let $v \mapsto \delta_{v,n}$ be the probabilistic projection constructed above, when we see $U$ and $V$ as metric spaces. Let $B = \bigcup_{n\in\mathbb{N}} B_n$ be the countable set used as the basis for the construction of $\delta_{v.n}$ for $v \in V$.

Our first move will be to construct the embedding $\pi^+$.
So let $f : X \to U$ and $y \in Y$ be given.
We will define $\pi^+(f)(y)$ separately for two cases.

***Case* 1.** The control $h_y$ is bounded by $\frac{1}{2}$.

Then let

$$\pi^+(f)(y) = \lim_{n\to\infty} \sum_{a\in A_n} \mu_{y,n}(a) \cdot f(a).$$

We will prove later that this is well defined.

***Case* 2.** The control $h_y$ is unbounded.

Let $z_y = n_y + \lambda_y$ be the unique real such that $h_y(z_y) = 1$, where $n_y \in \mathbb{N}$ and $0 \leq \lambda_y < 1$. $y \mapsto z_y$ is continuous when we are in this case. Let

$$\pi^+(f)(y) = (1 - \lambda_y) \sum_{a \in A_{n_y}} \mu_{y,n_y} \cdot f(a) + \lambda_y \sum_{a \in A_{n_y+1}} \mu_{y,n_y+1}(a) \cdot f(a).$$

***Claim 1.*** $\pi^+(f)(y)$ is well defined.

***Proof of Claim 1.*** This is only a problem in Case 1, where we have to prove that the limit exists. In this case, $y \sim \pi(x)$ for some $x$, and thus, whenever $\{a_n\}_{n \in \mathbb{N}}$ is a sequence from $\prod_{n \in \mathbb{N}} A_n$ such that $\mu_{y,n}(a_n) > 0$ for each $n$, we have that $x = \lim_{n \to \infty} a_n$, so $f(x) = \lim_{n \to \infty} f(a_n)$.
Then, using the compactness of $\prod_{n \in \mathbb{N}} A_n$, we see that

$$\forall \epsilon > 0 \exists n \forall m \geq n \forall a \in A_m(\mu_{y,m}(a) > 0 \Rightarrow d(f(x), f(a)) < \epsilon).$$

Given $\epsilon > 0$ and $n$ as above, for $m \geq n$ we have that

$$\begin{aligned} &d(f(x), \textstyle\sum_{a \in A_m} \mu_{y,m}(a) \cdot f(a)) \\ &\leq \textstyle\sum_{a \in A_m} d(\mu_{y,m}(a) \cdot f(x), \mu_{y,m}(a) \cdot f(a)) \\ &= \textstyle\sum_{a \in A_m} \mu_{y,m}(a) d(f(x), f(a)) \leq \epsilon, \end{aligned}$$

and we are through, the limit is $f(x)$.

***Claim 2.*** $\pi^+$ is continuous.

***Proof of Claim 2.*** We have to prove that $\pi^+$ is sequentially continuous, so let $f : X \to U$, $f_k : X \to U$ for each $k$, $y \in Y$ and $y_k \in Y$ for each $k$ such that

$$f = \lim_{k \to \infty} f_k$$

and

$$y = \lim_{k \to \infty} y_k.$$

We must show that $\pi^+(f)(y) = \lim_{k \to \infty} \pi^+(f_k)(y_k)$.

If $y \notin \overline{X}$, then for sufficiently large $k$ we have that $y_k \notin \overline{X}$, and in this case it is easily seen that the construction is continuous. It is actually computable by construction.

So assume that $y \in \overline{X}$. Then by assumption, if $a_k \in A_k$ with $\mu_{y_k,k}(a_k) > 0$ for each $k$, then $x = \lim_{k \to \infty} a_k$.
Since $f = \lim_{k \to \infty} f_k$ we have that

$$f(x) = \lim_{k \to \infty} f_k(a_k)$$

under the same assumptions, where $x \in X$ is the unique object such that $\pi(x) \sim y$. Then, using the same argument we used to show that $\pi^+(f)(y)$ is well defined for $y \in \overline{X}$, we have that

$$\pi^+(f)(y) = f(x) = \lim_{k\to\infty} \sum_{a\in A_k} \mu_{y_k,k}(a) \cdot f_k(a).$$

Since the sequences $\{y_k\}_{k\in\mathbb{N}}$ and $\{f_k\}_{k\in\mathbb{N}}$ are arbitrary, it also holds that

$$\pi^+(f)(y) = \lim_{n\to\infty} \sum_{a\in A_n} \mu_{y_{k_n},n}(a) \cdot f_{k_n}(a)$$

when $n \mapsto k_n$ is increasing and unbounded, but not necessarily strictly increasing.

By this remark, it is sufficient to prove that

$$\pi^+(f)(y) = \lim_{k\to\infty} \pi^+(f_k)(y_k)$$

for the two cases where all $y_k$ are in $\overline{X}$ and where none of them are in $\overline{X}$.

In the first case, we use that

$$\pi^+(f_k)(y_k) = \lim_{n\to\infty} \sum_{a\in A_n} \mu_{y_k,n}(a) \cdot f_k(a)$$

for each $k$. We let $n_1$ be such that

$$d\Big(\sum_{a\in A_m} \mu_{y_1,m}(a) \cdot f_1(a), \pi^+(f_1)(y_1)\Big) < \frac{1}{2}$$

for all $m \geq n_1$.

Then we let $n_2 > n_1$ be such that

$$d\Big(\sum_{a\in A_m} \mu_{y_2,m}(a) \cdot f_2(a), \pi^+(f_2)(y_2)\Big) < \frac{1}{4}$$

for all $m \geq n_2$, and so on.

We then slow down the sequence $\{k\}_{k\in\mathbb{N}}$ to the sequence $\{k_i\}_{i\in\mathbb{N}}$ by letting it be 1 until $i = n_1$, then letting it be 2 until $i = n_2$ and so on. Then

$$\pi^+(f)(y) = \lim_{i\to\infty} \sum_{a\in A_i} \mu_{y_{k_i},i}(a) \cdot f_{k_i}(a) = \lim_{k\to\infty} \pi^+(f_k)(y_k)$$

by construction.

In the second case, we just make use of the fact that

$$\lim_{k\to\infty} z_{y_k} = \infty$$

and the result follows from the limit space property.

This ends the construction of $\pi^+$ and the proof of its continuity, i.e. the proof of Claim 2.

Now, let $\overline{X \to U} = \{g : Y \to V \mid \forall a \in A(g(\pi(a)) \in U)\}$.
Then of course $\forall x \in X(g(\pi(x)) \in U)$ when $g \in \overline{X \to U}$, and $g$ may be projected to

$$\lambda x \in X g(\pi(x)) \in X \to U.$$

We must wait for the definition of the probabilistic projection before making further sense of this.
Our control will be

$$h_g(n+\lambda) = (1-\lambda)\cdot\sum_{i=0}^{n} d(g(\pi(a_i)), U) + \lambda\cdot\sum_{i=0}^{n+1} d(g(\pi(a_i)), U) + \frac{1}{2}(1-2^{-(n+\lambda)})$$

when $g : Y \to V$ is continuous.

We will now construct the finite sets supporting the probabilistic projection and prove the required properties.
Recall that $\{A_n\}$ and $\{B_n\}_{n\in\mathbb{N}}$ are the supports of the probabilistic projections of $Y$ to $X$ and of $V$ to $U$ resp.

Let $C_n^* = A_n \to B_n$, and let $\phi \in C_n^*$.
Let $f_\phi : X \to U$ be defined by

$$f_\phi(x) = \sum_{a\in A_n} \mu_{\pi(x),n}(a) \cdot \phi(a).$$

Let $C_n = \{f_\phi \mid \phi \in C_n^*\}$.

Now we will define the probabilistic projection $\mu^+$:
Let $g : Y \to V$ and let $f_\phi \in C_n^*$. Let

$$\mu^+_{g,n}(f_\phi) = \prod_{a\in A_n} \delta_{g(\pi(a)),n}(f(a)).$$

Since we construct $\mu^+$ using products of probability distributions, $\mu^+_{g,n}$ is itself a probability distribution.

Clearly, if $g \in \overline{X \to U}$, then

$$g \sim \pi^+(\lambda x \in X.g(\pi(x)))$$

since these two functions will be identical on $\overline{X}$, and then in particular on the $\pi$-image of $A$.
So assume that $g \in \overline{X \to U}$ and that $g = \lim_{n\to\infty} g_n$.

Assume further that $\mu^+_{g_n,n}(f_{\phi_n}) > 0$ for each $n$.
We must show that

$$\lambda x \in X.g(\pi(x)) = \lim_{n\to\infty} f_{\phi_n}.$$

Let $x \in X$ and $x = \lim_{n\to\infty} x_n$. We must show that

$$(*) \quad g(\pi(x)) = \lim_{n\to\infty} f_{\phi_n}(x_n).$$

We have that

$$f_{\phi_n}(x_n) = \sum_{a\in A_n} \mu_{\pi(x_n),n}(a) \cdot \phi_n(a).$$

Let $a_n \in A_n$ for each $n \in \mathbb{N}$.
Then, since $\mu^+_{g_n,n}(f_{\phi_n}) > 0$ we must have that

$$\delta_{g_n(\pi(a_n)),n}(\phi_n(a_n)) > 0$$

since the product probability would be zero otherwise.

Now, let us restrict ourselves to the sequences $\{a_n\}_{n\in\mathbb{N}}$ such that for all $n$ we have that $\mu_{x_n,n}(a_n) > 0$.
Then $x = \lim_{n\to\infty} a_n$ by assumption, so

$$g(\pi(x)) = \lim_{n\to\infty} g_n(\pi(a_n)).$$

But then $g(\pi(x)) = \lim_{n\to\infty} \phi_n(a_n)$ since $\phi_n$ has positive $g_n(\pi(a_n))$-probability for each $n$.

Then we finally apply Lemma 9.2 to see that $(*)$ holds.

We have now established the base case and the induction step needed to prove the following theorem:

**Theorem 9.1.** *Let $U_1, \ldots, U_n$ be complete and separable normed vector spaces that are subspaces of one space $V$. Let $\sigma$ be a type term in the type variables $u_1, \ldots, u_n$ and let $\sigma(U_1, \ldots, U_n)$ be the canonical interpretation in the category of limit spaces.*

*a) Each space $\sigma(U_1, \ldots, U_n)$ contains a dense countable set admitting probabilistic selection.*
*b) There are topological embeddings of $\sigma(U_1, \ldots, U_n)$ into $\sigma(V, \ldots, V)$ commuting with application and admitting probabilistic projections.*

**Remark 9.3.** Adding some notational details, we might include $\mathbb{N}$ in the set of base types. Then this theorem generalizes one of the theorems in [16]. In addition to be a generalization, the proof is carried out in the setting of limit spaces only, not using domain theory as we did in [16]. In many respects, we view the method of proof as important as the result itself.

## 9.4. Domain Representations and Density

When one works with an external notion of computability, i.e. transfers the notion from some representing structure, it is natural to try to establish results about how the representing structure relates to the space in question. One example is representations by domains and density theorems. In its simplest form, a density theorem states that the representing objects are dense in the underlying domain. In this section we will show that a density theorem for the standard domain representation of the set of continuous functions from one separable metric space to another will have strong implications.

There are certainly cases where we have an effective enumeration of a dense set, but where the canonical domain representation does not satisfy density. One lesson to learn is that the internal approach to density implicit in Section 3 may be as useful as using external approaches. By the way, all constructions in this section must be considered as extensional.

**Definition 9.6.** Let $X$ and $Y$ be topological spaces.
We say that $Y$ is *compactly saturated* over $X$ if whenever $C \subseteq E$ are compact subsets of $X$ and $g : C \to Y$ is continuous, then $g$ can be extended to a continuous $f : E \to Y$.

**Remark 9.4.** This property indicates that $Y$ in some sense is globally as connected as $X$ will be locally. What this means more precisely remains to be explored, but, for example, if there is a nontrivial path in $X$, then $Y$ will be path connected. Similar phenomena for higher dimensions indicate that "local lack of multidimensional holes" in $X$ implies the corresponding global lack of holes in $Y$.

In this section we will show that if $X$ and $Y$ are two separable complete metric spaces with domain representations $D_X$ and $D_Y$ in the style of Blanck [1], and the total objects are dense in the domain $D_X \to D_Y$, then $Y$ is compactly saturated over $X$.

We have to introduce some terminology:
Let $\langle X, d_X \rangle$ and $\langle Y, d_Y \rangle$ be the two spaces.
Let $\{\xi_k^X\}_{k \in \mathbb{N}}$ and $\{\xi_l^Y\}_{l \in \mathbb{N}}$ be prefixed enumerations of dense subsets of $X$ and $Y$ resp.

For $r \in \mathbb{Q}^+$, let $B^X_{k,r}$ and $B^Y_{l,r}$ be the open balls around $\xi^X_k$ and $\xi^Y_l$ resp. with radius $r$. Let $\overline{B}^X_{k,r}$ and $\overline{B}^Y_{l,r}$ be the corresponding closed balls.
The finitary objects in the domain representation used by Blanck [1] will be finite sets of such closed balls such that the distance between two centers do not exceed the sum of the corresponding radii. Thus we use finite sets of closed balls such that it is consistent that the intersection is nonempty. One finite set $\{B_1, \ldots, B_n\}$ is below another set $\{B'_1, \ldots, B'_m\}$ in the domain ordering if each $B'_i$ is formally the subset of some $B_j$, formally by comparing the radii and the distances between the centers.
The domains used are then the ideal completion of this ordering, and an ideal represents an element $x$ of $X$ (or $Y$) if $x$ is an element of all closed balls of the ideal, and any open ball around $x$ contains at least all closed ball in one of the elements of the ideal. The *total* ideals will be those representing elements in $X$ or $Y$, and an object in the domain $D_X \to D_Y$ is total if it maps total ideals to total ideals.

**Theorem 9.2.** *With the terminology above, if the total objects are dense in $D_X \to D_Y$, then $Y$ is compactly saturated over $X$.*

We will prove the theorem in several steps. Let $C \subseteq E$ and $g : C \to Y$ be as in the definition of compactly saturated. The rough idea is to use the density property to construct a sequence of continuous maps $f_n : X \to Y$ such that $\lim_{n\to\infty} f_n(x)$ exists on $E$ and will be a continuous extension of $g$ to $E$.

Let $\{B^X_k\}_{k\in\mathbb{N}}$ and $\{B^Y_l\}_{l\in\mathbb{N}}$ be enumerations of the two chosen bases for $X$ and $Y$, using $i_k, r_k$ and $j_l, s_l$ to denote the index of the center and the radius of each ball.
Throughout, we will let $D = g[C]$, the image of $C$ under $g$. $D$ is a compact subset of $Y$.

We will first construct a set $G$ of pairs of natural numbers such that

$$\{\langle \overline{B}^X_k, \overline{B}^Y_l \rangle \mid \langle k, l\rangle \in G\}$$

generates an ideal approximating $g$.
By recursion, we will define the finite set $G_n$ and the real $\delta_n > 0$.
We will refer to $G_n$ and to $\delta_n$ in the main construction. In the construction of the sequence of $G_n$s, we will also define the auxiliary $\Delta_n \subseteq \mathbb{N}$, and some other entities only needed in order to explain the recursion step.

Let $G_0 = \emptyset$, $\delta_0 = 1$ and $\Delta_0 = \emptyset$.
Assume that $G_n$, $\delta_n$ and $\Delta_n$ are constructed.

First, let $\Delta_{n+1} \subseteq \mathbb{N}$ be a finite set such that $\{B_l^Y \mid l \in \Delta_{n+1}\}$ is and open covering of $D$ and such that $s_l < 2^{-n}$ whenever $l \in \Delta_{n+1}$.
Let $\Pi_{n+1} = \{g^{-1}(\overline{B}_l^Y) \mid l \in \Delta_{n+1}\}$.
If $A, B \in \Pi_{n+1}$ and $A \cap B = \emptyset$, then $d_X(A, B) > 0$, since the sets $A$ and $B$ are closed subsets of $C$, and thus compact.
If $A \in \Pi_{n+1}$ and $B = \overline{B}_k^X$ for some $\langle k, l\rangle \in G_n$ and $\forall x \in A(d_X(x, \xi_{i_k}^X) > r_k)$, then

$$\inf\{d_X(x, \xi_{i_k}^X) - r_k \mid x \in A\} > 0.$$

This is because $A$ is compact. We call the latter number the *formal distance from $A$ to $\overline{B}_k^X$*.
Let $\pi_{n+1}$ be the least of the distances between disjoint elements of $\Pi_{n+1}$ and the formal distances between an element of $\Pi_{n+1}$ and a closed ball $\overline{B}_k^X$ for some $\langle k, l\rangle \in G_n$ as above.

Let $\Gamma_{n+1} \subset \mathbb{N}$ be a finite set such that

If $k \in \Gamma_{n+1}$, then $r_k < min\{\frac{\pi_{n+1}}{4}, \frac{\delta_n}{4}\}$.
If $k \in \Gamma_{n+1}$, then $B_k^X \cap C \neq \emptyset$
$\{B_k^X \mid k \in \Gamma_{n+1}\}$ is an open covering of $C$.
If $k \in \Gamma_{n+1}$, then $\overline{B}_k^X \cap C \subseteq g^{-1}(\overline{B}_l^Y)$ for some $l \in \Delta_{n+1}$.

The compactness of $C$ and the fact that $\{g^{-1}(B_l^Y) \mid l \in \Delta_{n+1}\}$ is an open covering of $C$ ensures that there exists a set $\Gamma_{n+1}$ like this.

We now let $\langle k, l\rangle \in G_{n+1}$ if $k \in \Gamma_{n+1}$, $l \in \Delta_{n+1}$ and $\overline{B}_k^X \cap C \subseteq g^{-1}(\overline{B}_l^Y)$. We let

$$\delta_{n+1} = d_X(C, X \setminus \bigcup\{B_k^X \mid k \in \Gamma_{n+1}\}).$$

By construction, $0 < \delta_{n+1} \leq \frac{\delta_n}{4}$.

Let $G = \cup_{n\in\mathbb{N}} G_n$.
Let $\hat{G}_n = \{\langle \overline{B}_k^X, \overline{B}_l^Y\rangle \mid \langle k, l\rangle \in G_n\}$ and let $\hat{G} = \cup_{n\in\mathbb{N}}\hat{G}_n$.

***Claim 1.*** Each $\hat{G}_n$ is a consistent set.

***Proof of Claim 1.*** Let $\langle k, l\rangle$ and $\langle k', l'\rangle$ be in $G_n$ such that

$$d_X(\xi_{i_k}^X, \xi_{i_{k'}}^X) \leq r_k + r_{k'}.$$

This is the formal consistency requirement on $\{\overline{B}_k^X, \overline{B}_{k'}^X\}$.
Since $\overline{B}_k^X \cap C$ is nonempty and contained in $g^{-1}(\overline{B}_l^Y) \in \Pi_{n+1}$ and $\overline{B}_{k'}^X \cap C$

is nonempty and contained in $g^{-1}(\overline{B}_{l'}^Y)$, there is $x \in g^{-1}(\overline{B}_l^Y)$ and $x' \in g^{-1}(\overline{B}_{l'}^Y)$ with

$$d_X(x, x') \le 2r_k + 2r_{k'} < \pi_{n+1}.$$

Then, by choice of $\pi_{n+1}$, we have that

$$g^{-1}(\overline{B}_l^Y) \cap g^{-1}(\overline{B}_{l'}^Y) \ne \emptyset,$$

and

$$\overline{B}_l^Y \cap \overline{B}_{l'}^Y \cap D \ne \emptyset.$$

Then

$$d_Y(\xi_{k_l}^Y, \xi_{k_{l'}}^Y) \le s_l + s_{l'},$$

which is the consistency requirement for $\{\overline{B}_l^Y, \overline{B}_{l'}^Y\}$.
Thus the right hand sides of two pairs in $\hat{G}_n$ are consistent when the left hand sides are consistent. Thus $\hat{G}_n$ is consistent, and Claim 1 is proved.

***Claim 2.*** $\hat{G}$ is consistent.
***Proof of Claim 2.*** Let $\langle \overline{B}_k^X, \overline{B}_l^Y \rangle \in \hat{G}_n$ and $\langle \overline{B}_{k'}^X, \overline{B}_{l'}^Y \rangle \in \hat{G}_m$ with $m < n$ and assume that

$$d_X(\xi_{i_k}^X, \xi_{i_{k'}}^X) \le r_k + r_{k'}.$$

Then we can argue in analogy with the proof of Claim 1 that there will be an

$$x \in g^{-1}(\overline{B}_l^Y) \cap \overline{B}_{k'}^X.$$

Since $x \in g^{-1}(\overline{B}_l^Y)$ and $x \in \overline{B}_{k'}^X \cap C \subseteq g^{-1}(\overline{B}_{l'}^Y)$ we have that

$$\overline{B}_l^Y \cap \overline{B}_{l'}^Y \ne \emptyset,$$

so

$$d_Y(\xi_{j_l}^Y, \xi_{j_{l'}}^Y) \le s_l + s_{l'}.$$

This ends the proof of Claim 2.

By construction, each $\hat{G}_n$ is an approximation to $g$ where $\hat{G}_n$ "decides" the value of $g(x)$ up to a precision of $2^{-n}$. Thus $\hat{G}$ will generate an ideal representing $g$.

The reason why we use tiny left hand sides as well is that we want to have some freedom in adjusting the construction of the extension $f$ of $g$ close to $C$.

We will now give the main construction. For each finitary $p : D_X \to D_Y$, i.e. finite consistent sets of pairs of closed balls, we let $f_p$ be a continuous, total extension.
We will construct a sequence $\{p_n\}_{n \in \mathbb{N}}$ of finitary objects, and we will let $f_n = f_{p_n}$.
$p_n$ will satisfy that $\cup_{i \le n} \hat{G}_n \sqsubseteq p_n$, which will ensure that we construct some extension of $g$.

For each $n$, let

$$C_n = \{x \in X \mid d_X(x, C) < \delta_n\}$$

and let $E_n = E \setminus C_n$. Let $p_0 = \emptyset$ and $f_0 = f_{p_0}$.
Assume that $p_n$ and $f_n$ are constructed, with $\cup_{i \le n} \hat{G}_i \subseteq p_n$.
Let

$$Y_n = \bigcup \{f_i[E] \mid i \le n\}.$$

Then $Y_n \subseteq Y$ is a compact set.
First, let $k_n$ be minimal such that

$$K_n = \{B_l^Y \mid l \le k_n \wedge s_l \le 2^{-n}\}$$

is an open covering of $Y_n$. Clearly $n \le m \Rightarrow k_n \le k_m$.
Let $L_n$ consist of all nonempty sets $f_n^{-1}(\overline{B}_l^Y) \cap E_n$ for $l \le k_n$. Let $\varepsilon_n$ be the least element in the set consisting of $\delta_n$, all distances between disjoint elements of $L_n$ and all formal distances between disjoint elements of $L_n$ and sets $\overline{B}_k^X$ for some $\langle k, l \rangle \in \bigcup_{i \le n} G_i$.
Let $m_n$ be the least number such that

$$M_n = \{B_k^X \mid k \le m_n \wedge r_k < \frac{\varepsilon_n}{4}\}$$

contains an open covering of each $A \in L_n$.

We then let $p_{n+1}$ consist of $\cup_{i \le n+1} \hat{G}_i$ together with the set $p'_{n+1}$ of all pairs $\langle \overline{B}_k^X, \overline{B}_l^Y \rangle$ such that $l \le k_n$, $\overline{B}_k^X \in M_n$ and

$$\overline{B}_k^X \cap f_n^{-1}(\overline{B}_l^Y) \cap E_n \neq \emptyset.$$

***Claim 3.*** $p_{n+1}$ is a consistent set.

***Proof of Claim 3.*** We have to prove that $p'_{n+1}$ is consistent with itself, with $\hat{G}_{n+1}$ and with $\cup_{i \le n} \hat{G}_i$.
Let $\langle \overline{B}_k^X, \overline{B}_l^Y \rangle$ and $\langle \overline{B}_{k'}^X, \overline{B}_{l'}^Y \rangle$ be elements of $p'_{n+1}$ such that

$$d_X(\xi_{i_k}^X, \xi_{i_{k'}}^X) \le r_k + r_{k'}.$$

Let $x \in \overline{B}_k^X \cap f_n^{-1}(\overline{B}_l^Y) \cap E_n$ and $x' \in \overline{B}_{k'}^X \cap f_n^{-1}(\overline{B}_{l'}^Y) \cap E_n$.
Then $d_X(x, x') \leq 2r_k + 2r_{k'} < \varepsilon_n$.
Thus $f_n^{-n}(\overline{B}_l^Y) \cap E_n$ and $f_n^{-1}(\overline{B}_{l'}^Y) \cap E_n$ intersect, so $\overline{B}_l^Y \cap \overline{B}_{l'}^Y \neq \emptyset$. Then

$$d_Y(\xi_{j_l}^Y, \xi_{j_{l'}}^Y) \leq s_l + s_{l'}.$$

This shows that $p'_{n+1}$ is consistent.

Now, let $\langle \overline{B}_k^X, \overline{B}_l^Y \rangle \in \hat{G}_{n+1}$ and $\langle \overline{B}_{k'}^X, \overline{B}_{l'}^Y \rangle \in p'_{n+1}$.
Since $\overline{B}_{k'}^X \cap C \neq \emptyset$ and $\overline{B}_{k'}^X \cap E_n \neq \emptyset$ and they both have radii $\leq \frac{\varepsilon_n}{4} \leq \frac{\delta_n}{4}$ the set $\overline{B}_k^X$ and $\overline{B}_{k'}^X$ must be formally disjoint, since we do not leap over the half distance between $C$ and $E_n$ in any of these balls.
Finally, let $\langle \overline{B}_k^X, \overline{B}_l^Y \rangle \in \hat{G}_i$ for some $i \leq n$ and $\langle \overline{B}_{k'}^X, \overline{B}_{l'}^Y \rangle \in p'_{n+1}$, and assume that

$$d_X(\xi_{i_k}^X, \xi_{i_{k'}}) \leq r_k + r_{k'}.$$

Now

$$A = f_n^{-1}(\overline{B}_{l'}^Y) \cap E_n \in L_n,$$

and since the distance between $\xi_{i_k}^X$ and $A$ is bounded by $r_k + 2r_{k'} < r_k + \varepsilon_n$, the formal distance between $A$ and $\overline{B}_k^X$ is $< \varepsilon_n$ Then $A \cap \overline{B}_k^X \neq \emptyset$.
Let $x \in A \cap \overline{B}_k^X$.
Since $\langle \overline{B}_k^X, \overline{B}_l^Y \rangle \in \hat{G}_i \subseteq p_n$, we will have that $f(x) \in \overline{B}_l^Y \cap \overline{B}_{l'}^Y$, so $d_Y(\xi_{j_l}^Y, \xi_{j_{l'}}^Y) \leq s_l + s_{l'}$.
This ends the proof of Claim 3.

***Claim 4***. Let $\langle \overline{B}_k^X, \overline{B}_l^Y \rangle \in p_n$ and let $m \geq n$.

a) If $\langle \overline{B}_k^X, \overline{B}_l^Y \rangle \in \cup_{i \leq n} G_n$ then

$$\overline{B}_k^X \subseteq f_m^{-1}(\overline{B}_l^Y).$$

b) If $\langle \overline{B}_k^X, \overline{B}_l^Y \rangle \in p'_n$, then

$$\overline{B}_k^X \cap E_n \subseteq f_m^{-1}(\overline{B}_l^Y).$$

***Proof of Claim 4***. a) is trivial since $\langle \overline{B}_k^X, \overline{B}_l^Y \rangle \in p_m$ for $m \geq n$.
b) is proved by induction on $m \geq n$ where the base case $n = m$ is trivial.
So let $x \in \overline{B}_k^X \cap E_n$ and assume as an induction hypothesis that $f_m(x) \in \overline{B}_l^Y$ where $l \leq k_n$.

We will show that $f_{m+1}(x) \in \overline{B}_l^Y$.
We have that $l \leq k_m$, so

$$f_m^{-1}(\overline{B}_l^Y) \cap E_m \in L_m$$

with

$$x \in f_m^{-1}(\overline{B}_l^Y) \cap E_m.$$

Then there will be a $\overline{B}_{k'}^X \in M_m$ such that $x \in B_{k'}^X$.
Then $\langle \overline{B}_{k'}^X, \overline{B}_l^Y \rangle \in p_{m+1}$, and consequently $f_{m+1}(x) \in \overline{B}_l^Y$.
This ends the proof of Claim 4.

***Claim 5***. For each $x \in E$, $\lim_{n\to\infty} f_n(x)$ exists.

***Proof of Claim 5***. We split the argument into two cases.
If $x \in C$, then $g(x) = \lim_{n\to\infty} \hat{G}_n(x)$ (in the sense of domains), and $\hat{G}_n(x) \sqsubseteq f_n(x)$ for each $n$, so

$$g(x) = \lim_{n\to\infty} f_n(x).$$

If $x \in E \setminus C$, let $\epsilon > 0$ be given. We choose $n$ so large that $2^{-n} < \epsilon$ and $x \in E_n$.
Then $f_n(x) \in Y_n$ and we pick one $B_l^Y \in K_n$ such that $f_n(x) \in B_l^Y$.
Then $x \in f_n^{-1}(\overline{B}_l^Y) \cap E_n \in L_n$ and then there is a $B_k^X \in M_n$ such that $x \in B_k^X$.
By construction then

$$\langle \overline{B}_k^X, \overline{B}_l^Y \rangle \in p'_{n+1}.$$

By Claim 4b), $f_m(x) \in \overline{B}_l^Y$ for all $m \geq n$.
This shows that $\{f_n(x)\}_{n\in\mathbb{N}}$ is a Cauchy sequence, so the limit exists.
This ends the proof of Claim 5.

Let $f(x) = \lim_{n\to\infty} f_n(x)$.

***Claim 6***. $f$ is continuous on $E$.

***Proof of Claim 6***. Let $x \in E$ and $\epsilon > 0$ be given.
If $x \in C$, choose $n$ so large that $2^{-n} < \epsilon$.
By construction of $\hat{G}_n$ there will be a pair $\langle \overline{B}_k^X, \overline{B}_l^Y \rangle \in \hat{G}_{n+1}$ such that $x$ is in the interior of $\overline{B}_k^X$.
By Claim 4a) we have that $f(y) \in \overline{B}_l^Y$ whenever $y \in \overline{B}_k^X$.
Thus, if $\delta > 0$ is so small that the open $\delta$-ball around $x$ is contained in $\overline{B}_k^X$, and then $d_Y(f(x), f(y)) \leq 2^{-n} < \epsilon$. This shows continuity in this case.

If $x \in E \setminus C$ we in addition choose $n$ such that $x \in E_n$.
We then use $p'_{n+1}$ to the same effect as we used $\hat{G}_{n+1}$ above, now applying Claim 4b).
This ends the proof of Claim 6 and the theorem.

*Concluding remarks*

This chapter can be seen as part two of a trilogy, where [18] is the first part. The third part is still in preparation [a]. There we will use external methods combined with the probabilistic approach in order to learn more about the typed hierarchy of total functionals over separable complete metric spaces in general.

## Acknowledgements

I am grateful to the anonymous referee for helpful suggestions on the exposition, and for pointing out several typos in the original manuscript.

## References

[1] J. Blanck, Domain representability of metric spaces, *Ann. Pure Appl. Logic.* **8**, 225–247, (1997).
[2] A. Edalat and P. Sünderhauf, A domain-theoretic approach to real number computation, *Theoret. Comput. Sci.* **210**, 73–98, (1998).
[3] M. H. Escardó, PCF extended with real numbers, *Theoret. Comput. Sci.* **162**(1), 79–115, (1996).
[4] M. H. Escardó, M. Hofmann, and T. Streicher, On the non-sequential nature of the interval-domain model of exact real number computation, *Math. Structures Comput. Sci.* **14**(6), 803–817, (2004).
[5] M. H. Escardó, Exhaustible sets in higher-type computation, *Logical Methods in Computer Science.* **4**(3), paper 3, (2008).
[6] J. E. Fenstad, *General Recursion Theory. An Axiomatic Approach.* Springer-Verlag, Berlin–Heidelberg, (1980).
[7] T. Grilliot, On effectively discontinuous type-2 objects, *J. Symbolic Logic.* **36**, 245–248, (1971).
[8] S. C. Kleene, Recursive functionals and quantifiers of finite types I, *Trans. Amer. Math. Soc.* **91**, 1–52, (1959).
[9] S. C. Kleene. Countable functionals. In ed., A. Heyting, *Constructivity in Mathematics*, pp. 81–100. North-Holland, Amsterdam, (1959).
[10] G. Kreisel. Interpretation of analysis by means of functionals of finite type. In ed. A. Heyting, *Constructivity in Mathematics*, pp. 101–128. North-Holland, (1959).

[a] Added in proof. The third part has now appeared in [19].

[11] G. Kreisel. Some reasons for generalizing recursion theory. In eds. R. O. Gandy and C. E. M. Yates, *Logic Colloquium 1969*, pp. 139–198. North-Holland, Amsterdam, (1971).

[12] M. Menni and A. Simpson, The largest topological subcategory of countably-based equilogical spaces, *Electr. Notes Theor. Comput. Sci.* (1999).

[13] R. Milner, Fully abstract models for typed $\lambda$-calculi, *Theoret. Comput. Sci.* **4**, 1–22, (1977).

[14] Y. N. Moschovakis. Axioms for computation theories - first draft. In eds., R. O. Gandy and C. E. M. Yates, *Logic Colloquium 1969*, pp. 199–255. North-Holland, Amsterdam, (1971).

[15] D. Normann. External and internal algorithms on the continuous functionals. In ed., G. Metakides, *Patras Logic Symposion*, pp. 137–144. North-Holland, Amsterdam, (1982).

[16] D. Normann, Comparing hierarchies of total functionals, *Logical Methods in Computer Science.* **1**(2), paper 4, (2005). Available at: `http://www.lmcs-online.org/ojs/viewarticle.php?id=95&layout=abstract` [Accessed October 2010]

[17] D. Normann, Computing with functionals – computability theory or computer science?, *Bull. Symbolic Logic.* **12**(1), 43–59, (2006).

[18] D. Normann. Internal density theorems for hierarchies of continuous functionals. In *CiE 2008: Proceedings of the 4th conference on Computability in Europe (Athens, Greece)*, pp. 467–475. Springer-Verlag, Berlin–Heidelberg, (2008).

[19] D. Normann, A rich hierarchy of functionals of finite types, *Logical Methods in Computer Science.* **5**(3), paper 11, (2009).

[20] G. Plotkin, LCF considered as a programming language, *Theoret. Comput. Sci.* **5**, 223–255, (1977).

[21] G. E. Sacks, *Higher Recursion Theory.* Springer-Verlag, Berlin–Heidelberg, (1990).

[22] M. Schröder. Admissible representations of limit spaces. In eds., J. Blanck, V. Brattka, P. Hertling, and K. Weihrauch, *Computability and Complexity in Analysis*, vol. 237, pp. 369–388. Informatik Berichte, volume 237, (2000).

[23] D. Scott. A type-theoretical alternative to ISWIM, CUCH, OWHY. Unpublished notes. Oxford, (1969).

[24] D. Scott, A type-theoretical alternative to ISWIM, CUCH, OWHY, *Theoret. Comput. Sci.* **121**, 411–440, (1993).

[25] V. Stoltenberg-Hansen and J. V. Tucker, Complete local rings as domains, *J. Symbolic Logic.* **53**, 603–624, (1988).

[26] J. V. Tucker and J. I. Zucker, Abstract versus concrete computation on metric partial algebras, *ACM Transactions on Computational Logic.* **5**, 611–668, (2004).

[27] K. Weihrauch, *Computable analysis.* Texts in Theoretical Computer Science, Springer Verlag, Berlin–Heidelberg, (2000).

# Chapter 10

# Recursive Functions: An Archeological Look

Piergiorgio Odifreddi
*Dipartimento di Matematica*
*Università degli Studi di Torino*
*10123, Torino, Italy*
*E-mail: piergiorgio.odifreddi@gmail.com*

## Contents



First of all, a disclaimer. I am not a historian. My interest in the development of Recursion Theory is not academic, but cultural. I want to know if and how the basic ideas and methods used in a restricted area of Logic derive from, or at least interact with, a wider mathematical and intellectual experience. I can only offer suggestions, not scholarly arguments, to those who share my interest.

For convenience, I refer to my book *Classical Recursion Theory* (see [19, 20]), CRT for short, for unexplained notation, references, and background.

## 10.1. Types of Recursion

The recursive functions take their name from the process of "recurrence" or "recursion", which in its most general numerical form consists in defining the value of a function by using other values of the same function. There are many different types of recursions, and among them the following are perhaps the most basic:

### 10.1.1. *Iteration*

The simplest type of recursion occurs when a given function is iterated. Technically, the $n$-th iteration of a function $f$ is defined as follows:

$$f^{(0)}(x) = x \quad f^{(n+1)}(x) = f(f^{(n)}(x)).$$

The first clause is needed to obtain $f^{(1)}(x) = f(x)$ from the second clause.

One of the earliest examples of iteration comes from the Rhind Papyrus, written about 1700 B.C., which gives as Problem 79 the following:

> In each of 7 houses are 7 cats; each cat kills 7 mice; each mouse would have eaten 7 ears of spelt (wheat); each ear of spelt would have produced 7 hekat (half a peck) of grain. How much grain is saved by the 7 house cats?

The solution amounts to computing the sixth term of a geometrical progression with first term 1 and multiplier 7, i.e. $f^{(6)}(7)$, with $f(x) = 7x$. The papyrus gives not only the correct answer (16,807), but also the sum of the first five terms of the progression (19,607).

A similar use of a geometrical progression comes from a medieval story about the origin of chess:

> According to an old tale, the Grand Vizier Sissa Ben Dahir was granted a boon for having invented chess for the Indian King, Shirham.
>
> Sissa addressed the King: "Majesty, give me a grain of wheat to place on the first square of the board, and two grains of wheat to place on the second square, and four grains of wheat to place on the third, and eight grains of wheat to place on the fourth, and so on. Oh, King, let me cover each of the 64 squares of the board."
>
> "And is that all you wish, Sissa, you fool?" exclaimed the astonished King.
>
> "Oh, Sire," Sissa replied, "I have asked for more wheat than you have in your entire kingdom. Nay, for more wheat that there is in the whole world, truly, for enough to cover the whole surface of the earth to the depth of the twentieth part of a cubit."[a]

[a]Reported in Newman [18].

Some version of the story was known to Dante, since he refers to it in the *Paradiso* (XXVIII, 92–93) to describe the abundance of Heaven's lights:

*eran tante, che 'l numero loro*
*più che 'l doppiar degli scacchi s'immilla.*
They were so many, that their number
piles up faster than the chessboard doubling.

As in the previous Egyptian problem, the solution amounts to computing the sum of the first 64 terms of a geometrical progression with first term 1 and multiplier 2, i.e.

$$\begin{aligned} 1 + 2 + 2^2 + \cdots + 2^{63} &= 2^{64} - 1 \\ &= 18,446,744,073,709,551,615. \end{aligned}$$

Coming closer to our times, an interesting use of iteration was made by Church [6] in the Lambda Calculus, which he had concocted as an alternative foundation for mathematics based on the notion of function and application, as opposed to set and membership. Church's idea was to represent the natural number $n$ in the Lambda Calculus as the binary operator $\overline{n}$ that, when applied to the arguments $f$ and $x$, produces the $n$-th iteration $f^{(n)}(x)$.

Apparently unnoticed by Church, the same idea had been proposed earlier by Wittgenstein [35], as follows:

6.02 And *this* is how we arrive at numbers. I give the following definitions

$$x = \Omega^{0\prime}x \text{ Def.},$$
$$\Omega'\Omega^{\nu\prime}x = \Omega^{\nu+1\prime} \text{ Def.}$$

So, in accordance with these rules, which deal with signs, we write the series

$$x, \quad \Omega'x, \quad \Omega'\Omega'x, \quad \Omega'\Omega'\Omega'x, \quad \ldots$$

in the following way

$$\Omega^{0\prime}x, \quad \Omega^{0+1\prime}x, \quad \Omega^{0+1+1\prime}x, \quad \ldots$$

[$\cdots$] And I give the following definitions

$$0 + 1 = 1 \text{ Def.},$$
$$0 + 1 + 1 = 2 \text{ Def.},$$
$$0 + 1 + 1 + 1 = 3 \text{ Def.},$$

(and so on)

> 6.021 A number is the exponent of an operation.

Even earlier, Peano [23] had suggested the same idea:

> Then, if $b$ is an $N$, by $a\alpha b$ we want to indicate what is obtained by executing the operation $\alpha$ on $a$, $b$ times in a row. Hence, if $a$ is a number, $a+b$ represents what is obtained by executing $b$ times on $a$ the operation $+$, that is the successor of $a$ of order $b$, i.e. the sum of $a$ and $b$. $[\cdots]$
> If $a$ and $b$ indicate two numbers, by their product $a \times b$ we will mean what is obtained by executing $b$ times on 0 the operation $+a$. $[\cdots]$
> If $a$ and $b$ indicate two numbers, by $a^b$ we will mean what is obtained by executing $b$ times on 1 the operation $\times a$.

Thus Peano, like Church but unlike Wittgenstein, saw that the definition of the numbers as iterators gives for free the representability of a number of functions obtained by iteration.

### 10.1.2. *Primitive recursion*

Primitive recursion is a procedure that defines the value of a function at an argument $n$ by using its value at the previous argument $n-1$ (see CRT, I.1.3).

Iteration is obviously a special case of primitive recursion, on the number of iterations. And so is the predecessor function, defined by

$$pd(n) = \begin{cases} 0 & \text{if } n = 0 \text{ or } n = 1 \\ pd(n-1)+1 & \text{otherwise.} \end{cases}$$

It is not immediate that the predecessor function can be reduced to an iteration, and hence is representable in the Lambda Calculus. It was Kleene [11] who saw how to do this, apparently during a visit to the dentist. Basically, $pd(n)$ is the second component of the $n$-th iteration of the function on pairs defined as

$$f((x,y)) = (x+1, x),$$

started on $(0,0)$.

More generally, it is possible to prove that any primitive recursion can be reduced to an iteration, in the presence of a coding and decoding mechanism (see CRT, I.5.10). This implies that all primitive recursive functions are actually representable in the Lambda Calculus, as proved by Kleene [12].

### 10.1.3. *Primitive recursion with parameters*

When defining a function of many variables by primitive recursion, all variables except one are kept fixed. Primitive recursion with parameters relaxes this condition, and it allows substitutions for these variables. Although apparently more general, this notion actually turns out to be reducible to the usual primitive recursion (see CRT, VIII.8.3.a).

One ancient example of a primitive recursion with parameters is the solution to the old problem known as the *Towers of Hanoi* or the *Towers of Brahma*:

> In the great temple of Benares, beneath the dome which marks the centre of the world, rests a brass-plate in which are fixed three diamond needles, each a cubit high and as thick as the body of a bee. On one of these needles, at the creation, God placed sixty-four disks of pure gold, the largest disk resting on the brass plate, and the others getting smaller and smaller up to the top one. This is the Tower of Brahma. Day and night unceasingly the priests transfer the disks from one diamond needle to another according to the fixed and immutable laws of Brahma, which require that the priest must not move more than one disk at a time and that he must place this disk on a needle so that there is no smaller disk below it. When the sixty-four disks shall have been thus transferred from the needle on which at the creation God placed them to one of the other needles, tower, temple, and Brahmins alike will crumble into dust, and with a thunderclap the world will vanish.[b]

The natural recursive solution is the following: to move $n$ disks from needle $A$ to needle $C$, first move $n-1$ disks from needle $A$ to needle $B$, then move one disk from needle $A$ to needle $C$, and then move $n-1$ disks from needle $B$ to needle $C$. More concisely:

$$move(n, A, C) = move(n-1, A, B) \wedge move(1, A, C) \wedge move(n-1, B, C).$$

Notice the use of $move(n-1, A, B)$ and $move(n-1, B, C)$, as opposed to $move(n-1, A, C)$, in the computation of $move(n, A, C)$, which makes this a primitive recursion with parameters (the value $move(1, A, C)$ does not count, being constant).

If we let $f(n)$ be the number of moves needed for $n$ disks provided by the previous solution, then

$$f(1) = 0 \qquad f(n+1) = 1 + 2f(n),$$

i.e.

$$f(n) = 1 + 2 + 2^2 + \cdots + 2^{n-1} = 2^n - 1,$$

[b]Reported in Rouse Ball [27].

and it is known that this is the least possible number of moves needed to solve the problem. In particular, according to the previous story, the doomsday will be reached after $2^{64}-1$ moves, i.e. the same number provided by the chessboard problem. If one correct move is made every second, for 24 hours a day and 365 days a year, the time required for the completion of the task would be of approximately 58 billion centuries.

### 10.1.4. *Course-of-value recursion*

When defining by primitive recursion a function at a given argument, only the value for the immediately preceeding argument can be used. Course-of-value recursion relaxes this condition, and it allows the use of any number of values for previous arguments. Although apparently more general, this notion actually turns out to be reducible to the usual primitive recursion (see CRT, I.7.1).

An early example of a course-of-value recursion was given by Leonardo da Pisa, also called Fibonacci, in his *Liber abaci*, written in 1202 and revised in 1228, when discussing the famous rabbit problem (*paria coniculorum*):

> How many pairs of rabbits can be bred in one year from one pair?
>
> A man has one pair of rabbits at a certain place entirely surrounded by a wall. We wish to know how many pairs can be bred from it in one year, if the nature of these rabbits is such that they breed every month one other pair, and begin to breed in the second month after their birth. Let the first pair breed a pair in the first month, then duplicate it and there will be 2 pairs in a month. From these pairs one, namely the first, breeds a pair in the second month, and thus there are 3 pairs in the second month. From these in one month two will become pregnant, so that in the third month 2 pairs of rabbits will be born. Thus there are 5 pairs in this month. From these in the same month 3 will be pregnant, so that in the fourth month there will be 8 pairs. [$\cdots$]

In the margin Fibonacci writes the sequence

$$1,\quad 2,\quad 3,\quad 5,\quad 8,\quad 13,\quad 21,\quad 34,\quad 55,\quad 89,\quad 144,\quad 233,\quad 377$$

and continues:

> You can see in the margin how we have done this, namely by combining the first number with the second, hence 1 and 2, and the second with the third, and the third with the fourth ... At last we combine the 10th with the 11th, hence 144 and 233, and we have the sum of the above-mentioned rabbits, namely 377, and in this way you can do it for the case of infinite numbers of months.

This provides the definition of the *Fibonacci sequence*:

$$f(0) = 0 \quad f(1) = 1 \quad f(n+2) = f(n) + f(n+1).$$

Notice the use of the two values $f(n)$ and $f(n+1)$ in the definition of $f(n+2)$, which makes this a course-of-value recursion.

The earliest record of a Fibonacci sequence is probably a set of weights discovered a few decades ago in Turkey, going back to around 1200 B.C. and arranged into a progression approximately equal to it (Petruso [24]). The sequence was also known in Egypt and Crete (Preziosi [25]), and it was used by the ancient and medieval Indians to define the metric laws of sanscrit poetry (Singh [31]).

### Double recursion

Primitive recursion can be used to define functions of many variables, but only by keeping all but one of them fixed. Double recursion relaxes this condition, and it allows the recursion to happen on two variables instead of only one. Although apparently more general, this notion actually turns out to be reducible in many cases (but not all) to the usual primitive recursion (see CRT, VIII.8.3.b and VIII.8.11).

The first use of a double recursion was made around 220 B.C. by Archimedes in his *Sand Reckoner* to solve the following problem:

> There are some, King Gelon, who think that the number of the sand is infinite in multitude; and I mean the sand not only which exists about Syracuse and the rest of Sicily, but also that which is found in every region whether inhabited or uninhabited. Again there are some who, without regarding it as infinite, yet think that no number has been named which is great enough to exceed this multitude. And it is clear that they who hold this view, if they imagined a mass made up of sand in other respects as large as the mass of the earth, including in it all the seas and the hollows of the earth filled up to a height equal to that of the highest of the mountains, would be many times further still from recognizing that any number could be expressed which exceeded the multitude of the sand so taken. But I will try to show you by means of geometrical proofs, which you will be able to follow, that, of the numbers named by me and given in the work which I sent to Zeuxippus,[c] some exceed not only the number of the mass of sand equal in magnitude to the earth filled up in the way described, but also that of a mass equal in magnitude to the universe.

To denote his large number, Archimedes fixes a number $a$ of units and defines the number $h_n(x)$ by a double recursion, on the cycle $x$ and the

[c] Archimedes is referring here to a work now lost.

period $n$, as follows:

$$h_0(x) = 1 \quad h_{n+1}(0) = h_n(a) \quad h_{n+1}(x+1) = a \cdot h_{n+1}(x),$$

so that

$$h_n(x) = (a^x)^n = a^{xn}.$$

Then he considers

$$h_a(a) = (a^a)^a = a^{(a^2)}$$

for the particular value $a = 10^8$, i.e. a myriad myriads (the myriad, i.e. 10,000, was the largest number for which the Greeks had a proper name). This takes him up to

$$(10^8)^{(10^{16})} = 10^{8 \cdot 10^{16}} \approx 10^{10^{17}},$$

which he calls "a myriad myriads units of the myriad-myriadesimal order of the myriad-myriadesimal period". This number, consisting of 80 million billions ciphers, remained the largest number used in mathematics until Skewes [32], who needed $10^{10^{10^{34}}}$ as a bound to the first place where the function $\pi(x) - li(x)$ first changes sign.

By an evaluation of the sizes of a grain of sand and of the then known universe, Archimedes gets an estimate of $10^{63}$ for the number of grains of sand needed to fill the universe, well below the bound above. It may be interesting to note that by using the values for the sizes of an electron ($10^{-18}$ meters) and of the currently known universe ($10^{35}$ light years), we get an estimate of $10^{207}$ for the number of electrons needed to fill the universe, still well below the bound above.

Archimedes' concludes his work as follows:

> I conceive that these things, King Gelon, will appear incredible to the great majority of people who have not studied mathematics, but that to those who are conversant therewith and have given thought to the question of the distances and sizes of the earth, the sun and moon and the whole universe, the proof will carry conviction. And it was for this reason that I thought the subject would not be inappropriate for your consideration.

## 10.2. The First Recursion Theorem

The so-called First Recursion Theorem (see CRT, II.3.15) provides a basic tool to compute values of functions which are solutions to recursive equations, implicitly defining functions by circular definitions involving the function itself.

The procedure is similar to a classical method to compute approximations to real numbers which are solutions to algebraic equations, implicitly defining real numbers by circular definitions involving the number itself. For example, consider the equation

$$x = 1 + \frac{1}{x}.$$

Then $x$ can be thought of as a fixed point of the function

$$f(x) = 1 + \frac{1}{x},$$

in the sense that

$$x = f(x).$$

To make $x$ explicit, we have at least two ways.

For example, we can transform the equation into the equivalent form

$$x^2 - x - 1 = 0,$$

and use the well-known formula for the solution to the second degree equation that was already known to the Babylonians around 2000 B.C., thus getting

$$x = \frac{1 \pm \sqrt{5}}{2}.$$

However, this works only for simple functions. Moreover, the solutions are not circular anymore, but are still implicit (the radical $\sqrt{5}$ still needs to be evaluated by other methods).

Alternatively, we can perform repeated substitutions of the right-hand side for $x$, thus obtaining a continuous function of the kind introduced in 1572 by Raffaele Bombelli in his *Algebra*:

$$x = 1 + \frac{1}{x} = 1 + \frac{1}{1 + \frac{1}{x}} = \cdots = 1 + \frac{1}{1 + \frac{1}{1 + \frac{1}{1 + \cdots}}}.$$

The infinite expression is built up as a limit of finite expressions, that provide approximations to the solution. More precisely, if we write $\frac{f(n+1)}{f(n)}$ for the $n$-th approximation, then

$$\frac{f(n+2)}{f(n+1)} = 1 + \frac{1}{\frac{f(n+1)}{f(n)}} = \frac{f(n) + f(n+1)}{f(n+1)},$$

i.e.

$$f(n+2) = f(n) + f(n+1).$$

In other words, $f$ is simply the Fibonacci sequence, and the approximations are given by the ratios of its successive terms:

$$\frac{2}{1} \quad \frac{3}{2} \quad \frac{5}{3} \quad \frac{8}{5} \quad \frac{13}{8} \quad \frac{21}{13} \quad \cdots.$$

This iterative method is the same underlying the proof of the First Recursion Theorem, and it has a long history.

### 10.2.1. *Differentiable functions*

An early appearance of the method is found in the Indian *Sulvasutra*, composed between 600 and 200 B.C. To compute numerical approximations to $\sqrt{2}$, the following recursive algorithm is proposed.

A first approximation is obtained by dissecting a rectangle of edges 1 and 2 (i.e. of area 2) into two squares of edge 1. One square is cut into two rectangles of short edge $\frac{1}{2}$, which are placed along the other square. The square of edge $1 + \frac{1}{2} = \frac{3}{2}$ has an area that exceeds 2 by a small square of edge $\frac{1}{2}$, thus producing an error equal to $\frac{1}{4}$.

A second approximation is obtained by subtracting from the square of edge $\frac{3}{2}$ giving the first approximation the error, i.e. two rectangular stripes of area $\frac{1}{8}$ and short edge $\frac{1}{8} \cdot \frac{2}{3} = \frac{1}{12}$. This produces a square of edge $\frac{3}{2} - \frac{1}{12} = \frac{17}{12}$, whose area differs from 2 by a small square of edge $\frac{1}{12}$, thus producing an error equal to $\frac{1}{144}$.

A third approximation is obtained by subtracting from the square of edge $\frac{17}{12}$ giving the second approximation the error, i.e. two rectangular stripes of area $\frac{1}{288}$ and short edge $\frac{1}{288} \cdot \frac{12}{17} = \frac{1}{408}$. This produces a square of edge $\frac{17}{12} - \frac{1}{408} = \frac{577}{408}$, which is the approximation to $\sqrt{2}$ given by the *Sulvasutra*, and is correct to 5 decimal places.

The procedure can be iterated as follows: Given an approximation $x_n$, we produce a new approximation

$$x_{n+1} = x_n - \frac{x_n^2 - 2}{2x_n},$$

where $x_n^2 - 2$ is the error of the $n$-th approximation, $\frac{x_n^2-2}{2}$ the area of each of the two rectangular stripes, and $\frac{x_n^2-2}{2x_n}$ their short edge.

If we let $f(x) = x^2 - 2$, then $f'(x) = 2x$ and $f(\sqrt{2}) = 0$. The previous recursive formula can thus be rewritten as

$$x_{n+1} = x_n - \frac{f(x_n)}{f'(x_n)}.$$

When generalized to any derivable functions, this becomes *Newton's formula* (1669) to approximate a zero of the given function by starting from a point $x_0$ sufficiently close to a zero and having a nonzero derivative.

In the case of the $f$ considered above, Newton's formula can be obtained directly by looking for an increment $h$ such that

$$f(x_n + h) = 0,$$

i.e.

$$(x_n + h)^2 - 2 = x_n^2 + 2x_n h + h^2 - 2 = 0.$$

By disregarding the quadratic term $h^2$ (which is the reason for the persistence of an error), we get

$$x_n^2 + 2x_n h = 2,$$

i.e.

$$h = -\frac{x_n^2 - 2}{2x_n}.$$

Similar proofs hold for any polynomial. In general, for an analytical function $f$ the increment is obtained from *Taylor's formula* (1715):

$$f(x+h) = f(x) + \frac{h}{1!}f'(x) + \frac{h^2}{2!}f''(x) + \cdots + \frac{h^n}{n!}f^{(n)}(x) + \cdots.$$

### 10.2.2. *Contractions*

When discussing the problem of consciousness, Royce [28] observed that an individual must have an infinite mental image of its own mind, since the image must contain an image of the image, which must contain an image of the image of the image, and so on.

Abstracting from the problem of consciousness, Royce presented a paradoxical metaphor that caught the fancy of the writer Jorge Luis Borges, who quoted it at least three times in his work with the following words:

> Imagine a portion of the territory of England has been perfectly levelled, and a cartographer traces a map of England. The work is perfect. There is no particular of the territory of England, small as it can be, that has not been recorded in the map. Everything has its own correspondence. The map, then, must contain a map of the map, that must contain a map of the map of the map, and so on to infinity.

The metaphor has been interpreted as a proof by contradiction that a perfect map is impossible, supporting the well-known aphorism of Korzybski [16]: "the map is not the territory".

Actually, from a mathematical point of view a perfect map that contains a copy of itself is not a *contradiction*, but rather a *contraction*, in the sense that it defines a function $f$ such that

$$|f(x) - f(y)| \leq c \cdot |x - y|,$$

for some $c$ such that $0 < c < 1$. Banach [2] has proved that a contraction on a complete metric space has a unique fixed point, and the proof is a typical iteration. Indeed, by induction,

$$|f^{(n+1)}(x) - f^{(n)}(x)| \leq c^n \cdot |f(x) - x|.$$

By the triangular inequality,

$$\begin{aligned} |f^{(n+m)}(x) - f^{(n)}(x)| &\leq \sum_{i<m} |f^{(n+i+1)}(x) - f^{(n+i)}(x)| \\ &\leq (\sum_{i<m} c^{n+i}) \cdot |f(x) - x|. \end{aligned}$$

Thus the sequence $\{f^{(n)}(x)\}_{n\in\omega}$ converges to a point $x_0$, and hence so does the sequence $\{f^{(n+1)}(x)\}_{n\in\omega}$. Since $f$ is continuous, the second sequence also converges to $f(x_0)$, which must then be equal to $x_0$. In other words, $x_0$ is a fixed point of $f$. Moreover, if $x_1$ is another fixed point, then

$$|x_0 - x_1| = |f(x_0) - f(x_1)| \leq c \cdot |x_0 - x_1|.$$

Since $c < 1$, it follows that $x_0 = x_1$, i.e. $x_0$ is the unique fixed point of $f$.

In the case of a perfect map, this means that there must be a point of the territory that coincides with its image on the map. Thus a perfect map is not the territory in general, but it is so in one (and only one) point.

### 10.2.3. *Continuous functions*

Banach's result was obtained as an abstraction of the technique of successive substitution developed in the XIX Century by Liouville, Neumann, and Volterra to find solutions to integral equations, in which an unknown function appears under an integral sign. A similar technique was used by Peano [22] to find solutions to systems of linear differential equations. In both cases an appropriate contraction is determined by the usual continuity and Lipschitz conditions, which ensure existence and uniqueness of the solution.

An extension of Banach's Fixed Point Theorem, for more special spaces but more general maps, was obtained by Brouwer [4], who proved that a continuous function of a convex compact subset of a Euclidean space on itself has a fixed point.

Brouwer's original proof determined the existence of a fixed point by contradiction, without actually exhibiting it (this was quite ironical, due to Brouwer's costructive philosophy). In the special case of a closed disk, Brouwer's proof amounted to the following: If a continuous function of a closed disk on itself had no fixed point, every point would be moved to a different point. By extending the vector determined by an argument and its image, we could associate to every point on the disk a point on the border. This would determine an impossible continuous deformation of the whole disk into the border.

However, a constructive version of Brouwer's Fixed Point Theorem for a continuous function on a closed square on itself can be obtained by the iteration technique, as follows: Suppose there is no fixed point on the border. Then the vector determined as above makes a complete turn while the point moves around the border. Divide the square into four equal squares. Either the vector vanishes on a point of the border on one of the squares, thus determining a fixed point of the given function, or there is at least one square on which the vector makes a complete turn while the point moves around the border, and the process can be started again. If no fixed point is found along the way, the process determines a sequence of telescopic squares which uniquely identifies a point. Since any neighborhood of the point contains vectors in every direction, by continuity the vector field must vanish at it, i.e. the process determines a fixed point.

In one dimension Brouwer's Fixed Point Theorem becomes a version of the Intermediate Value Theorem proved by Bolzano [3], according to which a continuous function on a closed interval that takes values respectively greater and smaller than $c$ on the extremes of the interval, must take value $c$ at some point of the interval. In this case, an intermediate value can be found by a bisection method similar to the above.

Even more abstract versions of Banach's theorem than Brouwer's were obtained by Knaster [15] and Tarski [34], who proved the existence of fixed points for any monotone function on a complete lattice. Abian and Brown [1] replaced complete lattices by chain-complete partial orderings, in which every chain of elements has a least upper bound. In particular, a chain-complete partial ordering has a least element $\perp$, since the empty chain must have a l.u.b.

Given a monotone function $f$ on a chain complete partial ordering, consider the following transfinite sequence of elements:

$$x_0 = \perp$$
$$x_{\alpha+1} = f(x_\alpha)$$
$$x_\beta = \text{the l.u.b. of } \{f(x_a)\}_{\alpha<\beta}, \text{ if } \beta \text{ is a limit ordinal.}$$

Since $f$ is monotone, this defines a chain, whose length cannot exceed the maximal length of chains on the given partial ordering. Then there is a largest element $x_{\alpha_0}$, otherwise the l.u.b. of the chain would be a larger element. And $f(x_{\alpha_0}) = x_{\alpha_0}$, otherwise $x_{\alpha_0}$ would not be the largest element of the chain. Moreover, $x_{\alpha_0}$ is the least fixed point, because every element of the chain is below any other fixed point, by induction.

It thus follows that any monotone function on a chain complete partial ordering has a least fixed point. If, moreover, $f$ is continuous (in the sense of preserving l.u.b.s $\bigsqcup$), then the fixed point is obtained in at most $\omega$ iterations, because

$$f(x_\omega) = f(\bigsqcup_{n\in\omega} x_n) = \bigsqcup_{n\in\omega} f(x_n) = \bigsqcup_{n\in\omega} x_{n+1} = x_\omega.$$

As an application, we can sketch a proof of the First Fixed Point Theorem of Kleene [14]. Consider the chain complete partial ordering consisting of the partial functions on the integers, ordered by inclusion. Since a recursive functional is monotone and continuous, it has a least fixed point $x_\omega$ by the theorem. Moreover, the least fixed point is recursive by the proof.

### 10.3. The Second Recursion Theorem

The so-called Second Recursion Theorem (see CRT, II.2.13) provides a basic tool to find explicit solutions to recursive equations, implicitly defining programs of recursive functions by circular definitions involving the program itself.

The procedure is the analogue of a classical method to find explicit definitions for functions implicitly defined by recursive equations. For example, consider the implicit definition of the Fibonacci sequence:

$$f(0) = 0 \quad f(1) = 1 \quad f(n+2) = f(n) + f(n+1).$$

To make $f$ explicit, we can use De Moivre's method (1718) of generating functions, and let

$$F(x) = f(0) + f(1) \cdot x + f(2) \cdot x^2 + \cdots + f(n) \cdot x^n + \cdots.$$

By computing

$$F(x) - F(x) \cdot x - F(x) \cdot x^2$$

we notice that most terms cancel out, since they have null coefficients of the form $f(n+2) - f(n+1) - f(n)$. We thus get

$$F(x) = \frac{x}{1 - x - x^2}.$$

By factoring the denominator, expanding the right-hand-side into a power series, and comparing it term by term to $F(x)$, we obtain the following explicit description for $f$:

$$f(n) = \frac{1}{\sqrt{5}} \left[ \left( \frac{1+\sqrt{5}}{2} \right)^n - \left( \frac{1-\sqrt{5}}{2} \right)^n \right].$$

This result, which uses $\frac{1 \pm \sqrt{5}}{2}$ to express the Fibonacci sequence, is the complement of the result proved above, which uses the Fibonacci sequence to approximate $\frac{1+\sqrt{5}}{2}$.

Kronecker [17] generalized the previous example to show that every linear recursive relation determines the coefficients of a power series defining a rational function. Conversely, every rational function can be expressed as a power series with coefficients satisfying a linear recursive relation.

The Second Recursion Theorem serves a similar purpose, by turning recursive programs which define functions by recursive calls, into programs for the same functions without any recursive call.

### 10.3.1. *The diagonal method*

The proofs of the Second Recursion Theorem and its variants (see CRT, II.2.10 and II.2.13) are elaborate and abstract forms of the diagonal method, which can be considered the most pervasive tool of Recursion Theory. Its essence is the following.

Given an infinite matrix $\{a_{ij}\}_{ij}$, we first transform the elements $a_{nn}$ on the diagonal by means of a switching function $d$, thus obtaining $d(a_{nn})$. If the switching function $d$ is never the identity on the elements of the matrix, then the transformed diagonal function is not a row of the matrix. More precisely, it differs on the $n$- th element from the $n$-th row.

Equivalently, if the transformed diagonal function is a row of the matrix, e.g. the $n$-th, then the switching function $d$ must be the identity on some element of the matrix. More precisely, it leaves the $n$-th element of the $n$-th row unchanged. In this form, the diagonal method provides a fixed point of the function $d$.

### 10.3.2. *The diagonal*

The first ingredient of the diagonal method is the consideration of the elements on the diagonal of an appropriate matrix.

This was done in a nontrivial way already by Archimedes in the *Sand Reckoner* discussed above, when stepping from the matrix $\{h_n(x)\}_{n,x}$ to the diagonal element $h_a(a)$.

In modern times, Du Bois Reymond has made a substantial use of diagonalization in his study of orders of infinity, reported in Hardy [10]. Basically, he defines an ordering based on domination (i.e. a function is greater than another if it is above it for almost all arguments), and classifies classes of functions by means of skeletons of fast growing functions obtained by starting with functions $f$ greater than the identity, iterating at successor stages, and diagonalizing at limit stages. More precisely, a function $f$ such that $f(x) \geq x$ for almost all arguments defines the following skeleton:

$$f_0(x) = f(x) \quad f_{\alpha+1}(x) = f_\alpha^{(x)}(x) \quad f_\alpha(x) = f_{\alpha_x}(x),$$

where in the last clause $\alpha$ is the limit of the ascending sequence of the ordinals $\alpha_x$ (the definition obviously depends on the choice of the ascending sequence.)

Today these skeletons have become standard in Complexity Theory, to classify complexity classes such as the primitive recursive functions (see CRT, VIII.8.10).

### 10.3.3. *The switching function*

The second ingredient of the diagonal method is the use of the switching function on the elements of the diagonal.

This was first done by Cantor [5], in his historical proof that the sets of natural numbers are more than the numbers themselves. By considering characteristic functions, the proof amounts to the observation that given a sequence $\{f_n\}_{n\in\omega}$ of 0,1-valued functions, the function

$$d(x) = 1 - f_x(x)$$

is 0,1-valued but not in the sequence, since it differs from $f_n$ on the argument $n$. The switching function, true to its name, is here the function that interchanges 0 and 1.

The same type of argument was used by Russell [29], to prove his celebrated paradox. This time we consider the set

$$R = \{x : \neg(x \in x)\}.$$

Then

$$x \in R \iff \neg(x \in x),$$

and thus

$$R \in R \iff \neg(R \in R),$$

contradiction. The switching function is now the negation operator that interchanges the truth values "true" and "false".

Russell's paradox was turned into a theorem by Curry [8], who proved the existence of fixed points for any $\lambda$-term in the Untyped Lambda Calculus, according to the following correspondence:

| Set Theory | Lambda Calculus |
|---|---|
| element | argument |
| set | term |
| membership | application |
| set formation {} | $\lambda$-abstraction |
| set equality | term equality |

If the term $N$ is supposed to correspond to negation, then the set $R$ corresponds to the term

$$C = \lambda x.\, N(xx).$$

By the reduction rules of the Lambda Calculus,

$$Cx = N(xx),$$

and thus

$$CC = N(CC).$$

However, this is not a contradiction, but rather a proof that $CC$ is a fixed point of $N$. In other words, in the Lambda Calculus there is no switching function, in the sense of a term that always changes its arguments.

Curry's Fixed Point Theorem is a version of the Recursion Theorems, and together with the representability of the predecessor function quoted above implies the representability of all recursive functions in the Lambda Calculus, as proved by Kleene [12] (see CRT, I.6.6.c).

### 10.3.4. *Selfreference*

In the last two arguments above, diagonalization takes the form of a selfreference. Indeed, the conditions "$x \in x$" in Russell's paradox can be read as: "$x$ belongs to itself". Similarly, the condition "$xx$" in Curry's theorem can be read as: "$x$ applied to itself".

Selfreference is obviously trivial in any language possessing the pronoun "I". The best known ancient reference is God's own description in *Exodus* (3.14): "I am that I am". However, this kind of selfreference is somewhat indirect, since the pronoun is a linguistic object that refers not to itself, but to the person who is pronouncing it. A better example is a phrase that talks of itself, for example: "This phrase consists of six words".

The first paradoxical selfreference was probably the Liar paradox, attributed to Eubulides (IV Century B.C.) in the form: "I am lying". A purely linguistic analogue is: "This phrase is false".

It is not paradoxical, instead, for a Cretian such as Epimenides (VI Century B.C.) to say: "All Cretians always lie". This phrase cannot be true, otherwise Epimenides would be a Cretian who is not always lying. Then it must be false, i.e. some Cretian does not always lie. It does not follow that such a Cretian is Epimenides. Nor would it follow, if he were, that the phrase is one of his truths. So being, the following comment by St. Paul in the *Epistle to Titus* (1.12) turns out to be even more cretin than it looks at first sight:

> For there are many unruly and vain talkers and deceivers, specially they of the circumcision: whose mouths must be stopped, who subvert whole houses, teaching things which they ought not, for filthy lucre's sake. One of themselves, even a prophet of their own, said, "The Cretians are always liars, evil beasts, slow bellies". This witness is true.

The Liar paradox had counteless versions in history. In particular, the original one-step selfreference was turned into a two-step one by Philip Jourdain in 1913 (following Buridan of the XIV Century), as follows:

> The following phrase is false.
> The previous phrase is true.

Finite $n$-steps versions are obtained in a similar fashion. An infinite diabolical version, as the name suggests, has been proposed by Yablo [36, 37]:

> All the following phrases are false.
> All the following phrases are false.
> ...

Suppose the first phrase is true. Then all the following ones are false, in particular the second. Moreover, all the remaining phrases are false, and hence the second one is true, contradiction. Then the first phrase is false, i.e. one of the following phrases is true, and a contradiction is reached as for the first. Thus the first phrase is contradictory. Similarly, so are all the remaining ones.

The turning point in these developments came with Gödel [9], who made an explicit reference to the Liar paradox in his paper. His main result can be stated as follows: Given any property $P$ weakly representable in a sufficiently strong formal system for Arithmetic, there is a sentence saying of itself that it has the property $P$ (see CRT, p. I.165). For the proof, consider an enumeration $\{\varphi_n\}_{n\in\omega}$ of the formulas with one free variable, the matrix

$$a_{ij} = \text{ the sentence "}\varphi_j\text{ has the property expressed by }\varphi_i\text{"}$$

and the switching function

$$d(\varphi) = \text{ the sentence "}\varphi\text{ has the property }P\text{".}$$

Since $P$ is weakly representable, the transformed diagonal sequence is still a row of the matrix, up to provable equivalence. Thus there is a $\varphi$ such that $d(\varphi)$ is provably equivalent to $\varphi$, i.e. $\varphi$ says of itself that it has the property $P$.

A first consequence is that truth cannot be weakly representable in any consistent and sufficiently strong formal system for Arithmetic. Otherwise so would be its negation, and the general result would give a contradictory sentence asserting its own negation, as in the Liar paradox. The unrepresentability of truth was obtained by Gödel before his Incompleteness Theorem, but he did not publish it. The result is thus usually attributed to Tarski [33].

A second consequence is that, since provability *is* weakly representable in any consistent and sufficiently strong formal system for Arithmetic, the general result gives a sentence asserting its own unprovability. From this one can easily obtain all the epochal results of Gödel [9], Rosser [26], and Church [7] (see CRT, pp. I.166–169).

By the same type of argument we can also prove the Second Recursion Theorem of Kleene [13], following Owings [21]. Given an effective transformation of programs $f$, consider an enumeration $\{\varphi_n\}_{n\in\omega}$ of the partial recursive unary functions, the matrix

$$a_{ij} = \text{ the function with program coded by }\varphi_i(j)$$

and the switching function

$$d(\varphi_e) = \text{ the function with program coded by } f(e).$$

Since $f$ is effective, the transformed diagonal sequence is still a row of the matrix. Thus there is an $e$ such that $d(\varphi_e)$ and $\varphi_e$ are the same function. Equivalently, the programs coded by $e$ and $f(e)$ compute the same function.

## References


[1] S. Abian and A. B. Brown, A theorem on partially ordered sets with applications to fixed-point theorems, *Can. J. Math.* **13**, 78–83, (1961).

[2] S. Banach, Sur les operations dans les ensembles abstraits et leurs applications aux equations integrales, *Fund. Math.* **3**, 7–33, (1922).

[3] B. Bolzano. Rein analytischer Beweise des Lehrsatzes, dass zwischen je zwey Werthen, die ein entgegengesetzes Resultat gewhren, wenigstens eine reelen Wurzel der Gleichen liege, Gottlieb Haase, Prague, (1817).

[4] L. Brouwer, Über Abbildungen von Mannigfaltigkeiten, *Math. Ann.* **71**, 97–115, (1911/12).

[5] G. Cantor, Über eine Eigenschaft des Inbegriffes aller reellen algebraischen Zahlen, *J. Reine Angew. Math.* **77**, 258–262, (1874).

[6] A. Church, A set of postulates for the foundation of logic (second paper), *Ann. of Math.* **34**, 839–864, (1933).

[7] A. Church, A note on the Entscheidungsproblem, *J. Symbolic Logic.* **1**, 40–41, (1936).

[8] H. B. Curry, The inconsistency of certain formal logics, *J. Symbolic Logic.* **7**, 115–117, (1942).

[9] K. Gödel, Über formal unentscheidbare Sätze der Principia Mathematica und verwandter Systeme I, *Monatsh. Math. Phys.* **38**, 349–360, (1931).

[10] G. H. Hardy, *Orders of infinity.* Cambridge University Press, Cambridge, (1910).

[11] S. C. Kleene, A theory of positive integers in formal logic, *Amer. J. Math.* **57**, 153–173, 219–244, (1935).

[12] S. C. Kleene, λ-definability and recursiveness, *Duke Math. J.* **2**, 340–353, (1936).

[13] S. C. Kleene, On notation for ordinal numbers, *J. Symbolic Logic.* **3**, 150–155, (1938).

[14] S. C. Kleene, *Introduction to Metamathematics.* Van Nostrand, New York, (1952).

[15] B. Knaster, Un théorème sur les fonctions d'ensembles, *Annales de la Société Polonaise de Mathématiques.* **6**, 133–134, (1928).

[16] A. Korzybski, *Science and Sanity.* Science Press, Lancaster, (1941).

[17] L. Kronecker, Zur Theorie der Elimination einer Variablen aus zwei algebraischen Gleichungen, *Monatsberichte der Kniglich Preussischen Akademie der Wissenschaften, Berlin.* pp. 535–600, (1881).

[18] J. Newman, *The World of Mathematics.* Simon and Schuster, New York, (1956).
[19] P. Odifreddi, *Classical Recursion Theory (Volume I).* North-Holland, Amsterdam, (1989).
[20] P. Odifreddi, *Classical Recursion Theory (Volume II).* North-Holland, Amsterdam, (1999).
[21] J. C. Owings, Jr., Diagonalization and the recursion theorem, *Notre Dame J. Formal Logic.* **14**, 95–99, (1973).
[22] G. Peano, Intégration par séries des équations différentielles linéaires, *Math. Ann.* **32**, 450–456, (1888).
[23] G. Peano, Sul concetto di numero, *Rivista di Matematica.* **1**, 87–102, 256–267, (1891).
[24] K. M. Petruso, Additive progressions in prehistoric mathematics: a conjecture, *Hist. Math.* **12**, 101–106, (1985).
[25] D. Preziosi, *Minoan Architectural Design: Formation and Signification.* Mouton, Berlin, (1983).
[26] B. J. Rosser, Extensions of some theorems of Gödel and Church, *J. Symbolic Logic.* **1**, 87–91, (1936).
[27] W. W. Rouse Ball, *Mathematical Recreations and Essays.* The MacMillan Company, New York, (1905).
[28] J. Royce, *The World and the Individual.* The MacMillan Company, New York, (1899).
[29] B. Russell, *The principles of mathematics.* Cambridge University Press, Cambridge, (1903).
[30] Yu. A. Shashkin, *Fixed Points. Amer. Math. Soc.*, (1991).
[31] P. Singh, The socalled Fibonacci numbers in Ancient and Medieval India, *Hist. Math.* **12**, 229–244, (1985).
[32] S. Skewes, On the difference $\pi(x) - li(x)$, *J. London Math. Soc.* **8**, 277–283, (1933).
[33] A. Tarski, Der Wahrheitsbegriff in der formalisierten Sprachen, *Studia Philosophica.* **1**(261–405), (1936).
[34] A. Tarski, A lattice-theoretical fixed-point theorem and its applications, *Pacific. J. Math.* **5**(285–309), (1955).
[35] L. Wittgenstein, Logisch-philosophische Abhandlung, *Ann. Naturphil.* **14**, 185–262, (1921).
[36] S. Yablo, Truth and reflection, *J. Philos. Logic.* **14**, 297–349, (1985).
[37] S. Yablo, Paradox without self-reference, *Analysis.* **53**, 251–252, (1993).

# Chapter 11

# Reverse Mathematics and Well-ordering Principles

Michael Rathjen* and Andreas Weiermann

*Department of Pure Mathematics, University of Leeds*
*Leeds LS2 9JT, UK*
*E-mail: rathjen@maths.leeds.ac.uk*

*Vakgroep Zuivere Wiskunde en Computeralgebra, Ghent University*
*Krijgslaan 281 - Gebouw S22, B9000 Gent, Belgium*
*E-mail: Andreas.Weiermann@ugent.be*

This chapter is concerned with generally $\Pi^1_2$ sentences of the form *"if X is well ordered then f(X) is well ordered"*, where $f$ is a standard proof theoretic function from ordinals to ordinals. It has turned out that a statement of this form is often equivalent to the existence of countable coded $\omega$-models for a particular theory $T_f$ whose consistency can be proved by means of a cut elimination theorem in infinitary logic which crucially involves the function $f$. To illustrate this theme, we shall focus on the well-known $\varphi$-function which figures prominently in so-called predicative proof theory. However, the approach taken here lends itself to generalization in that the techniques we employ can be applied to many other proof-theoretic functions associated with cut elimination theorems. In this paper we show that the statement *"if X is well ordered then $\varphi X0$ is well ordered"* is equivalent to $\mathbf{ATR}_0$. This was first proved by Friedman (see [7]) using recursion-theoretic and combinatorial methods. The proof given here is proof-theoretic, the main techniques being Schütte's method of proof search (deduction chains) [15], generalized to $\omega$ logic, and cut elimination for infinitary ramified analysis.

*Research of both authors was supported by Royal Society International Joint Projects award 2006/R3. The first author would like to thank the *Swedish Collegium for Advanced Study* in Uppsala for providing an excellent research environment for the completion of this paper.

## Contents



## 11.1. Introduction

The larger project broached in this paper is to present a general proof-theoretic machinery for investigating special kinds of $\Pi^1_2$ statements about well-orderings from a reverse mathematics point of view. These $\Pi^1_2$ statements are of the form

$$\mathbf{WOP}(f) \qquad \text{"}\textit{if } X \textit{ is well ordered then } f(X) \textit{ is well ordered}\text{"} \tag{11.1}$$

where $f$ is a standard proof-theoretic function from ordinals to ordinals. There are by now several examples of functions $f$ where the statement $\mathbf{WOP}(f)$ has turned out to be equivalent to one of the theories of reverse mathematics over a weak base theory (usually $\mathbf{RCA}_0$). The first example is due to Girard [9].

**Theorem 11.1.1.** (Girard, 1987) *Let* $\mathbf{WO}(\mathfrak{X})$ *express that* $\mathfrak{X}$ *is a well ordering. Over* $\mathbf{RCA}_0$ *the following are equivalent:*

*(i) Arithmetic Comprehension.*
*(ii)* $\forall\mathfrak{X}\,[\mathbf{WO}(\mathfrak{X}) \rightarrow \mathbf{WO}(2^{\mathfrak{X}})]$.

Recently two new theorems appeared in preprints [7, 11]. These results give characterizations of the form (11.1) for the theories $\mathbf{ACA}_0^+$ and $\mathbf{ATR}_0$, respectively, in terms of familiar proof-theoretic functions. $\mathbf{ACA}_0^+$ denotes the theory $\mathbf{ACA}_0$ augmented by an axiom asserting that for any set $X$ the $\omega$-th jump in $X$ exists while $\mathbf{ATR}_0$ asserts the existence of sets constructed by transfinite iterations of arithmetical comprehension. $\alpha \mapsto \varepsilon_\alpha$ denotes the usual $\varepsilon$ function while $\varphi$ stands for the two-place Veblen function familiar from predicative proof theory (cf. [15]). More detailed descriptions of $\mathbf{ATR}_0$

and the function $\mathfrak{X} \mapsto \varphi\mathfrak{X}0$ will be given shortly. Definitions of the familiar subsystems of reverse mathematics can be found in [17].

**Theorem 11.1.2.** (Montalban, Marcone, 2007) *Over* $\mathbf{RCA}_0$ *the following are equivalent:*

*(i)* $\mathbf{ACA}_0^+$.
*(ii)* $\forall\mathfrak{X}\,[\mathbf{WO}(\mathfrak{X}) \to \mathbf{WO}(\varepsilon_{\mathfrak{X}})]$.

**Theorem 11.1.3.** (Friedman) *Over* $\mathbf{RCA}_0$ *the following are equivalent:*

*(i)* $\mathbf{ATR}_0$.
*(ii)* $\forall\mathfrak{X}\,[\mathbf{WO}(\mathfrak{X}) \to \mathbf{WO}(\varphi\mathfrak{X}0)]$.

The proof of Theorem 11.1.3 uses rather sophisticated recursion-theoretic results about linear orderings and is quite combinatorial. Theorem 11.1.3 builds on a result from [6] to the effect that there is no arithmetic sequence of degrees descending by $\omega$-jumps. The latter result was then improved by Steel [18] to descent by Turing jumps: If $Q \subseteq \mathrm{Pow}(\omega) \times \mathrm{Pow}(\omega)$ is arithmetic, then there is no sequence $\{A_n \mid n \in \omega\}$ such that (a) for every $n$, $A_{n+1}$ is the unique set such that $Q(A_n, A_{n+1})$, (b) for every $n$, $A'_{n+1} \leq_T A_n$.

For a proof theorist, theorems 11.1.2 and 11.1.3 bear a striking resemblance to cut elimination theorems for infinitary logics. This prompted the first author of this paper to look for proof-theoretic ways of proving these results. The hope was that this would also unearth a common pattern behind them and possibly lead to more results of this kind. The project commenced with [2], where a purely proof-theoretic proof of Theorem 11.1.2 was presented. In this paper we shall give a new proof of Theorem 11.1.3. It is principally proof-theoretic, the main techniques being Schütte's method of proof search (deduction chains) [15] and cut elimination for ramified analysis. The general pattern, of which this paper provides a second example, is that a statement $\mathbf{WOP}(f)$ is often equivalent to a familiar cut elimination theorem for an infinitary logic which in turn is equivalent to the assertion that every set is contained in an $\omega$-model of a certain theory $T_f$.

To guide the reader through the paper we shall briefly sketch the main parts of the proof of Theorem 11.1.3, i.e., that (ii) implies (i). We start with the observation that $\mathbf{ATR}_0$ can be axiomatized over $\mathbf{ACA}_0$ via a single sentence of the form $\forall X(\mathbf{WO}(<_X) \to \forall Z \exists Y\, B_0(X,Y,Z))$ where $B_0(X,Y,Z)$ is an arithmetical formula (cf. Lemma 11.3.2). Thus to verify $\mathbf{ATR}_0$ it suffices to show that for every well-ordering $<_Q$ there exists an $\omega$-model of $\mathbb{M}$ of $\mathbf{ACA}_0$ which contains $Q$ such that $\mathbb{M} \models \forall Z \exists Y\, B_0(Q, Y, Z)$. To find $\mathbb{M}$ we employ Schütte's method of proof search from [15, II§4], which he used to prove the completeness theorem for first order logic (cf. [15, Theorem 5.7]). The method has to be extended to $\omega$-logic, though. Rather than working in the Schütte calculus of positive and negative forms we work in a Gentzen sequent calculus with finite sets of formulas, called sequents. Let $C$ be a sentence that axiomatizes arithmetic comprehension and let $D_Q(n)$ be the formula $n \in Q$ if the latter formula is true and $n \notin Q$ otherwise. The main idea is to start with the sequent $\{\neg\forall Z \exists Y\, B_0(Q, Y, Z), \neg C, \neg D_Q(0)\}$ and systematically apply the rules of $\omega$-logic for the second order sequent calculus backwards, giving rise to a tree of sequents $\mathcal{D}_Q$. One also has to add the formula $\neg D_Q(n)$ to all sequents generated in this way after $n$ steps.

There are two possible outcomes. If the tree $\mathcal{D}_Q$ is not well-founded then it contains an infinite path $\mathbb{P}$. Now define a set $M$ via

$$(M)_i = \{n \mid n \notin U_i \text{ occurs in } \mathbb{P}\}$$

and let $\mathbb{M} = (\mathbb{N}; \{(M)_i \mid i \in \mathbb{N}\}, +, \cdot, 0, 1, <)$. For a formula $F$, let $F \in \mathbb{P}$ mean that $F$ occurs in $\mathbb{P}$, i.e. $F \in \Gamma$ for some $\Gamma \in \mathbb{P}$. Let $U_0, U_1, U_2, \ldots$ be an enumeration of the free set variables. For the assignment $U_i \mapsto (M)_i$ one can then show that $F \in \mathbb{P} \;\Rightarrow\; \mathbb{M} \models \neg F$. Whence $\mathbb{M}$ is an $\omega$-model of $\mathbf{ACA}$ and $\mathbb{M} \models \forall Z \exists Y\, B_0(Q, Y, Z)$. Also note that $(M)_0 = Q$, thus $Q$ is in $\mathbb{M}$.

The other conceivable outcome is that $\mathcal{D}_Q$ is well-founded, i.e. all paths in $\mathcal{D}_Q$ are finite, and thus every maximal path ends in a sequent which contains a basic axiom. In other words $\mathcal{D}_Q$ is a proof tree and the Kleene-Brouwer ordering of this tree is some well-ordering $\tau$. The crucial step to perform next consists in envisaging $\mathcal{D}_Q$ as a skeleton of a proof tree in infinitary ramified analysis, dubbed $\mathbf{RA}^*$ in [15]. In actuality $\mathcal{D}_Q$ can be viewed as the skeleton of a proof of the empty sequent in $\mathbf{RA}^*$. As we can remove all cuts in this proof we end up with a cut free proof of the empty sequent. But this is impossible, and therefore $\mathcal{D}_Q$ cannot be well-founded. To be able to carry out the removal of all cuts we have to enlist the help of arithmetical transfinite induction, roughly up to the ordinal $\varphi\tau 0$. Hence this is the step where the principle $\forall \mathfrak{X}\, [\mathbf{WO}(\mathfrak{X}) \to \mathbf{WO}(\varphi\mathfrak{X}0)]$ makes its appearance in showing the direction $(ii) \Rightarrow (i)$ of Theorem 11.1.3.

## 11.2. The Ordering $\varphi\mathfrak{X}0$

Via simple coding procedures, countable well-orderings, and functions on them can be expressed in the language of second order arithmetic, $L_2$. Variables $X, Y, Z, \ldots$ are supposed to range over subsets of $\mathbb{N}$. Using an elementary injective pairing function $\langle,\rangle$ (e.g. $\langle n,m\rangle := (n+m)^2+n+1$), every set $X$ encodes a sequence of sets $(X)_i$, where $(X)_i := \{m \mid \langle i,m\rangle \in X\}$. We also adopt from [17], II.2 the method of encoding a finite sequence $(n_0, \ldots, n_{k-1})$ of natural numbers as a single number $\langle n_0, \ldots, n_{k-1}\rangle$.

**Definition 11.2.1.** Every set of natural numbers $Q$ can be viewed as encoding a binary relation $<_Q$ on $\mathbb{N}$ via $n <_Q m$ iff $\langle n,m\rangle \in Q$. The **field** of $Q$, $\mathrm{fld}(Q)$ is the set $\{n \mid \exists m\,[n <_Q m \vee m <_Q n]\}$.

We say that $Q$ is a **well-ordering** if $<_Q$ is a well-ordering, that is $<_Q$ is a linear ordering of its field and every non-empty subset $U$ of $\mathrm{fld}(Q)$ has a $<_Q$-least element.

**Definition 11.2.2.** Let $Q$ be a linear ordering with least element $0_Q$. Let $\varphi ua := \langle 0, \langle u,a\rangle\rangle$, $\mathbf{H} := \{\varphi ua \mid u,a \in \mathbb{N}\}$, $\mathbf{h}(\varphi ua) = u$ and $\mathbf{h}(b) = 0_Q$ if $b \notin \mathbf{H}$.

We introduce the ordering $\varphi Q0$ by inductively defining its field $\mathrm{fld}(\varphi Q0)$ and the ordering $<_{\varphi Q0}$:

(1) $0 \in \mathrm{fld}(\varphi Q0)$.
(2) $0 <_{\varphi Q0} \alpha$ if $\alpha \in \mathrm{fld}(\varphi Q0)$ and $\alpha \neq 0$.
(3) $\varphi u\alpha \in \mathrm{fld}(\varphi Q0)$ if $u \in \mathrm{fld}(Q)$, $\alpha \in \mathrm{fld}(\varphi Q0)$ and $\mathbf{h}(\alpha) \leq_Q u$.
(4) If $\alpha_1, \ldots, \alpha_n \in \mathrm{fld}(\varphi Q0) \cap \mathbf{H}$, $n > 1$ and $\alpha_n \leq_{\varphi Q0} \cdots \leq_{\varphi Q0} \alpha_1$, then

$$\alpha_1 + \ldots + \alpha_n \in \mathrm{fld}(\varphi Q0)$$

where $\alpha_1 + \ldots + \alpha_n := \langle 1, \langle \alpha_1, \ldots, \alpha_n\rangle\rangle$.
(5) If $\alpha_1 + \ldots + \alpha_n, \beta_1 + \ldots + \beta_m \in \mathrm{fld}(\varphi Q0)$, then

$$\begin{aligned} &\alpha_1 + \ldots + \alpha_n <_{\varphi Q0} \beta_1 + \ldots + \beta_m \text{ iff} \\ &n < m \wedge \forall i \leq n\, \alpha_i = \beta_i \text{ or} \\ &\exists i \leq \min(n,m)[\alpha_i <_{\varphi Q0} \beta_i \wedge \forall j < i\, \alpha_j = \beta_j]. \end{aligned}$$

(6) If $\alpha_1 + \ldots + \alpha_n \in \mathrm{fld}(\varphi Q0)$, $\varphi u\beta \in \mathrm{fld}(\varphi Q0)$ and $\varphi u\beta \leq_{\varphi Q0} \alpha_1$ then $\varphi u\beta <_{\varphi Q0} \alpha_1 + \ldots + \alpha_n$.
(7) If $\alpha_1 + \ldots + \alpha_n \in \mathrm{fld}(\varphi Q0)$, $\varphi u\beta \in \mathrm{fld}(\varphi Q0)$ and $\alpha_1 <_{\varphi Q0} \varphi u\beta$ then $\alpha_1 + \ldots + \alpha_n <_{\varphi Q0} \varphi u\beta$.

(8) If $\varphi u\alpha, \varphi v\beta \in \text{fld}(\varphi Q0)$, then

$$\begin{aligned} \varphi u\alpha <_{\varphi Q0} \varphi v\beta \text{ iff } & u <_Q v \wedge \alpha <_{\varphi Q0} \varphi v\beta \quad \text{or} \\ & u = v \wedge \alpha <_{\varphi Q0} \beta \quad \text{or} \\ & v <_Q u \wedge \varphi u\alpha <_{\varphi Q0} \beta. \end{aligned}$$

**Lemma 11.2.3.** ($\mathbf{RCA}_0$)

*(i) If $Q$ is a linear ordering then so is $\varphi Q0$.*
*(ii) $\varphi Q0$ is elementary recursive in $Q$.*

## 11.3. The Theory $\mathbf{ATR}_0$

**Definition 11.3.1.** Let $A(u, Y)$ be any formula. Define $H_A(X, Y)$ to be the formula which says that $<_X$ is a linear ordering and that $Y$ is equal to the set of pairs $\langle n, j \rangle$ such that $j$ is in the field of $<_X$ and $A(n, Y^j)$ where $Y^j = \{\langle m, i \rangle \mid i <_X j \wedge \langle m, i \rangle \in Y\}$. Intuitively $H_A(X, Y)$ says that $Y$ is the result of iterating $A$ along $<_X$.

$\mathbf{ATR}_0$ is the formal system in the language of second order arithmetic whose axioms consist of $\mathbf{ACA}_0$ plus all instances of

$$\forall X(\mathbf{WO}(<_X) \rightarrow \exists Y\, H_A(X, Y))$$

where $A$ is arithmetical.

**Lemma 11.3.2.** $\mathbf{ATR}_0$ *can be axiomatized over* $\mathbf{ACA}_0$ *via a single sentence*

$$\forall X(\mathbf{WO}(<_X) \rightarrow \forall Z \exists Y\, B_0(X, Y, Z)) \tag{11.2}$$

*where $B_0(X, Y, Z)$ is of the form $H_A(X, Y)$ for some arithmetical formula $A(u, Y, Z)$ with all free variables exhibited.*

***Proof.*** This is a standard result. One could for instance take $B_0(X, Y, Z)$ to mean that $Y$ is obtained from $Z$ by iterating the Turing jump operation along $<_X$ starting with $Z$; so $A(u, Y, Z)$ would actually be a $\Sigma^0_1$ (complete) formula. Another (shorter and citable) way of showing this is to use the fact that $\mathbf{ATR}_0$ is equivalent over $\mathbf{RCA}_0$ to the statement that every two well-orderings are comparable (see [17], Theorem V.6.8). The proof of the latter statement in $\mathbf{ATR}_0$ just requires an instance $H_A$ of said form (see the proof of [17, Lemma V.2.9]). □

**Definition 11.3.3.** Let $T$ be a theory in the language of second order arithmetic, $\mathrm{L}_2$. A *countable coded $\omega$-model of $T$* is a set $W \subseteq \mathbb{N}$, viewed as encoding the $\mathrm{L}_2$-model

$$\mathbb{M} = (\mathbb{N}, \mathcal{S}, +, \cdot, 0, 1, <)$$

with $\mathcal{S} = \{(W)_n \mid n \in \mathbb{N}\}$ such that $\mathbb{M} \models T$.

This definition can be made in $\mathbf{RCA}_0$ (see [17], Definition VII.2).

We write $X \in W$ if $\exists n\, X = (W)_n$.

## 11.4. Main Theorem

The main result we want to prove is the following:

**Theorem 11.4.1.** $\mathbf{RCA}_0 + \forall\mathfrak{X}\, [\mathbf{WO}(\mathfrak{X}) \to \mathbf{WO}(\varphi\mathfrak{X}0)]$ *proves* $\mathbf{ATR}_0$.

A central ingredient of the proof will be a method of proof search (deduction chains) pioneered by Schütte [15].

### 11.4.1. *Deduction chains in $\omega$-logic*

**Definition 11.4.2.**

(i) Let $U_0, U_1, U_2, \ldots$ be an enumeration of the free set variables of $\mathrm{L}_2$. For a closed term $t$, let $t^{\mathbb{N}}$ be its numerical value. We shall assume that all predicate symbols of the language $\mathrm{L}_2$ are symbols for primitive recursive relations. $\mathrm{L}_2$ contains predicate symbols for the primitive recursive relations of equality and inequality and possibly more (or all) primitive recursive relations. If $R$ is a predicate symbol in $\mathrm{L}_2$ we denote by $R^{\mathbb{N}}$ the primitive recursive relation it stands for. If $t_1, \ldots, t_n$ are closed terms the formula $R(t_1, \ldots, t_n)$ $(\neg R(t_1, \ldots, t_n))$ is said to be *true* if $R^{\mathbb{N}}(t_1^{\mathbb{N}}, \ldots, t_n^{\mathbb{N}})$ is true (is false).

(ii) Henceforth a **sequent** will be a finite set of $\mathrm{L}_2$-formulas *without free number variables.*

(iii) A sequent $\Gamma$ is **axiomatic** if it satisfies at least one of the following conditions:

(1) $\Gamma$ contains a true **literal**, i.e. a true formula of either form $R(t_1, \ldots, t_n)$ or $\neg R(t_1, \ldots, t_n)$, where $R$ is a predicate symbol in $\mathrm{L}_2$ for a primitive recursive relation and $t_1, \ldots, t_n$ are closed terms.

(2) $\Gamma$ contains formulas $s \in U$ and $t \notin U$ for some set variable $U$ and terms $s, t$ with $s^{\mathbb{N}} = t^{\mathbb{N}}$.

(iv) A sequent is **reducible** or a **redex** if it is not axiomatic and contains a formula which is not a literal.

**Definition 11.4.3.** For $Q \subseteq \mathbb{N}$ define

$$D_Q(n) = \begin{cases} \bar{n} \in U_0 & \text{if } n \in Q \\ \bar{n} \notin U_0 & \text{otherwise}. \end{cases}$$

For the proof of Theorem 11.4.1 it is convenient to have a finite axiomatization of arithmetic comprehension.

**Lemma 11.4.4.** $\mathbf{ACA}_0$ *can be axiomatized via a single* $\Pi^1_2$ *sentence* $\forall X\, C(X)$.

**Proof:** [17], Lemma VIII.1.5. □

**Definition 11.4.5.** Let $<_Q$ be a well-ordering. Let $B(U_i)$ be the formula $\exists Y\, B_0(U_0, Y, U_i)$ of Lemma 11.3.2.

A $Q$**-deduction chain** is a finite string

$$\Gamma_0, \Gamma_1, \ldots, \Gamma_k$$

of sequents $\Gamma_i$ constructed according to the following rules:

(i) $\Gamma_0 = \neg D_Q(0), \neg B(U_0), \neg C(U_0)$.

(ii) $\Gamma_i$ is not axiomatic for $i < k$.

(iii) If $i < k$ and $\Gamma_i$ is not reducible then

$$\Gamma_{i+1} = \Gamma_i, \neg D_Q(i+1), \neg B(U_{i+1}), \neg C(U_{i+1}).$$

(iv) Every reducible $\Gamma_i$ with $i < k$ is of the form

$$\Gamma_i', E, \Gamma_i''$$

where $E$ is not a literal and $\Gamma_i'$ contains only literals.
$E$ is said to be the **redex** of $\Gamma_i$.
Let $i < k$ and $\Gamma_i$ be reducible. $\Gamma_{i+1}$ is obtained from $\Gamma_i = \Gamma_i', E, \Gamma_i''$ as follows:

(1) If $E \equiv E_0 \vee E_1$ then

$$\Gamma_{i+1} = \Gamma_i', E_0, E_1, \Gamma_i'', \neg D_Q(i+1), \neg B(U_{i+1}), \neg C(U_{i+1}).$$

(2) If $E \equiv E_0 \wedge E_1$ then

$$\Gamma_{i+1} = \Gamma_i', E_j, \Gamma_i'', \neg D_Q(i+1), \neg B(U_{i+1}), \neg C(U_{i+1})$$

where $j = 0$ or $j = 1$.

(3) If $E \equiv \exists x\, F(x)$ then

$$\Gamma_{i+1} = \Gamma'_i, F(\bar{m}), \Gamma''_i, \neg D_Q(i+1), \neg B(U_{i+1}), \neg C(U_{i+1}), E$$

where $m$ is the first number such that $F(\bar{m})$ does not occur in $\Gamma_0, \ldots, \Gamma_i$.

(4) If $E \equiv \forall x\, F(x)$ then

$$\Gamma_{i+1} = \Gamma'_i, F(\bar{m}), \Gamma''_i, \neg D_Q(i+1), \neg B(U_{i+1}), \neg C(U_{i+1})$$

for some $m$.

(5) If $E \equiv \exists X\, F(X)$ then

$$\Gamma_{i+1} = \Gamma'_i, F(U_m), \Gamma''_i, \neg D_Q(i+1), \neg B(U_{i+1}), \neg C(U_{i+1}), E$$

where $m$ is the first number such that $F(U_m)$ does not occur in $\Gamma_0, \ldots, \Gamma_i$.

(6) If $E \equiv \forall X\, F(X)$ then

$$\Gamma_{i+1} = \Gamma'_i, F(U_m), \Gamma''_i, \neg D_Q(i+1), \neg B(U_{i+1}), \neg C(U_{i+1})$$

where $m$ is the first number such that $m \neq i+1$ and $U_m$ does not occur in $\Gamma_i$.

The set of $Q$-deduction chains forms a tree $\mathcal{D}_Q$ labeled with strings of sequents. We will first consider the case that $\mathcal{D}_Q$ is not well-founded. Then $\mathcal{D}_Q$ contains an infinite path $\mathbb{P}$. Now define a set $M$ via

$$(M)_i = \{t^{\mathbb{N}} \mid t \notin U_i \text{ occurs in } \mathbb{P}\}.$$

Set $\mathbb{M} = (\mathbb{N}; \{(M)_i \mid i \in \mathbb{N}\}, +, \cdot, 0, 1, <)$.

For a formula $F$, let $F \in \mathbb{P}$ mean that $F$ occurs in $\mathbb{P}$, i.e. $F \in \Gamma$ for some $\Gamma \in \mathbb{P}$.

**Claim.** Under the assignment $U_i \mapsto (M)_i$ we have

$$F \in \mathbb{P} \quad \Rightarrow \quad \mathbb{M} \models \neg F. \tag{11.3}$$

The Claim will imply that $\mathbb{M}$ is an $\omega$-model of **ACA**. Also note that $(M)_0 = Q$, thus $Q$ is in $\mathbb{M}$. The proof of (11.3) follows by induction on $F$ using Lemma 11.4.6 below. The upshot of the foregoing is that we can prove Theorem 11.4.1 under the assumption that $\mathcal{D}_Q$ is ill-founded for all sets $Q \subseteq \mathbb{N}$.

**Lemma 11.4.6.** *Let $Q$ be an arbitrary subset of $\mathbb{N}$ and $\mathcal{D}_Q$ be the corresponding deduction tree. Moreover, suppose $\mathcal{D}_Q$ is not well-founded. Then $\mathcal{D}_Q$ has an infinite path $\mathbb{P}$. $\mathbb{P}$ has the following properties:*

*(1)* $\mathbb{P}$ *does not contain literals which are true in* $\mathbb{N}$.
*(2)* $\mathbb{P}$ *does not contain formulas* $s \in U_i$ *and* $t \notin U_i$ *for constant terms* $s$ *and* $t$ *such that* $s^{\mathbb{N}} = t^{\mathbb{N}}$.
*(3) If* $\mathbb{P}$ *contains* $E_0 \vee E_1$ *then* $\mathbb{P}$ *contains* $E_0$ *and* $E_1$.
*(4) If* $\mathbb{P}$ *contains* $E_0 \wedge E_1$ *then* $\mathbb{P}$ *contains* $E_0$ *or* $E_1$.
*(5) If* $\mathbb{P}$ *contains* $\exists x F(x)$ *then* $\mathbb{P}$ *contains* $F(\bar{n})$ *for all* $n$.
*(6) If* $\mathbb{P}$ *contains* $\forall x F(x)$ *then* $\mathbb{P}$ *contains* $F(\bar{n})$ *for some* $n$.
*(7) If* $\mathbb{P}$ *contains* $\exists X F(X)$ *then* $\mathbb{P}$ *contains* $F(U_m)$ *for all* $m$.
*(8) If* $\mathbb{P}$ *contains* $\forall X F(X)$ *then* $\mathbb{P}$ *contains* $F(U_m)$ *for some* $m$.
*(9)* $\mathbb{P}$ *contains* $\neg B(U_m)$ *for all* $m$.
*(10)* $\mathbb{P}$ *contains* $\neg C(U_m)$ *for all* $m$.
*(11)* $\mathbb{P}$ *contains* $\neg D_Q(m)$ *for all* $m$.

***Proof.*** Standard. □

**Corollary 11.4.7.** *If* $\mathcal{D}_Q$ *is ill-founded then there exists a countable coded* $\omega$*-model of* $\mathbf{ACA}_0$ *containing* $Q$ *which satisfies* $\forall Z \exists Y\, B_0(Q, Y, Z)$.

The remainder of the paper will be devoted to ruling out the possibility that, whenever $Q$ is a well-ordering, $\mathcal{D}_Q$ can be a well-founded tree. This is the place where cut elimination for the infinitary proof system of ramified analysis, $\mathbf{RA}^*$ (see [15], part C), enters the stage. In a nutshell, the idea is that a well-founded $\mathcal{D}_Q$ gives rise to a derivation of the empty sequent (contradiction) in $\mathbf{RA}^*$ which can be ruled out by showing cut elimination for $\mathbf{RA}^*$ using transfinite induction up to $\varphi\mathfrak{X}0$, where $\mathfrak{X}$ is a well-ordering not much longer than $Q$. However, to simplify the technical treatment we first introduce an intermediate system $\Delta^1_1$-$\mathbf{CR}^Q_\infty$ based on the $\Delta^1_1$-comprehension rule and the $\omega$-rule. This theory basically coincides with Schütte's system $\mathbf{DA}^*$ (see [15], part C). It is not difficult to see that a well-founded $\mathcal{D}_Q$ can be viewed as a derivation of the empty sequent in $\Delta^1_1$-$\mathbf{CR}^Q_\infty$. The last step towards reaching a contradiction consists in embedding $\Delta^1_1$-$\mathbf{CR}^Q_\infty$ into $\mathbf{RA}^*$. Here we can basically follow [15] Theorem 22.14.

### 11.4.2. *The infinitary calculus* $\Delta^1_1$-$\mathbf{CR}^Q_\infty$

In what follows we fix $Q \subseteq \mathbb{N}$ such that $<_Q$ is a well-ordering. In the main, the system $\Delta^1_1$-$\mathbf{CR}^Q_\infty$ is obtained from $\mathbf{ACA}_0$ by adding the $\Delta^1_1$-comprehension rule, the $\omega$-rule and the basic diagram of $Q$. The language of $\Delta^1_1$-$\mathbf{CR}^Q_\infty$ is the same as that of $\mathbf{ACA}_0$ but the notion of formula comes enriched with set terms. Formulas and **set terms** are defined simultaneously. Literals are formulas. Every set variable is a set term. If $A(x)$ is a formula without set quantifiers (i.e. arithmetical) then $\{x \mid A(x)\}$ is a set term. If $P$ is a set term and $t$ is a numerical term then $t \in P$ and $t \notin P$ are formulas. The other formation rules pertaining to $\wedge, \vee, \forall x, \exists x, \forall X, \exists X$ are as per usual.

We will be working in a Tait-style formalization of the second order arithmetic with formulas in negation normal form, i.e. negations only in front of atomic formulas. Due to the $\omega$-rule there is no need for formulas with free numerical variables. Thus all sequents below are assumed to consist of formulas without free numerical variables.

**Axioms of $\Delta^1_1$-$\mathbf{CR}^Q_\infty$**

(i) $\Gamma, L$ where $L$ is a true literal.
(ii) $\Gamma, s \in U, t \notin U$ where $s^{\mathbb{N}} = t^{\mathbb{N}}$.
(iii) $\Gamma, s \in U_0$ if $s^{\mathbb{N}} \in Q$.
(iv) $\Gamma, s \notin U_0$ if $s^{\mathbb{N}} \notin Q$.

**Rules of $\Delta^1_1$-$\mathbf{CR}^Q_\infty$**

$$(\wedge)\ \frac{\Gamma, A \qquad \Gamma, B}{\Gamma, A \wedge B}.$$

$$(\vee)\ \frac{\Gamma, A_i}{\Gamma, A_0 \vee A_1} \text{ where } i \in \{0,1\}.$$

$$(\mathrm{Cut})\ \frac{\Gamma, A \qquad \Gamma, \neg A}{\Gamma}.$$

$$(\omega)\ \frac{\Gamma, F(\bar{n}) \quad \text{for all } n}{\Gamma, \forall x F(x)}.$$

$$(\exists_1)\ \frac{\Gamma, F(t)}{\Gamma, \exists x F(x)}.$$

$$(\forall_2)\ \frac{\Gamma, F(P) \quad \text{for all set terms } P}{\Gamma, \forall X F(X)}.$$

$$(\exists_2)\ \frac{\Gamma, F(P)}{\Gamma, \exists X F(X)} \text{ where } P \text{ is a set term.}$$

$(\Delta^1_1\text{-CR})\ \dfrac{\forall x[\forall Y A_0(x,Y) \leftrightarrow \exists Y A_1(x,Y)]}{\Gamma, \exists X \forall x[x \in X \leftrightarrow \forall Y A_0(x,Y)]}$ with $A_0, A_1$ arithmetical.

$(ST_1)\ \dfrac{\Gamma, A(t)}{\Gamma, t \in P}$ where $P$ is the set term $\{x \mid A(x)\}$.

$(ST_2)\ \dfrac{\Gamma, \neg A(t)}{\Gamma, t \notin P}$ where $P$ is the set term $\{x \mid A(x)\}$.

$\Delta^1_1$-$\mathbf{CR}^Q_\infty$ is a sequent calculus version of the system DA* of [15, §20]. The language of DA*, though, is based on the connectives $\bot, \forall, \rightarrow$ while $\Delta^1_1$-$\mathbf{CR}^Q_\infty$ has the connectives $\wedge, \vee, \forall, \exists, \neg$ and formulas are in negation normal form, i.e. the negation sign appears only in front of atomic formulas. The other main difference is that the deduction system of DA* is the Schütte calculus of positive and negative forms whereas $\Delta^1_1$-$\mathbf{CR}^Q_\infty$'s is the Gentzen sequent calculus.

**Lemma 11.4.8.** *We shall use* $\Delta^1_1$-$\mathbf{CR}^Q_\infty \vdash \Gamma$ *to convey that the sequent* $\Gamma$ *is derivable in* $\Delta^1_1$-$\mathbf{CR}^Q_\infty$. *Pivotal properties of* $\Delta^1_1$-$\mathbf{CR}^Q_\infty$ *we shall exploit are the following:*

*(a)* $n \in Q \Rightarrow \Delta^1_1\text{-}\mathbf{CR}^Q_\infty \vdash \bar{n} \in U_0$.
*(b)* $n \notin Q \Rightarrow \Delta^1_1\text{-}\mathbf{CR}^Q_\infty \vdash \bar{n} \notin U_0$.
*(c)* $\Delta^1_1\text{-}\mathbf{CR}^Q_\infty \vdash \mathbf{WO}(U_0)$.
*(d)* $\Delta^1_1\text{-}\mathbf{CR}^Q_\infty \vdash \exists Y H_A(U_0, Y)$ *for all arithmetical formulas* $A(u, Y)$ *having no other free numerical variables than* $u$.

**Proof:**

(a) and (b) are immediate by the axioms (iii) and (iv) of $\Delta^1_1$-$\mathbf{CR}^Q_\infty$.

(c) follows by (outer) transfinite induction on $<_Q$, crucially using the $\omega$-rule. This is standard but it seems to be a challenge to find a reference. Via the axioms (iii) and (iv), the role of $Q$ is played in $\Delta^1_1$-$\mathbf{CR}^Q_\infty$ by the variable $U_0$. Writing $s \in Q$ and $s <_Q t$ for $s \in U_0$ and $\langle s, t\rangle \in U_0$, respectively, we would like to show that $\Delta^1_1\text{-}\mathbf{CR}^Q_\infty \vdash \forall X(\mathit{Prog}_Q(X) \rightarrow \forall x\, x \in X)$, where $\mathit{Prog}_Q(U)$ stands for $\forall x[\forall y(y <_Q x \rightarrow y \in U) \rightarrow x \in U]$. It suffices to show

$$\Delta^1_1\text{-}\mathbf{CR}^Q_\infty \vdash \neg \mathit{Prog}_Q(U), \bar{n} \in U \tag{11.4}$$

for all $n$ for an arbitrary set variable $U$. To this end we proceed by induction on $Q$. Inductively assume that $\Delta^1_1\text{-}\mathbf{CR}^Q_\infty \vdash \neg \mathit{Prog}_Q(U), \bar{m} \in U$ holds for all $m <_Q n$. If $m <_Q n$ is false then $\langle m, n\rangle \notin Q$ and hence $\Delta^1_1\text{-}\mathbf{CR}^Q_\infty \vdash$

$\neg\bar{m} <_Q \bar{n}$. As a result, $\Delta^1_1\text{-}\mathbf{CR}^Q_\infty \vdash \neg Prog_Q(U), \neg\bar{m} <_Q \bar{n}, \bar{m} \in U$ holds for all $m$. Using $(\vee)$ inferences followed by an application of the $\omega$-rule, we get $\Delta^1_1\text{-}\mathbf{CR}^Q_\infty \vdash \neg Prog_Q(U), \forall y(y <_Q \bar{n} \to y \in U)$. As $\Delta^1_1\text{-}\mathbf{CR}^Q_\infty \vdash \bar{n} \notin Q, \bar{n} \in Q$, an inference $(\vee)$ (and weakening) yields

$$\Delta^1_1\text{-}\mathbf{CR}^Q_\infty \vdash \neg Prog_Q(U), \forall y(y <_Q \bar{n} \to y \in U) \wedge \bar{n} \notin Q, \bar{n} \in Q.$$

Hence via $(\exists_1)$ we arrive at

$$\Delta^1_1\text{-}\mathbf{CR}^Q_\infty \vdash \neg Prog_Q(U), \exists x[\forall y(y <_Q \bar{n} \to y \in U) \wedge x \notin Q], \bar{n} \in Q,$$

which is the same as $\Delta^1_1\text{-}\mathbf{CR}^Q_\infty \vdash \neg Prog_Q(U), \bar{n} \in Q$. Thus, by induction on $<_Q$, (11.4) follows.

(d) also follows by transfinite induction on $<_Q$ using $\Delta^1_1$-CR. A reference will be provided in Lemma 11.4.10. □

We shall need to measure the length of the previous derivations. For (c) and (d) the lengths of those derivations will be "longer" than $Q$, though not "much longer". Let $\tau$ be the ordinal giving the order-type of $Q$. It is easy to cook up a new ordering $Q^*$ in an elementary way from $Q$ corresponding to the ordinal $\omega^2 + \omega \cdot \tau + \omega$ in such a way that $\mathbf{RCA}_0$ suffices to prove $\mathbf{WO}(Q) \to \mathbf{WO}(Q^*)$ (see [9]). The rationale for the choice of $\omega^2 + \omega \cdot \tau + \omega$ is that it gives us enough elbow room for calibrating the lengths of the foregoing derivations.

From the standing assumption that $Q$ is a well-ordering we get that $Q^*$ is a well-ordering, too.

**Definition 11.4.9.** If $\alpha$ is an element of the field of $<_{Q^*}$, we use the notation $\Delta^1_1\text{-}\mathbf{CR}^Q_\infty \overset{\alpha}{\vdash} \Gamma$ to convey that the sequent $\Gamma$ is deducible in $\Delta^1_1\text{-}\mathbf{CR}^Q_\infty$ via a derivation of length $\leq \alpha$. More formally, this relation is defined by recursion on $\alpha$ as follows: $\Delta^1_1\text{-}\mathbf{CR}^Q_\infty \overset{\alpha}{\vdash} \Gamma$ holds if either $\Gamma$ is an axiom of $\Delta^1_1\text{-}\mathbf{CR}^Q_\infty$ or $\Gamma$ is the conclusion of a $\Delta^1_1\text{-}\mathbf{CR}^Q_\infty$-inference with premisses $(\Gamma_i)_{i\in I}$ such that for every $i \in I$ there exists $\beta_i <_{Q^*} \alpha$ with $\Delta^1_1\text{-}\mathbf{CR}^Q_\infty \overset{\beta_i}{\vdash} \Gamma_i$.

**Lemma 11.4.10.**

*(1)* $\Delta^1_1\text{-}\mathbf{CR}^Q_\infty \overset{0}{\vdash} D_Q(n)$ *for all* $n$ *with* $0$ *being the least element of* $Q$.

*(2)* $\Delta^1_1\text{-}\mathbf{CR}^Q_\infty \overset{\alpha}{\vdash} C(U)$ *for some* $\alpha \in$ *field*$(Q^*)$ *and all free set variables* $U$.

*(3)* $\Delta^1_1$-$\mathbf{CR}^Q_\infty \stackrel{\beta}{\vdash} \mathbf{WO}(U_0)$ *for some* $\beta \in$ *field*$(Q^*)$.

*(4)* $\Delta^1_1$-$\mathbf{CR}^Q_\infty \stackrel{\gamma}{\vdash} \exists Y\, H_A(U_0, Y)$ *for some* $\gamma \in$ *field*$(Q^*)$ *for all arithmetical formulas* $A(u, Y)$ *having no other free numerical variables than* $u$.

*(5)* $\Delta^1_1$-$\mathbf{CR}^Q_\infty \stackrel{\delta}{\vdash} B(U)$ *for some* $\delta \in$ *field*$(Q^*)$ *and all free set variables* $U$.

**Proof:** (1) is an immediate consequence of Lemma 11.4.8 (a) and (b). (2) follows since the rule $(\exists_2)$ gives arithmetical comprehension. (3) and (4) correspond to Lemma 11.4.8 (c) and (d), respectively. A detailed proof of (4) amounts to basically the same as that of [15, §21 Lemma 14]. (5) is an immediate consequence of (4). □

Recall that, by Corollary 11.4.7, there exists a countable coded $\omega$-model of $\mathbf{ACA}_0$ containing $Q$ and satisfying $\forall Z \exists Y\, B_0(Q, Y, Z)$ providing $\mathcal{D}_Q$ is ill-founded. Now let us assume that $Q$ is a well-ordering and that $\mathcal{D}_Q$ is well-founded. Then $\mathcal{D}_Q$ can be viewed as a deduction with **hidden cuts** involving formulas of the shape $\neg B(U_{i+1})$, $\neg C(U_{i+1})$ and $\neg D_Q(i+1)$. Note that by Lemma 11.4.10, $\Delta^1_1$-$\mathbf{CR}^Q_\infty \stackrel{0}{\vdash} D_Q(n)$, $\Delta^1_1$-$\mathbf{CR}^Q_\infty \stackrel{\alpha}{\vdash} C(U)$, and $\Delta^1_1$-$\mathbf{CR}^Q_\infty \stackrel{\gamma}{\vdash} B(U)$ for some $\alpha, \gamma \in$ field$(Q^*)$. Thus if $\Gamma$ is the sequent attached to a node $\tau$ of $\mathcal{D}_Q$ and $(\Gamma_i)_{i\in I}$ is an enumeration of the sequents attached to the immediate successor nodes of $\tau$ in $\mathcal{D}_Q$ then the transition $\frac{(\Gamma_i)_{i\in I}}{\Gamma}$ can be viewed as a combination of four inferences in $\Delta^1_1$-$\mathbf{CR}^Q_\infty$, the first one being a logical inferences and the other three being cuts. By interspersing $\mathcal{D}_Q$ with cuts and adding three cuts with cut formulas $\neg C(U_0)$, $\neg B(U_0)$ and $\neg D_Q(0)$ at the bottom we obtain a derivation $\tilde{\mathcal{D}}_Q$ in $\Delta^1_1$-$\mathbf{CR}^Q_\infty$ of the empty sequent. Since the preceding line of arguments can be done in $\mathbf{ACA}_0$ we arrive at the following:

**Corollary 11.4.11 ($\mathbf{ACA}_0$).** *If* $Q$ *is a well-ordering and* $\mathcal{D}_Q$ *is well-founded then there is a derivation* $\tilde{\mathcal{D}}_Q$ *in* $\Delta^1_1$-$\mathbf{CR}^Q_\infty$ *of the empty sequent.*

To finish the paper we thus have to show that the latter is impossible. This we shall do by embedding $\Delta^1_1$-$\mathbf{CR}^Q_\infty$ into a system $\mathbf{RA}^\infty$ defined below. Note that an upper bound for the length of $\tilde{\mathcal{D}}_Q$ is provided by $(\alpha+\gamma+\rho)\cdot 4$, where $\rho$ corresponds to the Kleene–Brouwer ordering on $\mathcal{D}_Q$.

## 11.5. Ramified Analysis $\mathbf{RA}^\infty$

The theories $\mathbf{RA}_\rho$ are designed to capture Gödel's notion of *constructibility* restricted to sets of natural numbers. They use ordinal indexed variables $X^\alpha, Y^\alpha, Z^\alpha, \ldots$ for $\alpha < \rho$, with the intended meaning that level 0 variables range over sets definable by numerical quantification, and level $\alpha > 0$ variables range over sets definable by numerical quantification and level $< \alpha$ set quantification. The proof-theoretic ordinal of $\mathbf{RA}_\alpha$ is $\varphi\alpha 0$. We are interested in an infinitary version of ramified analysis.

**Definition 11.5.1.** $\mathbf{RA}^\infty$ is basically the same system as $\mathbf{RA}^*$ in [15, §22]. One difference is that the language of $\mathbf{RA}^*$ is based on the connectives $\bot, \forall, \rightarrow$ while $\mathbf{RA}^\infty$ has $\wedge, \vee, \forall, \exists, \neg$ and formulas are in negation normal form, i.e. the negation sign appears only in front of atomic formulas. The other difference is that the deduction system of $\mathbf{RA}^*$ is the Schütte calculus of positive and negative forms whereas $\mathbf{RA}^\infty$'s is the Gentzen sequent calculus.

The formulas of $\mathbf{RA}^\infty$ do not have free numerical variables. Literals are formulas of the form $R(t_1, \ldots, t_n)$ and $\neg R(t_1, \ldots, t_n)$ with $R$ being a symbol for a primitive recursive relation and $t_1, \ldots, t_n$ being closed numerical terms.

$\mathbf{RA}^\infty$ uses ordinal indexed free set variables $U^\alpha, V^\alpha, W^\alpha, \ldots$ and bound set variables $X^\beta, Y^\beta, Z^\beta, \ldots$ with $\beta > 0$, where the ordinals are assumed to be elements of some countable well-ordering $R$.

The *set terms* and *formulas* together with their *levels* are generated as follows (cf. [15, §22]):

(1) Every literal is a formula of level 0.
(2) Every free set variable $U^\alpha$ is a **set term** of **level** $\alpha$.
(3) If $P$ is a set term of level $\alpha$ and $t$ is a numerical term, then $t \in P$ and $t \notin P$ are **formulas of level** $\alpha$.
(4) If $A$ and $B$ are formulas of levels $\alpha$ and $\beta$, then $A \vee B$ and $A \wedge B$ are formulas of level $\max(\alpha, \beta)$.
(5) If $F(0)$ is a formula of level $\alpha$, then $\forall x F(x)$ and $\exists x F(x)$ are formulas of level $\alpha$ and $\{x \mid F(x)\}$ is a **set term** of level $\alpha$.
(6) If $F(U^\beta)$ is a formula of level $\alpha$ and $\beta > 0$, then $\forall X^\beta F(X^\beta)$ is a formula of level $\max(\alpha, \beta)$.

**Definition 11.5.2.** The calculus $\mathbf{RA}^\infty_Q$

**Axioms**

$\Gamma, L$ where $L$ is a true literal.
$\Gamma, s \in U^\alpha, t \notin U^\alpha$ where $s^{\mathbb{N}} = t^{\mathbb{N}}$.
$\Gamma, s \in U_0$ if $s^{\mathbb{N}} \in Q$.
$\Gamma, s \notin U_0$ if $s^{\mathbb{N}} \notin Q$.

**Rules**

$(\wedge), (\vee)$, $(\omega)$, numerical $(\exists)$ and (Cut) as per usual.

$$(\exists^\alpha)\ \frac{\Gamma, F(P)}{\Gamma, \exists X^\alpha F(X^\alpha)} \qquad P \text{ set term of level} < \alpha.$$

$$(\forall^\alpha)\ \frac{\Gamma, F(P) \quad \text{for all set terms } P \text{ of level} < \alpha}{\Gamma, \forall X^\alpha F(X^\alpha)}.$$

$$(ST_1)\ \frac{\Gamma, F(t)}{\Gamma, t \in \{x \mid F(x)\}}.$$

$$(ST_2)\ \frac{\Gamma, \neg F(t)}{\Gamma, t \notin \{x \mid F(x)\}}.$$

**Definition 11.5.3.** The **cut rank** of a formula $A$ in $\mathbf{RA}_Q^\infty$, $|A|$, is defined as follows (cf. [15, §22]):

(1) $|L| = 0$ for arithmetical literals $L$.

(2) $|t \in U^\alpha| = |t \notin U^\alpha| = \omega \cdot \alpha$.

(3) $|B_0 \wedge B_1| = |B_0 \vee B_1| = \max(|B_0|, |B_1|) + 1$.

(4) $|\forall x B(x)| = |\exists x B(x)| = |t \in \{x \mid B(x)\}| = |t \notin \{x \mid B(x)\}| = |B(0)|+1$.

(5) $|\forall X^\alpha A(X^\alpha)| = |\exists X^\alpha A(X^\alpha)| = \max(\omega \cdot \gamma, |A(U^0)| + 1)$
where $\gamma$ is the level of $\forall X^\alpha A(X^\alpha)$.

By recursion on $\alpha$ we define the relation $\mathbf{RA}_Q^\infty \mid\frac{\alpha}{\rho} \Gamma$ as follows: $\mathbf{RA}_Q^\infty \mid\frac{\alpha}{\rho} \Gamma$ holds if either $\Gamma$ is an axiom of $\mathbf{RA}_Q^\infty$ or $\Gamma$ is the conclusion of an $\mathbf{RA}_Q^\infty$-

inference with premisses $(\Gamma_i)_{i\in I}$ such that for every $i \in I$ there exists $\beta_i < \alpha$ with $\mathbf{RA}_Q^\infty \mathrel{\vert\!\frac{\beta_i}{\rho}} \Gamma_i$ and, moreover, if this inference is a cut with cut formula $A$ then $|A| < \rho$.

The following three statements are proved in [15] for the system $\mathbf{RA}^*$. It is routine to transfer them to $\mathbf{RA}_Q^\infty$ since cut-elimination in a Schütte calculus of positive and negative is closely related to cut-elimination in sequent calculi. Moreover, the additional axioms pertaining to $Q$ do not impede the cut-elimination process.

**Theorem 11.5.4 (Cut-elimination I).**

$$\mathbf{RA}_Q^\infty \mathrel{\vert\!\frac{\alpha}{\eta+1}} \Gamma \quad\Rightarrow\quad \mathbf{RA}_Q^\infty \mathrel{\vert\!\frac{\omega^\alpha}{\eta}} \Gamma\,.$$

***Proof.*** Similar to [15, §22 Lemma 4]. □

**Theorem 11.5.5 (Cut-elimination II).**

$$\mathbf{RA}_Q^\infty \mathrel{\vert\!\frac{\alpha}{\omega^\rho}} \Gamma \quad\Rightarrow\quad \mathbf{RA}_Q^\infty \mathrel{\vert\!\frac{\varphi\rho\alpha}{0}} \Gamma\,.$$

***Proof.*** Similar to [15, Theorem 22.7]. □

For a formula $F$ of the language of $\Delta_1^1\text{-}\mathbf{CR}_\infty^Q$ let $F^\sigma$ be the result of replacing every bound variable $X$ by $X^\sigma$ and every free set variable by a set term of a level $< \sigma$. For $\Gamma = \{F_1, \ldots, F_n\}$ let $\Gamma^\sigma = \{F_1^\sigma, \ldots, F_n^\sigma\}$.

**Theorem 11.5.6 (Interpretation Theorem).**

$$\Delta_1^1\text{-}\mathbf{CR}_\infty^Q \mathrel{\vert\!\frac{\alpha}{}} \Gamma \quad\Rightarrow\quad \mathbf{RA}_Q^\infty \mathrel{\vert\!\frac{\omega\cdot\sigma+\omega+\omega\cdot\alpha}{\omega\cdot\sigma}} \Gamma^\sigma$$

*for all $\sigma$ of the form $\omega^\alpha \cdot \beta$ with $\beta \neq 0$.*

***Proof.*** This is basically the same as [15, Theorem 22.14]. □

There are different ways of formalizing infinite deductions in theories like **PA**. We just mention [16] and [8].

### 11.5.1. *Finishing the proof of the main theorem*

Recall that in order to finish the proof of Theorem 11.4.1 we want to show that $\mathcal{D}_Q$ is not well-founded whenever $Q$ is a well-ordering. By Corollary 11.4.11, if $Q$ is a well-ordering and $\mathcal{D}_Q$ is well-founded then there is a derivation $\tilde{\mathcal{D}}_Q$ in $\Delta_1^1\text{-}\mathbf{CR}_\infty^Q$ of the empty sequent. By the Interpretation Theorem 11.5.6 we would then get a derivation in $\mathbf{RA}_Q^\infty$ of the empty sequent. Using the principle $\mathbf{WO}(\mathfrak{X}) \to \mathbf{WO}(\varphi\mathfrak{X}0)$ we can then employ the cut-elimination Theorem 11.5.5 to obtain a cut-free derivation of the empty sequent in $\mathbf{RA}^{\infty}_{\tilde{Q}}$. But this is impossible.

From Corollary 11.4.7 we can thus conclude that for every well-ordering $\tilde{Q}$ there exists a countable coded $\omega$-model of $\mathbf{ACA}_0$ containing $\tilde{Q}$ and satisfying $\forall Z \exists Y\, B_0(\tilde{Q}, Y, Z)$. From this we would like to infer that for every well-ordering $Q$ and every set $Z_0$ there exists a set $Y$ such that $B_0(\tilde{Q}, Y, Z_0)$. We can do this by encoding $Q$ and $Z_0$ in a well-ordering $\tilde{Q}$ from which $Q$ and $Z_0$ can be retrieved in any $\omega$-model of $\mathbf{ACA}_0$ containing $\tilde{Q}$. One way of doing this is to define the new ordering $\tilde{Q}$ by letting

$$\begin{aligned}\langle n, m\rangle <_{\tilde{Q}} \langle n', m'\rangle \text{ iff } & [n = n' = 0 \wedge m <_{\tilde{Q}} m'] \vee \\ & [n = n' = 1 \wedge m, m' \in Z_0 \wedge m < m'] \vee \\ & [n = 0 \wedge n' = 1 \wedge m \in \mathrm{field}(Q) \wedge m' \in Z_0].\end{aligned}$$

Obviously $\tilde{Q}$ is a well-ordering, too, and any $\omega$-model $\mathbb{M}$ of $\mathbf{ACA}_0$ containing $\tilde{Q}$ will contain $Z_0$ as well. Moreover, $\mathbb{M} \models \exists Y\, B_0(\tilde{Q}, Z_0)$ implies $\mathbb{M} \models \exists Y\, B_0(Q, Z_0)$. Hence, in view of Lemma 11.3.2, we get $\mathbf{ATR}_0$, thereby finishing the proof of Theorem 11.4.1.

## 11.6. Finishing the Proof of Theorem 11.1.3

One direction of Theorem 11.1.3 follows from Theorem 11.4.1. The other direction is implicit in the proof of [15] Theorem 21.6.

## 11.7. Prospectus

The methodology exemplified in the proof of Theorem 11.1.3 should have many more applications. Every cut-elimination theorem in ordinal-theoretic proof theory potentially encapsulates a theorem of type 11.1.3. The first author has looked at two more examples and sketched proofs of the pertaining theorems. A familiar function from proof theory is the $\Gamma$-function where $\alpha \mapsto \Gamma_\alpha$ enumerates the fixed points of the $\varphi$-function. Since the proof of the next result has only been sketched we classify it as a conjecture.[a]

**Conjecture 11.7.1.** *Over* $\mathbf{RCA}_0$ *the following are equivalent:*

*(i) Every set* $X$ *is contained in a countable coded* $\omega$*-model of* $\mathbf{ATR}_0$.
*(ii)* $\forall \mathfrak{X}\, [\mathbf{WO}(\mathfrak{X}) \to \mathbf{WO}(\Gamma_{\mathfrak{X}})]$.

[a]Recently this conjecture has been proved in [13].

The direction (i)⇒(ii) follows from [12, 4.13,4.16].

For an example from impredicative proof theory one would perhaps first turn to the ordinal representation system used for the ordinal analysis of the theory $\mathbf{ID}_1$ of non-iterated inductive definitions, which can be expressed in terms of the $\theta$-function (cf. [5]). $\mathbf{ID}_1$ has the same strength as the subsystem of second order arithmetic based on bar induction, **BI** (cf. [4, 5, 14]). In Simpson's book the acronym used for **BI** is $\Pi^1_\infty$-$\mathbf{TI}_0$ (cf. [17, §VII.2]). In place of the function $\theta$ we prefer to work with simpler ordinal representations based on the $\psi$-function introduced in [3] or the $\vartheta$-function of [14]. For definiteness we refer to [14]. Given a well-ordering $\mathfrak{X}$, the relativized versions $\vartheta_{\mathfrak{X}}$ and $\psi_{\mathfrak{X}}$ of the $\vartheta$-function and the $\psi$-function, respectively, are obtained by adding all the ordinals from $\mathfrak{X}$ to the sets $C_n(\alpha, \beta)$ of [14, §1] and $C_n(\alpha)$ of [14, Definition 3.1] as initial segments, respectively. The resulting well-orderings $\vartheta_{\mathfrak{X}}(\varepsilon_{\Omega+1})$ and $\psi_{\mathfrak{X}}(\varepsilon_{\Omega+1})$ are equivalent owing to [14, Corollary 3.2].

Again, as the following statement has not been buttressed by a complete proof we formulate it as a conjecture.

**Conjecture 11.7.2.** *Over* $\mathbf{RCA}_0$ *the following are equivalent:*

*(i) Every set* $X$ *is contained in a countable coded* $\omega$*-model of* **BI**.
*(ii)* $\forall \mathfrak{X}\, [\mathbf{WO}(\mathfrak{X}) \rightarrow \mathbf{WO}(\psi_{\mathfrak{X}}(\varepsilon_{\Omega+1}))]$.

## References


[1] B. Afshari. *Proof-Theoretic Strengths of Hierarchies of Theories.* PhD thesis, University of Leeds, UK, (2008).

[2] B. Afshari and M. Rathjen, Reverse mathematics and well-ordering principles: a pilot study, *Ann. Pure Appl. Logic.* **160**, 231–237, (2009).

[3] W. Buchholz, A new system of proof–theoretic ordinal functions, *Ann. Pure Appl. Logic.* **32**, 195–207.

[4] W. Buchholz and K. Schütte, *Proof Theory of Impredicative Subsystems of Snalysis.* Bibliopolis, Naples, (1988).

[5] W. Buchholz, S. Feferman, W. Pohlers, and W. Sieg, *Iterated Inductive Definitions and Subsystems of Analysis.* Springer, Berlin, (1981).

[6] H. Friedman, Uniformly defined descending sequences of degrees, *J. Symbolic Logic.* **41**, 363–367, (1976).

[7] H. Friedman, A. Montalban, and A. Weiermann. Phi function. (Draft), (2007).

[8] H. Friedman and S. Sheard, Elementary descent recursion and proof theory, *Ann. Pure Appl. Logic.* **71**, 1–45, (1995).

[9] J.-Y. Girard, *Proof Theory and Logical Complexity.* vol. 1, Bibliopolis, Napoli, (1987).
[10] J. L. Hirst, Reverse mathematics and ordinal exponentiation, *Ann. Pure Appl. Logic.* **66**, 1–18, (1994).
[11] A. Marcone and A. Montalban. The epsilon function for computability theorists. (Draft), (2007).
[12] M. Rathjen, The strength of Martin-Löf type theory with a superuniverse. part i, *Arch. Math. Logic.* **39**, 1–39, (2000).
[13] M. Rathjen. $\omega$-models and well-ordering principles. To appear in *Foundational Adventures*, Proceedings in honor of Harvey Friedman's 60th birthday.
[14] M. Rathjen and A. Weiermann, Proof–theoretic investigations on Kruskal's theorem, *Ann. Pure Appl. Logic.* **60**, 49–88, (1993).
[15] K. Schütte, *Proof Theory.* Springer-Verlag, Berlin, Heidelberg, (1977).
[16] H. Schwichtenberg. Proof theory: Some applications of cut-elimination. In ed., J. Barwise, *Handbook of Mathematical Logic*, pp. 867–895. North Holland, Amsterdam, (1977).
[17] S. G. Simpson, *Subsystems of Second Order Arithmetic.* Springer-Verlag, Berlin, Heidelberg, (1999).
[18] J. Steel, Descending sequences of degrees, *J. Symbolic Logic.* **40**, 59–61, (1975).

# Chapter 12

# Discrete Transfinite Computation Models

Philip D. Welch
*School of Mathematics,*
*University of Bristol,*
*Bristol, BS8 1TW, UK*
*E-mail: p.welch@bristol.ac.uk*

Discrete computing models, such as that of the Turing machine or of Register machines, can be allowed to run transfinitely if suitable limit ordinal behavior is described. This chapter relates such models to classical accounts of higher type recursion theory, going back to Kleene. Using such models as a yardstick, one can analyse more modern models, such as computation in particular physical spacetimes, to give bounds on the complexity of such forms of computation. Computation on ordinals and set recursion are briefly discussed.

## Contents

## 12.1. Introduction

### 12.1.1. *The contents*

In the past few years there has been a resurgence of interest in discrete models of computation that are allowed to act *transfinitely*. Such conceptual machines act in simple steps or stages, and have as a paradigm the *standard Turing machine*. This, during its progress moves one cell at a time, to the left or the right along an unbounded tape that it is reading, and subsequently alters symbols, changes states and moves on. This paradigm has been with us for 70 years, and for much of this chapter we shall consider variants of such a device.

Our models will all be *discrete* acting computational digital models. We shall consider how Turing and other computing machines could possibly behave when allowed to perform *supertasks* (by which we mean that they are allowed to complete an infinite sequence of tasks or operations). Such a machine is usually envisaged with an unbounded tape. If supertasks are allowed then naturally the whole of that tape comes into play, and we can imagine that tape already having some characteristic function written on it. The machine can then act on that tape and is then essentially a computer acting *at a higher type*, namely that of infinite sequences of $0, 1$s, in other words, of real numbers.

Surprisingly, even if one restricts one's model to, say, a register machine model, where the registers are finite in number with natural number entries, then simply defined behaviour at transfinite limit stages of computation lead to quite powerful decision procedures. This chapter will look at these as well. Whereas at the finite level the power of Turing and register machines is the same, at the transfinite level they diverge markedly.

We shall not deal here with any machine that is, broadly speaking, an analog machine or computes in an analog fashion. Indeed apart from Section 2.1 we shall not be entering into any discussion of *physical* viabilities, feasibilities etc. We thus do not wish to discuss various machine proposals that could be seen to fall under the rubric we are setting ourselves, in that they seek to compute functions whilst being constructed to conform to some ambient physical theory or constraint. We have in mind models such as Davies [7], and the models of Beggs and Tucker [3] that attempt to compute any set of natural numbers, by some kinematic-based device, and thus do so within a fragment of Newtonian mechanics; nor do we consider the interaction of standard machines with physically based 'advice' functions or oracles, such as is done in [2]. We also shall not particularly consider *classical quantum computation*, as functions computed by such models are also Turing computable.

Section 2.1 is somewhat of an exception, in that we do consider a particular construction in detail due to Etesi and Németi [12], and to Hogarth [21], that allows Turing machines to be placed or arranged within particular spacetimes to allow for the algorithmic decision of $\Pi_1$ (and beyond) predicates without supertask phenomena. We can calculate somewhat precisely, the bounds to what can be computed in such models. We also recount the observation from [56], that *separability* of the spacetime manifold puts a universal countable bound on formal systems of computation within that spacetime (under some mild assumptions).

A major lacuna is that we do not consider finite automata on infinite graphs, or in particular on infinite binary trees. This would have fitted well in any discussion of discrete models, and is admittedly a serious omission. Lastly, this chapter is not about computation with an algebraic or structural flavour, and we include here, *inter alia*, the Blum, Shub, Smale model of computation on the reals ([4]). Although we do want to consider computation on the reals, $\mathbb{R}$ is considered primarily as unordered Baire space, or Cantor space, and there is not any other algebraic or structural feature that distinguishes the models considered here.

### 12.1.2. *Argument*

In the 1960s and 1970s much research was undertaken deriving ultimately from Kleene's theory ([26], [29]) of *recursion in higher types* (*Generalized Recursion Theory, Kleene Recursion*), and the pioneering work of Aczel, Gandy, Moschovakis amongst others, that would lead to *Spector classes*, the *theory of inductive definitions*, and the theory of *admissible sets*, (Barwise, Kripke, Platek).

Our motivation is that we wish to revisit some of these older theories and results and see how some recent activity in a class of models of computation fits in the older picture. Some of these later models are no more than familiar models with different style of inputs (register machines on ordinals say); others such as Infinite Time Turing Machine of Hamkins and Kidder ([15]) are versions of the standard model adapted to enable larger computations to be performed by allowing transfinite sequences of operations or stages; yet a third class simply consists of 'standard computation' placed in an unconventional framework (Turing machines stacked up, or regarded as inhabiting particular spacetimes). We want to see how these models fit in with our conceptions of recursion and computation formed in the earlier period.

We emphasise the logico-mathematical part of this, in particular the descriptive set theoretical descriptions. No apology is needed for this: to understand the model is to understand what the model produces: If this transcends that produced by finitary operations (in whatever form) then we are obliged to consider the underlying set-theoretical fundamentals. We consider further in Section 5 the connections of the computation models to the theory of inductive definitions (either monotone or not), and to subsystems of second order arithmetic. We do not consider this an accurate account on historical principles and we do not even claim to do justice to the concepts and individuals involved, but are merely taking a snapshot, as we rush past a fast-evolving subject.

## 12.2. Computation on Integers

We start ahistorically in terms of the published literature, but with a fact that surely must have been known to early recursion theorists such as Post: that allowing a Turing machine to at least run out indefinitely allows for the printing of the characteristic function not just of the halting problem (as a complete $\Sigma_1$-set) but of a $\Delta_2$-set. This has been called 'truth in the limit'. A Turing machine may answer any $\Pi_1$-question, *if* it is allowed $\omega$-many stages: given a recursive predicate $R(v_0, v_1)$ we may program a machine to investigate in turn $R(0, n), R(1, n), \ldots, R(k, n) \ldots$ in turn, and if for some $k$ $\neg R(k, n)$, then we require it to halt with a '0' for 'no' as output. If the machine does not halt, then were we able to 'transcend time' we could look back and say that the machine verified $\forall v_0 R(v_0, n)$. If we assume the machine has an output tape as well as a scratch tape, and if we assume the output tape starts out with every cell having a '1' written to it, we could dovetail all the queries $?\forall v_0 R(v_0, n)?$ for each $n$, and have the machine change a 1 to a 0 as soon as it verified that $\forall v_0 R(v_0, n)$ failed. Then after $\omega$ many stages, if we still could look at the output tape, we'd have written the characteristic function of the $\Pi_1$ set $A =_{\text{df}} \{n \mid \forall v_0 R(v_0, n)\}$. If we asked the question, *Is $A$ non-empty?* this is a $\Sigma_2$-query: $\exists v_1 \forall v_0 R(v_0, v_1)$? We cannot, in general, answer this, without allowing ourselves some further infinitary operation.

Putnam, much later, made the following definition:

**Definition 12.1. (Putnam [42])** $P$ is a *trial and error predicate* if and only if there is a general recursive function $f$ such that for every $x_1, \ldots, x_n$
$P(x_1, \ldots, x_n) \equiv \lim_{y \to \infty} f(x_1, \ldots, x_n, y) = 1$
$\neg P(x_1, \ldots, x_n) \equiv \lim_{y \to \infty} f(x_1, \ldots, x_n, y) = 0.$

Putnam asked, and answered, the question as to the complexity in the arithmetical hierarchy of such predicates: they are $\Delta_2 = \Sigma_2 \cap \Pi_2$ in the arithmetical hierarchy. We obtain the truth of $P(x_1, \ldots, x_n)$ 'in the limit' as $y \longrightarrow \infty$.

If we imagine the recursive function $f$ as being computed by a Turing machine $M$ writing its 0/1 output to a particular cell $C_0$ on its tape, the clauses above amount to a prescription of $M$'s behaviour on any input $x_1, \ldots, x_n$ that the contents of the cell $C_0$, after computing in turn $f(x_1, \ldots, x_n, 0)$, $f(x_1, \ldots, x_n, 1), \ldots$ , $f(x_1, \ldots, x_n, y), \ldots$, 'settles down' as $y$ increases: it must be either eventually 1/0 depending. We may thus rephrase Putnam as:

$P \subset \mathbb{N}^n$ *is a* trial and error predicate *if there is a Turing machine* $M_0$ *so that*
$P(x_1, \ldots, x_n) \Longleftrightarrow$ *the eventual value of* $M_0$*'s output cell on input* $n$ *is 1*
$\neg P(x_1, \ldots, x_n) \Longleftrightarrow$ *the eventual value of* $M_0$*'s output cell on input* $n$ *is 0.*

Continuing with this model for a moment, one sees that if there is in advance a fixed bound on the number of alternations that $M_0$ makes on the output cell $C_0$'s value, then that knowledge allows us to compute the characteristic functions of predicates in particular levels of the *difference hierarchy* of $\Sigma_1$ sets. Briefly: we say that $P(\vec{x})$ is in the $k$'th level of the difference hierarchy, if there are $\Sigma_1$ sets $Q_0, \ldots, Q_{2k-1}$ with $P(\vec{x}) \leftrightarrow \bigvee_{i \leq k}(Q_{2i-2}(\vec{x}) \wedge \neg Q_{2i-1}(\vec{x}))$; if we specify that $M'_0$ may only write to the cell $C_0$ at most $2k+1$ times, then $M'_0$ may decide $P(\vec{x})$. Note that it is the fixity of the value $k$ that determines the complexity within the class of Boolean combinations that is capable of being decided by such an arrangement. However even allowing $k$ to be unbounded, is not quite sufficient. The following can be shown:

**Fact**. *As long as the number of times that* $M'_0$ *can change its mind about the value of* $C_0$ *is a recursive function,* $f(\vec{x})$ *say, of the input* $\vec{x}$*, then still such predicates do not exhaust* $\Delta_2$. *To put it another way, there cannot be any recursive constraint on the number of alterations if the process is to decide all* $\Delta_2$ *predicates.*

There is the possibility of using the output of one Turing Machine as input to another. For functions on integers, this could be regarded as just composition of recursive functions. If we allow the output of a machine after $\omega$ many stages, either as a truth-in-the-limit operation or otherwise then we can display this as an infinitary operation on an infinite sequence of 0s and 1s (or whatever the alphabet of the machine is). How one does this, or under what preconditions one allows such models to be considered depends on how fastidiously one takes exception to 'supertasks', the latter being roughly defined here as a process that at some stage has completed infinitely many subtasks.

We now consider various mechanisms for this.

### 12.2.1. *Transcending the finite through stacking Turing machines*

One obvious objection to an infinitely running process is the 'Thomson Lamp' ( [52]) objection: if a switch has been thrown at times $\frac{1}{2}, \frac{2}{3}, \frac{3}{4}, \ldots, \frac{n}{n+1}, \ldots$ at what position is it at, or what value does it have, at time '1'? Similarly if a cell on the Turing machine tape has changed value infinitely often from 0 to 1 and back again at finite stages, what value should we allot it at 'time' $\omega$? Placing models in certain spacetimes neatly sidesteps this puzzle.

#### 12.2.1.1. *General relativistic models: Malament–Hogarth Spacetimes*

Pitowsky [40] gave an account of an attempt to define spacetimes in which the effect of infinitely many computational tasks could be seen by an observer $O_r$ in that spacetime. (By 'spacetime' we mean here a Hausdorff, paracompact, Riemannian manifold which is a solution to the Einsteinian GR equations – we refer the reader to [18].) He used the example of the, at the time, unresolved Fermat's Last Theorem but we can consider any other task that involves looking at a $\Pi^0_1$ question: for example, the consistency of the axioms of Peano Arithmetic, or Goldbach's Conjecture, both of which involve a simple universal quantifier $\forall n$ over a recursive predicate. Let us take the latter: another observer $O_q$ performs the tasks of checking that each even number in turn is the sum of two primes. This they do along their world-line $\gamma_1$, the proper time of which is infinite (as each calculation takes some finite unit of time). If they find a counterexample they send a signal to $O_p$ travelling along her worldline $\gamma_2$. The point of the example is to arrange that the proper time of $\gamma_2$ is finite, and has the whole of $\gamma_1$ in her chronological past. As Earman and Norton [10] mention, there are problems with this account not least that along $\gamma_1$ $O_p$ must undergo unbounded acceleration. Since then more sophisticated spacetimes due to Hogarth [21] and Etesi & Németi [12], have been devised. For the moment we follow formally [10] to define:

**Definition 12.2.** $\mathcal{M}=(M, g_{ab})$ is a *Malament–Hogarth (MH) spacetime* just in case there is a time-like half-curve $\gamma_1 \subset M$ and a point $p \in M$ such that $\int_{\gamma_1} d\tau = \infty$ and $\gamma_1 \subset I^-(p)$ (where $\tau$ denotes proper time and $I^-(p)$ the causal past of $p$).

This seemingly makes no reference to the word-line of the observer $\mathcal{O}_p$ travelling along their path $\gamma_2$, but they point out that there will be in any case such a future-directed timelike curve $\gamma_2$ passing through a point $q \in I^-(p)$ to $p$ such that $\int_{\gamma_2(q,p)} d\tau < \infty$, with $q$ chosen to lie in the causal future of the past endpoint of $\gamma_1$. The important point is that the *whole* of $\gamma_1$ lies in the chronological past of $\mathcal{O}_p$. As Hogarth showed in [21] such spacetimes are not globally hyperbolic, thus ruling out many standard spacetimes (such as Minkowski spacetime). (See [10] for a discussion on global hyperbolicity and a family of Penrosian Causal Censorship Hypotheses in this context – this is an interesting debate on how one might add extra axioms to GR to limit the types of spacetimes permissible – but would take this chapter too far off course.)

To obtain a spacetime as above, they take Minkowski spacetime $\mathcal{N}_0 = (\mathbb{R}^4, \eta_{ab})$ and choose a scalar field $\Omega$ which is everywhere equal to 1 outside of a compact set $C$, and which rapidly goes to $+\infty$ as the point $r$ is approached. The point $r$ is removed and the MH spacetime is then $\mathcal{N} = (\mathbb{R}^4 \backslash \{r\}, g_{ab})$, where $g_{ab} = \Omega^2 \eta_{ab}$. $\Omega$ and $\gamma_1$ can be chosen so that $\gamma_1$ is a timelike geodesic. This 'toy' spacetime is pictured on the left.

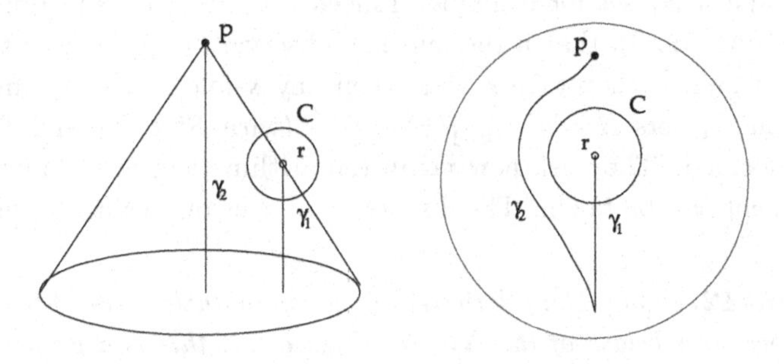


Figure 12.1. A 'toy' MH spacetime. On the right Hogarth's representation.

Earman and Norton discuss various possible spacetimes already in the literature that conform to being MH: Gödel spacetimes are MH but are causally 'vicious'; anti-de Sitter spacetime is MH, but fails a strong energy condition; Reissner–Nordstrom spacetime meets this, but as in all MH spacetimes there is divergent blue-shift of the signal to $\mathcal{O}_p$; further, of the unbounded amplification of signals that $\mathcal{O}_p$ may have to receive, etc., etc. Again this is beyond the terms of our discussion here, and apart from the rotating Kerr black hole solution of Etesi & Németi [12] to which we shortly turn, our aim is only to analyse the logico-mathematical possibilities inherent in these models.

#### 12.2.1.2. *Etesi & Németi's rotating black hole model*

The authors consider an observer sent axially into the region containing a rotating black hole of a certain size, and rotating at certain speeds – a Kerr solution. The first notable feature of such a black hole is that the primary singularity is ring-formed around the axis of rotation (see [18]). The observer, $\mathcal{O}_p$, is sent along the axis of rotation and receives signals sent from a Turing machine that is orbiting forever around the black hole. The machine is again looking for counterexamples, say, to a $\Pi_1$-predicate and will transmit one if it is found to $\mathcal{O}_p$. A clear desideratum for them is

*Assumption 1 'No swamping': it should not be the case that any part of the machinery or any observer should have to transmit or receive infinitely many signals.*

Initially, the orbiting machine and $\mathcal{O}_p$, should send and receive respectively, a single signal: the witness to the failure of the $\Pi_1$ predicate under inspection. They further remark though (their Proposition 2) that in fact the computational arrangement allows deciding queries $?n \in R?$ for sets $R$ slightly more complicated than $\Pi_1$ or $\Sigma_1$: $R$ can be taken as a union of a $\Sigma_1$ and a $\Pi_1$ set for example. Indeed they indicate an argument (at their Proposition 3), that if the machine-observer $\mathcal{O}_m$ is allowed to send $k$ different signals, (they take $k = 2$) then any $k$-fold Boolean combination of $\Sigma_1$ and $\Pi_1$ sets $R = \bigcap_{i<k-1}(S^i \cup P^i)$ (with $S^i \in \Sigma_1$ and $P^i \in \Pi_1$) can be decided. They ask how far in the arithmetical hierarchy this kind of argument can be taken. The discussion above concerning $\Delta_2$ predicates shows:

**Theorem 12.1.** [56] *The relations $R \subseteq \mathbb{N}$ computable in the Etesi–Németi model form a subclass of the $\Delta_2$ predicates of $\mathbb{N}$; this is a proper subclass if and only if there is a fixed finite bound on the number of signals sent to the observer $\mathcal{O}_p$.*

We have seen (at the Fact above) that for a $\Delta_2$ predicate $R$ there is no recursive bound on the input $n$ as to how many times the machine will have to change its mind concerning whether $n \in R$, and *a fortiori* no fixed in advance finite bound. Of course for checking whether any one $n$ is in $R$, finitely many signals will suffice; hence only if the architecture of the experiment allows for *potentially unboundedly* many signals, then (and only then) can $\Delta_2$ predicates be decided, still without breaking *Assumption 1*.

12.2.1.3. *Hogarth's arithmetically deciding spacetime regions*

Hogarth names a 'unit' or 'region' of spacetime that is capable of deciding a $\Pi_1$ question as above an '$\mathrm{SAD}_1$ spacetime or region', and as a shorthand denotes it by the right hand diagram above at Fig. 12.1. Hogarth in [22], (and in the later [23]) stacks up such regions to finite depths in order to answer $\Pi_n$ queries.

If a spacetime contains a sequence $\vec{O} = \langle O_j | j \geq 0 \rangle$ of non-intersecting open regions such that (1) for all $j \geq 0$ $O_j \subseteq I^-(O_{j+1})$ and (2) there is a point $p \in M$ such that $\forall j \geq 0$ $O_j \subseteq I^-(p)$ then $\vec{O}$ is said to form a *past temporal string* or just *string*. To decide membership in a $\Pi_2$-definable set of integers $P(n) \equiv \forall a \exists b Q(a, b, n)$ he then stacks up a string of regions taking each $O_j$ as a $\mathrm{SAD}_1$ region, each looking like the component of the right of Fig. 12.1, with $O_0$ being used to decide $\exists b Q(0, b, n)$. If this fails a signal is sent out to $\mathcal{O}_p$; but if this is successful, a signal is sent to $O_1$ to start to decide $\exists b Q(1, b, n)$ *etc.* Ultimately, putting this all together, again $\mathcal{O}_p$ receives a signal if $\neg P(n)$, or else knows after a finite interval that $P(n)$. It should be noted that

*Assumption 2 The open regions $O_j$ are disjoint*

and still that no observer or part of the machinery of the system has to send or receive infinitely many signals (thus the 'no swamping' assumption is kept). This whole region is then dubbed an '$\mathrm{SAD}_2$' spacetime.

An '$\mathrm{SAD}_{n+1}$' spacetime is defined accordingly as composed from an infinite string of (again disjoint) $\mathrm{SAD}_n$ regions $O_n$, again all in the past of some point $p$. (Earman and Norton [11] show that an $\mathrm{SAD}_1$ spacetime cannot decide $\Pi_2$ statements, and Hogarth [23] follows this up with the generalisation that $\mathrm{SAD}_j$ cannot decide $\Pi_{j+1}$ statements.) In Fig. 2.2 below, on the right is the underlying tree structure of an $\mathrm{SAD}_3$ region for computing queries of the form $?n \in A?$ for some $\Sigma_3$ set $A$: each large circle represents an $\mathrm{SAD}_2$ region (which in turn contains a string of infinitely many of the small circles – pictured by the terminal nodes of the tree

– representing the $SAD_1$ regions) which can be used for computing the answers to $\Sigma_2$ queries. Correspondingly, in Fig. 12.2, if each $O_n$ is a $SAD_j$ region then the diagram on the left is that of an $SAD_{j+1}$ region.

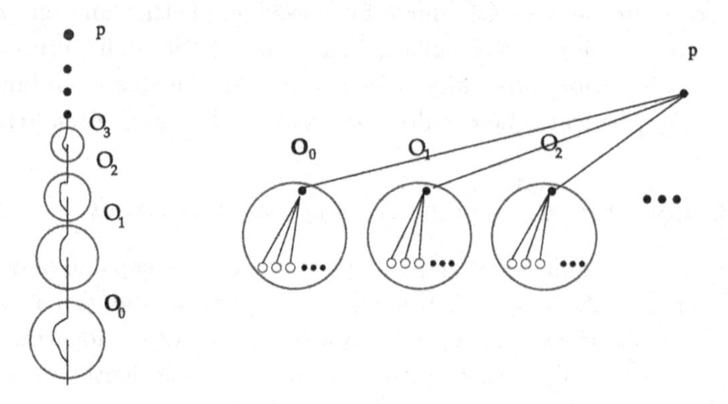


Figure 12.2. An $SAD_3$ region as a past temporal string of $SAD_2$ regions; and its tree representation (right).

In Fig.12.2, each region $O_n$ contains itself an $\omega$ sequence of $SAD_1$ regions which are shown in (the enlarged) circles of the tree.

We thus have that $\Pi_{n+1}$ questions can be decided by allowing nested stacks of suitable SAD regions to depth $n$ (if we define the depth of the simplest region in Fig. 12.1 as being 0). He then puts these altogether:

**Definition 12.3.** A spacetime $(M, g_{ab})$ is an *arithmetic deciding (AD) spacetime* just when it admits a past temporal string of disjoint open regions $\vec{O} = \langle O_j | j \geq 0 \rangle$ with each $O_j$ an $SAD_{j+1}$ region.

Of course there are many unresolved difficulties with this. There is a *recognition problem*: how, for example, could one ever recognise an AD spacetime region if such existed? Let alone one equipped with appropriate ranks of Turing machines coordinated and ready to go?

We now see how to go beyond Hogarth and observe that there is really no reason for us to stop at arithmetic. Hogarth has defined regions $SAD_{n+1}$ containing stacked $SAD_n$ subregions to a fixed finite depth $n$. He thus has used a subset of the class of *finite path trees* to label his regions. In the next definition $\mathbb{N}^{<\mathbb{N}}$ denotes the set of all finite sequences of natural numbers. $\mathbb{N}^{\mathbb{N}}$ denotes the set of all such infinite sequences from $\mathbb{N}$ to $\mathbb{N}$.

**Definition 12.4.** A *finite path tree* is any subtree $(T, <_T)$ of $\widetilde{T} = \langle \widetilde{T}, <_{\widetilde{T}} \rangle = \langle \mathbb{N}^{<\mathbb{N}}, \supseteq \rangle$where all branches under $<_T$ are of finite length.

We assign *ordinal ranks* to the nodes of a finite path tree (which are necessarily wellfounded) by recursion: the *rank of T* is then the rank of the empty sequence, (), the topmost node. A tree is in general infinitely splitting (a given node sequence node $u$ in $\mathbb{N}^{<\mathbb{N}}$ may have infinitely many immediate one step extensions), even though all branches are of finite length; hence ranks of nodes can in general be infinite, but of countable ordinal height (for an account of this and the following context, see e.g. [44] Section 15.2). Finite path trees in general can be used to describe the construction of the *Borel Sets* on spaces such as $\mathbb{N}^k \times (\mathbb{N}^{\mathbb{N}})^l$ for any $k, l < \omega$. A space, taken for simplicity as, $\mathbb{N} \times \mathbb{N}^{\mathbb{N}}$ has a topology constructed from *basic open sets* typically of the form $U_{\langle s,p \rangle} =_{\mathrm{df}} \{\langle s, x\rangle \in \mathbb{N} \times \mathbb{N}^{\mathbb{N}} \mid s \in \mathbb{N} \wedge \exists k \in \mathbb{N}(x \restriction k = p)\}$ where $p \in \mathbb{N}^{<\mathbb{N}}$.

**Definition 12.5.** (The Borel Hierarchy). (i) $X \subseteq \mathbb{N} \times \mathbb{N}^{\mathbb{N}}$ is in $\Sigma_0$ and in $\Pi_0$ if it is a basic open set in the above topology; (ii) $X \in \Pi_\xi$ iff $cX \in \Sigma_\xi$; (iii) $X \in \Sigma_\xi$ iff $X = \bigcup_n A_n$ where each $A_n \in \Pi_{\xi_n}$ for some $\xi_n < \xi$; a set $X$ is *Borel* if for some countable ordinal $\xi$ $X \in \Sigma_\xi$.

It is well known that this hierarchy is built up progressively through $\omega_1$ many stages (where $\omega_1$ is the first uncountable cardinal), and then no further sets are added (that is $\Sigma_{\omega_1} = \Pi_{\omega_1} = \Sigma_{\omega_1+1}$). Of particular interest is the *hyperarithmetical hierarchy* which is in one sense the constructive part of the Borel hierarchy. Here the construction of the Borel set is given by a *recursive finite path tree* (meaning the tree $T$ and its extension relation $<_T$ are given by computable functions) with a recursive assignment of recursively open sets to the bottommost rank 0 nodes, that is to the leaves of the tree. Membership then in an hyperarithmetic set of integers (that is taking $l = 0$ in the above) is given by testing a recursive protocol of queries. Already one construal of Hogarth's AD spacetime region is that it is capable in the above notation of answering questions concerning *some* unions of arithmetic sets, $S \in \Sigma_\omega$. Why 'some'? Because the description of the union $\bigcup_n A_n$ must be given to us in an effective, *i.e.* recursive way. The upshot is that for any hyperarithmetic set $H \subseteq \mathbb{N}$ there could be constructed a spacetime region $\mathrm{SAD}_H$ for which queries $?n \in H?$ could be answered. Such a region satisfies *Assumptions 1* and *2* and consists of SAD regions of smaller rank stacked according to the recursive finite path tree description for the construction of $H$.

A discussion and the details of the above can be found in [56]. Can a 'hyperarithmetically deciding spacetime' by analogy with Hogarth's AD deciding spacetime be constructed? It can, if we can enumerate those Turing programs that describe hyperarithmetic set building protocols.

**Proposition 12.1.** ([56]) *If* $\langle e_i \mid i \in \mathbb{N}\rangle$ *enumerates those indices of Turing programs that construct in the above sense hyperarithmetic sets* $S_{e_i}$, via *recursive trees, we may define a single MH 'hyperarithmetically deciding', HYPD, spacetime region in which any query of the form* $?n \in S_{e_i}?$ *can be answered in finite time.*

Here we piece together regions that are '$S_{e_i}$-deciding' just as the AD-deciding region is built. Given input $\langle i, n\rangle$ to an initial control machine, it activates $S_{e_i}$ and asks if $?n \in S_{e_i}?$ Of course this query will result in subqueries activating regions of lower rank down the tree coded by $e_i$ which are themselves $S_{e_j}$-deciding etc.

At this point the reader will, I think, object that the recognition problem has now got well out of hand: the collection of indices $\langle e_i \mid i \in \mathbb{N}\rangle$ enumerating hyperarithmetic set constructions is itself well beyond recursive or arithmetic, forming as it does a $\Pi^1_1$-complete set of integers. However it is worth emphasising that no machine in this tree array is itself performing 'supertasks' (i.e. performing infinitely many actions in its own proper time), but if it issues a signal to another process, it does so only once after a finite amount of its own proper time. It is simply that the overall tree no longer has a recursive description, and its ordinal rank is no longer a recursively given ordinal. We have not violated our two core assumptions. However the point should be that anthropomorphic considerations are being put aside and we are calculating what is feasible given the kind of techniques Hogarth contemplates. We have here what might be called a *hyperarithmetic computer*.

12.2.1.4. *A universal constant upper bound for any computation*

Nevertheless if we take this discussion to its logical conclusion, one might ask, How far could one possibly go building regions of higher and higher complexity without violating the core idea?

There is in any case a bound on the depth of any finite path tree to which we can assign MH regions without violating *Assumption 2*.

**Definition 12.6.** Let $\mathcal{M} = (M, g_{\rm ab})$ be a spacetime. We define $w(\mathcal{M})$ to be the least ordinal $\eta$ so that $\mathcal{M}$ contains no SAD region whose underlying tree structure has ordinal rank $\eta$.

Note that $0 \leq w(\mathcal{M}) \leq \omega_1$. Here a zero value $w(\mathcal{M})$ implies that $\mathcal{M}$ contains no SAD regions whatsoever, that is, is not MH; the upper bound is for the trivial reason that every finite path tree is a countable object and so cannot have uncountable ordinal rank.

**Proposition 12.2.** ([56]) *For any spacetime $\mathcal{M}$, $w(\mathcal{M}) < \omega_1$.*

***Proof.*** *Assumption 2* says that for different $\eta$ the different $\mathrm{SAD}_\eta$ components must occupy disjoint open regions $O_\eta$ of the manifold. However the manifold is separable (which follows from paracompactness and being Hausdorff). Let $X \subset M$ be a countable dense subset of $M$. Then each open region $O_\eta$ of $M$ contains members of $X$. As disjoint regions contain differing members of $X$ there can only be countably many such regions $O_\eta \subset M$, and therefore a countable bound. □

This is just the usual argument that separability of the real continuum $\mathbb{R}$ implies that any family of disjoint intervals of $\mathbb{R}$ must be countable. Consequently if $\mathcal{M}_{\mathrm{actual}}$ is *our* spacetime, (modelled using these basic assumptions) then $w(\mathcal{M}_{\mathrm{actual}})$ is a constant giving an upper bound to the complexity of MH regions, and so putative computations performable in $\mathcal{M}_{\mathrm{actual}}$. Dropping either of the Hausdorff or paracompactness properties from our list of properties of manifolds would seemingly result in unrecognisable (in current terms) 'spacetimes'. In short, although MH-spacetimes allow, at the most generous, for a reorganisation of any *countable* length computation (in some formalism, such as Turing machines) into one computation using trees of countable depth, this would be impossible for *uncountably* long (or many) computations whose stages occupy discrete spacetime regions. The same restriction would of course be true for any other system, or arrangement, of computations and is nothing to do with Hogarth style formalisations: this holds for any *separable* manifold and any generalised computation that requires a disjoint region of spacetime for each step or unit of computation. Somewhat more formally:

**Proposition 12.3.** *Let $\mathcal{M} = (M, g_{\mathrm{ab}})$ be a spacetime. Let $\mathcal{F}$ be some formal mechanism of computation, such that each computation step of the mechanism occupies a disjoint open neighbourhood of the manifold. Then there is a countable upper bound $w(\mathcal{M}, \mathcal{F})$ to the lengths of the $\mathcal{F}$-computations in $\mathcal{M}$.*

The proposition is not a completely precise mathematical statement, since we have not defined 'formal mechanism', but the point we hope should be clear. We have not specified 'step' or 'unit' but again this can mean a

Turing machine instruction step, or a cell unit. Anything that occupies a discrete interval in spacetime, whether it be an MH-spacetime, (so as to avoid supertask like phenomena), or in other spacetimes more generally with supertasks envisaged: one cannot in advance arrange the formalism to occupy uncountably many distinct open neighbourhoods. Hence the bound. If we are allowed to play God and are handed a separable manifold and (a set of integers coding) a countable ordinal $\alpha$, *in advance*, then indeed we could cook up an MH manifold to accommodate computations of that length (of proper time), using stacked Turing machines, or any other form of computational model $\mathcal{F}$. What we cannot have is one manifold $\mathcal{M}$ that will work for our chosen $\mathcal{F}$ for all countable $\alpha$.

### 12.2.2. *Allowing supertasks*

By means of using spacetime regions of a particular type the Etesi–Németi and Hogarth models avoid considering any supertasks, where any observer has performed infinitely many tasks in his or her chronological past. If we relax this constraint we may ask of our computing machines what they are capable of when given some well-defined behaviour in the transfinite. Amusingly even simple machines can perform a lot.

#### 12.2.2.1. *Punch-hole machines*

We first consider a simple kind of Turing machine. We envisage such a machine as having a tape, infinite in one direction, thus with a leftmost starting cell, and a read/write head traversing the tape in the usual fashion. The alphabet of the machine is simple: it consists of a blank and a '0'. The latter we can think of as a hole that the machine punches. Thus the machine can only write once, or punch a hole, in a cell; otherwise it 'reads' and moves a cell left or right in the usual fashion. The program or transition table for such a machine is simply that of an ordinary machine of this architecture.

We have to specify what happens at stage $\omega$ and subsequent stages. There are several possibilities: but let us say that we simply allow the machine to run for $\omega$ many stages, then consider what is on the tape (a potentially infinite sequence of holes and blanks) as input to be fed back into the machine at its starting state again for the next $\omega$ many stages. We thus reset the R/W head to first leftmost cell, and let it run the program afresh.

So, ignoring difficulties with 'hanging chads', such cells are usable once only. One easily sees that in $\omega$ steps again $\Delta_2$, or trial-and-error predicates are decidable. One may simply arrange that any new calculation extends beyond the scratch area of tape already used up. If one has a $\Delta_2$ predicate $P(v_0)$ it is not hard to arrange this so that a machine will punch holes (using additional blank gaps) on an output 'sub-tape' so that the correct final 0/1 value is recorded. Now we have allowed the possibility to reset the head to its starting position, and let the machine continue running. We let ourselves do and re-do this process, at every limit stage in time, pulling the head back to the start position and letting it work on the accumulated tape-full of punch-holes. Could we calculate more? For which predicates $P(n)$ of natural numbers $n$ is there a machine of this kind that halts for a given $n$ with the correct $P(n)\backslash\neg P(n)$ answer? These machines were first considered by Hamkins and Kidder but they discarded them as too weak, in favour of the *Infinite Time Turing Machine* to follow. Surprisingly perhaps, they can still calculate quite a lot: we have the following observation (due to S.D. Friedman and the author). Calling the above arrangement an 'infinite punch tape machine', it is not hard to demonstrate:

**Proposition 12.4.** *(i) Precisely the* arithmetical predicates *are decidable by infinite punch tape machines; (ii) any computation either halts by, or is in an infinite loop, by time* $\omega^2$.

12.2.2.2. *Infinite Time Register Machines (ITRMs)*

Koepke and Miller [34] consider the following register machine model. $M_N$ is a Shepherdson-Sturgis register machine (see [47], or as described in, e.g. Cutland [6]). $M_N$ has $N$ registers $R_i (i < N)$ each of which may contain a natural number. Suppose that the program under consideration has instruction set $\vec{I} = I_0, \ldots, I_q$. Let us say that at time $t$ $R_i$ contains $R_i(t) \in \mathbb{N}$, and that instruction $I(t)$ is about to be performed. We adopt a slightly more subtle behaviour than that for the punch-hole machines. We consider the state list $q_0, q_1, \ldots, q_p$ of the machine and at time $\lambda$ where $\lambda$ is any limit ordinal, we say that the machine will next perform the instruction numbered $I(\lambda) =_{\text{df}} \liminf_{\alpha \longrightarrow \lambda} I(\alpha)$ where $I(\alpha)$ is the instruction number about to be performed at time $\alpha$. This formulation has the rather pleasant effect of placing the machine at time $\lambda$, at the start of the outermost nested loop that it entered (if any) unboundedly often before time $\lambda$.

We have to assign register values, and here of course a register may have changed value infinitely often.

For $i < N$ we set:

*$\overline{R}_i(\lambda) =_{\rm df} \liminf_{\alpha\to\lambda} R_i(\alpha)$ and if this is finite we set $R_i(\lambda)= \bar{R}_i(\lambda)$. If infinite we set $R_i(\lambda) = 0$.*

It is this 'resetting' of a register that gives the model its strength. We may additionally consider such a machine to be able to consult an oracle: thus there is an instruction, so that if $Z \subseteq \mathbb{N}$, a register can be reset to 0 if $R_i(\alpha) \in Z$. Computations relative to an oracle $Z$ can be regarded in this manner as using the set $Z$ as an input; the infinite time available allows all of $Z$ to be consulted. We discuss the strength of this model below, but again surprisingly complicated predicates can be calculated.

### 12.2.2.3. *Infinite Time Turing Machines (ITTMs)*

This model, due to Hamkins and Kidder, awakened recent interest in transfinite computational models. It was designed in the 1990s but an account of them only appeared in [15] much later. We give a computationally equivalent version to that of their original model – and discuss the differences afterwards.

We go back to the punch-hole machine described above, but we now consider what to do if the machine is allowed to reuse cells. Clearly a cell may then be reused infinitely often and we must define a behaviour for it. We consider Turing machines with an alphabet of just three letters: 0,1 and B (for blank). We suppose that its standard program has states $q_0, \ldots, q_k$. The read/write head moves according to the description given by the usual transition table, but that also uses the *liminf of cell positions* that it has been reading at limit stages of time to calculate its position at a limit stage, (if this *liminf* of the read/write head positions is infinite at a limit stage, then the head is set back by fiat to the starting cell $C_0$). Further, we set the state $q(\lambda)$ at limit times to be the liminf of previous states. (This has the same effect of course of positioning the read/write head at outermost loops as it did for ITRMs.) If the cells of the machine are enumerated $\langle C_i \mid i \in \mathbb{N}\rangle$ with values at time $\nu$ denoted by $\langle C_i(\nu) \mid i \in \mathbb{N}\rangle$ then we set at limit time $\lambda$:

$$C_i(\lambda) = k \Longleftrightarrow \exists\alpha < \lambda\forall\beta < \lambda(\alpha < \beta \longrightarrow C_i(\beta) = k) \textit{for } k \in \{0, 1, \mathrm{B}\};$$
*Otherwise $C_i$ is set to a* 'B'.

Thus if the machine has changed its mind unboundedly often below $\lambda$ about the cell value then, this is set to a blank – for ambiguity if you will. Programs, are simply standard Turing machine ones, and may be enumerated as $\langle P_e \mid e \in \mathbb{N} \rangle$. If a particular program $P_e$ halts, then we can consider the contents of the tape the output of that machine. We may also prime the tape with an infinite string $x$ from the alphabet, and consider $P_e(x)$ to be the computation of the $e$'th program on input $x$.

Such machines can decide $\Pi^1_1$-predicates. We illustrate by means of the complete $\Pi^1_1$-predicate on integers: those $e \in \mathbb{N}$ so that the $e$'th (standard) recursive function, $\{e\} = f$, computes the characteristic function of a *wellorder* of $\mathbb{N}$. Given input $e$ the machine simulates the $e$'th standard Turing program and writes the output characteristic function on, say the cells of the tape $C_i$ where $i \equiv 0 (\text{mod}\, 3)$. The other cells are blank for scratch work. This takes $\omega$ many stages. When this is complete, the machine then checks the $\Pi_2$-condition of this characteristic function $f$ coding a *discrete linear ordering*. (This is of the form $\forall n \exists m R(n, m, f)$ and $R$ is recursive. This can be verified in $\omega$ steps, by starting with $n = 0$, recording '0' on the scratch tape cells $C_i$ where $i \equiv 1 (\text{mod}\, 3)$ , searching for an $m$, using the scratch tape cells $C_i$ where $i \equiv 2 (\text{mod}\, 3)$, for auxiliary calculation, then proceeding to $n = 1$ if successful, etc.) If this test is passed, we have an order $<_e$; we then need to test for wellorderedness. We wipe clean the scratch tape area, and search for the $<_e$'th least element of the ordering. We may do this by simply guessing a least element on a scratch tape, and then continually revising our guess $<_e$-downwards each time we find a lesser one. If after $\omega$ many steps we did not find such then we did not have a wellorder, and we can output a 0; if after $\omega$ many steps we only changed our minds finitely often, then we indeed located the $<_e$-least element, say it was 23, and we have it written concretely on the scratch tape. We now proceed through the code function $f$ and erase all mention of the element 23. This leaves us with a new code $f'$ of a discrete linear order written on the cells $\langle C_{3i} \mid i \in \mathbb{N} \rangle$, and we simply now repeat this process. There are only two outcomes: either at some point we arrived at the situation where the 'current' linear order is illfounded, and we discover this fact, by descending through it infinitely often looking for its least element, or else we end up emptying out the $C_{3i}$ cells for $i \in \mathbb{N}$ completely: after looking through this slice of the tape, and seeing it is empty (which takes a final $\omega$ many steps) we verify that it was truly wellfounded. If the order type of the ordering was $\alpha$ then this has been achieved in, rather roughly, $\omega + \omega.\alpha + \omega$ many steps.

Once the reader has convinced themselves of this, it is not hard to imagine programs that write out successfully on some slice of the scratch tape (which we might as well call 'output tape') those $e_i$ with $\{e_i\} \in WO$.

Moreover, one can also imagine a machine coding the *ordinal sum* of all the recursive ordinals $\alpha < \omega_1^{\mathrm{ck}}$ and outputting a code for that, i.e. $\omega_1^{\mathrm{ck}}$, on the output tape. However as Hamkins and Lewis showed, and we shall see later, this is only scratching the surface.

Suppose we denote by $P_e(n)$ the $e$'th computation on integer input of $n$, represented by an infinite string of $n$ 1s followed by an infinite string of 0s. Several natural questions arise:

**Q1** *What is* $0^\nabla = \{e | P_e(0)\downarrow\}$*? (The halting problem on integers).*

**Q2** *What are the halting times that arise? That is, if* $P_e(0)\downarrow$ *halts in* $\alpha$ *steps how large is* $\alpha$*?*

**Q3** *What are the decidable predicates? Where we say* $R(n)$ *is* semi-decidable *if there is some e so that* $R(n) \leftrightarrow P_e(n)\downarrow 1$*, and is* decidable *if both it and its complement are strongly semi-decidable.*

Hamkins and Lewis [15] first developed the theory of such machines, using the analogy of the standard Turing machine as a source of the notion of recursion: they note that there are versions of the Recursion Theorems, and the *Snm*-Theorem for this notion of computation and there is a universal machine with a universal program. Much of the standard development proceeds very smoothly but of course there are considerable differences: for Turing machines the whole run of a halting computation, the *snapshots* of the states and of tape's contents etc. can be encoded by an single integer; it is thus of the same *type* as the objects on which it operates. However for ITTMs, computations $P_e(n)$ on integer input, are in general a transfinite sequence $\mathcal{S} = \langle S_\beta \mid \beta \leq \alpha \rangle$ of snapshots of the cell values $\langle C_i(\beta) \mid i < \omega \rangle$ at each stage $\beta \leq \alpha$. A computation is then an infinite object and must be coded in this context by a set of integers or a 'real' number. (One uses reals that code wellorderings of length $\alpha + 1$ and attaches by pairing functions the snapshots to the nodes of the wellorder, together with any auxiliary information such as machine state etc. along the way). Computations are thus of different types from the integer inputs. A central representation of standard Turing machines comes via Kleene's $T$-predicate, yielding a canonical *Normal Form Theorem.* This theorem allows one to proceed effectively from $e$, and uniformly in $n$, from a halting computation of the form $P_e(n)\downarrow$ to a program $e'$ so that $P_{e'}(n)\downarrow$ will halt, and moreover produce an integer output which codes the whole *course-of-computation* that demonstrates $P_e(n)\downarrow$. For such a notion to work in this new area we should need the program $P_{e'}$ to be capable of producing the reals needed to code the potentially transfinitely many steps in calculations such as that of $P_e(n)\downarrow$. But are they capable of this? In short, *Is every ordinal length of a halting computation on an integer input capable of being itself 'written' or being the output of some other computation?* This must be true if we are to have a hope of producing a Normal Form theorem. Hamkins and Lewis called the halting time ordinals the *clockable* ordinals, and the question they asked is: *Is every clockable ordinal writable?*

The answer turns out, thankfully, to be affirmative, (it follows from the $\lambda, \zeta, \Sigma$-theorem below, [54]). We refer the reader not to the original papers, but to [57] for a later but somewhat tidier account of this theorem and the answers to the above three questions.

From this one gets the desired representation theorem (where $P_e(n)$ refers to ITTM computation).

**Theorem 12.2. (Normal Form Theorem I)** (Welch [55, 57]) $\forall e \exists e' \forall n \in \mathbb{N}$

*$P_e(n)\downarrow \longrightarrow (P_{e'}(n)\downarrow y$ where $y \in 2^{\mathbb{N}}$ codes a wellordered course-of-computation sequence for $P_e(x)\downarrow)$.*

*Moreover the map $e \longrightarrow e'$ is effective (in the usual Turing sense).*

There is a higher type version obtained by relativising all the results (now for $\lambda^x, \zeta^x$ etc.) above to real number inputs. Part (b) below is simply a variant form stated to be reminiscent of the Kleene $T$-predicate. We let $\varphi_e$ be the (partial) function computed by $P_e$.

**Corollary 12.1. (Normal Form Theorem II)** *(a) For any ITTM computable function $\varphi_e$ we can effectively find another ITTM computable function $\varphi_{e'}$ so that on any input $x$ from $2^{\mathbb{N}}$, if $\varphi_e(x)\downarrow$ then $\varphi_{e'}(x)\downarrow y \in 2^{\mathbb{N}}$, where $y$ codes a wellordered computation sequence for $\varphi_e(x)$. (b) There is* a universal predicate $\mathfrak{T}_1$ *which satisfies $\forall e \forall x$:*

$$P_e(x)\downarrow z \quad \leftrightarrow \quad \exists y \in 2^{\mathbb{N}}[\mathfrak{T}_1(e,x,y) \wedge Last(y) = z].$$

The effectivity is again established in the same way, noting that the input (whether $n \in \mathbb{N}$ or $x \in 2^{\mathbb{N}}$) does not affect the above description of an algorithm in any dynamic way.

However the proof that all clockable ordinals are writable proceeds via an analysis of how each single cell $C_i$ behaves during the course of a computation $P_e(n)$. In general cells may *stabilize* on some fixed value, or forever change value. The same is true for infinite sub-segments of the ITTM tape. Suppose we reserve cells $C_i$ ($i \equiv 2 \bmod 3$) for 'output' then we say that a real $y \in 2^{\mathbb{N}}$ is '*eventually computable*' if there is an ITTM computation $P_e(n)$ – which is not required to halt – but which has $y$ as the characteristic function of the output tape from some point in time onwards. The notion is then that a computation need not formally halt in order to 'produce' an output: it is sufficient that the output tape segment be stable from some point onwards. Hamkins and Lewis called an ordinal $\alpha$ *eventually writable*, if there was an eventually computable $y_\alpha \in 2^{\mathbb{N}}$ coding $\alpha$. Clearly we can consider any halting computation a special case of an eventually stable one, and thus if $\lambda$ is the supremum of all writable ordinals, and $\zeta$ the supremum of the eventually writable ordinals then $\lambda < \zeta$. Evaluating $\lambda, \zeta$ turned out to be tied up with calculating stabilisation points of cells $C_i$ in the universal machine calculations, and the following characterisation is possible.

**Theorem 12.3. (The $\lambda, \zeta, \Sigma$-theorem)** (Welch [57]) *(i) Any ITTM computation $P_e(n)$ on integers which halts, does so by time $\lambda$, the latter defined as the supremum of the writable ordinals;*
*(ii) any computation $P_e(n)$ with eventually stable output tape, will stabilise by time $\zeta$ the supremum of the eventually writable ordinals;*
*(iii) moreover $\zeta$ is the least ordinal so that there exists $\Sigma > \zeta$ with the property that*

$$L_\zeta \prec_{\Sigma_2} L_\Sigma;$$

*(iv) then $\lambda$ is the least ordinal satisfying:*

$$L_\lambda \prec_{\Sigma_1} L_\zeta.$$

We thus have a clear picture of the action of ITTM computations on integers. The machines run using very constructive rules, even for the limit stages, so their action is of course absolute to Gödel's constructible universe $L$. As [15] had noted, if an ITTM machine has its hands on a real $y$ coding an ordinal $\alpha$ then there is a standard Turing machine program for using that code to run a construction of the $L$-hierarchy 'along' that ordering $y$, thereby producing a real code for the $\alpha$'th level $L_\alpha$. Hence the tie up with $L$ is natural. A further observation on the $\lambda, \zeta, \Sigma$-Theorem is in order. The machine limit rules of *liminf* can be expressed in a $\Sigma_2$ way. If one has two levels of the $L$ hierarchy satisfying $L_{\zeta'} \prec_{\Sigma_2} L_{\Sigma'}$ then running the universal machine inside $L$ it is pretty much immediate that the machine's snapshots at time $\zeta'$ and $\Sigma'$ will be identical: this is what the elementarity entails. The machine will then either have halted, or, as one can show, has entered an eternally repeating loop (although the elementarity assumed is suggestive of this, in fact the latter still has to be shown). It turns out the pair $(\zeta, \Sigma)$ is the lexicographically least pair of ordinals where the universal machine has identical snapshots, and first enters an infinite loop.

What further seems to emerge from the proofs above, is that the primary notion here is not that of a 'halted computation' but of a 'stable computation': there are computations of the form $P_e(n)$ which do not formally halt, but eventually have a settled output tape, and thereafter just footle around for ever on their scratch tape areas. Halting is just a special case of *stabilising*, and this is borne out by the fact that we cannot fully analyse halting computations without analysing stabilising ones. Halting can be expressed by a $\Sigma_1$ statement in set theory ('There exists a real $y$ that successfully codes the course of computation of $P_e(n)$ with a last halting state'); this is at the basis of the $\Sigma_1$ characterisation of $\lambda$ in the $\lambda, \zeta, \Sigma$ theorem as $L_\lambda \prec_{\Sigma_1} L_\zeta$, *once* we have discovered $\zeta$. We may further establish theorems corresponding to those for halting computations.

The reader may have noticed that we seem to be avoiding discussion of the obvious fact that ITTMs can work on infinite input as well have infinite output: such computation is thus on one type up, on that of sets of integers, or reals themselves, rather than merely on integers. Before we turn to this we emphasise that Hamkins' and Kidder's original formulation of an ITTM immediately visualised such capabilities: their machine was devised as coming equipped with three infinite tapes, for input, scratch and output. A single read/write head surveyed a single cell from each of the three tapes simultaneously and according to its state and program, would write from an alphabet set of $\{0, 1\}$. At limit stages a cell $C_i$'s value was determined by taking the *limsup* of the previous cell values (there was no Blank character); the read/write head at limit stages would be brought back to the very first triplet of cells on the tapes, and the machine would enter a special 'limit state' $q_L$. The differences between this arrangement and that sketched above play no role in determining the classes of functions or sets computed (either on integers or on reals, which we are coming to): they are the same for either model. There are minor differences in calculating halting times, and in precisely which classes of ordinals are clockable often by an obvious factor of $\omega$ or so, but apart from these finer details there are functionally no differences between the models proposed. (This discussion does conceal one remark, that in fact, a one tape machine with two symbols cannot produce the same class of computable functions $f : \mathbb{R} \longrightarrow \mathbb{R}$. However for functions of type $f : \mathbb{R} \longrightarrow \mathbb{N}$ or $f : \mathbb{N} \longrightarrow \mathbb{N}$ a one tape two-alphabet machine turns out to be sufficient. For the wider class curiously a third character – which we have introduced here by the way of the Blank above – turns out to be necessary. See [17] for a discussion of this somewhat technical point and proofs of these results mentioned.)

## 12.3. Computation on Reals

Kleene developed an equational calculus for developing the notion of *recursion in a higher type object* (see [30], [26], [29]). The relevant type here is *Type 2*, the objects under consideration are functionals $\mathcal{I} : \mathbb{N}^{\mathbb{N}} \longrightarrow \mathbb{N}$. This generalised his earlier equational calculus for (standard) recursive functions that used (characteristic functions for Type 1) oracles, $I : \mathbb{N} \longrightarrow \mathbb{N}$. The intuitive notion of a functional $\mathcal{F}$ being computable relative to $\mathcal{I}$ is that we have some kind of machine that can take inputs in the form of (finite sequences) of integers and reals, $\vec{n}$, $\vec{x}$, and which is connected to some oracle/memory device that has access to the graph of $\mathcal{I}$ – itself a set of size the continuum. As the domain of $\mathcal{I}$ is $\mathbb{N}^{\mathbb{N}}$, the machine must compute a real $x$ to present to the oracle, which will return $\mathcal{I}(x)$. Thus an infinite amount of computation must be performed in some scratch/storage area before this oracle query can be launched. A *computation* will thus in general be of infinite length, but is perhaps better thought of as given by an infinite *tree* where, for example, there may be *infinite branching* nodes: below the call for $\mathcal{I}(x)$ will be the prior individual computations for $x(0), x(1), \ldots$. An illfounded tree, that is one with an infinite descending path, represents an undefined computation. There is some discussion of this in Rogers ([44], p. 406), where there is no oracle $\mathcal{I}$ discussed but where the allusion is to an "$\aleph_0$-mind' capable of forming such *generalised machine* computations. A crucial point is that a generalised computation step must only be allowed to take previously, inductively, defined generalised steps. The resulting notion is '*hyperarithmetic computability*'. (See also here [28], [27].)

With the addition of the oracle $\mathcal{I}$ it can thus be loosely characterised ([24]) as a model of computation in which the computational device has a

(i) *countably infinite memory;* and

(ii) *an ability to manipulate (search through, write to) that memory in finite time;* and optionally

(iii) *an ability to quiz an oracle $\mathcal{I}$ about that memory contents (in a single step).*

If the above is all done within the $e$'th program we call the above computation $\{e\}(\vec{n}, \vec{x}, \mathcal{I})$ which again, may or may not halt. The following functional is essential for developing much of the regular theory of relative recursiveness:

$$\mathcal{E}(x) = \begin{cases} 0 \text{ if } \exists n x(n) = 0; \\ 1 \text{ otherwise.} \end{cases}$$

The immediate import of this is that computation relative to the object $\mathcal{E}$ is closed under existential number quantification (for any $\mathcal{I}$ the class of relations semi-recursive in $\mathcal{I}$ is closed under universal number quantification). A second effect is that:

- If $A$ is an arithmetical set of reals then $A$ is recursive in $\mathcal{E}$.

More important consequences follow: if $\mathcal{I}$ is any functional such that $\mathcal{E}$ is recursive in $\mathcal{I}$, then we have the full *Ordinal Comparison Theorem* for stages of computation (see [39]) which is crucial for developing the theory of relations semi-recursive in a Type-2 functional. By 'relation' in the next theorem, we mean any predicate $R(\vec{n}, \vec{x}) \subseteq \mathbb{N}^k \times (\mathbb{N}^{\mathbb{N}})^l$ for $k, l \in \mathbb{N}$. We thus arrive at the notion of *Kleene Recursion.*

**Theorem 12.4. (Kleene)** *The hyperarithmetic relations are precisely those recursive in $\mathcal{E}$.*

*The $\Pi^1_1$ relations are precisely those semi-recursive in $\mathcal{E}$.*

If we are considering relative recursion of a set of reals $A \subseteq \mathbb{R}$ in a set of reals $B$ (which we may identify with its characteristic function oracle $\mathcal{I}_B$) we may denote such:

$x \in A \simeq \{e\}(x, B, \mathcal{E}) \downarrow 1$

and say that '$A$ is recursive in $B$' if $\{e\}$ gives a function total on inputs $x$, and then one has appropriate versions of the above theorem relativised to $B$. There is an appropriate notion of *Kleene Degree*:

**Definition 12.7. (Kleene degrees)** Let $A, B \subseteq \mathbb{R}$; we say that

$A \leq_K B$ iff there is an index $e$ and a real $y$ so that

$$\text{for any } x \in \mathbb{R}\ (x \overset{\in}{\underset{\notin}{}} A \longleftrightarrow \{e\}(x, y, B, \mathcal{E}) \downarrow {}^1_0);$$

$A$ is *Kleene-semi-recursive in $B$* iff there is an index $e$ and a real $y$ so that

$$\text{for any } x \in \mathbb{R}\ (x \in A \longleftrightarrow \{e\}(x, y, B, \mathcal{E}) \downarrow 1).$$

The presence of the fixed real $y$ ensures that the degree class of $B$ contains continuum many sets of reals $A$; moreover the degree of $B$, being thus closed under continuous pre-images, forms a so-called *Wadge degree.* In general a computation evolves its own tree structure as it grows, according to its instruction set. But one can think of $y$ as also contributing to some part of the computational tree structure. In this case, as $y$ is allowed to vary, we see that $0_K$ contains $\varnothing, \mathbb{R}$, and in fact consists of the *Borel sets.* $0'_K$ (the $K$-degree of a complete Kleene semi-recursive set of reals) contains WO, the set of reals coding wellorders, and so a complete $\Pi^1_1$ set of reals. In fact it consists of all the co-analytic, so precisely the $\mathbf{\Pi}^1_1$, sets.

It is possible to give a set theoretical description of Kleene recursion in a relation $B$ and $\mathcal{E}$. In what follows, $\omega_{1\,\mathrm{ck}}^{B,y,x}$ denotes the least ordinal which does not have a real code recursive in $(B, x, y)$; it turns out that the wellfounded computation tree of a converging Kleene recusion will have rank less than $\omega_{1\,\mathrm{ck}}^{B,y,x}$. This is the basis of the following characterisation: we only need to look inside a model with enough ordinals – namely $\omega_{1\,\mathrm{ck}}^{B,y,x}$ – to see whether the computation tree is wellfounded. Moreover, in admissibility theory wellfoundedness of any relation inside a transitive admissible set is actually a $\Sigma_1$-notion. Here $\mathcal{L}_{\in,\dot{X}}$ is the language of set theory augmented by a predicate symbol $\dot{X}$ – to be interpreted by $B$.

**Lemma 12.1.** *$A \leq_K B$ iff there are $\Sigma_1$-formulae in $\mathcal{L}_{\in,\dot{X}}$ $\varphi_1(\dot{X}, v_0, v_1)$, $\varphi_2(\dot{X}, v_0, v_1)$, and there is $y \in \mathbb{R}$, so that for any $x \in \mathbb{R}$*

$$x \in A \Longleftrightarrow L_{\omega_1^{B,y,x}}[B, y, x] \models \varphi_1[B, y, x] \Longleftrightarrow L_{\omega_1^{B,y,x}}[B, y, x] \models \neg\varphi_2[B, y, x].$$

Thus to determine whether $x \in A/x \notin A$ we perform $\Sigma_1$-searches through the least admissible set $L_{\omega_1^{B,y,x}}[B, y, x]$ relative to $B$ containing $y, x$. As intimated equivalence with the former definition comes about through the original (relativised) theorem of Kleene (Theorem 12.4) and the theory of admissible sets (*cf.* [1]).

The generalised theory of recursion in higher types was much investigated and developed in the late 1960s and 1970s, with a history too rich to go into here, with names such as Gandy, Moschovakis, Sacks, Grilliot, Fenstad, Normann, Moldestad, Harrington prominent. The recursion relative to the single operator $\mathcal{E}$ is in one respect merely illustrative, being the historical example from the earlier days and papers ([26], [29], [30]) of the Kleene Equational Calculus. The reader may consult Hinman [20] for an overall development of the theory, Fenstad [13] for an attempt to present an axiomatic approach to general computation theories and the latter Part D of Sacks [45] for the further development in relation to set recursion.

Mention must now be made of the connections to the theory of *inductive definitions* and here more particularly to the theory of *Spector classes*. The latter is a general unifying theory of definability developed by Moschovakis in [39]. We consider here just pointclasses $\Gamma \subseteq \mathbb{N}^k \times (\mathbb{N}^{\mathbb{N}})^l$ (for any $k, l < \omega$) and use the notation that $\breve{\Gamma} = \{\neg R : R \in \Gamma\}$. $\exists^{\mathbb{N}}$, $\forall^{\mathbb{N}}$ represent natural number quantifiers as opposed to $\exists^{\mathbb{N}^{\mathbb{N}}}$, $\forall^{\mathbb{N}^{\mathbb{N}}}$ over elements of $\mathbb{N}^{\mathbb{N}}$.

**Definition 12.8.** A *Spector class* of pointsets $\Gamma \subseteq \mathbb{N}^k \times (\mathbb{N}^{\mathbb{N}})^l$ for any $k, l$, is a collection that is (i) closed under $\cap, \cup$, number quantification: $\exists^{\mathbb{N}}$,

$\forall^{\mathbb{N}}$; closed under (standard) recursive substitutions, has a universal set $U$ indexing by $\mathbb{N}$ all members of $\Gamma$ and lastly has the Prewellordering property:

PW: For any $P \in \Gamma$ there is $\sigma : P \longrightarrow \lambda$ for some ordinal $\lambda$ with the property that there are relations: $x \leq_0^\sigma \in \Gamma$, $x \leq_1^\sigma y \in \breve{\Gamma}$ so that:

$$P(y) \Rightarrow (\forall x[P(x) \wedge \sigma(x) \leq \sigma(y)] \Longleftrightarrow x \leq_0^\sigma y \Longleftrightarrow x \leq_1^\sigma y).$$

It would be impossible to give a full exposition of the import of Spector pointclasses here, but suffice it to say that the definition above encapsulates a fundamental unifying approach to the theory of inductive definability. Familiar Spector pointclasses are $\Pi_1^1$ and $\Sigma_2^1$ but there are many others. For $\Sigma_n^1$ or $\Pi_n^1$ the existence of the prewellordering property depends on the surrounding set theory in which one works. We shall only be discussing Spector pointclasses within $\Delta_2^1 = \Pi_2^1 \cap \Sigma_2^1$. The Kleene recursion theory then throws up a canonical example of a Spector pointclass: the Kleene semi-recursive (in $\mathcal{E}$) sets are precisely the $\Pi_1^1$ sets.

The type of formalism on the right-hand side equivalences of Lemma 12.1 in fact is also one way of defining Spector classes within the $\Delta_2^1$ pointclass.

### 12.3.1. *ITTM computations on reals*

If we now return to the ITTM model we shall see that it fits very nicely into this overall general theory. We have a choice to make here. At **Q3** above we called the *decidable* predicates, those where a characteristic function of the predicate could always be computed by a *halting* computation. It is natural, particularly given the machine nature of the origins of the notion, to think of *halting* as somehow fundamental, and therefore it is this that should be used to characterise 'decidability'. However here we are adopting the position that the fundamental feature of the ITTMs is the $\Sigma_2$ nature of the limit rule for the cell values, and the concomitant phenomenon of their having stabilised output without halting; it was indeed from this stabilising and looping times, from $\zeta$ and $\Sigma$, that we could characterise the halting times. The halting computations are for this purpose to be regarded as the special sub-class of 'fully stabilised' computations: halting is just a special kind of stabilisation (sic). This position is further strengthened when we consider below its relation to previous notions of higher type recursion.

We stated the Normal Form Theorems in the stronger, halting, version, as these would be more familiar to the reader, but there are equally well Normal Form Theorems which are verbatim as above but with $\downarrow$ replaced by $\uparrow$ throughout. The viewpoint here is that the strongly stabilising, i.e. halting, computations should probably be thought to give rise to a notion of *strong decidability* (and *strong semi-decidability*) whilst the stabilising computations correspond to the notion of *decidability* and *semi-decidability*. However most papers distinguish the 'stabilising' form, with the adverb 'eventually', used in the form: 'eventually (semi)-decidable predicates' or adjectively as in 'eventual ITTM degrees' etc. This is established enough that it would egregious to go against it here.

However for notions from higher type recursion theory, one says in general that a class of 'semi-decidable sets are those semi-recursive in a $\mathcal{F}$' where the latter $\mathcal{F}$ is some higher type functional. *Then*, for the appropriate $\mathcal{F}$ for ITTMs, actually '*semi-recursive in* $\mathcal{F}$' would correspond to the stabilising behaviour rather than the halting one. This would also accord with the usage inherited from Kleene Recursion. We shall call here then 'ITTM-semi-recursive' those predicates where membership facts can be represented as the stable output of some program, and thus corresponding to 'eventually semi-decidable' in the literature. (For different classes of machines such as the $\Sigma_n$-machines mentioned below, the notion of 'output' becomes somewhat more rarified, but these too one we would like to think of as providing mathematical classes of sets that are generalised (semi-)recursive in some way.)

It is a conceptually simple adjustment to have within an ITTM program an oracle call that requests of some oracle $B \subseteq \mathbb{R}$ (here $2^{\mathbb{N}}$) whether the current contents of the scratch tape, $y \in 2^{\mathbb{N}}$, is an element of $B$, and receive a 0/1 reply. Thus computation relative to an oracle for sets of reals is unproblematic. We again adopt the same notation that $P_e^B(x) \downarrow y$ if the $e$'th machine with oracle $B$, on input $x \in 2^{\mathbb{N}}$ halts with output $y \in 2^{\mathbb{N}}$. Changing the arrow to $P_e^B(x) \uparrow y$ indicates that *eventually* $y$ is written to the output tape, and remains there unchanging from some point on. (We have to have some other notation such as '$P_e^B(x) \mid$' for when the computation diverges or is undefined.)

We first give the integer version.

**Definition 12.9.** (i) A set of integers $x$ is *ITTM-semi-recursive* in a set y if and only if:

$$\exists e \forall n \in x\, [P_e^y(n) \uparrow 1 \longleftrightarrow n \in x \ ];$$

(ii) A set of integers $x$ is *ITTM-recursive* in a set $y$ if and only if:

$$\exists e \forall n \in x\, [P_e^y(n) \uparrow 1 \leftrightarrow n \in x \wedge P_e^y(n) \uparrow 0 \leftrightarrow n \notin x].$$

We may write $x \preceq^{\infty} y$ for the reducibility ordering.

Equivalently: $x$ is ITTM-recursive in $y$ if both $x$ and $\neg x$ are ITTM-semi-recursive in $y$ (since if the latter holds it is easy to amalgamate the two programs into a single program $P_e$ with the effect of the Definition. The relation $\preceq^{\infty}$ is in the class $\Delta^1_2$. There is a natural prewellordering that arises on computations $P_e$ establishing membership in some set $x$: put $n \prec m$ if the computation $P_e(n) \uparrow 1$ stabilises to an output of 1 before that of $P_e(m) \uparrow 1$ does. The relation $\prec$ is itself ITTM-semi-recursive (think of the universal machine that observes the simulated copies of computation sequences of $P_e$ for various $n$ – eventually it itself will stabilise into seeing that $P_e(n)$ stabilises before $P_e(m)$) and thus we can establish the prewellordering property very easily.

There is a natural notion of *complete ITTM-semi-recursive set of integers*:

**Definition 12.10.** $\tilde{x} =_{\text{df}} \{e \mid P_e(0) \uparrow\}$ – the complete set of *stable indices.*

The following tells us what this set is by way of a set theoretic characterisation. We regard $x \rightarrowtail \tilde{x}$ as an analogy to the hyperjump operation.

**Theorem 12.5.** (Welch [54]) *$\tilde{x}$ is (Turing-)recursively isomorphic to the $\Sigma_2$-theory of $\langle L_{\zeta^x}[x], \in, x\rangle$. In particular $\tilde{0}$ is recursively isomorphic to the $\Sigma_2$-theory of $\langle L_\zeta, \in\rangle$.*

This should be compared with Kleene's result that his notation system set $\mathcal{O}$ – a complete $\Pi^1_1$ set of integers coding indices of wellfounded finite path trees – is in fact (Turing-)recursively isomorphic to the $\Sigma_1$-truth set of $\langle L_{\omega_1^{\text{ck}}}, \in\rangle$. Indeed Klev has defined in [31] an extension of Kleene's $\mathcal{O}$ to an $\mathcal{O}^{++}$, that mirrors exactly Kleene's original definition as a tree (indeed the tree is literally an extension of Kleene's). By the above, it is thus to the complete $\Sigma_2(L_\zeta)$ set what $\mathcal{O}$ is to $\Sigma_1(L_{\omega_1^{\text{ck}}})$.

The following is the natural version for real computation:

**Definition 12.11.** A set of reals $A$ is *ITTM-semi-recursive* in a set of reals $B$ if and only if:

$$\exists e \forall x \in 2^{\mathbb{N}} \left[P_e^B(x) \uparrow 1 \leftrightarrow x \in A\right];$$

(ii) A set of reals $A$ is *ITTM-recursive* in a set of reals $B$ if and only if:

$$\exists e \forall x \in 2^{\mathbb{N}} \left[P_e^B(x) \uparrow 1 \leftrightarrow x \in A \wedge P_e^B(x) \uparrow 0 \leftrightarrow x \notin A\right].$$

**Definition 12.12.** $A \leq^\infty B$ iff for some $e \in \omega$, for some $y \in \mathbb{R}$ : $A$ is ITTM-recursive in $(y, B)$.

Notice in the above that we have included a parameter real $y$ to ensure the closure under continuous preimages as before. This will ensure we have *Wadge pointclasses* and that the ensuing notion of $\leq^\infty$*-degree* with the degree ordering induced, will be wellfounded. The structure of this degree ordering is dependent on the ambient set theory – we shall not go into this now, but under the assumption of 'sufficient Determinacy' (that of two-person perfect information games of sufficient complexity in their payoff sets) we shall have that the degrees are wellordered; under the assumption of $V = L$ the ordering of $\leq^\infty$ degrees is very different, and below the complete $\leq^\infty$-semi-recursive set of reals there are plenty of $\leq^\infty$-incomparable sets (and hence Post's problem has a rich positive solution); whilst under 'sufficient determinacy' assumptions, there are no intermediate degrees at all. This was to be expected, and serves only to confirm the position of the pointclass of ITTM-semi-recursive sets as one within the totality of the Wadge ordering of all reasonable pointclasses of sets of reals. See [55] for a further discussion and results.

By analogy with Kleene recursion we have:

**Lemma 12.2.** *$A \leq^\infty B$ iff there are $\Sigma_2$-formulae in $\mathcal{L}_{\in,\dot{X}}$ $\varphi_1(\dot{X}, v_0, v_1)$, $\varphi_2(\dot{X}, v_0, v_1)$ and $y \in \mathbb{R}$, so that for all $x \in \mathbb{R}$*

$$x \in A \iff L_{\zeta^{B,y,x}}[B, y, x] \models \varphi_1[B, y, x] \iff L_{\zeta^{B,y,x}}[B, y, x] \models \neg\varphi_2[B, y, x].$$

The lemma then identifies structures in which we can look to ascertain the outcomes of our ITTM computations relative to a set of reals $B$, say. By way of analogy with $\zeta$, the ordinal $\zeta^{B,y,x}$ is the least that is not ITTM-$(B, x, y)$-recursive. It is thus also least such that $L_{\zeta^{B,y,x}}[B, y, x]$ has a proper $\Sigma_2$-elementary end-extension.

## 12.4. Computation on Ordinals, and Ordinal Length Machines

In the 1970s the theory of *α-recursion* coming out of the *meta-recursion* of the 1960s reached its highest stage of development. The observation that an enumeration of a $\Pi^1_1$-complete set of integers was very naturally effected, not in $\omega$, but in $\omega_1^{\mathrm{ck}}$ (the least non-recursive ordinal) steps led to a discussion on the role of hyperarithmetic *vis à vis* finite. In meta-recursion the motivation was to have a generalisation of recursion theory where infinitely long computations converged. Initially the emphasis had been on using an analogy between finite/recursive/recursively enumerable to yield a notion of meta-finite/meta-recursive/meta-r.e. In the latter the integers would be replaced by recursive ordinals, and a meta-r.e. set was a set of recursive ordinals whose *indices* formed a $\Pi^1_1$ set. Meta-recursive sets would be those that were both meta-r.e. and co-meta-r.e. The notion that replaced finiteness, was that of meta-finiteness which was to be identified with a set of ordinals together with a hyperarithmetic index set. In particular the domain of computation had now changed: instead of $\omega$ it would become $\omega_1^{\mathrm{ck}}$. (See, e.g., the discussion in [45] Part V for an account of this development.) This was not the first generalisation of recursion theory to ordinals: Takeuti [51] had replaced 'recursive enumerability' by a scheme equivalent to $\Sigma_1$-definability and was the first to generalise recursion theory from natural numbers to ordinals. There were a number of developments from Kleene's equational calculus to include ordinal valued functions in equations: Machover [38], Levy [37], Tugué [53], Kripke [36], Platek [41] all had such calculi. The latter two involved what emerged as a primary notion, that of an *admissible ordinal* with the concomitant axiomatisation of *admissible set theory* as a fragment of full ZFC set theory. Platek had the notion of an *admissible set*. From one perspective, it seems pointless to split the distinction between an 'equational calculus' and an abstract 'machine' (if there is one to split). Platek, though, seems to have had in mind, or at least the picture of, an ordinal register machine of some sorts, which we shall turn to these later.

### 12.4.1. *Ordinal length tapes*

Since the Hamkins–Kidder machines can construct levels of the Gödel $L$-hierarchy up to a certain stage (below the level $\Sigma$ alluded to above) it is a natural generalisation to think up behaviours for machines with tape not an $\omega$ sequence of cells, but longer. Indeed why not consider a sequence of cells $C_\alpha$ for $\alpha$ any ordinal? Both Koepke and Dawson independently, and at roughly around the same time, came up with the idea of ordinal length tape machines, equipped with *liminf* rules to locate read/write heads and instruction numbers within a program list. One allows the head to move left, but again must specify if the head is over a limit cell $C_\lambda$ what default action to do if the machine is asked to move left one cell. As sets can be coded by sets of ordinals (assuming the Axiom of Choice) we have some means of dealing with, or representing sets on tapes. If a machine runs and produces a sequence of 0s and 1s on a tape, then again, if of the right form, we can say that the machine is *producing* (codes for) sets. We may label these machines as *Ordinal Time Turing Machines (OTM).* Dawson [8] formulated an *Axiom of Computability* that states that every set is computable, in that there is a program that produces (not necessarily halting) at some point a code for that set. He then proves that the computable sets form a transitive class satisfying the ZF axioms together with AC. A condensation lemma on the elements appearing in a table of a long computation then produces the Generalised Continuum Hypothesis. As the construction of the machine and its action is completely absolute in character, we can imagine the machine running inside Gödel's constructible universe $L$, performing the same actions with the same outcomes. Since $L$ is the minimal transitive class model of ZF, then of course the machine is producing precisely the construcible sets.

Koepke gave a detailed description in [32], [33] of the organisation of such results, and whereas Dawson was considering codes for sets running on an everlasting machine, Koepke considers *halting computations* starting from an input tape with marks for finitely many ordinals. Koepke then shows in detail that a *Bounded Truth function* for $L$ is computable. He then has:

**Theorem 12.6.** (Koepke [32]) *A set $x$ is computable from a finite set of ordinal parameters if and only if it is a member of the constructible hierarchy.*

He then proceeds to derive GCH again using this analysis. During the 1970s Silver produced a description of the constructible hierarchy using, what came to be called 'Silver Machines'. Silver's motivation was to avoid R.B. Jensen's 'fine-structural' description of $L$, which Jensen had used to great effect in establishing fundamental properties both of $L$, and of the universe of all sets. An account of Silver's method is in [9], Part IX. The 'machine' nature of the description is essentially that of an extremely slowed production of constructible sets, and owes more to a desire to have as simple as possible method of set construction, rather than a perspective with a mechanical a flavour. Silver convincingly made use of his theory by producing a fine-structure free proof of an important combinatorial principle of $L$ called $\Box$, due to Jensen. The ordinal length tape Turing machine model held out hope that another different proof of $\Box$ might be possible using the machine's description. That hope has not been realised, and it seems that despite the smoothness of set construction at successor steps, the infinitary nature of the limit rule mitigates against certain construction principles that seem common to most proofs of □ to date, so maybe this approach would seem difficult.

Nevertheless, the description of the constructible sets, now adds a further method of describing $L$ besides the two originally due to Gödel, and to those of Jensen and Silver.

12.4.1.1. *$\alpha$-length tapes*

Rather than take ON length tapes, it would be possible to consider computation using the above machines but with the length of tape, and perhaps time, restricted to, say, suitable ordinals $\alpha$, such as initial ordinals or cardinal numbers. There would indeed be nothing against this: one could produce, say, just the hereditarily countable members of $L$ by allowing only computations that took countable lengths of time. For restricting to computations not of cardinal length, some closure considerations come into effect. In order to have effective methods of combining even very elementary processes on sets, one should require that ordinals be sufficiently closed to enable this, and something such as closure under the primitive recursive set functions (*cf.* [9], p.100) would be suitable.

The notion of *admissible ordinal* stands out, not least because of the development of *$\alpha$-recursion theory* in the 1960s and 1970s. We have briefly mentioned the origins of this theory at the beginning of this section. The motivation for its development was indeed a theoretical one: to lift from $\mathbb{N}$ the theory of recursion to other domains. The closure of an admissible ordinal was soon seen to result in a powerful theory of sets that when axiomatised gives essentially a reduced form of ZF, with the scheme of Replacement restricted to $\Sigma_1$ instances, and that of Comprehension to $\Delta_1$. An *admissible set* was then a model of this theory, and $\langle L_{\omega_1^{ck}}, \in\rangle$ is the least transitive model of this theory (if one includes the axiom of infinity). An account of this development is contained in [45].

One could therefore simply restrict an ordinal length tape machine to an admissible ordinal length $\alpha$, and consider calculations of length at most $\alpha$ in time.

Does one get back precisely the theory of $\alpha$-recursion theory? Does 'computably enumerable' correspond to $\alpha$-r.e., and if so does the machine approach give any new slant on the old results from the 1970s such as the Sacks–Simpson theorem [46] that there are incomparable $\alpha$-r.e. sets neither (weakly) $\alpha$-recursive in the other; or the Shore Splitting and Density theorems [48], [49]? These are matters still under investigation. Dawson ([8]) has established for a notion of what he calls *uniform $\alpha$-computation* that indeed one has the Sacks–Simpson and Shore Density results.

### 12.4.2. *Ordinal Register Machines*

We now turn to full blooded finite register machines with the capability of ordinal entries. Again, such machines are allowed to run transfinitely using an ordinary register arrangement, and finite instruction set, with a suitable *liminf* rules for *register values* and *instruction numbers* at limit ordinal $\lambda$ lengths of time. We have mentioned that one (unpublished by Platek) approach yielded an equational calculus for ordinal recursion up to $\omega_1^{\mathrm{ck}}$, Siders and Koepke [35], consider register machines with a stack and remarkably even a machine with finitely many registers allows one to calculate a bounded truth predicate for $L$. One thus can represent $L$ both using Ordinal Register Machines (ORM) and Ordinal Time Turing Machines (OTM).

As for the ITTMs one has notions of *clockable ordinal* (one for which an ORM or OTM halts on, say, 0 input) and *writable ordinal* (one for which a code can be written: this is easier to formulate for an OTM: a code can be written literally on the tape; for an ORM one simply has the machine halt with the ordinal in, say, the first register). For both these notions it is easier than for ITTMs to conclude that $\gamma$ the supremum of the clockable ordinals is that of the writable ordinals. In [16] it is explicitly shown how to convert calculations from an ORM to an OTM and vice versa.

Using ordinal register machines with values up to the admissible ordinal $\gamma$ Hamkins and Miller have used priority arguments to produce a Friedberg–Muchnik like solution to Post's problem [16] for ORMs: they produce ORM-enumerable but incomparable sets $A, B \subset \gamma$ that are below the appropriate notion of jump.

**Definition 12.13.** Let $P_e$ be the $e$'th ORM program, the (weak) jump is the set

$$0^{\diamond} = \{e \in \mathbb{N} \mid P_e(0) \downarrow\}.$$

Although neither [16] nor [35] make the following characterisation, it appears reasonable to argue that the ordinal $\gamma$ in fact is recognisable by set theorists as the first $\Sigma_1$-*stable ordinal* $\sigma$. This is defined to be the least ordinal $\sigma$ so that $\langle L_\sigma, \in\rangle \prec_{\Sigma_1} \langle V, \in\rangle$, that is, $L_\sigma$ is an elementary substructure of the universe of all sets, but only for $\Sigma_1$ sentences expressible using parameters from $L_\sigma$. (See [20] p. 412 for an equivalent definition in terms of $\infty$-partial recursive functions.) Consequently if any ORM (or OTM) halts on integer input (or indeed any input less than $\gamma$) then the length of that computation must be also less than $\sigma$, as this halting assertion is itself a simple $\Sigma_1$-statement in the language of set theory. (Moreover anything output by such a machine must also clearly be an ordinal less than $\sigma$ by the same reasoning.) Hence $\gamma \leq \sigma$. To see that $\sigma \leq \gamma$ observe that in the $L$ hierarchy, new $\Sigma_1$ sentences become true in $L_\delta$ for arbitrarily large ordinals $\delta < \gamma$. Now given a true $\Sigma_1$ sentence in the language of set theory, run an ORM (or OTM) program to search for that ordinal $\delta$, and then halt. This task must take more than $\delta$ (but also less than $\sigma$) steps. Hence $\sigma = \gamma$. One then obtains:

**Proposition 12.5.** $0^{\diamond}$ *is recursively isomorphic to the* $\Sigma_1$*-truth set of* $\langle L_\sigma, \in\rangle$.

One can compare this with the statement that the standard Turing halting set is recursively isomorphic to the $\Sigma_1$-truth set of $\langle L_\omega, \in\rangle$ where $L_\omega = \text{HF}$ the class of hereditarily finite sets. A similar result holds (with the appropriate formulations) for OTMs for the same reasons.

## 12.5. Theoretical Machine Strength

We consider finally the *theoretical strengths* of the various types of mechanisms discussed here. We have answered, in one fashion at least, the capabilities of the machines in the Malament–Hogarth spacetimes, and the Etesi–Németi model in particular. It is also clear that the ON-length tape machines are full ZFC-machines that are capable of producing Gödel's constructible universe.

For the intermediate machine models we have mentioned, one could simply be satisfied by seeing at which level of complexity the machines can answer queries concerning predicates. One can however somewhat more formally, formulate a theory in which the behaviour of the machine can be represented, and one may then calibrate this theory, not necessarily proof theoretically, but at least as a theory within other theories, for example as a subsystem of second order analysis, much as is done in the *Reverse Mathematics Program* (see [50]). The discussion becomes somewhat technical, but for the logician, interesting.

Towards analysing the ITTMs we first look at connections to certain kinds of *quasi-inductive definitions* that were defined earlier, at least in one form, by Burgess in [5].

Let $\Gamma : \mathcal{P}(\omega) \to \mathcal{P}(\omega)$ be any arithmetic operator (that is '$n \in \Gamma(X)$' is arithmetic; we emphasise that $\Gamma$ need be neither monotone nor progressive). We define the following iterates of $\Gamma : \Gamma_0(X) = X; \Gamma_{\alpha+1}(X) = \Gamma(\Gamma_\alpha(X)); \Gamma_\lambda(X) = \liminf_{\alpha\to\lambda} \Gamma_\alpha(X) = \bigcup_{\alpha<\lambda} \bigcap_{\lambda>\beta>\alpha} \Gamma_\beta(X)$. Following [5], we say that $Y \subseteq \omega$ is *arithmetically quasi-inductive* if for some such $\Gamma, Y$ is (1-1) reducible to$\Gamma_{\mathrm{On}}(\varnothing)$. Any such definition has a least countable $\xi = \xi(\Gamma)$ with $\Gamma_\xi(\varnothing) = \Gamma_{\mathrm{On}}(\varnothing)$. If we let $\zeta$ denote the supremum of all such $\xi(\Gamma)$, then we have that the $\zeta$ defined here is none other than the $\zeta$ defined above relating to ITTMs. In fact the ITTMs give an example of a *recursive quasi-inductive operator* that is *complete* for all *arithmetic quasi-inductive operators*. (Think: a universal ITTM can be programmed to mimick any *arithmetic* quasi-inductive operator.) Hence the same class of sets arises, it turns out, if one restricts to simply recursive $\Gamma$.

For any such arithmetic quasi-inductive operator $\Gamma$ let us now define the *repeat pair* of $\Gamma$ on a starting set $X$, as the lexicographically least pair $(\zeta(\Gamma, X), \Sigma(\Gamma, X))$ with $\Gamma_\zeta(X) = \Gamma_\Sigma(X)$.

**Definition 12.14.** AQI is the sentence: *'For every arithmetic operator* $\Gamma$, *for every* $X \subseteq \mathbb{N}$, *there is a wellordering* $W$ *with a repeat pair* $(\zeta(\Gamma, X), \Sigma(\Gamma, X))$ *in* Field($W$)*'.* If an arithmetic operator $\Gamma$ acting on $X$ has a repeat pair, we say that $\Gamma$ *converges* (with input X).

Then AQI can be formulated in second order number theory, and essentially is asserting that there are sufficient wellorderings for every operator on every input set $X$ to converge. One may ask:

Q: *What is the strength of* $\mathsf{ACA}_0 + \mathsf{AQI}$?

(The choice of $\mathsf{ACA}_0$, *arithmetical comprehension*, as a base theory is somewhat arbitrary. We refer the reader to [50] in what follows for all notions concerning these axiom systems, and determinacy hypotheses etc.) We could have equivalently reformulated a version of AQI which mentioned instead looping points of ITTMs, but this would turn out to be equivalent, as we have intimated. $\Pi^1_3\mathsf{CA}_0$ is sufficient to prove there are $\beta$-models of $\mathsf{ACA}_0 + \mathsf{AQI}$.

**Theorem 12.7.** ([58]) *(i)* $\Pi^1_3\mathsf{CA}_0$, $\mathsf{ACA}_0+\mathsf{AQI}$ *and* $\Pi^1_2\mathsf{CA}_0$ *are in descending order of strength in that each theory proves the existence of* $\beta$*-models of the next.*

*More precisely, and in the same sense:*

*(ii)* $\Delta^1_3\mathsf{CA}_0+\mathbf{\Sigma}^0_3$-Determinacy, $\mathsf{ACA}_0+\mathsf{AQI}$, and $\Delta^1_3\mathsf{CA}_0$ *are similarly in strictly descending order of strength.*

*Determinacy* makes an appearance here, since this theorem is the outcome of an attempt to generalise the theorem of Solovay (see [25]) that strategies for $\Sigma^0_2$-games appear at the level $\sigma^1_1$ of the $L$ hierarchy, the closure ordinal for $\Sigma^1_1$-monotone inductive definitions. (In turn, as is well known, strategies for $\Sigma^0_1$-games appear at $\omega_{1\,\mathrm{ck}}$ the closure ordinal for $\Pi^1_1$-monotone inductive definitions. We are thus trying to link AQIs, or ITTMs to strategies for certain infinite games.) Thus AQIs are close to, but not equivalent to, $\Sigma^0_3$-Determinacy. That they are stronger than $\Sigma^0_2$-Determinacy, is because $\sigma^1_1 < \zeta$. Moreover, letting 'Bool($\Sigma^0_2$)' denote Boolean combinations of $\Sigma^0_2$ sets, the constructible rank of the height of the least $\beta$-model of $\Pi^1_2\mathsf{CA}_0$ (as shown by Möllerfeld and Heinatsch [19]) where strategies for Bool($\Sigma^0_2$) games are to be found, is less than $\zeta$, and in fact is again a 'writable' ordinal less than $\lambda$, in the sense of ITTMs. This shows that the assertion that any ITTM halts or loops, is stronger than Bool($\Sigma^0_2$)-Determinacy.

That $\Delta^1_3\mathsf{CA}_0 + \mathsf{AQI}$ is stronger than $\Delta^1_3\mathsf{CA}_0$ in the above sense should be plausible in that the universal ITTM on input $\varnothing$ will go into a loop at the 'repeat pair' ordinals $\zeta$ and $\Sigma$ where $L_\zeta \prec_{\Sigma_2} L_\Sigma$. However it is easy to see from this $\Sigma_2$-extendability property that any such $L_\zeta$ is a $\Sigma_2$-admissible set (where we now require the admissible set to additionally be a model of $\Sigma_2$-Replacement) and is also a union of such. The reals of such a model then form a $\beta$-model of $\Delta^1_3\mathsf{CA}_0$.

*Connections to ordinal analysis*

The notion of '$\Sigma_2$-extendibility' of a model, that is of having a proper $\Sigma_2$-elementary end extension, would seem *prima facie*, to be connected to any attempt to prove a generalisation of Rathjen's ordinal analysis ([43]) of $\Pi^1_2\mathsf{CA}_0$ that could be lifted to $\Pi^1_3\mathsf{CA}_0$. In the former proof, chains of arbitrary but finite length of $\Sigma_1$-end extensions in the constructible hierarchy of the form $L_{\xi_1} \prec_{\Sigma_1} L_{\xi_1} \prec_{\Sigma_1} \cdots \prec_{\Sigma_1} L_{\xi_n}$ are analysed. (Note that the least $\beta$-model of $\Pi^1_2\mathsf{CA}_0$ consists of $\mathcal{P}(\mathbb{N}) \cap L_{\xi_\omega}$ where $\xi_\omega$ is the least ordinal with $L_{\xi_\omega}$ a union of an infinite tower of $\Sigma_1$ substructures.) To analyse $\Pi^1_3\mathsf{CA}_0$ in a similar way would require lifting the '1' above to '2' and looking at arbitrarily long chains of $\Sigma_2$-extensions. It would seem then that any ordinal analysis of $\Pi^1_3\mathsf{CA}_0$ would first have to go through an analysis of $\mathsf{AQI}$, the latter being but the very first step in this linkage.

*$\Sigma_n$- or 'hypermachines.'*

The notion of *liminf* is essentially a two quantifier alternation: *'there exists a time such that for all later times...'*. It is possible to enquire whether there are other types of limit rule that bring about different notions of computation, or different classes of computable function. One attempt to consider this question is a result of [54] which shows that, amongst all possible $\Sigma_2$-rules the *liminf* (or equivalently the *limsup*) rule is complete, or the most general. This is entirely unsurprising: if the universal ITTM machine can produce the constructible hierarchy up to $L_\Sigma$, there is little else for it to do. Further any other $\Sigma_2$-rule would itself produce looping behaviour between $L_\zeta$ and $L_\Sigma$.

One may thus broaden the enquiry and look for more complex rules. It is possible to develop a $\Sigma_3$-machine which incorporates a $\Sigma_3$-limit rule cf. [14]; the essential idea is that instead of taking a liminf along all ordinals one takes a liminf using only those ordinals that already bound the reappearances of earlier (shifted) snapshots. One thus has in some sense a *dynamic limit rule* in that the behaviour at a limit rule depends more formally on the tapes' prior contents. One then has the analogous result that a universal machine program would then have identical snapshots at the least pair $(\zeta(3), \Sigma(3))$ where $L_{\zeta(3)} \prec_{\Sigma_3} L_{\Sigma(3)}$to mirror the earlier $\lambda$-$\zeta$-$\Sigma$ theorem at $\Sigma_2$. After $\Sigma(3)$ it then returns to the previous snapshot at $\zeta(3)$ and thereafter repeats forever. It is possible to generalise this to higher quantificational levels $\Sigma_n$ with the snapshot/looping behaviour at the appropriate pair $(\zeta(n), \Sigma(n))$ lexicographically least with $L_{\zeta(n)} \prec_{\Sigma_n} L_{\Sigma(n)}$. However showing these facts is more technical, and is reliant more on the underlying set theory; it thus perhaps has decreasingly less of an appeal to intuitions concerning machine computation. This is explained in [14].

*ITRMs on integers*

The ITRMs of Miller and Koepke (12.2.2.2) with entries restricted to natural numbers turn out to be pleasantly strong. It is possible show that such machines are $\Pi^1_1$-complete, in that for any $\Pi^1_1$ set $A$ there is a program on an ITRM, that correctly accepts or rejects $n$ depending on whether $n$ is or is not in $A$. It can thus, for example, decide for which indices $e$ the $e$'th (standard) Turing function $\{e\}$ is the characteristic function of a wellorder or not. Moreover it can be shown that the strength of the machine strictly increases with the number $N$ of registers. It is possible with $2N$ registers to simulate an $N$ register machine, whilst giving as output integer codes of those programs on $N$ registers that halt. A corollary is that there can be no such *universal machine.* Here we let $P_{e,N}$ denote the $e$'th ITRM program for an $N$ register machine.

**Definition 12.15.** Let ITRM be the axiom scheme that states for each $N \in \mathbb{N}$ that halting sets for $N$-register machines exists:

'For any $N \in \mathbb{N}, K_N =_{\mathrm{df}} \{e | P_{e,N}(\vec{0}) \downarrow\}$ exists'.

One then obtains (with $\mathsf{RCA}_0$ as the *recursive comprehension scheme*):

**Theorem 12.8.** $\mathsf{RCA}_0 \vdash \mathsf{ITRM} \longleftrightarrow \Pi^1_1\mathsf{CA}_0$.

Reverse mathematics has shown that a wealth of theorems can be proven in the system $\Pi^1_1\mathsf{CA}_0$. As a sample we have the following, in which we assume the ITRM is equipped with an *oracle set* $Z \subseteq \mathbb{N}$ (and a register operation to query it):

**Theorem 12.9.** *Let* $T \subseteq \{\sigma \mid \sigma \in {}^{<\mathbb{N}}\mathbb{N}\}$ *be a set of sequences which form* a tree. *Then if* $Z \subseteq \mathbb{N}$ *codes* $T$ *(via some recursive coding), the* perfect kernel *of* $T$ *is ITRM-computable in the oracle* $Z$.

(By the perfect kernal we mean the maximal subtree whose branches form a perfect set, that is without isolated points.) Thus, as with much of this kind of study, a seemingly simple model in fact turns out to be rather powerful.

## Acknowledgements

The author would like to gratefully acknowledge grants from the EPSRC and the Templeton Foundation during the writing of this chapter. He would also like to thank the Kurt Gödel Research Center, Vienna for its hospitality also during this period.

## References

[1] K. J. Barwise, *Admissible Sets and Structures.* Perspectives in Mathematical Logic, Springer Verlag, Berlin–Heidelberg, (1975).

[2] E. Beggs, J. F. Costa, B. Loff, and J. V. Tucker. Oracles and advice as measurement. In eds., C. S. Calude *et al.*, *Unconventional Computing*, vol. 5204, *Lecture Notes in Computer Science*, pp. 33–50. Springer, Berlin, (2008).

[3] E. Beggs and J. V. Tucker, Can Newtonian systems, bounded in space, time, mass and energy compute all functions?, *Theoret. Comput. Sci.* **371**, 4–19, (2007).

[4] L. Blum, M. Shub, and S. Smale, On a theory of computation and complexity over the real numbers, *Notices Amer. Math. Soc. (N. S.).* **21**(1), 1–46, (1989).

[5] J. P. Burgess, The truth is never simple, *J. Symbolic Logic.* **51**(3), 663–681, (1986).

[6] N. Cutland, *Computability: an Introduction to Recursive Function Theory.* Cambridge University Press, Cambridge, 1980).
[7] E. B. Davies, Building infinite machines, *British J. Philos. Sci.* **52**(4), 671–682, (2001).
[8] B. Dawson. *Ordinal time Turing computation.* PhD thesis, Bristol, (2009).
[9] K. Devlin, *Constructibility.* Perspectives in Mathematical Logic, Springer-Verlag, Berlin–Heidelberg, (1984).
[10] J. Earman and J. D. Norton, Forever is a day: Supertasks in Pitowsky and Malament–Hogarth spacetimes, *Philos. Sci.* **60**, 22–42, (1993).
[11] J. Earman and J. D. Norton. Infinite pains: the trouble with supertasks. In eds., A. Morton and S. Stich, *Benacerraf and his critics*, vol. xi, *Philosophers and their critics.* Blackwell, Oxford, (1996).
[12] G. Etesi and I. Németi, Non-Turing computations via Malament–Hogarth space-times, *Internat. J. Theoret. Phys.* **41**(2), 341–370, (2002).
[13] J. E. Fenstad, *General Recursion Theory: an Axiomatic Approach.* Perspectives in Mathematical Logic, Springer, Berlin–Heidelberg–New York, (1980).
[14] S. D. Friedman and P. D. Welch, Hypermachines, *to appear in the J. Symbolic Logic.* (2010).
[15] J. D. Hamkins and A. Lewis, Infinite time Turing machines, *J. Symbolic Logic.* **65**(2), 567–604, (2000).
[16] J. D. Hamkins and R. Miller, Post's problem for ordinal register machines: an explicit approach, *Ann. Pure Appl. Logic.* **160**(3), 302–309, (2009).
[17] J. D. Hamkins and D. Seabold, Infinite time Turing machines with only one tape, *MLQ Math. Log. Q.* **47**(2), 271–287, (2001).
[18] S. W. Hawking and G. F. R. Ellis, *The Large Scale Structure of Space-Time.* Cambridge University Press, Cambridge, (1973).
[19] C. Heinatsch and M. Möllerfeld. Determinacy in second order arithmetic. In eds., S. Bold, B. Löwe, T. Räsch and J. van Bentham, *Foundations of the Formal Sciences V*, Studies in Logic, pp. 143–155, College of London Publications, (2007).
[20] P. Hinman, *Recursion-Theoretic Hierarchies.* $\Omega$ Series in Mathematical Logic, Springer, Berlin, (1978).
[21] M. Hogarth, Does general relativity allow an observer to view an eternity in a finite time?, *Found. Phys. Lett.* **5**(2), 173–181, (1992).
[22] M. Hogarth. Non-Turing computers and non-Turing computability. In *PSA: Proceedings of the Biennial Meeting of the Philosophy of Science Association Vol. 1*, pp. 126–138. (1994).
[23] M. Hogarth, Deciding arithmetic using *SAD* computers, *British J. Philos. Sci.* **55**, 681–691, (2004).
[24] K. Hrbacek and S. Simpson. On Kleene degrees of analytic sets. In eds., J. Barwise, H. J. Keisler and K. Kunen, *Proceedings of the Kleene Symposium*, Studies in Logic, pp. 347–352. North-Holland, Amsterdam, (1980).
[25] A. S. Kechris. On Spector classes. In eds., A. S. Kechris and Y. N. Moschovakis, *Cabal Seminar 76-77*, vol. 689, *Lecture Notes in Mathematics Series*, pp. 245–278. Springer, Berlin, (1978).
[26] S. C. Kleene, Recursive quantifiers and functionals of finite type I, *Trans.*

*Amer. Math. Soc.* **91**, 1–52, (1959).
[27] S. C. Kleene. Turing-machine computable functionals of finite type I. In *Proceedings 1960 Conference on Logic, Methodology and Philosopy of Science*, pp. 38–45. Stanford University Press, California, (1962).
[28] S. C. Kleene, Turing-machine computable functionals of finite type II, *Proc. Lond. Math. Soc.* **12**, 245–258, (1962).
[29] S. C. Kleene, Recursive quantifiers and functionals of finite type II, *Trans. Amer. Math. Soc.* **108**, 106–142, (1963).
[30] S. C. Kleene. Recursive functionals and quantifiers of finite types revisited. In *Generalized Recursion Theory II, Proceedings 2nd Scandinavian Logic Symposium, Oslo, 1977*, vol. 94, *Studies in Logic and Foundations of Mathematics*, pp. 185–222, North-Holland, Amsterdam, New York, (1978).
[31] A. Klev, *Magister thesis*, (August 2007).
[32] P. Koepke, Turing computation on ordinals, *Bull. Symbolic Logic.* **11**, 377–397, (2005).
[33] P. Koepke and M. Koerwien, Ordinal computations, *Math. Structures Comput. Sci.* **16**(5), 867–884, (2006).
[34] P. Koepke and R. Miller. An enhanced theory of infinite time register machines. In eds., A. Beckmann *et al.*, *Logic and the Theory of Algorithms*, vol. 5028, *Springer Lecture Notes Computer Science*, pp. 306–315. Springer, Berlin, (2008).
[35] P. Koepke and R. Siders. Computing the recursive truth predicate on ordinal register machines. In eds. A. Beckmann et al., *Logical Approaches to Computational Barriers*, Computer Science Report Series, p. 21. Swansea, (2006).
[36] S. Kripke, Transfinite recursion on admissible ordinals I,II, *J. Symbolic Logic.* **29**, 161–162, (1964).
[37] A. Levy, Transfinite computability (abstract), *Notices Amer. Math. Society.* **10**, 286, (1963).
[38] M. Machover, The theory of transfinite recursion, *Bull. Amer. Math. Soc.* **67**, 575–578, (1961).
[39] Y. N. Moschovakis, *Elementary Induction on Abstract Structures.* vol. 77, *Studies in Logic*, North-Holland, Amsterdam, (1974).
[40] I. Pitowsky, The physical Church–Turing thesis and physical computational complexity, *Iyyun.* **39**, 81–99, (1990).
[41] R. Platek. *Foundations of Recursion Theory.* PhD thesis, Stanford, California, (1966).
[42] H. Putnam, Trial and error predicates and the solution to a problem of Mostowski, *J. Symbolic Logic.* **30**, 49–57, (1965).
[43] M. Rathjen, An ordinal analysis of parameter-free $\Pi^1_2$ comprehension, *Arch. Math. Logic.* **44**(3), 263–362, (2005).
[44] H. Rogers, Jr., *Theory of Recursive Functions and Effective Computability.* McGraw-Hill, New York, (1967).
[45] G. E. Sacks, *Higher Recursion Theory.* Perspectives in Mathematical Logic, Springer Verlag, Berlin–Heidelberg, (1990).
[46] G. E. Sacks and S. G. Simpson, The $\alpha$-finite injury method, *Ann. Math.*

*Logic.* **4**, 343–367, (1972).

[47] J. Shepherdson and H. Sturgis, Computability of recursive functionals, *J. ACM.* **10**, 217–255, (1963).

[48] R. A. Shore, Splitting an $\alpha$ recursively enumerable set, *Trans. Amer. Math. Soc.* **204**, 65–78, (1975).

[49] R. A. Shore, The recursively enumerable $\alpha$-degrees are dense, *Ann. Math. Logic.* **9**, 123–155, (1976).

[50] S. G. Simpson, *Subsystems of Second Order Arithmetic.* Perspectives in Mathematical Logic, Springer, Berlin, (1999).

[51] G. Takeuti, On the recursive functions of ordinal numbers, *J. Math. Soc. Japan.* **12**, 119–128, (1960).

[52] J. Thomson, Tasks and supertasks, *Analysis.* **15**(1), 1–13, (1954-55).

[53] T. Tugué, On the partial recursive functions of ordinal numbers, *J.. Math. Soc. Japan.* **16**, 1–31, (1964).

[54] P. D. Welch, Eventually infinite time Turing degrees: infinite time decidable reals, *J. Symbolic Logic.* **65**(3), 1193–1203, (2000).

[55] P. D. Welch. Post's and other problems in higher type supertasks. In eds., B. Löwe, B. Piwinger and T. Räsch, *Classical and New Paradigms of Computation and their Complexity hierarchies, Papers of the Conference Foundations of the Formal Sciences III*, vol. 23, *Trends in logic*, pp. 223–237. Kluwer, Dordrecht, (2004).

[56] P. D. Welch, Turing Unbound: The extent of computations in Malament-Hogarth spacetimes, *British J. Philos. Sci.* **15**(4), 659–674, (2008).

[57] P. D. Welch, Characteristics of discrete transfinite Turing machine models: halting times, stabilization times, and normal form theorems, *Theoret. Comput. Sci.* **410**, 426–442, (2009).

[58] P. D. Welch, Weak systems of determinacy and arithmetical quasi-inductive definitions, *arXiv: 0905.4412, to appear in the J. Symbolic Logic.* (2010).